ADVANCES IN
GEOPHYSICS

VOLUME 18A

Advances in
GEOPHYSICS

Edited by

H. E. LANDSBERG

Institute for Fluid Dynamics and Applied Mathematics
University of Maryland, College Park, Maryland

J. VAN MIEGHEM

Royal Belgian Meteorological Institute
Uccle, Belgium

Editorial Advisory Committee

BERNARD HAURWITZ R. STONELEY
ROGER REVELLE URHO A. UOTILA

INTERNATIONAL UNION OF THEORETICAL AND APPLIED MECHANICS
AND INTERNATIONAL UNION OF GEODESY AND GEOPHYSICS

TURBULENT DIFFUSION IN ENVIRONMENTAL POLLUTION

Proceedings of a Symposium held at Charlottesville, Virginia,
April 8–14, 1973

Edited by

F. N. FRENKIEL

Naval Ship Research and Development Center
Bethesda, Maryland

R. E. MUNN

Atmospheric Environment Service
Toronto, Ontario, Canada

VOLUME 18A

1974

Academic Press • New York   San Francisco   London
A Subsidiary of Harcourt Brace Jovanovich, Publishers

COPYRIGHT © 1974, BY ACADEMIC PRESS, INC.
ALL RIGHTS RESERVED.
NO PART OF THIS PUBLICATION MAY BE REPRODUCED OR
TRANSMITTED IN ANY FORM OR BY ANY MEANS, ELECTRONIC
OR MECHANICAL, INCLUDING PHOTOCOPY, RECORDING, OR ANY
INFORMATION STORAGE AND RETRIEVAL SYSTEM, WITHOUT
PERMISSION IN WRITING FROM THE PUBLISHER.

ACADEMIC PRESS, INC.
111 Fifth Avenue, New York, New York 10003

United Kingdom Edition published by
ACADEMIC PRESS, INC. (LONDON) LTD.
24/28 Oval Road, London NW1

LIBRARY OF CONGRESS CATALOG CARD NUMBER: 52-12266

ISBN 0-12-018818-X

PRINTED IN THE UNITED STATES OF AMERICA

CONTENTS

LIST OF CONTRIBUTORS vii
FOREWORD xi
PREFACE xiii

Turbulent Atmospheric Diffusion: The First Twenty-Five Years, 1920–1945. By J. KAMPÉ DE FÉRIET 1
Limitations of Gradient Transport Models in Random Walks and in Turbulence. By STANLEY CORRSIN 25
Addendum: Random Walks on Markovian Binary Velocity Fields. By G. S. PATTERSON, JR. 61
Height of the Mixed Layer in the Stably Stratified Planetary Boundary Layer. By JOOST A. BUSINGER AND S. P. S. ARYA 73
Transport of Heat across a Plane Turbulent Mixing Layer. By HEINRICH E. FIEDLER 93
Thermodynamic Model for the Development of a Convectively Unstable Boundary Layer. By D. J. CARSON AND F. B. SMITH 111
Vertical Component of Turbulence in Convective Conditions. By S. J. CAUGHEY AND C. J. READINGS 125
Surface Layer of the Atmosphere in Unstable Conditions. By ROBERT R. LONG 131
Numerical Simulation of Lagrangian Turbulent Quantities in Two and Three Dimensions. By RICHARD L. PESKIN 141
The Eulerian–Lagrangian Relationship Resulting from a Turbulence Model. By HJALMAR FRANZ 165
Computational Modeling of Turbulent Transport. By JOHN L. LUMLEY and BEJAN KHAJEH-NOURI 169
Modeling the Atmospheric Boundary Layer. By J. C. WYNGAARD, O. R. COTÉ, and K. S. RAO 193
Instantaneous Velocity and Length Scales in a Turbulent Shear Flow. By P. J. SULLIVAN 213
Numerical Computation of Turbulent Shear Flows. By STEVEN A. ORSZAG and YIH-HO PAO 225
Energy Cascade in Large-Eddy Simulations of Turbulent Fluid Flows. By A. LEONARD 237
Statistical Analysis of Wall Turbulence Phenomena. By Z. ZARIĆ . 249
Characteristics of Turbulence within an Internal Boundary Layer. By R. A. ANTONIA and R. E. LUXTON 263

Structure of the Reynolds Stress and the Occurrence of Bursts in the Turbulent Boundary Layer. By W. W. WILLMARTH and S. S. LU 287

Some Features of Turbulent Diffusion from a Continuous Source at Sea. By K. F. BOWDEN, D. P. KRAUEL, and R. E. LEWIS ... 315

Diffusion of Turbidity by Shear Effect and Turbulence in the Southern Bight of the North Sea. By JACQUES C. J. NIHOUL 331

Investigation of Small-Scale Vertical Mixing in Relation to the Temperature Structure in Stably Stratified Waters. By GUNNAR KULLENBERG... 339

Turbulent Diffusion of Heat and Momentum in the Ocean. By CARL H. GIBSON, LUIS A. VEGA, and ROBERT BRUCE WILLIAMS 353

Turbulent Diffusion and Beach Deposition of Floating Pollutants. By G. T. CSANADY.. 371

Estimation of the Diffusion Coefficient of Thermal Pollution. By K. TAKEUCHI and S. ITO 383

Activities in, and Preliminary Results of, Air–Sea Interactions Research at I.M.S.T. By MICHEL COANTIC and ALEXANDRE FAVRE 391

Eddy Diffusion Coefficients in the Planetary Boundary Layer. By F. K. WIPPERMANN 407

Lagrangian–Eulerian Time-Scale Relationship Estimated from Constant Volume Balloon Flights Past a Tall Tower. By JAMES K. ANGELL 419

Effect of Molecular Diffusion on the Structure of a Turbulent Density Interface. By P. F. LINDEN and P. F. CRAPPER 433

Renormalization for the Wiener–Hermite Representation of Statistical Turbulence. By W. C. MEECHAM........................ 445

SUBJECT INDEX 457

LIST OF CONTRIBUTORS

Numbers in parentheses indicate the pages on which the authors' contributions begin.

JAMES K. ANGELL, *Air Resources Laboratories, National Oceanographic and Atmospheric Administration, Silver Spring, Maryland, U.S.A.* (419)

R. A. ANTONIA, *Department of Mechanical Engineering, University of Sydney, Sydney, Australia* (263)

S. P. S. ARYA, *Department of Atmospheric Sciences, University of Washington, Seattle, Washington, U.S.A.* (73)

K. F. BOWDEN, *Oceanography Department, University of Liverpool, Liverpool, England* (315)

JOOST A. BUSINGER, *Department of Atmospheric Sciences, University of Washington, Seattle, Washington, U.S.A.* (73)

D. J. CARSON, *Boundary Layer Research Branch, Meteorological Office, Bracknell, Berkshire, England* (111)

S. J. CAUGHEY, *Meteorological Research Unit, R.A.F. Cardington, Bedford, U.K.* (125)

MICHEL COANTIC, *Institut de Mécanique Statistique de la Turbulence, Marseille, France* (391)

STANLEY CORRSIN, *Department of Mechanics and Materials Science, The Johns Hopkins University, Baltimore, Maryland, U.S.A.* (25)

O. R. COTÉ, *Air Force Cambridge Research Laboratories LYB, Bedford, Massachusetts, U.S.A.* (193)

P. F. CRAPPER,[1] *Department of Applied Mathematics and Theoretical Physics, University of Cambridge, Cambridge, England* (433)

G. T. CSANADY,[2] *Department of Mechanical Engineering, University of Waterloo, Waterloo, Canada* (371)

ALEXANDRE FAVRE, *Institut de Mécanique Statistique de la Turbulence, Marseille, France* (391)

HEINRICH E. FIEDLER, *Institut für Thermo- und Fluiddynamik Technische Universität, Berlin, Germany* (93)

HJALMAR FRANZ, *Deutsches Hydrographisches Institut, Hamburg, Germany* (165)

CARL H. GIBSON, *Department of Applied Mechanics and Engineering Sciences and the Scripps Institution of Oceanography, University of California, San Diego, La Jolla, California, U.S.A.* (353)

[1] *Present address*: Department of Mechanical Engineering, University of Toronto, Toronto, Canada.
[2] *Present address*: Woods Hole Oceanographic Institution, Woods Hole, Massachusetts, U.S.A.

LIST OF CONTRIBUTORS

S. Ito, *Japan Meteorological Association, Tokyo, Japan* (383)

J. Kampé de Fériet, *Université de Lille, Lille, France* (1)

Bejan Khajeh-Nouri, *The Pennsylvania State University, University Park, Pennsylvania, U.S.A.* (169)

D. P. Krauel,[3] *Oceanography Department, University of Liverpool, Liverpool, England* (315)

Gunnar Kullenberg, *Institute of Physical Oceanography, University of Copenhagen, Copenhagen, Denmark* (339)

A. Leonard,[4] *Department of Mechanical Engineering, Stanford University, Stanford, California, U.S.A.* (237)

R. E. Lewis,[5] *Oceanography Department, University of Liverpool, Liverpool, England* (315)

P. F. Linden, *Department of Applied Mathematics and Theoretical Physics, University of Cambridge, Cambridge, England* (433)

Robert R. Long, *Department of Mechanics and Materials Science, The Johns Hopkins University, Baltimore, Maryland, U.S.A.* (131)

S. S. Lu,[6] *Department of Aerospace Engineering, University of Michigan, Ann Arbor, Michigan, U.S.A.* (287)

John L. Lumley, *The Pennsylvania State University, University Park, Pennsylvania, U.S.A.* (169)

R. E. Luxton,[7] *Department of Mechanical Engineering, University of Sydney, Sydney, Australia* (263)

W. C. Meecham, *School of Engineering and Applied Science, University of California, Los Angeles, California, U.S.A.* (445)

Jacques C. J. Nihoul, *Institut de Mathématique, Université de Liège, Belgium* (331)

Steven A. Orszag, *Flow Research, Inc., Kent, Washington, U.S.A.* (225)

Yih-Ho Pao, *Flow Research, Inc., Kent, Washington, U.S.A.* (225)

G. S. Patterson, Jr., *Advanced Study Program, National Center for Atmospheric Research, Boulder, Colorado, U.S.A.* (61)

Richard L. Peskin, *Department of Mechanical, Industrial and Aerospace Engineering, Geophysical Fluid Dynamics Program, Rutgers University, New Brunswick, New Jersey, U.S.A.* (141)

K. S. Rao, *Air Force Cambridge Research Laboratories LYB, Bedford, Massachusetts, U.S.A.* (193)

---

[3] *Present address*: Bedford Institute of Oceanography, Dartmouth, Nova Scotia, Canada.
[4] *Present address*: NASA Ames Research Center, Moffett Field, California, U.S.A.
[5] *Present address*: I.C.I. Brixham Laboratory, Devon, England.
[6] *Present address*: Department of Power Mechanical Engineering, National Tsing Hua University, Hsinchu, Taiwan.
[7] *Present address*: Department of Mechanical Engineering, University of Adelaide, Adelaide, South Australia.

C. J. READINGS, *Meteorological Research Unit, R.A.F. Cardington, Bedford, U.K.* (125)

F. B. SMITH, *Boundary Layer Research Branch, Meteorological Office, Bracknell, Berkshire, England* (111)

P. J. SULLIVAN, *Department of Applied Mathematics, The University of Western Ontario, London, Ontario, Canada* (213)

K. TAKEUCHI,[8] *Osaka District Meteorological Observatory, Osaka, Japan* (383)

LUIS A. VEGA, *Department of Applied Mechanics and Engineering Sciences, and the Scripps Institution of Oceanography, University of California, San Diego, La Jolla, California, U.S.A.* (353)

ROBERT BRUCE WILLIAMS, *NATO Saclant ASW Research Center, La Spezia, Italy* (353)

W. W. WILLMARTH, *Department of Aerospace Engineering, University of Michigan, Ann Arbor, Michigan, U.S.A.* (287)

F. K. WIPPERMANN, *Institut für Meteorologie der Technische Hochschule Darmstadt, Darmstadt, Germany* (407)

J. C. WYNGAARD, *Air Force Cambridge Research Laboratories LYB, Bedford, Massachusetts, U.S.A.* (193)

Z. ZARIĆ, *Department of Thermal Physics, Boris Kidrič Institute, University of Belgrade, Yugoslavia* (249)

---

[8] *Present address*: Meteorological Research Institute, Tokyo, Japan.

FOREWORD

The Editors of *Advances in Geophysics* present with great satisfaction the Proceedings of the 1973 IUTAM–IUGG Symposium on Turbulent Diffusion in Environmental Pollution. Volumes 18A and 18B are a sequel to Volume 6, which contained the first symposium; though published 16 years ago, a number of these papers are still widely cited. There is little doubt that the contents of the present volumes contain equally outstanding contributions. The organizers of the Symposium were singularly successful in attracting a worldwide group of authorities in the field and deserve for this the gratitude of the interested scientific public. We also want to thank the volume editors for their unselfish labors in collecting and processing of the manuscripts. Both Dr. François N. Frenkiel and Dr. R. E. Munn did an outstanding job for which the readers should give them full credit.

A second reason why we are particularly pleased to see these volumes appear is the current need to present all available knowledge in the field in a world threatened by pollution. Admittedly we are still far from the goal of complete scientific solution to both theoretical and practical problems in the field. Yet there has been great progress in understanding and not inconsiderable success in use of whatever insight has been gained. Moreover, as these volumes show, there is a close alliance between pure and applied science that has helped in the past and bodes well for the future. It is our hope, nay our conviction, that the papers in these volumes will provide considerable stimulus for further research in an intriguing but still intractable field.

H. E. LANDSBERG
J. VAN MIEGHEM

PREFACE

The second IUTAM-IUGG Symposium on Turbulent Diffusion in Environmental Pollution was held at the University of Virginia in Charlottesville, Virginia, April 8 to 14, 1973. The Symposium was sponsored jointly by the International Union of Theoretical and Applied Mechanics (IUTAM) and the International Union of Geodesy and Geophysics (IUGG). The scope covered by the Symposium was similar to that of the IUTAM-IUGG Symposium on Atmospheric Diffusion and Air Pollution held in Oxford, England in 1958[1] and has some similarity to the symposia at Marseille, France in 1961[2] and Kyoto, Japan in 1966[3] organized by the two Unions. The two IUGG Associations particularly concerned with the scope of the Symposium are the International Association of Meteorology and Atmospheric Physics (IAMAP) and the International Association of the Physical Sciences of the Ocean (IAPSO).

The main purpose of the Symposium was to survey our present knowledge of turbulent diffusion as related to environmental pollution. The scope included theoretical and experimental studies of turbulent diffusion in fluid dynamics, atmospheric physics, and physical oceanography; the relation between the Eulerian and Lagrangian statistics of turbulent structure; the dispersion and deposition of pollutants; turbulent mixing of chemically reacting species; large-scale and global dispersion of pollutants and predictability of pollutant concentration patterns; and some basic aspects of their mathematical and physical modeling. In one sense, the Symposium was an IUTAM-IUGG response to the 1972 United Nations Conference on the Human Environment held in Stockholm, surveying the frontiers of knowledge in the field of turbulent diffusion in environmental pollution.

Participation in the Symposium was by invitation, and the selection of papers and invited participants was made by the Scientific Committee appointed by the two International Unions. About 130 participants and observers took part in the Symposium representing 17 countries (Australia, Belgium, Bulgaria, Canada, Denmark, France, Fed. Rep. Germany, India,

---

[1] "Atmospheric Diffusion and Air Pollution," Proceedings of a IUTAM-IUGG Symposium held at Oxford, England, August 24-29, 1958 (F. N. Frenkiel and P. A. Sheppard, eds.). Advances in Geophysics, Vol. 6. Academic Press, New York, 1959.

[2] "Fundamental Problems in Turbulence and Their Relation to Geophysics," Proceedings of a IUGG-IUTAM Symposium held at Marseille, France, September 4-9, 1961, Journal of Geophysical Research, Vol. 67, No. 8, 1962.

[3] "Boundary Layers and Turbulence," Proceedings of a IUGG-IUTAM Symposium held at Kyoto, Japan, September 19-24, 1966 (K. F. Bowden, F. N. Frenkiel, and I. Tani, eds.). The Physics of Fluids, Suppl., Vol. 10, No. 9, Part II, 1967.

Japan, Kenya, Netherlands, Norway, Sweden, UK, USA, USSR, and Yugoslavia). Among the invited participants were several younger scientists selected by the Scientific Committee. Sixty-two formal papers were presented; of these, 54 were shorter papers and 8 were invited papers. Extensive discussion followed most of the papers and appreciable time for informal discussion was arranged during the Symposium.

The studies of turbulent diffusion and more particularly the use of the statistical theory of turbulence in these studies originates, to a major extent, with the work on diffusion by continuous movement by G. I. Taylor published in 1921. This study has been an important source of inspiration for both theoretical and experimental research in fluid dynamics and atmospheric physics. An introductory paper presented by J. Kampé de Fériet at the Symposium was devoted to the discussion of some of the contributions to the study of turbulent diffusion and of atmospheric turbulence during the first 25 years after the work originating with Taylor. While turbulent diffusion involves the use of the Lagrangian approach, most of the experimental measurements of turbulence as well as the studies of the statistical theory of turbulence are based on the Eulerian approach. The relation between the Eulerian and Lagrangian approaches is thus of particular importance in such studies and still remains an object of challenge to both theoretical and experimental investigations, as indicated in a paper by S. Corrsin surveying the second 25 years. Some recent contributions are based on experiments involving the modeling of turbulent phenomena using high-speed computer methods. These studies lead to some relations between the Eulerian and Lagrangian characteristics of turbulence for the computer models which result in better insight into the nature of the laws governing turbulent diffusion. This aspect of modeling of turbulence phenomena was discussed in several papers.

While some of the basic knowledge has been developing, the many more practical problems of atmospheric and oceanic pollution require solutions. Several papers on turbulence in the planetary boundary layer, on shear flow phenomena, and on numerical and wind-tunnel simulation of such flows were presented. The interaction between the viewpoints of geophysicists and fluid dynamicists in these discussions has been of particular interest and deserves to be continuously encouraged. The experimental studies of geophysicists involve considerably larger scale phenomena than most of the studies of fluid dynamicists. Although fluid dynamic studies may involve very precise measuring methods (such as hot-wire anemometry), some of the more complex fluid dynamic phenomena may often be lost in such studies. The use of high-speed computer methods to analyze experimental measurements of turbulent phenomena now make it possible to resolve some of the hot-wire measurements and to observe what could well be referred to as the micrometeorology of wind-tunnel turbulence. Several papers contained results of measurements in wind tunnels describing some aspects of the structure of turbulence in the boundary layer. In addition to the statistical characteristics, experimental wind-tunnel studies of flow in turbulent shear layers involve the observation of well-ordered flow structures including bursts and turbulent intermittencies. Studies of such phenomena and of their similarity in the laboratory and in nature appear to be of particular importance.

One of the sessions was devoted to the studies of turbulent diffusion in the sea and in stratified waters. A report was also presented on some preliminary studies of air–sea interactions in a large wind–water tunnel.

Several papers were concerned with atmospheric pollution and with mathematical modeling of urban pollution. It should be noted that the first use of mathematical modeling of urban pollution reported in 1957 and based on the statistical theory of turbulence was only mentioned in the first IUTAM–IUGG Symposium held in Oxford; however, during the second IUTAM–IUGG Symposium there were several papers presented on this subject. The use of mathematical modeling of pollution has now become more practical, but several fundamental questions about the methodology of simulation modeling require further studies. Of particular importance is the possible application of the statistical theory of turbulence to such studies and the availability of appropriate data on the structure of turbulence. Discussions of the laboratory simulation of atmospheric turbulence and the comparison between turbulent diffusion in the wind tunnel and in the atmosphere appeared in several papers. Such studies appear to be of special interest and could involve the use of mathematical modeling of the laboratory experiments whose main purpose would be to improve the methodology of simulation modeling. While such studies have not been reported during the Symposium, they appear to be quite promising.

Two sessions of the Symposium were organized in cooperation with the IAMAP Commission on Air Chemistry and Global Pollution and were devoted to regional and global pollution studies. Both experimental and theoretical results of such studies were reported.

Following the Symposium banquet, Thomas F. Malone, Secretary-General of the Scientific Committee on Problems of the Environment (International Council of Scientific Unions) presented an address to the Symposium participants.

The IUTAM–IUGG Scientific Committee included A. Favre (France); K. Grasshoff (Fed. Rep. Germany); J. O. Hinze (Netherlands); C. E. Junge (Fed. Rep. Germany); M. Landahl (Sweden); R. E. Munn (Canada); A. M. Obukhov (USSR); E. Palm (Norway); H. A. Panofsky (USA); F. Pasquill (UK); C. H. B. Priestley (Australia); and F. N. Frenkiel (USA), Chairman.

A local organizing committee, consisting of members of the faculty of the University of Virginia—D. Barnes, R. A. Lowry, G. A. McAlpine, J. B. Morton, J. E. Scott, Jr., S. F. Singer, and R. A. Kuhlthau, Chairman—assisted in on-site arrangements and acted as hosts throughout the meeting.

The American Geophysical Union, the U.S. Academy of Sciences–National Research Council, and the University of Virginia cooperated in the organizaton of the Symposium.

The arrangements made for the participants and observers attending the Symposium contributed much to the success of the Symposium. The local arrangements committee and W. C. Phillips were particularly helpful in this regard. The American Geophysical Union and more specially A. F. Spilhaus, Jr., Executive Director and Mrs. Cynthia Beadling contributed greatly to several aspects of the organization of the Symposium.

Financial support from the Environmental Protection Agency, National Science Foundation (Atmospheric Sciences Section), Office of Naval Research, U.S. Atomic Energy Commission, in addition to the support provided by the two Unions, is gratefully acknowledged.

F. N. FRENKIEL
R. E. MUNN

TURBULENT ATMOSPHERIC DIFFUSION: THE FIRST TWENTY-FIVE YEARS, 1920–1945

J. KAMPÉ DE FÉRIET

Université de Lille, Lille, France

1.

When the Scientific Committee invited me to deliver this lecture, their decision was, I suppose, motivated by the conviction that nobody could be better qualified to talk about the past than a man of the past! How much the archaeologists would learn about the Stone Age if they could, at one of their meetings, listen to a Stone Age Man! But there is a danger: the Stone Age Man would talk only about his own cavern and thus give a biased account of the whole story . . . . I am afraid that unfortunately this is what may happen in my lecture: theoretical and experimental researches on turbulent diffusion in the atmosphere are not known to me by library files only; during the twenty-five years from 1920 to 1945 I did in fact follow them day by day and sometimes took part in some of them; looking at the field from my own window, I am able to overemphasize some details and to neglect some others. I apologize for the distortions that my souvenirs could possibly introduce in some parts of the picture.

2.

Under the title *diffusion* one describes a large class of phenomena, concerning various types of scattering of many kinds of materials in fluids, gaseous or liquid; diffusion in the atmosphere can be easily observed in everyday life; when smoke puffs out of a chimney, particles of unburned coal, water droplets, and hot gases spread out; when the wind blows in irregular gusts, it carries puffs of smoke to and fro through a wide wake, which looks like a grey dark snake over the sky, undulating more or less rapidly; the concentration of the wake is diminishing continuously as the puffs propagate further, and finally the wake disappears completely. The fact

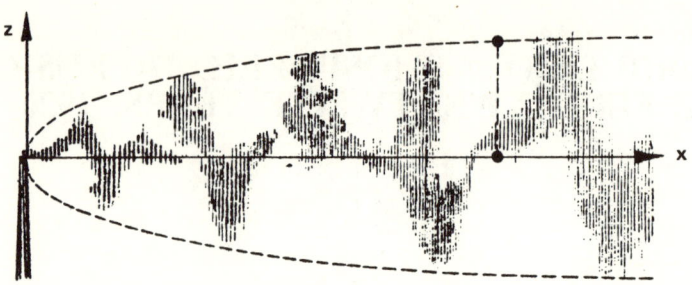

Fig. 1

that this scattering phenomenon depends on the wind structure, on the frequency and intensity of the gusts, has been recognized for a long time; if one observes the same chimney in the early hours, before sunrise, when the wind blows slowly and smoothly, the smoke looks like a narrow ribbon, propagating very quietly, sometimes to very great distances, before disappearing in the atmosphere. The diffusion of smoke by wind gusts is a typical example of turbulent atmospheric diffusion; the word *turbulent* specifies clearly that the main cause of the diffusion is attributed to the fluctuations of the wind velocity and that molecular diffusion, ever-present in any fluid motion, contributes a negligible part to the observed phenomenon.

Turbulent diffusion is clearly a particular aspect of the large class of facts described under the general heading of turbulence; but it must be pointed out that during the development that took place in the early history of this field, both theoretical as well as experimental progress of our knowledge on turbulent diffusion have played a foremost role in the development of our ideas about turbulence. For instance, it was from his observations of the diffusion of a colored streamline in a pipe that Osborne Reynolds (1883) made the fundamental distinction between two types of flow in a pipe and introduced the "Reynolds number," the nondimensional parameter which has proved to be of paramount importance; the pioneering work of Sir Geoffrey Taylor (1915a) on turbulence started with his interpretation of the observations on the diffusion of warm air over a cold sea on the Great Bank of New Foundland.

3.

Among the different types of turbulent diffusion the case of atmospheric diffusion is specially difficult, because a well-adapted theoretical frame has been lacking during the period we have to consider here; the theory of turbulence was first developed in the case of a homogeneous (even most of the time isotropic) turbulence in an incompressible fluid; these conditions are approximately satisfied in a wind tunnel, but are very far from the real situation in the atmosphere; near the ground the structure of the wind is highly inhomogeneous and the vertical gradient of temperature (leading to important variations of the density) plays a foremost role.

Thus one can only marvel at the number and importance of the results that have been obtained during this pioneering period. The only theoretical tools available then were the Reynolds' equations established in 1895 for the turbulent motion of an incompressible fluid (Reynolds, 1895); the case of a compressible fluid was, in a somewhat superficial ways touched upon for the first time by Keller and Friedmann (1924); the complete equation taking account of the temperature fluctuations in a gas was given, for the first time as far as I know, in 1937 in my paper (Kampé de Fériet, 1937).

It was J. Boussinesq (1877), who as early as 1872 pointed out that the flows which we call now "turbulent," are much too complicated to be completely described in all their details; using the Eulerian point of view, every quantity $f(x, y, z, t)$, like pressure p, temperature T, density ρ, components u, v, w of the velocity, has to be put in the form

$$f(x, y, z, t) = \bar{f}(x, y, z, t) + f'(x, y, z, t)$$

$\bar{f}$ being the mean value and f' the fluctuation of this quantity; only the mean value $\bar{f}$ can be measured. The main problem for the foundation of a theory is thus to devise a set of equations containing only the mean values of the velocity, pressures, etc.

The situation is exactly the same as ignoring the individual motions of the molecules, when we consider a gas or a liquid as a continuous medium; the velocity $u(x, y, z, t)$, $v(x, y, z, t)$, $w(x, y, z, t)$ of the continuous medium is the mean value of the velocities (U_n, V_n, W_n) of the molecules present at time t in the neighborhood of (x, y, z); the Navier–Stokes equations take account of the fluctuations of the velocity of the molecules $(u - U_n, v - V_n, w - W_n)$ by the viscosity terms.

The fundamental step was made by Reynolds (1895), who obtained a new set of equations by averaging the Navier–Stokes equations; the influence of the fluctuations on the mean motion $(\bar{u}, \bar{v}, \bar{w})$ is expressed by the tensor of the Reynolds stresses:

$$\tau_{xx} = -\rho \overline{u'^2}, \quad \tau_{xy} = -\rho \overline{u'v'}, \quad \tau_{xz} = -\rho \overline{u'w'}$$
$$\tau_{yx} = -\rho \overline{v'u'}, \quad \tau_{yy} = -\rho \overline{v'^2}, \quad \tau_{yz} = -\rho \overline{v'w'}$$
$$\tau_{zx} = -\rho \overline{w'u'}, \quad \tau_{zy} = -\rho \overline{w'v'}, \quad \tau_{zz} = -\rho \overline{w'^2}$$

The exact meaning of the process by which the mean value $\bar{f}$ is obtained is extremely important for correct use of the Reynolds equations; research on this subject has developed along two different lines.

a. Mathematical Point of View

Averaging the Navier–Stokes equations, Reynolds makes use of some properties of the mean; it has been pointed out by Oseen in 1930 that neither a time average over a finite interval of time nor a space average over a bounded volume has the properties used by Reynolds. In 1935, I proved that the Reynolds equations are logical consequences of the Navier–Stokes equations if and only if the mean satisfies the following conditions:

$$(R_1) \quad \overline{f+g} = \bar{f} + \bar{g}$$

$$(R_2) \quad \overline{\alpha f} = \alpha \bar{f}, \quad \alpha \equiv \text{const}$$

$$(R_3) \quad \overline{\bar{f}g} = \bar{f}\bar{g}$$

(R_2 and R_3 imply $\bar{\bar{f}} = \bar{f}$, thus $\overline{f'} = 0$)

$$(R_4) \quad \overline{\partial f/\partial x} = \partial \bar{f}/\partial x, \quad \overline{\partial f/\partial y} = \partial \bar{f}/\partial y,$$
$$\overline{\partial f/\partial z} = \partial \bar{f}/\partial z, \quad \overline{\partial f/\partial t} = \partial \bar{f}/\partial t$$

This remark has lead since 1949 to a large number of purely mathematical papers by Garrett Birkhoff, Gian Carlo Rota, John Sopka, John Kelley, J. B. Miller, and many others; except for trivial cases, it is extremely difficult (Kampé de Fériet, 1946, 1949) to satisfy conditions (R_1)–(R_4); to obtain a large class of solutions, one has to use a statistical interpretation, considering a function $f(x, y, z, t)$ as a *sample* of a random function $f(x, y, z, t, \omega)$, a Gibbsian *ensemble* of flows, ω being chosen at random in a given probability space $(\Omega, \mathscr{S}, P)$; with this interpretation the mean is considered as the statistical average (mathematical expectation):

$$\bar{f}(x, y, z, t) = \int_\Omega f(x, y, z, t, \omega) \, dP$$

In Section 6 we will come back to this interpretation, which started at the end of the period, slowly penetrated all theoretical research, and gave many important results after 1945.

b. Experimental Point of View

Much more fruitful has been the discussion of the experimental data collected about atmospheric turbulence: the clarification of the notion of the *scale of turbulence*, which emerged around 1930, is of paramount importance for understanding turbulent atmospheric diffusion.

Every instrument is able to measure fluctuations in a fluid if they last more than a time characteristic of this instrument; while in a given flow some instruments measure a constant velocity, some others will manifest very rapid fluctuations around the mean velocity. Two recordings of wind velocity by a cup anemometer and a hot-wire anemometer look at first glance very different. The gusts lasting less than a few seconds completely disappear in the cup anemometer recording; on the contrary they appear as very sharp peaks on the hot-wire recording. Magnan (1931; Huguenard et al., 1924, 1928), who was the first, in 1929, to use systematically the hot-wire anemometer in the atmosphere near the ground, has detected fluctuations of the wind velocity, which were completely smoothed by the cup anemometer; to give only one example, on the seashore (isle de Ré) he has observed a wind velocity increasing from 12 to 21 m/sec in a time of $\frac{1}{3}$ sec.

In the kinetic theory the velocity fluctuations about their mean value (the velocity of the continuous medium) are attributed to the molecules; in a turbulent flow it is much more difficult to give a precise formulation; the role of the molecules has to be played by "cells" or "balls" of some sort, which one calls "eddies." It is remarkable that the first paper of Sir Geoffrey Taylor (1915a) on atmospheric turbulence has as its title, "Eddy diffusion in the atmosphere"; but one "eddy" cannot have a sharp definition, like a molecule of the kinetic theory; when at the point x, y, z at time t one records a velocity fluctuation

$$u'(x, y, z, t), \quad v'(x, y, z, t), \quad w'(x, y, z, t)$$

this is interpreted as the passing of one eddy, a ball of fluid rotating around a center, this center moving with the mean velocity

$$\bar{u}(x, y, z, t), \quad \bar{v}(x, y, z, t), \quad \bar{w}(x, y, z, t)$$

But the analogy must not be pressed to much; the molecules—at least in the naïve view of classical kinetic theory are permanent objects and their energy can change only by collisions; the balls, rolling around themselves, which are called eddies, have no sharp boundaries and their energy is continually changing; nevertheless these rather fuzzy objects gave a significant meaning to the "scale of turbulence" which is fundamental for any interpretation of turbulent atmospheric diffusion. The scale is roughly defined by the diameter of the eddies, whose influence is predominant on the scattering of the material under consideration. Dedebant and Wehrlé (1935) have placed great emphasis on this point of view; for instance, in the diffusion of the smoke of a chimney the significant eddies might have an average diameter of one meter (Dobson, 1919); for the diffusion of pollen over the Baltic sea, this average diameter might be several hundred meters (Schmidt, 1925); for the diffusion over the whole earth of the particles blown up by the Krakatoa, eddies figuring the general circulation of the atmosphere were responsible.

It was only when spectral analysis (in time and space) of the turbulent energy of the wind was finally introduced that a precise definition of the scale of turbulence became possible, but it must be stressed that the rather loose description of an eddy has proved to be the key opening for the interpretation of many aspects of turbulent atmospheric diffusion.

The most complete series of measurements in this early period were made by the British "Wind Structure Subcommittee," over a large flat plain at Cardington (U.K.); seven poles (one of them 150 ft high) were equipped with Dines anemometers, a modified type of Pitot tube, giving much finer recordings of the wind velocity than a cup anemometer; temperature was measured by electric thermometers on the ground and at the top of the highest pole; the range occupied by the poles was 4000 × 700 ft. The book published by Giblett (1932) contains a considerable amount of data; from the discussion of C. Durst it results that the eddies, over this plain, in the layer $0 < z < 150$ m are essentially of two types. *Type I:* Friction eddies, identical in structure, but much larger than eddies observed in the boundary layer near a wall in a wind tunnel; the axes of these eddies are oriented at random, in all directions; their diameter never exceeds 100 ft. If the mean temperature gradient

$$\beta = -d\bar{T}/dz$$

is overadiabatic, $\beta > \beta_0$, where $\beta_0 = 1°C/100$ m is the *dry adiabatic lapse-rate*, these eddies are very abundant. On the contrary, if the gradient is subadiabatic ($\beta < \beta_0$) and in particular when there is a temperature inversion ($\beta < 0$), they are rare and are very rapidly damped. *Type II:* Convection eddies are like the well-known Benard cells developing in a liquid over a heated wall; their axes are horizontal; their diameter in the wind direction is between 3000 and 8000 ft, and normally to this direction, between 600 and 2000 ft.

Many other interesting results are given in this report about the repartition of the turbulent energy of the wind in the direction of the mean wind velocity and normally to this direction. Scrase (1930) has carried out with a three-dimensional vane a series of measurements giving some insight on the fluctuations of the wind velocity at 1.5 and 19 m above the ground. As an example, for one series, when the mean velocity at $z = 1.5$ m was $u = 4.70$ m/sec, he gives the values

	$\overline{|u'|}$	$\overline{|v'|}$	$\overline{|w'|}$	
$z = 1.5$ m	57	66	43	cm/sec
$z = 19$ m	55	40	31	cm/sec

he has also computed the values of the corresponding Reynolds stresses:

$$-\rho\overline{u'w'} = 3.62, \quad -\rho\overline{w'v'} = 1.50, \quad -\rho\overline{v'u'} = 0.58 \quad \text{(cgs)}$$

4.

Turbulent atmospheric diffusion is due to the turbulent fluctuations of the wind velocity; but the fluctuations can be defined only if one knows the mean motion, thus the whole theoretical frame depends on knowledge of the mean motion.

As a first approximation, in the lower layer of the atmosphere over a flat country, one has made the following hypotheses:

(a) The mean motion is horizontal and depends only on the altitude above the ground:

$$\bar{u} = \bar{u}(z), \qquad \bar{v} = \bar{v}(z), \qquad \bar{w} = 0$$

(b) The density of the air is constant all over the layer:

$$\rho = \bar{\rho} = \text{const}$$

(this is a very crude assumption, neglecting the temperature fluctuations and assuming that the thickness of the layer is very small).

(c) The Reynolds stresses depend only on z.

Thus the Reynolds equations reduce to

$$(2\omega \cos \varphi)\rho\bar{v} = \frac{\partial \bar{p}}{\partial x} - \frac{d}{dz}\left(\rho v \frac{d\bar{u}}{dz} - \overline{\rho u'w'}\right)$$

$$(-2\omega \cos \varphi)\rho\bar{u} = \frac{\partial \bar{p}}{\partial y} - \frac{d}{dz}\left(\rho v \frac{d\bar{v}}{dz} - \overline{\rho v'w'}\right)$$

$$(2\omega \sin \varphi)\rho\bar{v} = \frac{\partial \bar{p}}{\partial z} + \rho g + \frac{d}{dz}\left(\overline{\rho w'^2}\right)$$

where ω is the angular velocity of the Earth's rotation and φ the latitude. In most of the studies one still oversimplifies, neglecting the Earth's rotation and the horizontal pressure gradient; thus the rotation of the mean wind velocity (Eckmann spiral) disappears and one can take $\bar{v} = 0$ (the mean motion is parallel to the x axis at any altitude z); then from the simplified Reynolds equations one deduces

(4.1) $$\rho v \frac{d\bar{u}}{dz} - \overline{\rho u'w'} = \text{const} = \tau_0$$

(4.2) $$\overline{\rho v'w'} = \text{const} = 0$$

(4.3) $$\bar{p} + \rho g z + \overline{\rho w'^2} = \text{const} = \bar{p}_0 \qquad \text{(mean pressure on the ground)}$$

because, the ground being assimilated to a flat plate, one has

$$\overline{\rho u'w'} = 0, \quad \overline{\rho v'w'} = 0, \quad \overline{\rho w'^2} = 0 \quad \text{for} \quad z = 0$$

Thus the horizontal stress on the ground reduces to the viscous stress:

$$\tau_0 = \rho v \left.\frac{d\bar{u}}{dz}\right|_{z=0}$$

In this sublayer neglecting the Reynolds stress the mean velocity profile is given by

$$\bar{u} = (\tau_0/\rho v)z$$

But out of this sublayer the Reynolds stress becomes much larger than the viscous stress, which can be neglected,

$$\tau_0 = -\overline{\rho u'w'}$$

The essential step is made when one introduces [using one idea first expressed by Boussinesq in 1872 (see Boussinesq, 1877)] the turbulent viscosity coefficient $\gamma^*(z)$ defined by

(4.4) $$\gamma^*(z)\, d\bar{u}/dz = -\overline{u'w'}$$

whence

(4.5) $$\tau_0 = \rho\gamma^*(z)\, d\bar{u}/dz$$

The turbulent viscosity coefficient $\gamma^*(z)$ [often designated by the symbol K, the *Austauschgrösse* for Schmidt (1925)] had a leading role in the research about turbulent atmospheric diffusion during the period 1920–1945.

Equation (5) shows clearly that there is a connection between the mean velocity profile $\bar{u}(z)$ and the turbulent viscosity coefficient $\gamma^*(z)$.

To the degree of approximation assumed here, the density ρ being constant, one can take advantage of the mean velocity profiles corresponding to turbulent flows of an incompressible fluid over a flat plate.

Around 1925 a power law was first proposed:

$$\bar{u}(z) = \bar{u}(z_1)(z/z_1)^p$$

Giblett (1932), using the measurements made at Cardington found for the exponent p:

$p = 0.01$ if $\beta > \beta_0$

$p = 0.60$ for large temperature inversions $\beta < 0$

But in 1930 the logarithmic law was discovered for the mean velocity profile, a law which was tested in the laboratory in many experiments and agreed with with great accuracy; Prandtl (1932) and von Kármán (1930) gave theoretical justifications of this law based respectively on the (very controversial) idea of the *Mischungsweg* and on general similarity considerations (which seemed quite plausible). Prandtl's formula reads

(4.6) $$\bar{u}(z) = 5.75(\tau_0/\rho)^{1/2} \log(30z/k)$$

where k is a length characteristic of the roughness of the plate. To this velocity profile corresponds the linear law for the turbulent velocity coefficient

(4.7) $$\gamma^*(z) = 0.174(\tau_0/\rho)^{1/2}z$$

A considerable amount of experimental research has been devoted to the verification of the logarithmic law in the lower layer of the atmosphere. One of the most complete series of measurement is due to Rossby and Montgomery (1935); they use the formula

(4.8) $$\bar{u}(z) = 5.75(\tau_0/\rho)^{1/2} \log[(z + z_0)/z_0]$$

which has the advantage over (4.6) of giving $\bar{u}(0) = 0$. When the temperature gradient is adiabatic, $\beta = 1°C/100$ m, they found rather good agreement, taking the values (cm)

$z_0 = 0.25$ over snow

$z_0 = 0.50$ over grass

$z_0 = 4.00$ over a calm sea

Paeschke (1937) introduces two constants z_0 and z'_0

(4.9) $$\bar{u}(z) = 5.75(\tau_0/\rho)^{1/2} \log[(z - z'_0)/z_0]$$

His measurements, done with a hot-wire anemometer, are the most precise and were done over a great number of varied grounds. He gives the following values (cm)

$z'_0 = 3$	$z_0 = 0.5$	over snow
$z'_0 = 20$	$z_0 = 3.2$	over short grass
$z'_0 = 30$	$z_0 = 3.9$	over high grass
$z'_0 = 45$	$z_0 = 6.7$	over field of beets.

It must be pointed out that if $z > 1$ m, the three formulas (4.6), (4.8), and (4.9) practically coincide.

We did ourselves a series of measurements during World War II (unpublished) in the layer $0 < z < 10$ m over the Hamada, a very flat plateau in the Sahara. The Hamada extends over hundreds of miles, thus the homogeneity of the flow of air was exceptionally good, the boundary conditions on the ground being uniform over a great distance. The influence of the temperature gradient over the mean velocity profile was clearly marked. The logarithmic law was well verified when the temperature gradient was overadiabatic $\beta > \beta_0$. During the strong temperature inversions occurring before dawn (due to the intense radiation of the ground), there was no mean velocity profile at all; the wind velocity at $z = 1$ m could be much greater, during several minutes, than the velocity at $z = 10$ m. The exchanges between two contiguous layers being promptly damped, each layer was flowing independently of the others.

As a rule the experimental values of $\gamma^*(z)$ are deduced from the observed mean velocity profile $\bar{u}(z)$ through (4.5); I know few computations of $\overline{u'w'}$ from the recordings of u' and w'. In Giblett (1932) there is quite a number of results giving the mean value γ_m^* of $\gamma^*(z)$ for the layer less than 150 ft above the ground. A statistical table summarizes the average values of γ_m^*, tabulated as a function of the mean temperature gradient β and the mean wind velocity $\bar{u}$ at 150 ft. For high wind velocities (25–29 mi/hr), γ_m^* is not very sensitive to β; on the contrary for low wind velocities (10–14 mi/hr), γ_m^* decreases in the ratio of 1/50 when β changes from superadiabatic to an inversion.

From its definition (4.4), γ^* is a function of the velocity fluctuations u', w'; thus it must depend on the scale of turbulence, corresponding to the observed flow; a rather large amount of experimental work is available: Dobson (1919) diffusion of smoke from a chimney; Roberts (1923) and Lehmann (1933) diffusion of acetylene near the ground; Frenkiel (1947) various chemical pollutants; Taylor (1915a) diffusion of warm air over a cold sea; Schmidt (1925) pollen scattering; Dobson and Taylor (see Haude, 1933) balloons; etc. Richardson (1920b) has made the most extensive series of experiments using chimney smokes (even from a ship) and many chemical fumes (NH_4Cl, P_2H_5, etc.). The smallest value of γ^* observed has been 78 cm^2 sec^{-1} at $z = 30$ cm above lawn in almost still air and the greatest 1.6×10^6 cm^2 sec^{-1} at $z = 250$ m above a forested hill with a wind velocity $\bar{u} = 16$ m sec^{-1}. From a general discussion of the known results Richardson (1926), suggests the orders of magnitude shown in Table I.

TABLE I

Scale	Diameter of eddies	γ^* (cm^2 sec^{-1})
Molecular		0.2
Aerology	100 m	10^5
Synoptic meteorology	100 km	10^8
General circulation	10,000 km	10^{11}

5.

If the mean motion is unsteady, one must add to the left-hand member of the Reynolds equations the mean accelerations $\overline{\gamma_x}$, $\overline{\gamma_y}$, $\overline{\gamma_z}$, thus assuming that
$$\bar{u} = \bar{u}(z, t), \qquad \bar{v} = 0, \qquad \bar{w} = 0$$
and making otherwise the same assumptions as in Section 4; Eq. (4.1) is replaced by

(5.1) $$\rho \frac{\partial \bar{u}}{\partial t} = \frac{\partial}{\partial z}\left(\rho v \frac{\partial \bar{u}}{\partial z} - \rho \overline{u'w'}\right)$$

neglecting the molecular viscosity, defining γ^* by (4.4) and dividing by the constant $\rho > 0$, one obtains

(5.2) $$\frac{\partial \bar{u}}{\partial t} = \frac{\partial}{\partial z}\left[\gamma^*(z) \frac{\partial \bar{u}}{\partial z}\right]$$

If and only if $\gamma^*(z)$ is a constant, this equation reduces to the classical diffusion equation (sometimes called the Fick equation) which is identical with the equation given by Fourier for the conduction of heat in a rigid body (heat equation); practically all the applications reduce to this particular case, supposing implicitly that the layer of atmosphere considered is small enough to neglect the variation of $\gamma^*(z)$. Let us note that

$$\rho \frac{\partial}{\partial z}\left[\gamma^*(z) \frac{\partial \bar{u}}{\partial z}\right]$$

represents *the gain of momentum* of the layer z by unit volume of fluid. This gain is precisely produced by the velocity fluctuations w'; particles of the neighboring layers constantly penetrate in the layer z and their action is expressed by the Reynolds stress $\tau(z) = -\rho \overline{u'w'}$. From this point of view, Eq. (5.2) expresses the diffusion of momentum, under the particular assumptions made about the flow.

Let us now consider diffusion in an incompressible fluid of a material in equilibrium with the fluid; the velocity of a particle of the material is the same as the velocity of the surrounding fluid; moreover the material is neither created nor destroyed; let us call $s(x, y, z, t)$ the density of the material. Averaging the condition

$$ds/dt = 0$$

one obtains the general diffusion equation for a conservative material in equilibrium with an incompressible fluid:

(5.3) $$\frac{\partial \bar{s}}{\partial t} + \frac{\partial}{\partial x}(\overline{u}\overline{s}) + \frac{\partial}{\partial y}(\overline{v}\overline{s}) + \frac{\partial}{\partial z}(\overline{w}\overline{s}) = -\frac{\partial}{\partial x}(\overline{u's'}) - \frac{\partial}{\partial y}(\overline{v's'}) - \frac{\partial}{\partial z}(\overline{w's'})$$

Making the following assumptions:

(a) $\bar{u} = \bar{u}(z,t)$, $\quad \bar{v} = 0$, $\quad \bar{w} = 0$;
(b) $\bar{s} = \bar{s}(z,t)$;
(c) $\overline{u's'}$, $\overline{v's'}$, $\overline{w's'}$ are constant in the horizontal plane;

one obtains

(5.4) $$\partial \bar{s}/\partial t = -\partial/\partial z(\overline{w's'})$$

Now introducing a turbulent diffusion coefficient

(5.5) $$\gamma_s^*(z,t)\,\partial \bar{s}/\partial z = -\overline{w's'}$$

we can write the diffusion equation

(5.6) $$\frac{\partial \bar{s}}{\partial t} = \frac{\partial}{\partial z}\left(\gamma_s^* \frac{\partial \bar{s}}{\partial z}\right)$$

It is now clear that (5.2) corresponds to the particular case

$$s = \rho u$$

It must be kept in mind that the equilibrium of a material with the surrounding atmosphere is very seldom satisfied; if the density of the material is different from the air density, every particle of the material has a relative vertical velocity, ascendant if it is lighter than air, descendant if it is heavier, and Eq. (5.6) has to be modified.

For heat diffusion, one must take into account the variation of heat produced by a change of volume of the fluid. The turbulent heat diffusion equation reads

(5.7) $$\frac{\partial \bar{T}}{\partial t} = \frac{\partial}{\partial z}\left[\gamma_T^*\left(\frac{\partial \bar{T}}{\partial z} + \beta_0\right)\right]$$

where $\gamma_T^* \partial \bar{T}/\partial z = -\overline{w'T'}$ and $\beta_0 \equiv$ adiabatic lapse rate $= 1°C/100$ m. Let us note that (5.7) is a particular case of the general heat turbulent diffusion given in Kampé de Fériet (1937).

Now arises a fundamental question: does γ_s^* depend on the material? Is it true that $\gamma_s^* = \gamma^*$? The question did not receive a final answer in the period we are considering; there has been a lot of controversy; but, pro or con, a final argument has never been produced; there were only some hints that γ_s^* was dependent, even very highly, on the material: for instance G. I. Taylor, from a few observations in the ocean deduced that the diffusion coefficient is 19 times greater for heat than for salinity; but, as far as I know, no such conclusion has been finally established for the diffusion of different materials in the atmosphere.

The influence of the gradient of temperature

$$\beta(z) = -\partial \bar{T}/\partial z$$

on the structure of atmospheric turbulence has been known for some time; an elementary proof, sketched in Fig. 2, shows how a particle having a

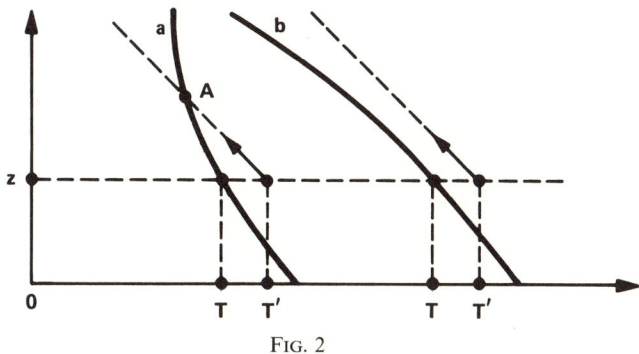

Fig. 2

temperature T' greater than the temperature $\bar{T}(z)$ of the layer, starts an ascending motion, which is damped when $\beta(z) < \beta_0$ (a), or accelerated when $\beta(z) > \beta_0$ (b). Richardson (1920a), Taylor (1931), and Prandtl (1929, 1932) have examined the problem of the damping of small perturbations of a shear flow having a mean velocity $\bar{u}(z)$; all three conclude that the damping depends only on the value of the nondimensional parameter

$$\gamma(z) = \frac{g[\beta_0 - \beta(z)]}{\bar{T}(z)(\partial \bar{u}/\partial z)^2}$$

the perturbations are damped only if $\gamma > \gamma_0$. Turbulence could appear only if the perturbations are amplified, when $\gamma < \gamma_0$. But they disagree on the critical value of γ_0: for Taylor, $\gamma_0 = \frac{1}{4}$; for Richardson, $\gamma_0 = 1$; and for Prandtl, $\gamma_0 = 2$. This shows that, at that time, these types of problems were not ripe for a final solution.

During our research on the Hamada in the Sahara (December 1939–March 1940) we had an opportunity to observe how the night temperature inversion is transformed rapidly in a large superadiabatic lapse rate and to detect the curious effect of this transformation on atmospheric diffusion in the layer $0 < z < 10$ m near the ground. During the night the ground is radiating strongly; the air being dry and the sky clear, only a small fraction of this radiation is coming back and the ground is cooling rapidly; the ground which at noon may have a temperature of 40°C comes very close to the freezing point just before dawn and the air in contact with the ground becomes very cool; a strong inversion of temperature occurs then in the

atmosphere, extending frequently up to 1500 m; for instance,
$$\bar{T}(100\text{ m}) - \bar{T}(0) = +10°C$$
Immediately after sunrise the ground, absorbing the solar radiation, warms quickly; on the average, between February 15 and March 15, the temperature increase of the ground was 6.5°C per hour, with a maximum of 9.2°C on March 3; the air in contact with the ground becomes warmer, the temperature takes the shape 1; the layer $0 \leq z \leq h$ is unstable and the inversion starts at $z = h$. Gradually the unstable layer becomes thicker and the inversion begins at a higher level (Fig. 3); for instance, 30 min after

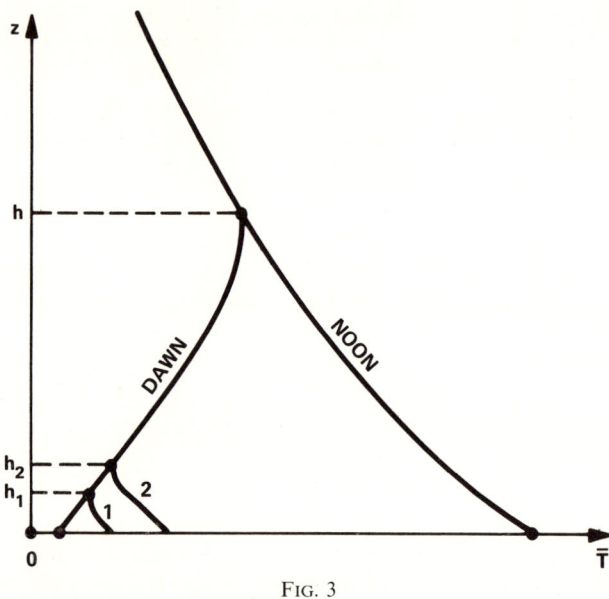

Fig. 3

sunrise h is of the order of magnitude of 1 m.

This gradual change of the mean temperature distribution $\bar{T}(z, t)$ has a remarkable effect on atmospheric diffusion near the ground. A chemical pollutant was continuously emitted from a line of sources near the ground; before dawn, the wind being light, the cloud of pollutants was flowing quietly over the ground, like a steady ribbon, for miles; but after sunrise, the basis of the cloud separated more and more from the ground, still flowing quietly, but now over a layer of clear air.

In order to compare the theory with experiment, an expression of the integral of the heat diffusion equation (5.7) was evaluated as a function of the initial temperature distribution in the atmosphere (Kampé de Fériet 1942):
$$\bar{T}(z, 0) = f(z)$$

and of the evolution of the temperature of the ground:

$$\bar{T}(0, t) = g(t)$$

supposing that $\gamma_T^*(z, t)$ had a constant value γ_m^*. Taking as a numerical example

$$f(z) = -\beta z + T_0, \qquad g(t) = Bt + T_0$$

for $\gamma_m^* = 200$ cm² sec⁻¹, the computations gave the basis of the inversion at $h = 1$ m for $t = 38$ min 20 sec after sunrise. The general diffusion equation (5.3) has been considered under much less restrictive conditions than (a)–(c), leading to Eq. (5.6). In particular, let us suppose

$$\bar{u} = \text{const}, \qquad \bar{v} = \bar{w} = 0$$

and let us introduce the three turbulent **diffusion coefficients** defined by

(5.8a) $\qquad \gamma_{s,x}^* \, \partial \bar{s}/\partial x = -\overline{u's'}$

(5.8b) $\qquad \gamma_{s,y}^* \, \partial \bar{s}/\partial y = -\overline{v's'}$

(5.8c) $\qquad \gamma_{s,z}^* \, \partial \bar{s}/\partial z = -\overline{w's'}$

the diffusion equation now reads

(5.9) $\qquad \dfrac{\partial \bar{s}}{\partial t} + \bar{u}\dfrac{\partial \bar{s}}{\partial x} = \dfrac{\partial}{\partial x}\left(\gamma_{s,x}^* \dfrac{\partial \bar{s}}{\partial x}\right) + \dfrac{\partial}{\partial y}\left(\gamma_{s,y}^* \dfrac{\partial \bar{s}}{\partial y}\right) + \dfrac{\partial}{\partial z}\left(\gamma_{s,z}^* \dfrac{\partial \bar{s}}{\partial z}\right)$

If one assumes that the diffusion is isotropic and that

$$\gamma_{s,x}^* = \gamma_{s,y}^* = \gamma_{s,z}^* = \gamma_s^* = \text{const}$$

then the diffusion equation reduces to

(5.10) $\qquad \dfrac{\partial \bar{s}}{\partial t} + \bar{u}\dfrac{\partial \bar{s}}{\partial x} = \gamma_s^* \nabla^2 \bar{s}$

For the integrals referring to an instantaneous point source, a continuous point source, and a continuous line source, we refer to Frenkiel (1949).

6.

All the theoretical research briefly described in the preceding sections use Eulerian variables, whereas the Lagrangian particle displacement is a rather natural variable in studying dispersion. If so much use of the Eulerian point of view has been made, this is due to the fact that the Navier–Stokes equations and their statistical translation, the Reynolds equations, are more straightforward than their Lagrangian counterpart; but it is obvious that in order to study the scattering of particles, the consideration of their trajectories should be very illuminating.

The introduction in turbulence diffusion theory of the Lagrangian point of view was made in G. I. Taylor's (1921) famous paper, "Diffusion by continuous movements." He considers what could be called a "linear homogeneous turbulent flow":

$$\bar{u} = 0, \qquad \bar{v} = 0, \qquad \bar{w} = 0$$

$$u'(x, y, z, t) = u'(t), \qquad v'(x, y, z, t) = w'(x, y, z, t) = 0$$

using for the mean $\bar{f}$ a time average, he supposes

(6.1) $$\overline{u'(t)^2} = U_0^2 \qquad \text{independent of time}$$

and introduces the correlation coefficient of the turbulent velocity

(6.2) $$\overline{u'(t)u'(t+h)} = U_0^2 R(h)$$

the independence of the time t of $\overline{u'(t)u'(t+h)}$ is a definition of the homogeneity of turbulence, (6.1) being then a particular case of (6.2). Now considering the displacement of a particle, starting from the origin at time $t = 0$:

$$x(t) = \int_0^t u'(s)\, ds, \qquad y(t) = z(t) = 0$$

the main result of G. I. Taylor is the formula

(6.3) $$\overline{x(t)^2} = U_0^2 \int_0^t \int_0^t R(s_2 - s_1)\, ds_1\, ds_2$$

One can compute the dispersion of the particles if one knows the correlation of the turbulent velocity $u'(t)$; from (6.3) he obtains this remarkable asymptotic expression:

(6.4) $$\overline{x(t)^2} = At \quad \text{for } t \text{ large}$$

From this formula G. I. Taylor deduces a result that has been the foundation of one of the most popular experimental methods of measuring γ^* in atmospheric diffusion. Let us assume that in the neighborhood of a chimney (Fig. 1):

(a) $\bar{u} = \text{const}, \qquad \bar{v} = \bar{w} = 0$
(b) $u' = v' = 0$
(c) $w'(t)$ satisfies (6.2)

A particle of smoke emitted by the chimney at the time t_0 will have trajectory

$$x(t) = \bar{u}(t - t_0), \qquad y(t) = 0, \qquad z(t) = \int_0^t w'(s)\,ds$$

Thus, from (6.4), if $t - t_0$ is large enough,

$$\overline{z(t)^2} = A(t - t_0) = Ax(t)/\bar{u}$$

But the percentage of smoke contained in the interval $[-z(t), +z(t)]$ being proportional to $\overline{z(t)^2}$, it is clear that the domain of the plane O, z containing a given percentage of smoke (e.g., 90 %) is a parabola. Moreover, Richardson (1920b) has shown that $A = 2\gamma^*$; thus any determination of the parameter p of the diffusion parabola

(6.5) $$z^2 = 2pX$$

gives directly the value of $\gamma^* = p\bar{u}$. Many experimental determinations of γ^* [Dobson (1919), Lehmann (1933), Frenkiel (1947), Richardson (1920b)] are based on this remark.

G. I. Taylor made another step of great importance, introducing the idea of the energy spectrum of his linear homogeneous turbulence, defined as the Fourier transform of the correlation:

(6.6) $$f(\omega) = (2/\pi) \int_0^{+\infty} \cos(\omega s) R(s)\,ds$$

The amount of kinetic energy coming from the band

$$\omega' \leq \lambda \leq \omega''$$

is equal to

$$U_0^2 \int_{\omega'}^{\omega''} f(\lambda)\,d\lambda$$

One would never admire enough how his marvelous physical intuition, using rather elementary mathematical tools, lead Taylor (1938) to open this new road, along which Wiener (1930, 1933) has developed his powerful "generalized harmonic analysis," solving problems which were completely out of reach of the classical Fourier harmonic analysis. At the time Taylor (1921) was writing his paper, probability theory was in its infancy: the first rigorous definition of a random function was made two years later by Wiener (1923) in his pioneering paper on Brownian motion; there are many obvious similarities between turbulence and Brownian motion, the most striking being the identity of Taylor's asymptotic expression (4) and the famous Einstein formula for the dispersion of particles; it became more and more clear that the theory of random functions, which has mostly developed after 1930, was the right tool to use in the theory of turbulence; as we pointed out in discussing the Reynolds equations, it is much more convenient to define the mean value by a statistical average than by a time (or space) average. Taking advantage of the progress recently made by A. Khintchine, A. Kolmogoroff, N. Wiener, P. Lévy, we applied (Kampé de Fériet, 1939) the theory of stationary random functions to the linear homogeneous turbulence of G. I. Taylor. We consider a Gibbsian ensemble of velocity fields:

$$u(t, \omega), \quad v(t, \omega) = w(t, \omega) = 0$$

$u(t, \omega)$ being a *stationary random function continuous in quadratic mean*:

(6.7) $$\overline{u(t, \omega)} = 0$$

(6.8) $$\overline{u(t, \omega)u(t + h, \omega)} = U_0^2 R(h)$$

(6.9) $$\lim_{h \to 0} R(h) = R(0) = 1$$

A. Khintchine has proved that (6.9) implies the representation of the correlation by a Fourier–Stieltjes integral

(6.10) $$R(h) = \int_0^{+\infty} \cos(\lambda h) \, d\mathscr{F}(\lambda)$$

the function $\mathscr{F}$ being nondecreasing, continuous to the left:

$$\mathscr{F}(\lambda - 0) = \mathscr{F}(\lambda)$$

$\mathscr{F}(\lambda) = 0$ for $\lambda \leq 0$ and $\mathscr{F}(+\infty) = 1$. This function represents the *integrated energy spectrum* because the mean kinetic energy (per unit of mass) of the turbulent fluctuations corresponding to the wave numbers λ such that $\lambda_1 \leq \lambda < \lambda_2$ is precisely equal to $U_0^2[\mathscr{F}(\lambda_2) - \mathscr{F}(\lambda_1)]$. This spectrum generalizes Taylor's spectrum f. In the particular case when $\mathscr{F}$ is absolutely continuous, one has $f(\lambda) = d\mathscr{F}/d\lambda$ (pure band spectrum). In the general case, a line spectrum could be superposed on the band spectrum, the line λ_0 contributing a finite amount of energy: $U_0^2[\mathscr{F}(\lambda_0 + 0) - \mathscr{F}(\lambda_0)]$. The spectrum does not contain any line if and only if

$$\lim_{H \to +\infty} (1/H) \int_0^H R(h)^2 \, dh = 0$$

The dispersion of particles moving along the Ox axis with the turbulent velocity $u(t, \omega)$ is given by

(6.11) $$\overline{x(t, \omega)^2} = 2U_0^2 \int_0^t (t - s)R(s) \, ds$$

Taylor double integral being replaced by a more tractable simple integral;

one has also

(6.12) $$\overline{x(t,\omega)^2} = 2U_0 \int_0^{+\infty} \frac{1-\cos\lambda t}{\lambda^2} d\mathscr{F}(\lambda)$$

From this formula one deduces that the increases of dispersion with time t depends essentially on the behavior of the spectrum at very small wave numbers; if there exists a wave number $\alpha > 0$ such that $\mathscr{F}(\alpha) = 0$ (no energy corresponding to wave numbers $\lambda \leq \alpha$), then the dispersion is bounded:

$$\overline{x(t,\omega)^2} \leq 4U_0^2/\alpha^2, \quad \text{for all } \bar{t} > 0$$

One can also refine the asymptotic formula (6.4): If $hR(h)$ is absolutely integrable in $[0, +\infty]$, one has

(6.13) $$\overline{x(t,\omega)^2} = 2U_0^2[At - B + \eta(t)]$$

where:

$$A = \int_0^{+\infty} R(h)\,dh, \qquad B = \int_0^{+\infty} hR(h)\,dh$$

the function $\eta(t)$ tending toward 0 when $t \to +\infty$.

These few results, abstracted from Kampé de Fériet (1939), show clearly enough what powerful tools the new theory of random functions has supplied to theoretical research on turbulent diffusion; nevertheless it must be kept in mind that in the experiments one is always working on *one sample* of the random function and using time averages, like Taylor; only an implicit use of some kind of ergodic theorem could justify the use of formulas established for statistical averages in computations based on time averages.

Another major difficulty arises from the lack of connection between the Eulerian and the Lagrangian points of view in the statistical theory of turbulence. Even from a practical, purely computational point of view, it was urgent to classify and relate together the innumerable definitions used for the correlation and the spectrum; a thorough investigation and comparison has been made by Frenkiel (1948) (the work was completed in 1942 but published after the war, due to the circumstances), which has, at the same time, given several analytic representations well adapted to the various types of experimental correlations and spectra.

Nevertheless it must be stressed that despite the difficulties of interpretation we have mentioned, the theory has been a safe basis for experimentation, and formula (6.13) has been an invaluable guide in research on turbulent diffusion; as an example let us take the research done at l'Institut de Mécanique des Fluides de Lille in 1936–1938 (Kampé de Fériet, 1938b). Small soap bubbles (3–4 mm diameter) were emitted at point E of the horizontal wind tunnel (Fig. 4); the relative velocity of the bubbles with respect

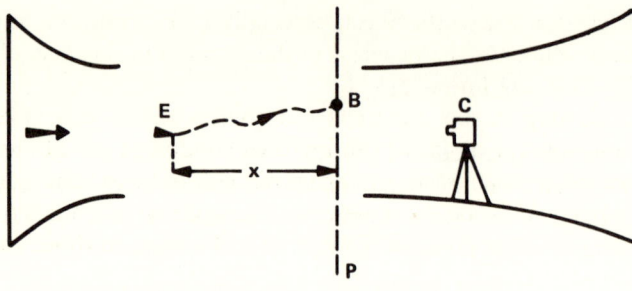

Fig. 4

to the air was less than 1 or 2 cm/sec; a plane P perpendicular to the mean velocity was lighted and the passage of the soap bubbles through this plane was recorded by the camera C. From 1000 to 2000 bubbles were emitted in each experiment (see the picture of a cloud of 1568 bubbles in Frenkiel, 1949, Fig. 8); the dispersion σ^2 of the bubbles from the center of the cloud was computed; Martinot-Lagarde has proved that the experimental results were well represented by the formula

$$\sigma^2 = 2 \frac{(\overline{v'^2})^{1/2}}{U_0} L(x - \alpha L)$$

where U_0 denotes the mean velocity, $\overline{v'^2}$ the dispersion of the transverse turbulent fluctuations, L the length characteristic of the scale of turbulence, and α a nondimensional coefficient depending on $R(h)$. Assuming $R(h) = \exp(-kh^2)$, α had the value 0.64; during the experiments $(\overline{v'^2})^{1/2}/U_0$ was varied from 0.030 to 0.041, depending on the grid used in the wind tunnel.

7.

Richardson has injected a great number of new ideas in research on turbulent diffusion; in particular Richardson (1926) considered the "relative" dispersion of two particles; if we call $D(t)$ the distance of these particles, he has suggested the empirical "law for diffusivity":

$$d\overline{D(t)^2}/dt = C[\overline{D(t)^2}]^{2/3}$$

It is very interesting to note that much later a demonstration of this law was given by Obukhov (1941); but we shall not expand further on this point because Obukhov's proof is based on the paper of Kolmogoroff (1941), containing new and profound views on *locally isotropic turbulence*; this paper was a turning point in the development of the theory of turbulence and had many applications to atmospheric turbulence and as a consequence to turbulent atmospheric diffusion. Published in Russia in 1941, Kolmogoroff's paper was known to the West, especially to France, only after 1945; thus it did not influence the progress of the research in this field during the period 1920–1945 as I know it.

At the beginning of this lecture, I did confess that I was able only to describe the landscape as seen from my own window; I am afraid that the picture made from this narrow vantage point was not clear enough to show the vivid colors, which could carry to you the sense of youthful enthusiasm, typical of the people working in the field of turbulence, in those old days of 1920–1945.

REFERENCES

Becker, R. (1930). Untersuchung über den Feinbau des Windes mittels räumlicher Windfahnenfelder. *Beitr. Phys. Freien Atmos.* **16**, 271–288.
Best, A. C. (1935). Transfer of heat and momentum in the lowest layers of the atmosphere. *Meteorol. Off. Geophys. Mem., London* No. 65.
Boussinesq, J. (1877). Essai sur la théorie des eaux courantes (1872). *Mem. Savants Etrangers Acad., Paris.*
Budel, A. (1932). Eine photogrammetrische Methode zum Studium der Strömung und Austauschvorgänge. *Beitr. Phys. Freien Atmos.* **20**, 9–17.
Budel, A. (1933). Individuelle Bewegungen kleiner Luftmassen. *Beitr. Phys. Freien Atmos.* **20**, 214–219.
Davies, E. L., and Sutton, O. G. (1931). The present position of the theory of turbulent motion in the atmosphere. *Quart. J. Roy. Meteorol. Soc.* **57**, 405–413.
de Backer, S. (1934). Sur la turbulence de l'air atmosphérique. *Bull. Cl. Sci., Acad. Roy. Belg.* **20**, 125–131.
Dedebant, G., and Wehrle, P. (1935). Le rôle de l'échelle en météorologie. *C. R. Congr. Avanc. Sci., 68th, Paris.*
Dobson, G. M. B. (1919). Measurement of turbulence in the atmosphere by the spreading of a smoke trail. *Aeronaut. Res. Commun. R. M.* No. 67.
Frenkiel, F. N. (1946). Étude de la diffusion turbulence par une méthode colorimétrique. *C. R. Acad. Sci.* **22**, 1331–1333.
Frenkiel, F. N. (1947). Mesure de la diffusion turbulente du vent naturel dans le voisinage du sol. *C. R. Acad. Sci.* **224**, 98–100.
Frenkiel, F. N. (1948). Etude statistique de la turbulence: spectres et correlations. *Off. Nat. Etudes Rech. Aeronaut., Rapp. Tech.* No. 34. [Engl. transl.: *NACA (Nat. Adv. Comm. Aeronaut.) Tech. Memo.* No. 1436 (1958).]
Frenkiel, F. N. (1949). Turbulent diffusion. *Proc. U.S. Nav. Ordnance Lab. Res. Symp., White Oak, Md.* No. 1136, pp. 67–86.
Friedmann, A. (1924). Ueber atmosphärische Wirbel und die Turbulenz des Windes. *Beitr. Phys. Freien Atmos.* **11**, 154–163.
Giblett, M. A. (1932). The structure of wind over level country. *Meteorol. Off. Geophys. Mem., London* No. 54.
Haude, W. (1933). Temperatur und Austausch der bodennahen Luft über einer Wüste. *Beitr. Phys. Freien Atmos.* **21**, 129–143.
Hesselberg, T. (1928). Untersuchungen über die Gesetze der ausgeglichenen atmosphärischen Bewegungen. *Geofys. Publ.* **5**, No. 4.
Hesselberg, T., and Bjorkdal, E. (1929). Ueber das Verteilungsgesetz der Windunruhe. *Beitr. Phys. Freien Atmos.* **15**, 121–133.

Huguenard, E., Magnan, A., and Planiol, P. (1924). Contribution à l'étude du vent. *Bull. Lab. Morphol. Ec. Htes Etud.* No. 1.
Huguenard, E., Magnan, A., and Planiol, P. (1928). Etude sur les accélérations et les vitesses angulaires du vent naturel. *Bull. Serv. Tech. Aeronaut.* No. 49.
Hummel, E. (1931). Vergleichende Untersuchungen der Böigkeit des Windes. *Beitr. Phys. Freien Atmos.* **17**, 264–278.
Johnson, N. K. (1929). A study of the vertical gradient of temperature in the atmosphere near the ground. *Meteorol. Off. Geophys. Mem.*, London No. 46.
Kampé de Fériet, J. (1934). L'état actuel du problème de la turbulence. *Sci. Aerienne* **3**, 9–34.
Kampé de Fériet, J. (1935a). L'état actuel du problème de la turbulence. *Sci. Aerienne* **4**, 12–52.
Kampé de Fériet, J. (1935b). La turbulence du vent dans les couches inférieures de l'atmosphère. *Rev. Quest. Sci.* **27**, 18–63.
Kampé de Fériet, J. (1936). La turbulence atmosphérique. *Journees Tech. Int. Aeronaut.*, Paris pp. 431–467.
Kampé de Fériet, J. (1937). Sur les équations de la diffusion thermique par turbulence. *Ann. Soc. Sci. Bruxelles, Ser. 1* **57**, 67–72.
Kampé de Fériet, J. (1938a). Recherches sur la turbulence atmosphérique au centre national de vol sans moteur de la banne d'ordanche. *Int. Studienkomm. Motorlosen Flug, ISTUS Mitteilungbl.* **6**, 9–13.
Kampé de Fériet, J. (1938b). Some researches on turbulence. *Appl. Mech. Proc. Int. Congr.*, 5th, Cambridge, Mass. pp. 352–355.
Kampé de Fériet, J. (1939). Les fonctions aléatoires stationnaires et la théorie statistique de la turbulence homogène. *Ann. Soc. Sci. Bruxelles, Ser. 1* **59**, 143–194.
Kampé de Fériet, J. (1942). Sur l'effacement de l'inversion de température après le lever du soleil dans les couches basses de l'atmosphère. *Meteorologie* No. 40, pp. 137–149.
Kampé de Fériet, J. (1946). La notion de moyenne dans les équations du mouvement turbulent d'un fluide. *Congr. Int. Mec. Appl.*, 6th, Paris.
Kampé de Fériet, J. (1949). Sur un problème d'algèbre abstraite posé par la définition de la moyenne dans la théorie de la turbulence. *Ann. Soc. Sci. Bruxelles, Ser. 1* **63**, 156–172.
Keller, L., and Friedmann, A. (1924). Differentialgleichungen für die turbulente Bewegung einer kompressibelen Flüssigkeit. *C. R. Congr. Int. Mec. Appl.*, 1st, Delft pp. 395–405.
Köhler, H. (1932). Ein kurzes Studium des Austausches auf Grund des Potenzgesetzes. *Beitr. Phys. Freien Atmos.* pp. 91–105.
Kolmogoroff, A. N. (1941). The local structure of turbulence in incompressible fluid for very large Reynolds numbers. *Dokl. Akad. Nauk SSSR* **30**, 301–305.
Lehmann, P. (1933). Eine Azytilenmethode zur Ermittlung des Turbulenzgrades der Luft. *Z. Angew. Meteorol.* **50**, 54–57.
Magnan, A. (1931). Méthodes de mesure des variations rapides du vent. *Veroeff. Forsch. Inst. Rhoen Rossiten Ges.* No. 4, pp. 92–101.
Martinot-Lagarde, A. (1934). Sur un anémomètre peu sensible aux changements de direction du vent. *C. R. Acad. Sci.* **198**, 338.
Obukhov, A. M. (1941). Energy distribution in the spectrum of turbulent flow. *Izv. Akad. Nauk SSSR, Ser. Geogr. Geo Fiz.* **5**, 453.
Paeschke, W. (1937). Experimentelle Untersuchungen zum Rauhigkeits- und Stabilitätsproblem. *Beitr. Phys. Freien Atmos.* **24**, 163.
Prandtl, L. (1929). Einfluss stabilisierender Kräfte auf die Turbulenz. *Vortr. Gebiete Aerodyn.*, Aachen pp. 1–17.
Prandtl, L. (1932). Meteororologische Anwendung der Strömungslehre. *Beitr. Phys. Freien Atmos.* **19**, 188–202.
Prandtl, L., and Tollmien, W. (1924). Die Windverteilung über dem Erdboden errechnet aus den Gesetzen der Rohrströmung. *Z. Geophys.* **1**, 47–55.

Reynolds, O. (1883). An experimental investigation of the circumstances which determine whether the motion of water shall be direct or sinuous and of the law of resistance in parallel channels. *Phil. Trans. Roy. Soc. London, Ser. A* **174**, 935–982.

Reynolds, O. (1895). On the dynamical theory of incompressible viscous fluids and the determination of the criterion. *Phil. Trans. Roy. Soc. London, Ser. A* **186**, 123–164.

Richardson, L. F. (1920a). The supply of energy from and to atmospheric eddies. *Proc. Roy. Soc. Ser. A* **97**, 354.

Richardson, L. F. (1920b). Some measurements of atmospheric turbulence. *Phil. Trans. Roy. Soc. London, Ser. A* **221**, 28.

Richardson, L. F. (1925). Turbulence and vertical temperature near trees. *Phil. Mag.* **49**, 81.

Richardson, L. F. (1926). Atmospheric diffusion on a distance neighbour graph. *Proc. Roy. Soc., Ser. A* **110**, 709–737.

Roberts, P. F. T. (1923). The theoretical scattering of smoke in a turbulent atmosphere. *Proc. Roy. Soc., Ser. A* **104**, 640–654.

Rossby, C. G., and Montgomery, R. B. (1935). The layer of frictional influence in wind and ocean currents. *Mass. Inst. Technol. Meteorol.* No. 3(3).

Sauvegrain, J. (1934). Sur une méthode d'étude de la structure du vent au voisinage du sol. *C. R. Journees Mec. Fluides, Lille*.

Schmidt, W. (1925). "Der Massenaustausch in freier Luft und verwandte Erscheinungen." Hamburg.

Schmidt, W. (1929). Die Struktur des Windes. *Wien. Sitzungsber. M. N. Kl., Abt. 2* **138**, 85.

Scrase, F. J. (1930). Some characteristics of eddy motion in the atmosphere. *Meteorol. Off. Geophys. Mem., London* No. 52.

Sutton, O. G. (1932). Eddy diffusion in the atmosphere. *Proc. Roy. Soc., Ser. A* **135**, 143–165.

Taylor, G. I. (1915a). Eddy motion in the atmosphere. *Phil. Trans. Roy. Soc. London, Ser. A* **125**, 1–26.

Taylor, G. I. (1915b). Skin friction of wind at the earths surface. *Proc. Roy. Soc., Ser. A* **92**, 196.

Taylor, G. I. (1917). Phenomena connected with the turbulence in the lower atmosphere. *Proc. Roy. Soc., Ser. A* **94**, 137.

Taylor, G. I. (1921). Diffusion by continuous mouvements. *Proc. London Math. Soc.* **20**, 196–211.

Taylor, G. I. (1927). Turbulence. *Quart. J. Roy. Meteorol. Soc.* **53**, 201–213.

Taylor, G. I. (1931). Effect of variation in density on the stability of superposed streams of fluid. *Proc. Roy. Soc., Ser. A* **132**, 499–523.

Taylor, G. I. (1932). The transport of vorticity and heat through fluids, in turbulent motion. *Proc. Roy. Soc., Ser. A* **135**, 685–705.

Taylor, G. I. (1935). Statistical theory of turbulence. *Proc. Roy. Soc., Ser. A* **151**, 421–478.

Taylor, G. I. (1938). The spectrum of turbulence. *Proc. Roy. Soc., Ser. A* **164**, 476–490.

von Kármán, T. (1930). Mechanische Ähnlichkeit und Turbulenz. *Appl. Mech., Proc. Int. Congr., 3rd, Stockholm* pp. 85–93.

von Kármán, T., and Howarth, L. (1938). On the statistical theory of isotropic turbulence. *Proc. Roy. Soc., Ser. A* **164**, 192–215.

Wenk, F. (1930). Aenderung der Stabilität atmosphärischer Schichtungen bei adiabatischer Hebung. *Beitr. Phys. Freien Atmos.* **16**, 298–304.

Wiener, N. (1923). Differential space. *J. Math. Phys. (Cambridge, Mass.)* **2**, 131–174.

Wiener, N. (1930). Generalized harmonic analysis. *Acta Math.* **55**, 117–258.

Wiener, N. (1933). "The Fourier Integral and Some of Its Applications." Cambridge Univ. Press, London and New York.

Wiener, N., Paley, R. E. A. C., and Zygmund, A. Z. (1933). Notes on random functions. *Math. Z.* **37**, 647–668.

LIMITATIONS OF GRADIENT TRANSPORT MODELS IN RANDOM WALKS AND IN TURBULENCE

STANLEY CORRSIN

*Department of Mechanics and Materials Science
The Johns Hopkins University, Baltimore, Maryland 21218, U.S.A.*

1. INTRODUCTION

Possibly starting with de St. Venant (1843), certainly with Boussinesq (1877), turbulent transport (e.g., of momentum, heat, or contaminant) has been modeled by more or less ad hoc analogy to the simplest molecular transport, i.e., by linear mean gradient models. Virtually all turblent transport theories, such as the mixing length theories of Taylor (1915, 1932, 1935), Prandtl (1925), and von Kármán (1930), the "homologous turbulence" theory of von Kármán (1937), the dimensionally inspired free shear flow theory of Prandtl (1942), the more continuous interaction model of Nevzgliadov (1945), and most of the more elaborate higher moment approaches of recent decades (see, e.g., the papers in the symposia proceedings edited by Kline et al., 1969; and by the staff of the NASA Langley Res. Cent., 1973) assume a linear gradient transport model for one or more properties. Originally it was applied directly to mean velocity, temperature, and concentration. More recently it has been applied to transport of turbulent energy, mean square temperature and concentration fluctuations, and even (seemingly inappropriately) to represent the energy transfer due to the work done by fluctuating static pressure gradient forces.

The application of the second moment differential equations began with Reynolds (1895) and turbulent energy balance, followed after a long interval by von Kármán (1937) and Taylor (1938a) with the differential equation for the balance of mean square vorticity fluctuation, and Chou (1945) with the shear stress balance. Since that time, more moment equations have been introduced, e.g., Rotta (1951a,b) with dissipation rate and integral scale balances, Corrsin (1952, 1953) with temperature fluctuation, gradient fluctuation, and heat flux balances.

Since any finite collection of statistical moment equations generated from a nonlinear field equation (such as the Navier–Stokes equation) must have a number of moments greater than the number of equations ("the closure problem," a phrase introduced to turbulence possibly by Kraichnan), the object of most theoretical basic research and analytical practical technology in the area is to discover/devise whatever number of additional relations is needed to render the problem determinate. For technological purposes (including meteorological and oceanographic forecasting), we want primarily some additional assumptions which are self-consistent, violate no scientific principles, and "work"; for research, we want assumptions which not only work but also have some a priori plausibility (even if discovered a posteriori), and contribute enough explanation to be called a "theory."

Simple gradient transport assumptions are virtually the only kind used for practical calculations in turbulent transport in 1973, in spite of the fact that it has long been realized (see, e.g., the textbook of Bosworth, 1952) that a gradient transport model requires (among other things) that the characteristic scale of the transporting mechanism [mean free path in gas kinetics; Lagrangian velocity integral time scale multiplied by root-mean-square velocity, in turbulence (Taylor, 1921, 1938b)], must be small compared with the distance over which the mean gradient of the transported property changes appreciably. From time to time (e.g., Batchelor, 1950; Corrsin, 1957) it has been pointed out that nearly all traditional turbulent transport problems violate this requirement, yet papers introducing or using gradient transport models proliferate, and contain no hint that they must often be wrong in principle. Surprisingly, they sometimes can be made to yield good agreement with experiment, although this "success" may be attributable to the number of empirical constants (sometimes an empirical function) available. There has been an astonishing lack of amazement at the partial success of these models.

The purpose of this paper is to identify some conditions which are necessary (though not sufficient) for the validity of a gradient transport model in a simple mean-free-path type of random transport process, essentially a random walk. Presumably, more complex random convective transport phenomena are subject to analogous necessary conditions in order that gradient transport be a plausible approximation, and we shall see how well these conditions are fulfilled in some standard turbulent transport problems. It will be concluded that the partial success of gradient transport models in turbulence is largely fortuitous, and certainly surprising.

Doubtless the "generalized" effects of inhomogeneity and nonstationarity indicated in this account have been presented in the kinetic theory literature many decades ago, and in considerably more rigorous fashion. Unfortunately, there has not been sufficient time to search for them, so I have been unable to ascribe proper credit to writers who deserve it. The monograph of

Chapman and Cowling (1970) and the article of Grad (1958), for example, will provide more rational starting points for some kinds of generalized analyses. Long ago, Chapman (1928) and others pointed out that the approach to be used here can lead to errors in higher order kinetic theory of gases. Since our goals are primarily heuristic, we simply accept the hazards.

2. Mathematical Restrictions Leading to the Gradient Transport Approximation

Suppose we are interested in the spatial transport rate of a scalar field whose mean "concentration" is $\bar{\Gamma}(z, t)$. z is a spatial coordinate (we restrict to one dimension for simplicity) and t is time. Restricting also to a conserved property, we can write an expression for the mean "flux" rate $\bar{F}(z, t)$ of $\bar{\Gamma}$ from all space at coordinate positions $< z$ to all space positions $> z$, viz.

$$(1) \qquad (\partial/\partial t) \int_{-\infty}^{z} \bar{\Gamma}(z', t) \, dz' = -\bar{F}(z, t).$$

In a sense this defines $\bar{F}$. To deduce the familiar differential equation expressing this conservation, we differentiate Eq. (1) with respect to z:

$$(2) \qquad \partial \bar{\Gamma}/\partial t = -\partial \bar{F}/\partial z.$$

In a very general transport process, $\bar{F}(z, t)$ will depend on what went on at all places in space, and at all times up to the present, t. Avoiding specific assumptions about transport mechanism, we may indicate this by saying that $\bar{F}$ is a "functional"[1] of $\bar{\Gamma}$ and of the statistical properties of the velocity, and the concentration fluctuation field $\gamma(z, t_1) \equiv \Gamma(z, t_1) - \bar{\Gamma}(z, t_1)$, for time $t_1 \leq t$, e.g.,[2]

$$(3) \qquad \bar{F}(z, t) = \int_{-\infty}^{t} dt' \int_{-\infty}^{\infty} dz' \, \Phi_{\Gamma}\left[\bar{\Gamma}(z', t'), M^{(n)}(z, z', t, t'), z - z', t - t'\right],$$

---

[1] By *functional* we mean a function which, at a single value of the independent variable, depends upon the behavior of another function, either (a) at a different value of the independent variable or (b) over a finite range of the independent variable. Examples:

(a) $A(x) = \exp[B(x^2 + a)]$; (b) $A(x) = \int_{x-K}^{x} B^{1/2}(x') \, dx'$.

It is the second class of functional which is more likely in convective transport problems.

[2] As a simple illustration, imagine a process in which each space point "emits" $\bar{\Gamma}$ at a mean rate proportional to local $\bar{\Gamma}$, with half traveling in each direction at the constant speed c. Then the flux is obviously the following functional of $\bar{\Gamma}$:

$$\bar{F}(z, t) = K \int_{0}^{\infty} [\bar{\Gamma}(z - c\tau, t - \tau) - \bar{\Gamma}(z + c\tau, t - \tau)] \, d\tau.$$

Evidently, this has a qualitative resemblance to radiative transport.

where $M^{(n)}$ represents various moments of velocity and concentration fluctuations. In Eq. (3) Φ_Γ is a function of $\bar{\Gamma}$ and the $M^{(n)}$, expressing the appropriate consequences of the physical (or mathematical) mechanism which causes the transport. $\bar{F}$ is a functional of Φ_Γ.

In very general cases, Φ_Γ itself may be a functional of $\bar{\Gamma}$ and the $M^{(n)}$'s. For example, if random convection is dominant (as in some turbulent cases), Φ_Γ must depend on the history of the material point displacements Δ (or velocities):

(4) $$\Phi_\Gamma = \chi_\Gamma\{\bar{\Gamma}(z'', t''), M_\Delta^{(n)}(z''', t'''), \ldots\},$$

with $-\infty \leq z'' \leq \infty$, $-\infty \leq z''' \leq \infty$, $-\infty \leq t'' \leq t$, $-\infty \leq t''' \leq t$. $M_\Delta^{(n)}$ denotes the nth moment of displacements.

If the transporting mechanism is fairly *local* in space and time, we can expect the much simpler possibility that $\bar{F}(z, t)$ be expressible as an ordinary function of the concentration at the same point in space–time, and of its first few derivatives:

(5) $$\bar{F}(z, t) = fn(\bar{\Gamma}, \bar{\Gamma}_z, \bar{\Gamma}_t, \bar{\Gamma}_{zz}, \bar{\Gamma}_{tt}, \bar{\Gamma}_{zt}, \ldots, z, t).$$

The z, t dependence is included to reflect possible dependence on moments of displacement and concentration fluctuation fields (the latter generally unknown, of course).

If the mechanism is further restricted to be independent of the value of $\bar{\Gamma}$ itself, to be homogeneous and stationary, and to be a linear function of the $\bar{\Gamma}$ derivatives,

(6) $$\bar{F}(z, t) = K_1 \bar{\Gamma}_z + K_2 \bar{\Gamma}_t + K_3 \bar{\Gamma}_{zz} + K_4 \bar{\Gamma}_{zt} + K_5 \bar{\Gamma}_{tt} + \cdots.$$

The K_i are constants which are to be evaluated from detailed information on the mechanism, e.g., the displacement field and γ-field statistics. The letter subscripts on variables denote partial differentiation.

Equation (6) simplifies still further if we require the character of mechanism to be *symmetric* in space, so that only odd order $\bar{\Gamma}$ spatial derivatives are allowed.[3] If we restrict to random convective processes with large enough transporting speeds, all time derivative effects are negligible, and Eq. (6) reduces to

(7) $$\bar{F}(z, t) = K^{(1)} \bar{\Gamma}_z + K^{(2)} \bar{\Gamma}_{zzz} + \cdots.$$

---

[3] This does not mean that the actual transport events are necessarily symmetric; it means merely that the process is invariant under a reflection of the z axis. An even order derivative would not reverse sign under such a reflection, yet $\bar{F}$ must reverse its sign.

Transport processes in which the effects of all derivations higher than the first can be neglected give finally

(8) $$\bar{F}(z, t) = -D\bar{\Gamma}_z \quad [D > 0]$$

which might be called "Fourier/Fick transport," but may be more clearly identified by a generic name such as "simple gradient transport." Of course, some classes of terms may be ruled out by other considerations, such as the second law of thermodynamics, self-consistency conditions, and, in systems of more than one space dimension, appropriate invariance constraints.

A familiar process that illustrates the simplification from Eq. (7) to Eq. (8) is Brownian motion or rarefied gas molecular kinetics (see, e.g., Einstein, 1926).

3. The Simplest "Kinetic Theory" or Random Walk Model; Size Limitation on Homogeneous Mean Free Path

If transport of $\bar{\Gamma}$ is effected by statistically homogeneous and stationary random motion of particles having "mean free path" l, with root-mean-square particle speed V so large that no measurable changes occur in $\bar{\Gamma}(z, t)$ during the "mean free time" $\tau = l/V$, then the mean flux of $\bar{\Gamma}$ across any fixed observation plane at z can be estimated roughly by assuming all particles to travel with the rms speed over the mean distance. We choose the plane z at the midpoint of a mean free path (or lattice step), and assume that each particle begins its free path journey tagged with the $\bar{\Gamma}$ level of the emigration lamina, and that it gives up the excess or deficiency in $\bar{\Gamma}$ immediately upon arriving at its immigration lamina (Fig. 1). (See, e.g., Jeans,

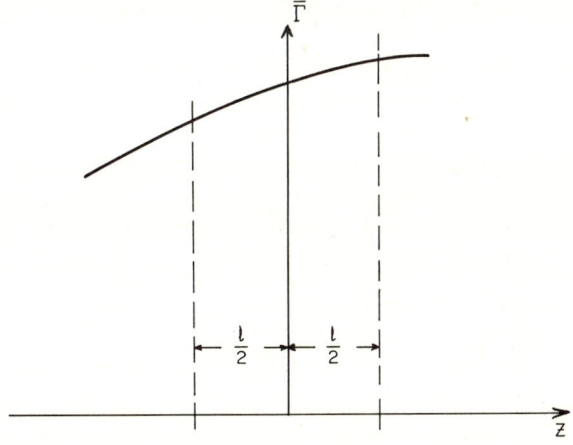

Fig. 1. Schema for rough estimate of mean flux of $\bar{\Gamma}$ over a "mean free path," l.

1940; Loeb, 1934; these texts consider only the primary term in the following analysis.)

If $n/2$ particles per unit time cross unit area in each direction, the net mean flux of $\bar{\Gamma}$ is

$$\bar{F} = \frac{1}{2}nV\left\{\bar{\Gamma}\left(z - \frac{l}{2}\right) - \bar{\Gamma}\left(z + \frac{l}{2}\right)\right\}. \tag{9}$$

If l is small enough relative to distances over which $\bar{\Gamma}$ varies appreciably, both $\bar{\Gamma}$ terms are expressible by the first few terms of Taylor series; then the flux is

$$\bar{F}(z, t) = -nV\left\{\frac{l}{2}\bar{\Gamma}_z + \frac{1}{3!}\left(\frac{l}{2}\right)^3 \bar{\Gamma}_{zzz} + \cdots\right\}. \tag{10}$$

All even derivatives have canceled out. The first term is the gradient transport model, with diffusivity

$$D \equiv \tfrac{1}{2}nVl. \tag{11}$$

The balance differential equation for a conserved property is then

$$\bar{\Gamma}_t = nV\left\{\frac{l}{2}\bar{\Gamma}_{zz} + \frac{1}{3!}\left(\frac{l}{2}\right)^3 \bar{\Gamma}_{zzzz} + \cdots\right\}. \tag{12}$$

The simple gradient transport approximation is applicable when the second and higher order terms are negligible with respect to the first, e.g.

$$|\bar{\Gamma}_{zzz}/\bar{\Gamma}_z|(l^2/24) \ll 1 \quad (\text{for } \bar{F}), \tag{13}$$

or

$$|\bar{\Gamma}_{zzzz}/\bar{\Gamma}_{zz}|(l^2/24) \ll 1 \quad (\text{for } \partial \bar{F}/\partial z).$$

Then Eq. (12) reduces to

$$\bar{\Gamma}_t = D\bar{\Gamma}_{zz}, \tag{14}$$

the simplest diffusion equation.

Near extrema and inflection points there may, however, be local regions where Eq. (13) is violated. Does this mean that Eq. (14) cannot be used across such regions? A pragmatic, ad hoc answer seems to be that Eq. (14) *can* be used, provided the aberrant region is very small compared with the domain over which $\bar{\Gamma}(z, t)$ manifests most of its variations, or is at locations containing only negligible amounts of $\bar{\Gamma}$.

To emphasize purely spatial considerations, we look at a mean concentration profile $\bar{\Gamma}(z)$, which is independent of time. If, for example, we are able to maintain (e.g., by remote boundary conditions, by chemical reaction, by radiative heating, or by mathematical fiat) the distribution

(15) $$\bar{\Gamma}(z) = \exp\{-z^2/2\sigma^2\},$$

then $\bar{F} \sim -\bar{\Gamma}_z$ provided that Eq. (13) is satisfied. The symbol $\sim$ denotes proportionality. For Eq. (15) the first part of Eq. (13) is

(16) $$(l/\sigma)^2 \ll 24/[3 - (z/\sigma)^2].$$

Even for a mean free path value as large as $l = 0.2\sigma$, this condition is fulfilled over a z range containing most of $\bar{\Gamma}$. The ratio of fourth to second derivative terms in $\bar{F}_z$ is

(17) $$\frac{1 - 2(z/\sigma)^2 + \frac{1}{3}(z/\sigma)^4}{1 - (z/\sigma)^2} \frac{l^2}{8\sigma^2},$$

which is infinite at the inflection points $z_I = \pm\sigma$, hence violates Eq. (13) in that neighborhood.

A Taylor series expansion for $\bar{\Gamma}_{zz}$ about z_I shows that the pathological interval, $\delta_{(10)}$, over which

$$|\bar{\Gamma}_{zz}/\bar{\Gamma}_{zzzz}|(24/l^2) \leq 10,$$

for example, is

(18) $$\delta_{(10)}/\sigma \doteq 0.6(l/\sigma)^2,$$

which is very small if $l \ll \sigma$.

In this section we see that the "size limitation on l" is also a uniformity condition on $\bar{\Gamma}_z$.

4. Rapidly Changing Mean Field

The simple kinetic theory/random walk model discussed in the previous section contains a time restriction, viz. the mean field $\bar{\Gamma}(z, t)$ changes negligibly during one mean free time $\tau = l/V$. If that restriction is relaxed, the flux at time t depends on the $\bar{\Gamma}$ field that existed at earlier time $t - \tau$. The Taylor series expansion must then be carried out in time as well as in distance.

In fact, it is a straightforward exercise to generalize this procedure by also allowing the transport speed V and length l to be functions of position and time. The series expansion for such a general case is carried out in the Appendix as far as the second order "correction" terms. But in this section and the following one only first order correction terms are presented.

With V and l constant, but $\bar{\Gamma} = \bar{\Gamma}(z, t)$, the mean flux expression expanded in series as far as the second derivative terms is

(19) $$\bar{F}(z, t) = -\tfrac{1}{2}nVl\{\bar{\Gamma}_z - \bar{\Gamma}_{zt}\tau + \cdots\}.$$

The $\bar{\Gamma}_{zz}$ and $\bar{\Gamma}_{tt}$ contributions balance individually between particles crossing z from the two directions.

For a simple conservative process, in which $\bar{\Gamma}$ changes only because of the diffusion, a first estimate of $\bar{\Gamma}_{zt}$ can be obtained merely by differentiating Eq. (14) with respect to z. With $D = \tfrac{1}{2}nVl$, this approach shows that a sufficient condition for the negligibility of the $\bar{\Gamma}_{zt}$ term is equivalent to that for negligibility of the $\bar{\Gamma}_{zzz}$ term.

On the other hand, when the evolution of $\bar{\Gamma}(z, t)$ is in part due to another process, such as a chemical reaction, then the negligibility of the $\bar{\Gamma}_{zt}$ term is not equivalent to a spatial condition. Instead, we require that

(20) $\qquad \tau|\bar{\Gamma}_{zt}/\bar{\Gamma}_z| \ll 1, \quad \text{or} \quad \tau|\bar{\Gamma}_{zzt}/\bar{\Gamma}_{zz}| \ll 1.$

5. Inhomogeneity and Nonstationarity of the Transport Mechanism

As remarked in the preceding section, this simplest model can also be generalized to allow for both inhomogeneity and nonstationarity of the random transport mechanism, e.g. of the rms particle speed and of the mean free path, $V = V(z, t)$ and $l = l(z, t)$.

An obvious ad hoc first attempt at generalization is to allow variability of the diffusivity:

(21) $\qquad \bar{F}(z, t) = -D(z, t)\bar{\Gamma}_z,$

where $D = \tfrac{1}{2}nVl$. To see when this variability of D can be neglected, we substitute Eq. (21) into Eq. (2), to get a generalized diffusion equation,

(22) $\qquad \bar{\Gamma}_t = D\bar{\Gamma}_{zz} + D_z\bar{\Gamma}_z,$

so the condition, written in suggestive form, is

(23) $\qquad |D_z/D| \ll |(\bar{\Gamma}_z)_z/\bar{\Gamma}_z|.$

The validity of Eq. (21) as a second approximation is, however, not obvious. For example, the result of the power series expansions of the three functions $\bar{\Gamma}$, V, and l in the steady but *inhomogeneous* case (see the Appendix) is

(24) $\qquad \bar{F} = -\tfrac{1}{2}nl\{V\bar{\Gamma}_z + \bar{\Gamma}V_z\},$

to the first order in "correction." To discover conditions sufficient for the viability of Eqs. (21) and (22), we substitute Eq. (24) into the balance equation. Then the generalized diffusion equation is

(25) $\qquad \bar{\Gamma}_t = \tfrac{1}{2}n\{lV\bar{\Gamma}_{zz} + (lV)_z\bar{\Gamma}_z + lV_z\bar{\Gamma}_z + l_zV_z\bar{\Gamma} + lV_{zz}\bar{\Gamma}\}.$

Sufficient conditions for the negligibility of the last three terms compared with the second one on the right-hand side are the individual inequalities

(26a) $$|V_z/V| \ll |l_z/l|,$$

(26b) $$|l_z/l + V_z/V| \ll |\bar{\Gamma}_z/\bar{\Gamma}|,$$

(26c) $$\left|\frac{V_z}{V_{zz}} + \frac{l_z}{l}\cdot\frac{V}{V_{zz}}\right|\left|\frac{\bar{\Gamma}_z}{\bar{\Gamma}}\right| \gg 1.$$

The left-hand side of Eq. (26b) is $|D_z/D|$.

Our tentative conclusion is that a generalized diffusion equation in the form of Eq. (22) may not be self-consistent. This random walk analysis is far from rigorous, but it should at least make us less sanguine about the common practice of using simple, variable diffusivity models.[4]

It is interesting that the z dependences of V and of l influence the flux in different ways, with l_z entering less in lower order terms (see the Appendix).

The effects of rapid *time dependences* of $l(t)$ and $V(t)$ can add time derivative effects of both to expression (19) for flux in a rapidly varying mean field. The individual series terms linear in time (see the Appendix) can be grouped to give a single correction term for nonstationarity:

(27) $$\bar{F}(z, t) = -\tfrac{1}{2}n\{Vl\bar{\Gamma}_z - (Vl\bar{\Gamma}_z)_t\tau\}.$$

The corresponding generalized diffusion equation is

(28) $$\bar{\Gamma}_t = \tfrac{1}{2}n\{Vl\bar{\Gamma}_{zz} - (Vl\bar{\Gamma}_{zz})_t\tau\}.$$

The first terms on the right-hand sides of Eqs. (25) and (28) are the simple gradient transport terms. The remaining terms (along with the relationship $l = V\tau$) will suggest conditions that must be fulfilled for them to be negligible.

6. Transition Probability and Another Generalized Diffusion Equation

For random, convective, transport processes in which the particle velocity is uncorrelated from one step to the next (an example of a "Markov process"), the probability density function[5] (p.d.f.) P of particle position Z

---

[4] Loeb (1934, Sect. 72 and footnote on p. 255) shows a case in which Eq. (22) may be viable in two-species gas diffusion. However, he does not raise the question of Eq. (25).

[5] The probability density function is defined by $P(Z, t)\,dZ$ = probability that at time t, $Z \le Z(t) \le Z + dZ$. This (common) notation is actually confusing because it uses the same symbol for the random variable in "physical space" and for the possible value in "probability space" which the random variable can take on. A more explicit notation is illustrated by saying that the probability density function $P_Z(s, t)$ of $Z(t)$ is defined as follows: $P_z(s, t)\,ds$ is the probability that at time t, $s \le Z \le s + ds$.

at time $t + \theta$ can be expressed as an integral over the p.d.f. at any earlier time t, weighted with the "transition probability" T that the particle moves a distance Δ during the time difference θ[6]:

$$P(Z, t + \theta) = \int_{-\infty}^{\infty} P[Z - \Delta(\theta), t] T[\Delta(\theta)] \, d\Delta. \tag{29}$$

For the basic initial value problem of random convective diffusion of a material point, $P(z, t)$ is proportional to the (ensemble) mean concentration $\bar{\Gamma}(z, t)$, so we can associate this with our previous discussions of diffusion. The function form of the transition probability contains "the physics" of the process.

It is often convenient to have a differential equation in place of Eq. (29), and the more or less standard procedure (Chandrasekhar, 1943) is to use the "Kramers–Moyal expansion." This means to expand the left-hand side as a power series in θ and the integrand as a power series in Δ. The outcome is a generalized diffusion equation

$$\frac{\partial P}{\partial t} = \sum_{m=1}^{\infty} \frac{(-1)^m}{m!} \frac{\partial^m}{\partial z^m} [A_m P(z, t)]. \tag{30}$$

The A_m are moments of P which need not concern us here.

Harking back to Eq. (12), we might seize upon this form as an invitation to try a sequence of approximations corresponding to various truncations of the series. Unfortunately for that plan, Pawula (1967) has shown that *only the $m = 2$ truncation is self-consistent*. This is the so-called "Fokker–Planck equation," with drift velocity chosen equal to zero in our case.

The nonviability of a particular "higher order approximation," e.g., with $m = 4$ (we put $A_1 = A_3 = 0$, to ensure a symmetric diffusive process, and A_2 and A_4 constant for simplicity),

$$\bar{\Gamma}_t = D\bar{\Gamma}_{zz} + D_1 \bar{\Gamma}_{zzzz}, \tag{31}$$

can be seen pragmatically from the Fourier Transform with respect to z:

$$\tilde{\Gamma}_t(k, t) = -Dk^2 \tilde{\Gamma} + D_1 k^4 \tilde{\Gamma}. \tag{32}$$

k is the wave number. Equation (12) suggests that in a random walk or kinetic theory application, both D and D_1 are positive. The solution to Eq. (32) corresponding to the "elementary solution" of Eq. (31) (i.e., for Dirac function initial form) is

$$\tilde{\Gamma}(k, t) \sim \exp\{(-Dk^2 + D_1 k^4)t\}, \tag{33}$$

which obviously misbehaves at large enough k or t.

---

[6] See, for example, Einstein (1926), Chandrasekhar (1943), Wang and Uhlenbeck (1945). The equation is called by a variety of names, including those of Chapman, Kolmogorov, and Smoluchowski.

7. Continuous Limits of Formal Random Walks

The simple diffusion equation [Eq. (14)] is a particular continuous limit of the simplest formal random walk process, a connection apparently made first in essence by Rayleigh (1894) in a different context.

The simple wave equation is a different continuous limit of the random walk process. In a sense this is implied by the more general work of Goldstein (1951), but seems to have been explicitly stated first in unpublished reports by Michelson (1954), and by Davies *et al.* (1954).

Motivated by the wish to subsume in a single model the highly correlated "small time" dispersion effect of turbulent motion with the uncorrelated "large time" effect, Goldstein (1951) devised a random walk limiting procedure which led to the so-called "telegraph equation." A similar hyperbolic model [the diffusion equation, Eq. (14), is parabolic] had been proposed also be Lyapin (1948, 1950; see also the paper of Monin, 1955). Without the details of the general difference equation for the random walk, a sense of the three different limits alluded to above can be given briefly. We consider the elementary, one-dimensional walk on a regular lattice. Each step length is δZ and has duration δt. Suppose that the probability that two successive steps are in the same direction is p; the probability of a reversal must, of course, be $1 - p$.

The *diffusion equation limit* follows from choosing

(34) $$p = \tfrac{1}{2}; \quad (\delta Z)^2/\delta t \to 2D \quad \text{as} \quad \delta Z \to 0,$$

where D is a nonzero and noninfinite constant, essentially the diffusivity. Then the difference equation (which is not shown here) becomes

(35) $$P_t = DP_{zz}.$$

The *wave equation limit* follows from choosing

(36) $$p = 1; \quad \delta Z/\delta t \to V \quad \text{as} \quad \delta Z \to 0,$$

where V is a nonzero and noninfinite constant, essentially the wave speed. Then

(37) $$P_{tt} = V^2 P_{zz}.$$

The *telegraph equation limit* follows from choosing

(38) $$\left. \begin{array}{c} p \to 1 \quad \text{with} \quad (1-p)/\delta t \to 1/2T \\ \text{and} \quad \delta Z/\delta t \to V \end{array} \right\} \quad \text{as} \quad \delta Z \to 0,$$

where V and T are nonzero and noninfinite constants, essentially wave speed and relaxation time, respectively. Then

$$P_{tt} + T^{-1}P_t = V^2 P_{zz}. \tag{39}$$

The exact solutions to Eqs. (35), (37), and (39) need not be repeated here, but their contrasting traits can be inferred from the qualitative sketches in Fig. 2. For "small times" the telegraph solution is wavelike; for "large times" it is diffusionlike. It is this combination of asymptotic attributes which resembles turbulent diffusion (Taylor, 1921). The initial conditions here are a Dirac function for $P(z, 0)$ and, for the equations which require an addition condition, a finite $P_{tt}(z, 0)$. Figure 3 shows the solution of the telegraph equation for a normal (Gaussian) initial distribution,

$$P(z, 0) = \exp\{-\tfrac{1}{4}z^2\}, \tag{40}$$

and for two different dimensionless times. The solid curve is for $t/T = 2$, the dashed one for $t/T = 5$. Both ordinates are scaled so that $P_{\max} = 1$. Here the wavelike propagation of the initially localized $P(z, 0)$ function has nearly disappeared by the time $t = 5T$. The abcissa widths are nearly the same because t is the same for the two curves; T was varied.

The solution to Eq. (39), even with a Dirac function initial condition, implied for Fig. 2, is a bit complicated (Goldstein, 1951; Monin and Yaglom, 1965, 1971), but ordinary differential equations for the second moments are easily deduced from the three transport equations, Eqs. (35), (37), and (39). They are, writing D as $V^2 T$,

$$d\sigma^2/dt - 2V^2 T = 0, \tag{41}$$

$$d^2\sigma^2/dt^2 - 2V^2 = 0, \tag{42}$$

and

$$d^2\sigma^2/dt^2 + T^{-1}\, d\sigma^2/dt - 2V^2 = 0. \tag{43}$$

With initial conditions $\sigma(0) = 0$ [for all] and $\ddot{\sigma}(0) \to V$ [like turbulence (Taylor, 1921), for Eqs. (42) and (43)], the solutions are[7]

$$\sigma^2 = 2V^2 Tt, \tag{44}$$

$$\sigma^2 = V^2 t^2, \tag{45}$$

---

[7] Of course, a great variety of $\sigma(t)$ behaviors can be achieved with the diffusion equation by making the diffusivity a function of time. Suppose Eq. (35) is generalized to $P_t = D(t)P_{zz}$. Taylor (1938) pointed out that, at least for the unsheared case, turbulent dispersion from a point source can be empirically described this way, with $D(t) = \tfrac{1}{2} d\,\overline{Z^2}/dt$, where Z is the fluid point displacement. Solutions to the differential equation are obtained simply by observing that the $D(t)$ can be absorbed into a rescaled time variable, $d\theta \equiv D(t)\, dt$, which gives the classical diffusion equation. Only the initial conditions pose a slight challenge.

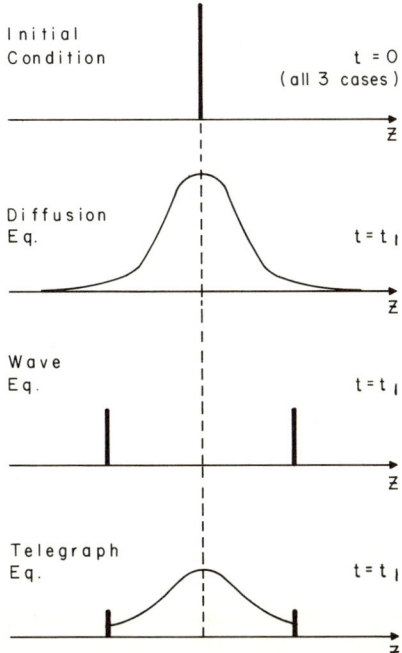

FIG. 2. Qualitative traits of the mean concentration profiles given by the solutions of Eqs. (35), (37), and (39)—all with a Dirac function at the origin as initial condition.

and

(46) $$\sigma^2 = 2V^2 Tt - 2V^2 T^2(1 - e^{-t/T}).$$

Clearly Eqs. (44) and (45) are the two asymptotic forms of Eq. (46).

Before leaving this discussion, it may be helpful to identify a sense in which turbulent diffusion may be considered "wavelike" for small enough time intervals after a fluid point has been "tagged" at a prescribed space position. There may be two or three rationalizations for the adjective *wavelike*. One which has been mentioned from time to time is rather like the following:

No contaminant dispersing from a point source by turbulence is observed to reach infinite distance in infinitesimal time. But the (parabolic) diffusion equation requires that behavior of its solutions; therefore, turbulent transport cannot be parabolic. The apparently bounded shapes of smoke plumes in atmospheric turbulence are the observational justification. The statement is correct, but is literally correct for molecular diffusion as well; the energy of

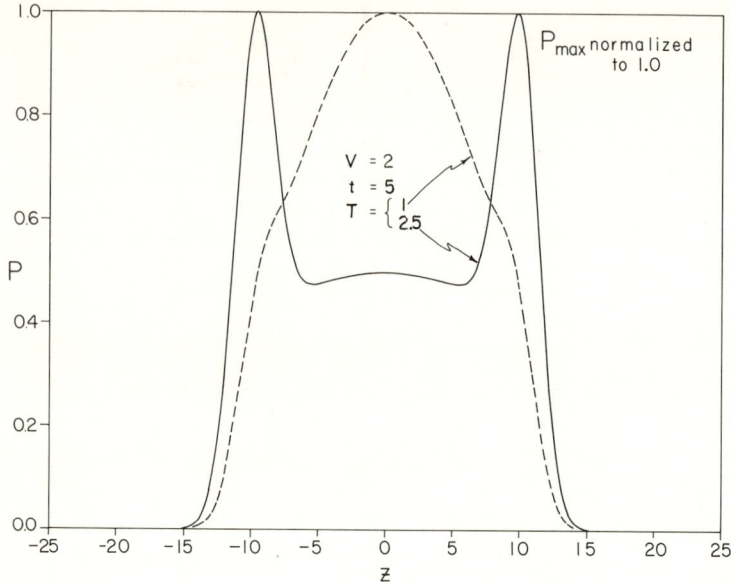

FIG. 3. Solution of the telegraph equation with an origin-centered Gaussian function as initial condition. The solid curve corresponds roughly to the configuration at the bottom of Fig. 2 [and the initial distribution is about as broad as one of the peaks here]; the dashed curve corresponds to a more nearly diffusive (less wavelike) telegraph equation.

any finite system is finite, so there are no infinite speeds (even without invoking relativistic questions). The normal (Gaussian) solution for a point source in molecular diffusion is, of course, just an approximation, albeit a good one for times larger than a few mean free times. Appropriate generalizations for photon and molecular diffusion and Brownian motion started with Fock (1926). A brief historical outline is given by Monin and Yaglom (1965, 1971).

A second sense in which turbulent diffusion is "wavelike" for small times is that the concentration field propagates along essentially straight lines. This happens simply because fluid particle paths are differentiable at least twice (the accelerations are finite). Therefore, neglecting simultaneous molecular diffusion, the "characteristics" of the relevant partial differential equation, $\Gamma_t + \mathbf{u} \cdot \nabla \Gamma = 0$, are straight (in each "realization" of a "Gibbsian ensemble"), a trait of simple wave phenomena. In his modeling of turbulent diffusion by the hyperbolic telegraph equation, Monin (1955, 1956) describes the system as displaying "diffusion with finite velocity" (see also Girgidov, 1973).

A third sense is basically the same as the second one: for $t \to 0$, the standard deviation of contaminant dispersed from a point source by turbulence behaves like $\sigma \to w't$ (Taylor, 1921), which is identical with Eq. (45), the wave equation result.

Another concept may be appended to complicate at least the semantics of this discussion: when a simple wave beam (like a pencil of light) propagates through a random medium, the average intensity does not spread linearly. For example, if the inhomogeneity scale is much larger than the wavelength, each ray (i.e., characteristic) performs a (continuous) random walk, and the "large t" intensity spread is "diffusionlike."

8. Continuous Limit of a Random Walk That Possesses both Eulerian and Lagrangian Statistics

A principal shortcoming of the telegraph equation as a turbulent diffusion model is that it possesses no Eulerian (i.e., spatial) statistical properties, whereas Eulerian coordinates are those in which most experimental investigations and theoretical analysis of turbulent motion are conveniently carried out. This is, in fact, a major deficiency of all classical random walk processes as simulations of turbulent diffusion.

To overcome this shortcoming, Lumley and Corrsin (1959) introduced the random walk process we have come to call the "walk on a random field." In these processes, we prescribe not a stochastic algorithm for the peregrinations in time of a particle on a blank lattice, but rather the statistical properties of a velocity field on a field of spatial and temporal intervals. These velocity field properties are naturally Eulerian. It is then a straightforward matter to compute the random trajectories of "material" points on these fields; the associated statistical properties are most directly Lagrangian.

Patterson and Corrsin (1966) reported some computer experiments on rather general binary examples, particularly comparing the Lagrangian velocity autocorrelation function with the Eulerian velocity correlation function in space–time. Binary fields were used for simplicity and economy, but there is no conceptual difficulty in doing the same thing for continuous fields when enough computing capacity is available, and this generalization has recently been carried out by Liu and Thompson (1973).

In the special case of binary Eulerian fields generated by two Markov chains, one for space and one for time, Patterson (1966 and this volume, p. 61) corrected the solution of Lumley and Corrsin, and also devised a continuous limit procedure which converted the difference equations to a generalized telegraph equation. His work is presented in these proceedings, so a brief outline will suffice here.

The Eulerian binary velocity field has uniform velocity over distances at least δZ and over times at least δt. If p is the probability that neighboring spatial domains of size δZ have equal velocities at the same t, while q is the probability that a point in space retains the same velocity over successive time intervals δt, then Patterson sets the following limiting process:

$$
(47) \quad \left.\begin{array}{l} p \to 1 \quad \text{with} \quad (1-p)/\delta Z \to 1/2L \\ q \to 1 \quad \text{with} \quad (1-q)/\delta t \to 1/2T \\ \text{and} \quad \delta Z/\delta t \to V \end{array}\right\} \quad \text{as} \quad \delta Z \to 0.
$$

Clearly the limits are like Goldstein's, but the process is Eulerian. L and T have been defined as *Eulerian* length and time scales.

Patterson's generalized telegraph equation for the probability density function of the particle walks which can occur on an ensemble of these random fields is

$$
(48) \quad \beta_t + \frac{1}{2T}\beta + \frac{V}{4LT}\left(1 + \frac{VT}{L}\right)P_t
$$

$$
= \frac{1}{8T^2}\int_0^t \left\{ A(t-\tau)\beta(z,\tau) + \frac{V}{8LT}\left[1 + \frac{VT}{L}\right]B(t-\tau)P_\tau \right\} d\tau,
$$

where

$$(49a) \quad \beta(z,t) \equiv P_{tt} + 1/T(1 + VTL^{-1})P_t - V^2 P_{zz},$$

a kind of "telegraph operator" on P,

$$(49b) \quad A(t) \equiv (4T/t)\exp(-t/2T)I_1(t/2T),$$

and

$$(49c) \quad B(t) \equiv \int_0^\infty A(\tau)A(t-\tau)\,d\tau.$$

I_1 is the modified Bessel function of first order. For $L = \infty$, this reduces to the Goldstein case $\beta = 0$, as it should. In the Patterson equation, the line $z = Vt$ is singular, as is the analogous line in the telegraph case. Equation (48) has not yet been solved, although Patterson has developed a simpler form by Laplace transformation.

9. A Random Walk with Large Steps

The wavelike character of a simple random walk at "small time" or with "large mean free path" (i.e., step length not very much smaller than a characteristic length of the mean concentration field), is dramatically demonstrated by the *binary* velocity case. To make the problem more "physical," the initial concentration field is taken as

(50) $$\Gamma(z, 0) = 3 \exp\{-z^2/2\sigma_0^2\}$$

instead of a Dirac function. To give smooth curves for $\bar{\Gamma}(z, t)$, the ensemble of lattices is given a uniform distribution of phases, i.e. the lattice positioning is different (relative to the origin) in each realization, such that over the ensemble of realizations the lattice points are uniformly distributed over one step length ($\equiv$ one lattice spacing). Figure 4 shows the outcome (which

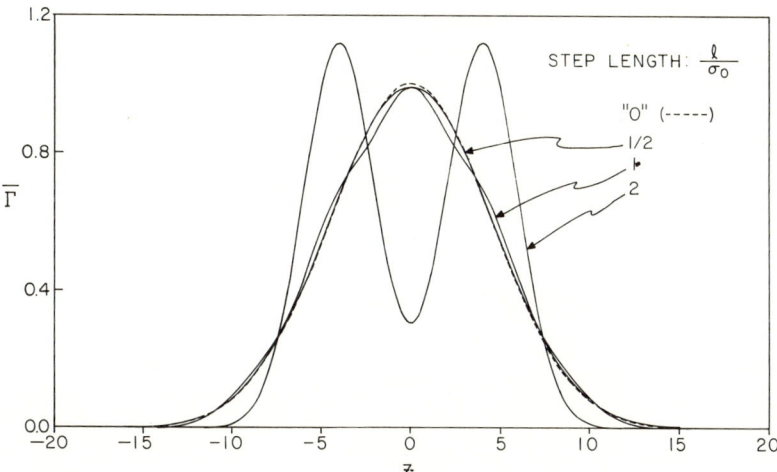

FIG. 4. Mean concentration profiles resulting from dispersion of an initial Gaussian profile by simple binary random walks whose step lengths l have different ratios to the standard deviation σ_0 of the initial profile.

could, of course, be written out analytically) for three ratios l/σ_0 of step length to standard deviation of initial field, compared with the diffusion equation solution $l/\sigma_0 \to 0$. To maintain the same degree of dispersal for all cases, the number of steps is chosen as

(51) $$n = 16(\sigma_0/l)^2,$$

thus ranging from 4 for the longest steps to 64 for the shortest.

The resemblance between the $l/\sigma_0 = 2$ case ($n = 4$) and one of the telegraph equation dispersion solutions in Fig. 3 is evident.

The large-step-length random walk shows far less dramatic results *if the probability density function of velocities is not binary but normal* (Gaussian), because in that case it can be shown that *an initially normal (or Dirac) concentration field remains always normal*, no matter how large a value we choose for l/σ_0. A similar situation occurs during dispersion in isotropic turbulence. It is found empirically that the fluid velocity fluctuations are

Gaussian within the accuracy of measurement (see, e.g., Batchelor, 1953), and this leads to Gaussian concentration distributions about a single tagging point. Since the elementary solution of the simple diffusion equation is a Gaussian function, it is essential to remind ourselves that the occurrence of Gaussian distributions does *not* ensure that the process can be described by a simple diffusion equation. On the other hand, by allowing the diffusivity to be a function of time (see footnote 7 to Section 7), we should be able to devise a diffusion equation whose solution is a spreading Gaussian function with fairly general growth history.

As another example, consider a random walk with a Gaussian p.d.f. of step length

$$p_l(l) = (2\pi)^{-1}\sigma_l^{-1} \exp\{-l^2/2\sigma_l^2\}. \tag{52}$$

If the initial concentration field is a Dirac function,

$$\Gamma(z, 0) = \delta(z), \tag{53}$$

then it can be shown that after n times steps

$$\bar{\Gamma}(z, n) = (2\pi)^{-1/2}\sigma_n^{-1} \exp\{-z^2/2\sigma_n^2\}, \tag{54}$$

with

$$\sigma_n = n^{1/2}\sigma_l. \tag{55}$$

With equal time steps, Eq. (55) is equivalent to

$$\sigma \sim t^{1/2}. \tag{56}$$

This behavior is appropriate to solutions of the simple diffusion equation. Yet the system certainly does not fulfill a condition like Eq. (13), which is necessary if a process is in fact describable by the simple diffusion equation.

One of the qualitative differences between transport processes whose scale is *not* much smaller than the characteristic length of the mean concentration profile and classical transport processes, which fulfill such a condition, is that the former can apparently transport properties against a local concentration gradient.[8] In turbulent transport, this possibility has led to such apparent paradoxes as "reversed" kinetic energy transformation, wherein energy is fed locally from turbulence into mean flow, instead of the other way around (Erian and Eskinazi, 1969).

---

[8] There has not been time to calculate an illustration. Any concentration distribution with an unsymmetrical maximum or minimum should serve as an appropriate example. Parenthetically, this constant density phenomenon should not be confused with the "countergradient vertical heat flux" observed in the planetary atmospheric boundary layer (see, e.g., Deardorff, 1972).

10. Implications of the Foregoing Conditions for Gradient Modeling of Turbulent Transport

The types of conditions mentioned as necessary for the gradient transport model or approximation can be listed here, omitting the rather general discussion of Section 2:

(a) the transport mechanism length scale must be much smaller than the distance over which the curvature of the mean transported field gradient changes appreciably [Eq. (13)];

(b) the transport mechanism time scale must be much smaller than the time during which the mean transported field gradient changes appreciably [Eq. (20)];

(c) the transport mechanism length scale must be essentially constant over a distance of a length scale (merely for the concept to be meaningful), and over a distance for which the mean transported field changes appreciably [Eq. (26b)];

(d) the transport mechanism velocity must be appreciably more uniform than the length scale [Eq. (26a)].

A rather complex condition involving nonuniformities of mean field, length scale and velocity [Eq. (26c)] is not as simple to interpret.

(e) the restrictions on time dependences of mean field, length scale, and velocity combine (in the random walk) to a collective condition, obvious from the form of Eq. (28), which can be described as follows: the relative local accumulation of $\bar{\Gamma}$ must be very small during one time scale of the transport mechanism. In a flowing turbulence, for example, this should be established in a frame traveling with the mean flow.

Although we have discussed and illustrated these necessary conditions in the context of mean free path kinetic theory or random walk problems, it seems likely that qualitatively analogous forms apply to general random convective transport as well, including turbulence. Therefore, we shall estimate analogous expressions for a turbulent boundary layer, as an illustration.

Earlier, more casual, studies gave grounds for pessimism about the validity in principle of gradient transport models. It has been pointed out before (Corrsin, 1957) that a characteristic length of the turbulent momentum transfer in a boundary layer is twice as large as the "momentum thickness" of the layer, and it has been known for decades that the two-point velocity correlation function has measurable value across nearly the full extent of a pipe flow (Taylor, 1936), a jet (Corrsin, 1943), and a wake (Townsend, 1949). The archival literature is replete with data showing, either directly or indirectly, for both scalar and momentum transport, that the mean gradients vary considerably over distances comparable to the length scales characterizing the "eddies."

As a more recent illustration, we can use the turbulent boundary layer data of Blackwelder and Kovasznay (1972), by estimating a length characteristic of momentum transport. A plausible choice would be the product of Lagrangian integral time scale of turbulent velocity and rms value of the same velocity. Unfortunately, there are no data available on Lagrangian velocity autocorrelation function in any of the traditional turbulent shear flows, but we might speculate that the Eulerian integral time scale in a frame convected downstream at the speed for maximum correlation at each distance will be the same order of magnitude as the Lagrangian time scale (the isotropic case has been studied: Corrsin, 1963; Shlien and Corrsin, 1974).

From the Blackwelder/Kovasznay data, we can estimate the integral time scale of the turbulent shear stress (momentum transport) covariance in the "optimum" convected frame. This turns out to be[9]

$$(57) \qquad T_s \approx \begin{Bmatrix} 4.2\delta/\bar{U} \\ 6.8\delta/\bar{U} \end{Bmatrix} \quad \text{at} \quad z/\delta = \begin{Bmatrix} 0.20 \\ 0.45 \end{Bmatrix}.$$

z is wall distance, δ is a boundary layer thickness defined by $\bar{U}(x, \delta) \equiv 0.99 \bar{U}_\infty$, where x is downstream location, and $\bar{U}_\infty$ is free stream velocity. The symbol z is used in place of their y in order to introduce a meteorological flavor.

We should like to know whether the mean velocity gradient $\partial \bar{U}/\partial z$ changes appreciably over a z distance

$$(58) \qquad L_s \equiv w'_L T_s,$$

where w'_L is the r.m.s. Lagrangian turbulent velocity component normal to the wall. There is also no information on w'_L and, although we know that it is equal to w' (the Eulerian rms component) only in a stationary, homogeneous turbulence (Lumley, 1962), we assume the rough equality. Then

$$(59) \qquad L_s \approx w' T_s.$$

Blackwelder and Kovasznay (1972) and Kovasznay et al. (1970) give only u' data, so we estimate w' by assuming that the ratio w'/u' measured by Klebanoff (1955) is applicable. This gives

$$(60) \qquad L_s \approx \begin{Bmatrix} 0.23\delta \\ 0.31\delta \end{Bmatrix} \quad \text{at} \quad z = \begin{Bmatrix} 0.20\delta \\ 0.45\delta \end{Bmatrix}.$$

[9] The empirical function chosen to fit the envelope of their R_{uw} data at $z/\delta = 0.45$ is misprinted in Fig. 6. Dr. Blackwelder (private communication) suggests 7.93 in place of 4.75 and 1.39 in place of 0.61.

In the boundary layer region between 0.2δ and 0.45δ, an estimate of the mean velocity gradient difference over a distance L_s gives

$$(61) \qquad \frac{\bar{U}_z(z) - \bar{U}_z(z + L_s)}{\frac{1}{2}\{\bar{U}_z(z) + \bar{U}_z(z + L_s)\}} \approx 0.7,$$

a 70 % difference in mean gradient over one characteristic transport scale! Although this is rather large, the operant dimensionless criterion analogous to Eq. (13) is

$$(62) \qquad |\bar{U}_{zzz}/\bar{U}_z|(L_s^2/24) \ll 1.$$

At $z/\delta = 0.2$ and 0.45, this number is 0.12 and 0.04, respectively. We conclude that in the outer part of the boundary layer, this condition is well satisfied.

There may be regions in some meteorological or oceanographic flows which also fulfill this condition.

In recent years there has been a systematic effort to generate laboratory flows which satisfy Eq. (13) over relatively large distances (Wiskind, 1962; Rose, 1966; Champagne et al., 1970). It is hoped that these experiments will gradually be generalized to include small amounts of mean gradient variation.

It is, of course, much easier to postulate a homogeneous turbulent transport situation (e.g., Corrsin, 1952; Reis, 1952; Burgers and Mitchner, 1953a,b; Craya, 1958; Deissler, 1961; Fox, 1964; Courseau and Loiseau, 1972a,b) than to generate it in the laboratory.

We can supplement the *mean field homogeneity condition* which relates to condition (a) above with a look at the *mean field time stationarity condition*, which is a kind of time scale limitation, condition (b). With momentum transport time scale T_s already estimated for a turbulent boundary layer, we can ask whether the mean velocity gradient changes appreciably in this time interval, observing in the convected frame. The analog of Eq. (20) is

$$(63) \qquad |\bar{U}_{zt}/\bar{U}_z|(T_s/2) \ll 1 \quad \text{or} \quad |\bar{U}_{zzt}/\bar{U}_{zz}|(T_s/2) \ll 1$$

in the convected frame. Estimating just the first of these from the Blackwelder/Kovasznay data, we find 0.005 and 0.009 at $z/\delta = 0.20$ and 0.45, respectively. Condition (63) is well satisfied.

The third and fourth conditions [(c) and (d)] listed at the start of this section specify homogeneity of the length and velocity which characterize the transport mechanism itself. We look at the boundary layer data for T_s and w' variations between $z/\delta = 0.20$ and 0.45; from the data of Kovasznay et al. (1970), and of Blackwelder and Kovasznay and of Klebanoff, we can estimate Table I.

TABLE I

z/δ	$\bar{U}/\bar{U}_\infty$	$u'/\bar{U}_\infty$	$w'/\bar{U}_\infty$	$(\bar{U}/\delta)T_s$	L_s/δ
0.20	0.74	0.068	0.041	4.2	0.23
0.45	0.85	0.056	0.038	6.8	0.31

The combined homogeneity condition on L_s and w' which is analogous to Eq. (26b) is

$$\bar{U} \left|\frac{\partial \bar{U}}{\partial z}\right|^{-1} \left|\frac{1}{L_s}\frac{\partial L_s}{\partial z} + \frac{1}{w'}\frac{\partial w'}{\partial z}\right| \ll 1. \quad (64)$$

With the derivatives estimated crudely by differences in Table I, the left-hand side of (64) turns out to be 1.6, so this inequality is seriously violated.

On the other hand, the data in the table indicate that a relation analogous to Eq. (26a) is moderately well fulfilled.

The collective stationarity condition in an appropriately moving frame [condition (e), above] follows directly from Eq. (28) as

$$\frac{\bar{U}\,|\partial/\partial x(w'L_s\,\partial^2\bar{U}/\partial z^2)|\,L_s/w'}{w'L_s\,|\partial^2\bar{U}/\partial z^2|} \ll 1. \quad (65)$$

If the nonstationarity effect, the second term on the right-hand side of Eq. (28) is moderately small compared with the principal (gradient transport) term, we can approximate this "correction" in a way which will make the estimation simpler. Neglecting the correction term entirely, and differentiating the resulting diffusion equation with respect to t, we have

$$\bar{\Gamma}_{tt} \approx \tfrac{1}{2} n (V\bar{l}\,\bar{\Gamma}_{zz})_t, \quad (66)$$

which estimates the correction term in Eq. (28) as

$$-\bar{\Gamma}_{tt}\tau, \quad (67)$$

an expression which would look more at home on the left-hand side. In any case, the analogous boundary layer condition simpler than Eq. (65) is

$$\frac{|\partial/\partial x(\bar{U}\,\partial\bar{U}/\partial x)|\,(L_s/w')}{|\partial \bar{U}/\partial x|} \ll 1. \quad (68)$$

With the same boundary layer data, plus the assumption that the mean velocity profile can be roughly approximated by $\bar{U}/\bar{U}_\infty \approx (z/\delta)^{1/7}$, the left-hand side of the latter inequality comes out to 0.15. We infer that the transport is quasi-stationary in a frame moving with the mean speed.

In summary of this section, we have found that, of several homogeneity and (convected frame) stationarity conditions which are necessary for the applicability of a gradient transport model in turbulence, one is strongly violated in the central region of a turbulent boundary layer: that requiring cross-stream uniformity of the length scale and rms velocity fluctuations. Other homogeneity conditions are moderately well satisfied for $z/\delta \geq 0.2$. The stationarity conditions are fairly well satisfied.

The reason that the relatively large mean gradient difference across one characteristic transport length [70 %; see Eq. (61)] does not cause a major violation, is that the second derivative is excluded by mechanism symmetry requirements, so the "correction" term is proportional to the mean third derivative and the cube of the length scale ratio.

11. Directional Constraint in Gradient Transport; the Turbulent Diffusivity Tensor

The simplest n-dimensional generalization of Eq. (8), the mean flux of a transported property which undergoes simple gradient transport, is

$$\bar{\mathbf{F}}(\mathbf{x}, t) = -D \, \nabla \bar{\Gamma}. \tag{69}$$

The scalar character of the diffusivity obviously implies an isotropic transport mechanism.

The well-known, nonisotropic generalization of Eq. (69) requires that the diffusivity be a second rank tensor (rather than a vector), in order that $\bar{\mathbf{F}}$ remain a vector:

$$\bar{\mathbf{F}} = -\mathbf{D} \cdot (\nabla \bar{\Gamma}). \tag{70}$$

It was noted quite a while ago (e.g., by Richardson, 1920) that the turbulent diffusivity (of heat, for example) has a different value in different directions, in turbulent shear flow. Richardson suggested, in effect, that

$$\bar{F}_1 = -D_1 \, \partial \bar{\Gamma}/\partial x_1; \qquad \bar{F}_2 = -D_2 \, \partial \bar{\Gamma}/\partial x_2; \qquad \bar{F}_3 = -D_3 \, \partial \bar{\Gamma}/\partial x_3. \tag{71}$$

A review of the applications of this form as applied to a turbulent diffusivity has been given by Monin and Yaglom (1965, 1971), and it need not be pursued here. An obvious difficulty of Eq. (71) appears when it is written in correct form, i.e. Eq. (70), which recognizes that a nonisotropic diffusivity must be a tensor of at least second order. Recognition that Eq. (71) is actually an inner product of a second rank tensor and a vector, in a particular coordinate system, highlights Eq. (71) as an assumption on the principal axis directions of the diffusivity tensor. It assumes that they coincide with the mean flow direction and the normal to the boundary (if, for example, we apply it to a boundary layer flow):

$$(72) \qquad D_{ik} = \begin{pmatrix} D_1 & 0 & 0 \\ 0 & D_2 & 0 \\ 0 & 0 & D_3 \end{pmatrix}.$$

Apparently it was Batchelor (1949), in his tensorial generalization of Taylor's (1921) theory of "diffusion by continuous movements," who introduced the appropriate diffusivity form [Eq. (70)] into the description of turbulent diffusion. Lettau (1952) independently pointed out that in a shear flow there is no a priori reason to expect the off-diagonal terms in Eq. (72) to be zero.

Monin and Yaglom have also outlined the history of the (relatively recent) work with turbulent diffusivity as a more general tensor. Therefore, we shall focus here primarily on concepts.

In turbulent transport, we know from the mean conservation equation for a scalar field, $\Gamma(\mathbf{x}, t) = \bar{\Gamma}(\mathbf{x}, t) + \gamma(\mathbf{x}, t)$, that the mean turbulent flux is (in cartesian tensor notation)

$$(73) \qquad \bar{F}_i = \overline{\gamma u_i}$$

(Kampé de Fériet, 1937), so[10]

$$(74) \qquad \overline{\gamma u_i} = -D_{ik} \, \partial \bar{\Gamma} / \partial x_k,$$

which might be regarded as the definition of **D**. This form shows at once a necessary (though not sufficient) directional condition: the vector $\overline{\gamma \mathbf{u}}$ must be parallel to the vector $-\mathbf{D} \cdot (\nabla \bar{\Gamma})$. Parenthetically, we note that a scalar diffusivity [Eq. (69)] cannot be appropriate unless $\overline{\gamma \mathbf{u}}$ is parallel to $-\nabla \bar{\Gamma}$.

Yaglom (1969; also Gee and Davies, 1964) has pointed out that, in a shear flow $[\bar{U}_1(x_3), 0, 0]$ with $\partial \bar{U}_1/\partial x_3 > 0$, (a) we should expect D_{13} and D_{31} to be negative, (b) $D_{31} \neq D_{13}$, i.e., there is no reason to expect the diffusivity tensor to be symmetric (see also, Gee, 1967). From turbulent transport rate data in a neutrally stable atmospheric boundary layer with mean temperature $\bar{\Gamma}$ essentially a function of x_3 only (Zubkovsky and Tsvang, 1966) he computed

$$(75) \qquad D_{13} \approx -3 D_{33}.$$

---

[10] The analogous "turbulent kinematic viscosity" must be a fourth rank tensor:

$$-\overline{u_i u_k} = v_{ikjl}[\partial \bar{U}_j/\partial x_l + \partial \bar{U}_l/\partial x_j] \quad \text{(See, e.g., Hinze, 1959.)}$$

Furthermore, assuming that the hypothetical configuration $\bar{\Gamma} = \bar{\Gamma}(x_1)$ in a boundary layer would yield $\overline{\gamma u_1}/\overline{\gamma' u_1'} \approx \overline{\gamma u_3}/\overline{\gamma' u_3'}$, an experimental result for $\bar{\Gamma} = \bar{\Gamma}(x_3)$, he offered the conjecture

$$(76) \qquad D_{31} \approx -\tfrac{1}{3} D_{11}.$$

If we speculate that $D_{11} \approx 4.5 D_{33}$ in a boundary layer [based on the fact that at $x_3/\delta = 0.45$, the Blackwelder/Kovasznay (1972) data give $\overline{u_1^2} T_{11} \approx 4.5 \overline{u_3^2} T_{33}$, where these T's are component Eulerian time scales in the frame convected with the correlation peak], then we find that Yaglom's suggestions correspond to

$$(77) \qquad D_{31} \approx \tfrac{1}{2} D_{13}.$$

Later in this section we shall estimate some D components from data taken under better defined conditions, in the laboratory.

First, it may be helpful to recall Yaglom's (1969) qualitative explanation of the off-diagonal components of D_{ik}.

The 1-component of the mean flux vector, Eq. (70), is

$$(78) \qquad \bar{F}_1 = -D_{11} \, \partial \bar{\Gamma}/\partial x_1 - D_{13} \, \partial \bar{\Gamma}/\partial x_3,$$

where we restrict to two dimensions for simplicity. At first glance it may seem paradoxical that transport of $\bar{\Gamma}$ in the x_1 direction can be proportional to the mean gradient along x_3, the axis perpendicular to x_1. A simple example can illustrate the possibility. Suppose that at an arbitrary time the $\bar{\Gamma}$ field has a constant mean gradient (Fig. 5a). The material field undergoes random displacements in the plane. It is essentially obvious that the random particle displacements will convect $\bar{\Gamma}$ down the gradient on the average. In particular, the random x_1 displacements will transport $\bar{\Gamma}$ down the $\partial \bar{\Gamma}/\partial x_1$ gradient, and the random x_3 displacements will transport $\bar{\Gamma}$ down the $\partial \bar{\Gamma}/\partial x_3$ gradient. If X_1 and X_3 (the x_1 and x_3 displacements) are uncorrelated, that is the only kind of transport which will occur. However, if $\overline{X_1 X_3} \neq 0$, x_3 displacements, which by definition are along the $\partial \bar{\Gamma}/\partial x_3$ component direction, on the average will transport $\bar{\Gamma}$ parallel to the x_1 axis as well as down the $\partial \bar{\Gamma}/\partial x_3$ gradient.

Figure 5b clearly illustrates the cross-effect. Here $\partial \bar{\Gamma}/\partial x_3' = 0$, yet there will be mean transport of $\bar{\Gamma}$ parallel to the x_3' axis. It is also clear from Fig. 5b that if $\overline{X_1 X_3} > 0, D_{31} > 0$, and if $\overline{X_1 X_3} < 0, D_{31} < 0$ [Eq. (80)].

If a cross-correlated random walk example were worked out with $\overline{X_1^2} = \overline{X_3^2}$, it would give $D_{31} = D_{13}$. This is obvious from a comparison of Fig. 5b with the same case with x_1' and x_3' axes interchanged.

Having seen why the off-diagonal terms of the diffusivity tensor arise, we may turn to the question of their experimental determination in a turbulent flow for which the gradient transport model is assumed to be satisfactory.

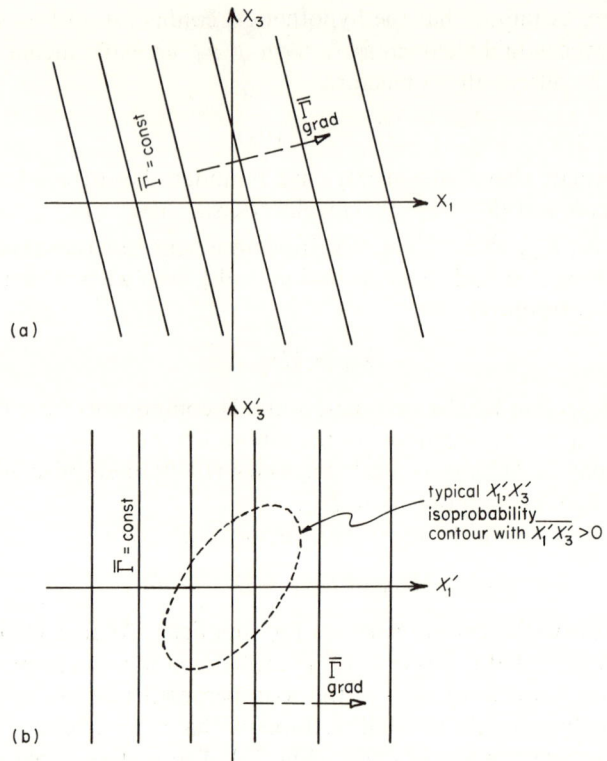

FIG. 5. Qualitative sketch to show how correlation of orthogonal displacement components can cause mean convective transport of contaminant $\bar{\Gamma}$ in a direction perpendicular to the mean gradient of $\bar{\Gamma}$. (a) A local $\bar{\Gamma}$ field with gradient in arbitrary direction. (b) The same field with cartesian coordinates reoriented for simplicity; a typical isoprobability contour for correlated particle displacement components, as sketched, yields mean transport in the x_3 direction (as well as the x_1 direction), even though the mean $\bar{\Gamma}$ gradient has no component along x_3. This means that $D_{31} \neq 0$.

This means that the mean shear, turbulence, mean concentration gradient, and concentration fluctuation fields should all be approximately homogeneous and stationary. With Eq. (74), the mean conservation equation for $\bar{\Gamma}$ is

(79) $$\frac{\partial \bar{\Gamma}}{\partial t} + \bar{U}_k \frac{\partial \bar{\Gamma}}{\partial x_k} = \frac{\partial}{\partial x_i}\left[D_{ij} \frac{\partial \bar{\Gamma}}{\partial x_j} \right].$$

What measurements are needed to establish the nine components of D? We have only four equations, so data at a single point will not suffice. We can set up several cases, or we can evaluate Eqs. (74) or (79) at several locations in a gently nonuniform case.

It is important to interject the remark that if D is constant, the resulting simplification in the diffusive term of Eq. (79) shows that only the symmetric part of **D** influences the transport; a term like $\partial^2 \bar{\Gamma}/\partial x_1 \partial x_3$ is multiplied by $(D_{13} + D_{31})$.

If it were experimentally feasible in a prescribed flow field, we could successively set up $\bar{\Gamma}$ fields whose mean gradients lie along cartesian axis directions. With measurement of $\overline{\gamma \mathbf{u}}$ and $\nabla \bar{\Gamma}$ in each case, we could determine the components of **D** directly.

Johnson's (1957, 1959) experiment on the mean and fluctuating temperature fields during boundary layer heat transfer from a warm wall provides the data for determination of one relation between **D** components. The example is two dimensional in the mean, so we write only x_1 (streamwise) and x_3 (normal to wall) components. In addition to Eq. (78) for $\bar{F}_1$,

$$\bar{F}_3 = -D_{31} \partial \bar{\Gamma}/\partial x_1 - D_{33} \partial \bar{\Gamma}/\partial x_3. \tag{80}$$

The mean temperature gradient $\nabla \bar{\Gamma}$ is dominantly along $-x_3$, so we estimate

$$\nabla \bar{\Gamma} \doteq |\nabla \bar{\Gamma}| \begin{bmatrix} 0 \\ -1 \end{bmatrix}. \tag{81}$$

The heat flux vector is

$$\bar{\mathbf{F}} = \overline{\gamma \mathbf{u}} = |\overline{\gamma \mathbf{u}}| \begin{bmatrix} -0.915 \\ 0.40 \end{bmatrix}. \tag{82}$$

We note in passing that $-\overline{\gamma \mathbf{u}}$ is not even approximately parallel to $\nabla \bar{\Gamma}$, so a scalar turbulent diffusivity is out of the question.

A tensor diffusivity must satisfy the directional condition implied by Eq. (70); $-\mathbf{D} \cdot (\nabla \bar{\Gamma})$ must be parallel to $\bar{\mathbf{F}}$. This inner product gives

$$-\mathbf{D} \cdot (\nabla \bar{\Gamma}) \sim \begin{bmatrix} -D_{13} \\ -D_{33} \end{bmatrix}. \tag{83}$$

Comparing Eq. (83) with (82), we obtain

$$D_{13} \doteq -2.3 D_{33}, \tag{84}$$

which is much like Yaglom's estimate [Eq. (75)].

Unfortunately, the orientation of isothermal surfaces so nearly parallel to the boundary prevents any corresponding estimate of D_{31}.

Until appropriate Lagrangian data are available, estimates such as

$$D_{11} \approx \overline{u_1^2} T_{11} \quad \text{and} \quad D_{33} \approx \overline{u_3^2} T_{33} \tag{85}$$

may suffice for the diagonal components. Also, the estimate that $D_{11} \approx 4.5 D_{33}$ is even more appropriate here than in the atmospheric boundary layer, since the numerical values come from a laboratory boundary layer flow.

Laboratory data have long been available for a different flow configuration, in which the mean temperature gradient vector covers more than a full 90° range of directions, the warm turbulent jet (e.g., Corrsin and Uberoi, 1950).

Fifteen diameters from the round jet nozzle, the fields are fairly well developed. At a radial distance, $r \equiv r_{1/2}$, where the velocity $\bar{U}_1$ is half that on the axis, the mean temperature gradient vector is very nearly radial, toward the axis:

$$\nabla \bar{\Gamma} = |\nabla \bar{\Gamma}| \begin{bmatrix} 0 \\ -1 \end{bmatrix}. \tag{86}$$

The heat flux vector is

$$\bar{\mathbf{F}} = \overline{\gamma \mathbf{u}} = |\overline{\gamma \mathbf{u}}| \begin{bmatrix} 0.24 \\ 0.97 \end{bmatrix}, \tag{87}$$

and

$$-\mathbf{D} \cdot (\nabla \bar{\Gamma}) \sim \begin{bmatrix} D_{13} \\ D_{33} \end{bmatrix}, \tag{88}$$

which must be parallel to $\bar{\mathbf{F}}$, whence

$$D_{13} \approx \tfrac{1}{4} D_{33}. \tag{89}$$

The subscript 3 now denotes the radial direction r. This contrasts sharply with the boundary layer [Eq. (84)] in both sign and magnitude. The difference in sign is simply a consequence of the 180° difference in temperature gradient direction relative to velocity gradient. An explanation for the difference in relative magnitude is not obvious.

On the jet axis the mean temperature gradient is along $-x_1$, 90° different from Eq. (86), but $\overline{\gamma u_3} = 0$ there, and the foregoing procedure leads only to

$$D_{31} = 0. \tag{90}$$

Fairly close to the jet axis, at $r = 0.4 r_{1/2}$,

$$\nabla \bar{\Gamma} = |\nabla \bar{\Gamma}| \begin{bmatrix} -0.23 \\ -0.97 \end{bmatrix}, \tag{91}$$

and

$$\bar{\mathbf{F}} = \overline{\gamma \mathbf{u}} = |\overline{\gamma \mathbf{u}}| \begin{bmatrix} 0.28 \\ 0.96 \end{bmatrix}. \tag{92}$$

In passing, we note the near parallelism of vectors $-\nabla \bar{\Gamma}$ and $\bar{\mathbf{F}}$. This is presumably coincidental, but reminds us that scalar diffusivity and viscosity models work better for free shear flows than for wall flows.

$$\tag{93} -\mathbf{D} \cdot (\nabla \bar{\Gamma}) \sim \begin{bmatrix} 0.23 D_{11} + 0.97 D_{13} \\ 0.23 D_{31} + 0.97 D_{33} \end{bmatrix}.$$

Since this must be parallel to $\overline{\gamma \mathbf{u}}$,

$$\tag{94} D_{13} - 0.069 D_{31} \approx -0.24 D_{11} + 0.29 D_{33}.$$

We could estimate a ratio of D_{11} to D_{33} from space–time covariance function data on u_1 and u_3, if it were available. A brief search of likely journals has encountered only u_1 data.

Lacking the necessary facts, we can speculate. First, we try the guess that

$$\tag{95} D_{11} \approx 2 D_{33},$$

a less drastic inequality than in the boundary layer. This transforms Eq. (94) to

$$\tag{96} D_{13} - 0.069 D_{31} \approx -0.19 D_{33}.$$

Unless $|D_{31}|$ is considerably larger than $|D_{13}|$, we infer that

$$\tag{97} D_{13} \approx -\tfrac{1}{5} D_{33},$$

which is opposite in sign from Eq. (83), the value a short radial distance away! Since D_{11} and D_{33} probably change relatively slowly with radial position, Eqs. (89) and (97) indicate a rapid change in D_{13}—another warning that the gradient transport model for turbulence may not be viable.

If, instead of neglecting the D_{31} term in Eq. (80), we assume that D_{13} has the same value as at $r_{1/2}$ [Eq. (89)], then Eq. (96) gives

$$\tag{98} D_{31} \approx 6.4 D_{33},$$

suggesting a strongly unsymmetric **D**. There is little basis for guessing whether Eq. (98) is more plausible than Eq. (97).

A generalization of Batchelor's (1949) work to include the effect of uniform mean velocity gradient (extending the analysis of Corrsin, 1953; Riley and Corrsin, 1974) may soon give at least semiquantitative insight into the diffusivity tensor unsymmetry for the case of homogeneous turbulent shear flow.

12. Concluding Remarks

By qualitative analogy with random walk transport, analyzed without rigor, we have inferred a collection of conditions that may be necessary (though not sufficient) for the applicability of simple gradient transport models in turbulence. These conditions stipulate degrees of homogeneity and stationarity of the mean field being transported and of the turbulence properties central to the transport mechanism. One or more of these conditions appear to be violated in each of the traditional turbulent flow boundary value problems, such as boundary layer and jet. The largest error in principle would seem to arise from the effects of inhomogeneity in transporting mechanism, e.g., the turbulence scale and rms velocity, as illustrated by Eq. (64).

Is there a simple generalization which may suffice for interim applications while basic research proceeds? Perhaps correction terms analogous to those in Eq. (25) could be added. A second-order correction term like that in Eq. (9) may not be appropriate unless the multitude of other second-order corrections are included as well. These can be inferred in a straightforward (but tedious) Taylor series expansion in space and time, allowing mean concentration $\bar{\Gamma}$, length scale l, and transport velocity scale V all to vary with position and time. This would be an extension of the calculation in the Appendix.

It is also interesting to ask whether a model for turbulent transport can be approached from the other length-scale limit, i.e., analogous to radiative transport in an "optically thin" medium—a direction which has also occurred to Spalding.[11] The difficulties are, of course, enormously greater than in the radiation problem, the most obvious reasons being the *continuous* interaction in turbulent motion, and the fact that the instantaneous speed of random turbulent transport is not orders of magnitude larger than the mean motion of the medium; it is, in fact, often smaller. The formulation which most nearly resembles radiative transport is the "backward diffusion" approach of Corrsin (1952, 1972; see also Batchelor, 1949), in which statistical moments at a point in space–time $(\mathbf{x}, t)$ are expressed as integrals over the (dispersed) prior locations of the ensemble of fluid particles which arrive at $\mathbf{x}$ at time t (one particle in each "realization").

Finally, it must be remembered that there are many alternative approaches to turbulent transport which are related to neither a gradient transport approximation (and its direct generalizations) nor to a hypothetical, long-path, radiative transport analogy. "Closure" schemes for hierarchies of turbulence moment equations have been based on a variety of truncated expansion procedures which often invoke no explicit physical or phenomenological images, but are more or less ad hoc, based sometimes on the intuitive belief that higher order correlation coefficients tend to be very much smaller than lower order ones. It is possible that one of these more formal closure methods will eventually succeed.

[11] D. B. Spalding, comment in NASA (1973).

APPENDIX

Power Series for Random Walk Transport Rate

Suppose a passive scalar field $\Gamma(z, t)$ undergoes transport in the mean, due to the random convective action of indelibly tagged particles (tagged with the "Γ substance") which perform random walks. Suppose also that both the mean step length $l(z, t)$ and the root-meansquare speed $V(z, t)$ are smooth functions of position and time. We ignore density changes merely for algebraic simplicity. They, of course, may be important in some applications.

Under hypotheses like those in the crudest kinetic theory of gases (see, e.g., Loeb, 1934), the mean net transport rate ("flux") of $\bar{\Gamma}(z, t)$ across the plane z in the positive direction is something like

(A.1)
$$\bar{F} = \tfrac{1}{2}n\{V(z - \tfrac{1}{2}l_-, t - \tau_-)\bar{\Gamma}(z - \tfrac{1}{2}l_-, t - \tau_-) \\ - V(z + \tfrac{1}{2}l_+, t - \tau_+)\bar{\Gamma}(z + \tfrac{1}{2}l_+, t - \tau_+)\}.$$

(See Fig. 6.) In Eq. (A.1), n is number density and $\tau \equiv l/V$ is "mean free time." The symbols with $+$ and $-$ subscripts are defined by

(A.2) $\quad l_- \equiv l(z - \tfrac{1}{2}l_-, t - \tau_-), \quad \tau_- \equiv \tau(z - \tfrac{1}{2}l_-, t - \tau_-),$

$\quad\quad\quad l_+ \equiv l(z + \tfrac{1}{2}l_+, t - \tau_+), \quad \tau_+ \equiv \tau(z + \tfrac{1}{2}l_+, t - \tau_+).$

The arguments of the V's and Eqs. (A.2) indicate the inhomogeneity and nonstationarity of the transporting mechanism.

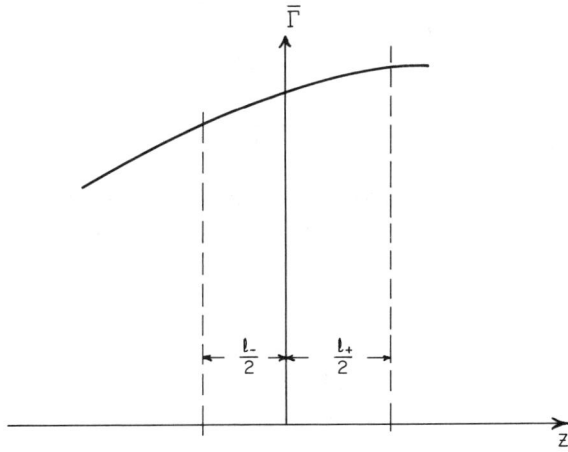

FIG. 6. Schema for estimate of mean flux of $\bar{\Gamma}$ over a mean free path when the latter is a function of position (and possibly time as well), $l = l(z, t)$.

In order to simplify Eq. (A.1) from a functional form to a function form all evaluated at (z, t), we suppose that each function in it can be approximated by its first few terms in a Taylor series in $l/2$ and τ about (z, t). An interesting facet is that an expansion to second order in l and τ requires expansion of the argument functions, l_-, l_+, τ_- and τ_+ as well.

A typical function in Eq. (A.1) will suffice as illustration:

(A.3)
$$V(z - \tfrac{1}{2}l_-, t - \tau_-) = V(z, t) + V_z(z, t)\left(-\frac{l_-}{2}\right) + V_t(z, t)(-\tau_-)$$
$$+ \frac{1}{2!} V_{zz}(z, t)\left(-\frac{l_-}{2}\right)^2 + \frac{1}{2!} V_{tt}(z, t)(-\tau_-)^2$$
$$+ V_{zt}(z, t)\left(-\frac{l_-}{2}\right)(-\tau_-) + \cdots.$$

But

(A.4)
$$l_- = l(z, t) + l_z(z, t)\left(-\frac{l_-}{2}\right) + l_t(z, t)(-\tau_-) + \frac{1}{2!} l_{zz}(z, t)\left(-\frac{l_-}{2}\right)^2$$
$$+ \frac{1}{2!} l_{tt}(z, t)(-\tau_-)^2 + l_{zt}(z, t)\left(-\frac{l_-}{2}\right)(-\tau_-) + \cdots.$$

and

(A.5)
$$\tau_- = \tau(z, t) + \tau_z(z, t)\left(-\frac{l_-}{2}\right) + \tau_t(z, t)(-\tau_-) + \frac{1}{2!} \tau_{zz}(z, t)\left(-\frac{l_-}{2}\right)^2$$
$$+ \frac{1}{2!} \tau_{tt}(z, t)(-\tau_-)^2 + \tau_{zt}(z, t)\left(-\frac{l_-}{2}\right)(-\tau_-) + \cdots.$$

In effect, Eqs. (A.4) and (A.5) must be substituted successively into themselves (so that they become series in l and τ), then into (A.3). The result, to second order, is

(A.6)
$$V(z - \tfrac{1}{2}l_-, t - \tau_-) \doteq V - V_z(l/2) - V_t\tau + V_z(l_z/2)(l/2) + V_z(l_t/2)\tau$$
$$+ V_t\tau_z(l/2) + V_t\tau_t\tau + \tfrac{1}{2}V_{zz}(l/2)^2 + \tfrac{1}{2}V_{tt}\tau^2$$
$$+ V_{zt}(l/2)\tau.$$

where all quantities are evaluated at (z, t).

Substituting Eq. (A.6) and the corresponding expressions for the other velocity and the two mean concentration terms into Eq. (A.1), we arrive at

(A.7)

$$(2/n)\bar{F}(z, t)$$
$$= -lV\bar{\Gamma}_z - l\bar{\Gamma}V_z + lV\bar{\Gamma}_t\tau_z + l\bar{\Gamma}V_t\tau_z + l\tau V_z\bar{\Gamma}_t + l\tau V_t\bar{\Gamma}_z + V\tau l_t\bar{\Gamma}_z$$
$$+ \bar{\Gamma}\tau l_t V_z + l\tau V\bar{\Gamma}_{zt} + l\tau\bar{\Gamma}V_{zt}.$$

The symmettry in $\bar{\Gamma}$ and V is a reflection of that in Eq. (A.1).

With the convention $\tau = l/V$, this can be written as an expansion in l. After some algebra,

(A.8)

$$(2/n)\bar{F}$$
$$= -lV\bar{\Gamma}_z - l\bar{\Gamma}V_z + ll_z\bar{\Gamma}_t + (ll_z\bar{\Gamma}/V)V_t + ll_t\bar{\Gamma}_z + (ll_t\bar{\Gamma}/V)V_z + l^2\bar{\Gamma}_{zt}$$
$$+ (l^2\bar{\Gamma}/V)V_{zt}.$$

For a steady state, homogeneous transport mechanism $(l_t = V_t = l_z = V_z = 0)$, to $O(l^2)$,

(A.9) $\qquad (2/n)\bar{F} = -lV\bar{\Gamma}_z + l^2\bar{\Gamma}_{zt},$

which is Eq. (19).

For steady state field and mechanism $(\bar{\Gamma}_t = l_t = V_t = 0)$, to $O(l^2)$,

(A.10) $\qquad (2/n)\bar{F} = -lV\bar{\Gamma}_z - l\bar{\Gamma}V_z,$

which is Eq. (24).

It should be remarked again that any specific physical or mathematical case may have material conservation constraints which will have to be taken into account.

Acknowledgments

I should like to thank Michael Karweit for critical discussion and computation, James Riley and Lyle Horn for constructive reading of the manuscript, and the National Science Foundation (Engineering Mechanics Section) and Office of Naval Research (Fluid Dynamics Program) for financial support.

References

Batchelor, G. K. (1949). *Proc. Roy. Soc., Ser. A* **195**, 513.
Batchelor, G. K. (1950). *J. Aeronaut. Sci.* **17**, 441.
Batchelor, G. K. (1953). "The Theory of Homogeneous Turbulence." Cambridge University Press, London and New York.
Blackwelder, R. F., and Kovasznay, L. S. G. (1972). *Phys. Fluids* **15**, 1545.
Bosworth, R. C. L. (1952). "Heat Transfer Phenomena." Wiley, New York.
Boussinesq, J. (1877). *Mem. Pres. Div. Savants Acad. Sci. Paris* **23**, 46.
Burgers, J. M., and Mitchner, M. (1953a). *Proc. Kon. Ned. Akad. Wetensch., Ser. B* **56**, Part I, 228.
Burgers, J. M., and Mitchner, M. (1953b). *Proc. Kon. Ned. Akad. Wetensch., Ser. B* **56**, Part II, 343.
Champagne, F. H., Harris, V. G., and Corrsin, S. (1970). *J. Fluid Mech.* **41**(1), 81.
Chandrasekhar, S. (1943). *Rev. Mod. Phys.* **15**, 1.
Chapman, S. (1928). *Phil. Mag., J. Sci.* **5**, 630.
Chapman, S., and Cowling, T. G. (1970). "The Mathematical Theory of Non-Uniform Gases," 3rd Ed. (with D. Burnett). Cambridge Univ. Press, London and New York.
Chou, P. Y. (1945). *Quart. Appl. Math.* **3**, 38.
Corrsin, S. (1943). *Nat. Adv. Comm. Aeronaut., Rep.* **ACR 3L23**. [Reissued as Wartime Rep. No. W-94 (1946).]
Corrsin, S. (1952). *J. Appl. Phys.* **23**, 113.
Corrsin, S. (1953). *Proc. Iowa Thermodyn. Symp., State Univ. Iowa, Iowa City* p. 5.
Corrsin, S. (1957). *In* "Naval Hydrodynamics," Proc. 1st Symp. (R. Sherman, ed.), Publ. No. 515, pp. 373-400. Nat. Acad. Sci.—Nat. Res. Counc., Washington, D.C.
Corrsin, S. (1963). *J. Atmos. Sci.* **20**, 115.
Corrsin, S. (1972). *Phys. Fluids* **15**, 986.
Corrsin, S., and Uberoi, M. S. (1950). *Nat. Adv. Comm. Aeronaut., Rep.* **998**. [*NACA Tech. Note* **1865** (1949).]
Courseau, P., and Loiseau, M. (1972a). *C. R. Acad. Sci., Ser. B* **274**, 1582, 1655.
Courseau, P., and Loiseau, M. (1972b). *C. R. Acad. Sci., Ser. B* **275**, 1199, 1247.
Craya, A. (1958). *Publ. Sci. Tech. Min. Air (Fr.)* **345**.
Davies, R. W., Diamond, R. J., and Smith, T. B. (1954). Tech. Rep. No. 201. Amer. Inst. Aerol. Res., Pasadena, California.
Deardorff, J. W. (1972). *J. Geophys. Res.* **77**, 5900.
Deissler, R. G. (1961). *Phys. Fluids* **4**, 1187.
de St. Venant, J.-C. B. (1843). *C. R. Acad. Sci.* **17**.
Einstein, A. (1926). "Investigations on the Theory of the Brownian Movement" (Engl. transl., ed. with notes by R. Fürth). Menthuen, London. (Reprinted by Dover, New York.)
Erian, F. F., and Eskinazi, S. (1969). *Phys. Fluids* **12**, 1988.
Fock, V. A. (1926). *Tr. Gos. Opt. In-ta* **4**, 34.
Fox, J. (1964). *Phys. Fluids* **7**, 562.
Gee, J. H. (1967). *Quart. J. Roy. Meteorol. Soc.* **93**, 237.
Gee, J. H., and Davies, D. R. (1964). *Quart. J. Roy. Meteorol. Soc.* **90**, 478.
Girgidov, A. D. (1973). *Izv. Akad. Nauk SSSR, Fiz. Atmos. Okeana* **9**, 91.
Goldstein, S. (1951). *Quart. J. Mech. Appl. Math.* **4**, 129.
Grad, H. (1958). *In* "Handbuch der Physik" (S. Flügge, ed.), Vol. 12, p. 26. Springer-Verlag, Berlin and New York.

Hinze, J. O. (1959). "Turbulence: An Introduction to Its Mechanism and Theory." McGraw-Hill, New York.
Jeans, J. H. (1940). "An Introduction to the Kinetic Theory of Gases." Cambridge Univ. Press, London and New York.
Johnson, D. S. (1957). *J. Appl. Mech.* **24**, 2. (*Trans. ASME* **79**.)
Johnson, D. S. (1959). *J. Appl. Mech.* **26**, 325. (*Trans. ASME* **81**)
Kampé de Fériet, J. (1937). *Ann. Soc. Sci. Bruxelles, Ser. I* **57**, 67.
Klebanoff, P. S. (1955). *NACA Rep.* **1247**. [*NACA Tech. Note* **3178** (1954).]
Kline, S. J., Morkovin, M. V., Sovran, G., and Cockrell, D. J., eds. (1969). *Proc. Conf. Comp. Turbulent Boundary Layers.* Mech. Eng. Dep., Stanford Univ., Stanford, California.
Kovasznay, L. S. G., Kibens, V., and Blackwelder, R. F. (1970). *J. Fluid Mech.* **41**, 283.
Lettau, H. (1952). *Geophys. Res. Pap.* No. 19, 437.
Liu, H. T., and Thompson, R. O. R. Y. (1973). "Lagrangian Statistics for Isotropic Turbulence in One Dimension," Unpubl. Rep., Contrib. No. 3012. Woods Hole Oceanogr. Inst., Woods Hole, Massachusetts.
Loeb, L. B. (1934). "The Kinetic Theory of Gases," 2nd Ed. McGraw-Hill, New York.
Lumley, J. L. (1962). *Colloq. Int. Cent. Nat. Rech. Sci.* **108**, 17.
Lumley, J. L., and Corrsin, S. (1959). *Advan. Geophys.* **6**, 179.
Lyapin, Y. S. (1948). *Meteor. Gidrol.* **5**, 13.
Lyapin, Y. S. (1950). *Tr. Gl. Geofiz. Observ.* **19**, 175.
Michelson, I. (1954). Tech. Rep. No. 193. Amer. Inst. Aerol. Res., Pasadena, California.
Monin, A. S. (1955). *Izv. Akad. Nauk. SSSR, Ser. Geofiz.* **3**, 234.
Monin, A. S. (1956). *Izv. Akad. Nauk SSSR, Ser. Geofiz.* **12**, 1461.
Monin, A. S., and Yaglom, A. M. (1965). "Statistical Hydromechanics," Vol. 1. Nauka, Moscow.
Monin, A. S., and Yaglom, A. M. (1971). "Statistical Fluid Mechanics," Vol. 1 (Engl. Ed. of Monin and Yaglom, 1965; updated, augmented, and revised by the authors; ed. by J. L. Lumley). MIT Press, Cambridge, Massachusetts.
NASA Langley Res. Cent. Staff, eds. (1973). "Free Turbulent Shear Flows." *NASA Spec. Publ.* **NASA SP-321**.
Nevzgliadov, V. G. (1945). *J. Phys. USSR* **9**, 235.
Patterson, G. S., Jr. (1966). Ph.D. Thesis, Johns Hopkins Univ., Baltimore, Maryland.
Patterson, G. S., Jr., and Corrsin, S. (1966). *In* "Dynamics of Fluids and Plasmas" (S. I. Pai, ed.), p. 275. Academic Press, New York.
Pawula, R. F. (1967). *Phys. Rev.* **162**, 186.
Prandtl, L. (1925). *Z. Angew. Math. Mech.* **5**, 136.
Prandtl, L. (1942). *Z. Angew. Math. Mech.* **22**, 241.
Rayleigh, Lord (Strutt, J. W.). (1894). "The Theory of Sound," 2nd Ed., Sect. 42a. Macmillan, New York. (Reprinted by Dover, New York, 1945.)
Reis, F. B. (1952). Ph.D. Thesis, Mass. Inst. Technol., Cambridge, Massachusetts.
Reynolds, O. (1895). *Phil. Trans. Roy. Soc. London, Ser. A* **186**, 123.
Richardson, L. F. (1920). *Proc. Roy. Soc., Ser. A* **97**, 354.
Riley, J. J., and Corrsin, S. (1974). *J. Geophys. Res.* **79**, 1768.
Rose, W. G. (1966). *J. Fluid Mech.* **25**, 97.
Rotta, J. (1951a). *Z. Phys.* **129**, 547 (I).
Rotta, J. (1951b). *Z. Phys.* **131**, 51 (II).
Shlien, D. J., and Corrsin, S. (1974). *J. Fluid Mech.* **62**, 255.
Taylor, G. I. (1915). *Phil. Trans. Roy. Soc. London, Ser. A* **215**, 1.
Taylor, G. I. (1921). *Proc. London Math. Soc.* **20**, 196.
Taylor, G. I. (1932). *Proc. Roy. Soc., Ser. A* **135**, 685.

Taylor, G. I. (1935). *Proc. Roy. Soc., Ser. A* **151**, 421.
Taylor, G. I. (1936). *Proc. Roy. Soc., Ser. A* **157**, 537.
Taylor, G. I. (1938a). *Proc. Roy. Soc., Ser. A* **164**, 15.
Taylor, G. I. (1938b). *Appl. Mech., Proc. Int. Congr., 5th* (J. P. Den Hartog and H. Peters, eds.), pp. 294–310. Wiley, New York.
Townsend, A. A. (1949). *Proc. Roy. Soc., Ser. A* **197**, 124.
von Kármán, T. (1930). *Nachr. Ges. Wiss. Goettingen, Math.-Phys. Kl.* p. 58.
von Kármán, T. (1937). *J. Aeronaut. Sci.* **4**, 131.
Wang, M. C., and Uhlenbeck, G. E. (1945). *Rev. Mod. Phys.* **17**, 323.
Wiskind, H. K. (1962). *J. Geophys. Res.* **67**, 3033.
Yaglom, A. M. (1969). *In* "Fluid Dynamics Transactions" (W. Fiszdon, P. Kucharczyk and W. J. Prosnak, eds.), Vol. 4, p. 801. Inst. Fundam. Tech. Res., Polish Acad. Sci., Warsaw.
Zubkovsky, S. L., and Tsvang, L. R. (1966). *Izv. Akad. Nauk SSSR, Fiz. Atmos. Okeana* **2**, 1307.

Addendum[1]

RANDOM WALKS ON MARKOVIAN BINARY VELOCITY FIELDS

G. S. PATTERSON, JR.[2]

Advanced Study Program.
National Center for Atmospheric Research, Boulder, Colorado 80302, U.S.A.

1. INTRODUCTION

Binary random walks in one or more dimensions have a long history of being useful in giving insight into various physical phenomena—principally as heuristic examples. The best known use of the random walk is in the problems of molecular diffusion and Brownian motion (Einstein, 1905; von Smoluchowski, 1916; Chandrasekhar, 1943), although similar walks had been used by Rayleigh (1880) in a sound wave problem. In the simplest one-dimensional case, the walking particle moves to the right or left a distance dx with equal probability, and arrives a time dt later. If a limiting process is used whereby $dx, dt \to 0$, but in such a way that $(dx)^2/dt \to D$, a constant, then the probability distribution for the walking particles $P(x, t)$ satisfies the diffusion equation with diffusivity D.

(1) $$\partial P(x, t)/\partial t = D\, \partial^2 P(x, t)/\partial x^2$$

While the diffusion equation is a satisfactory representation of molecular diffusion, it is not completely satisfactory for the case of turbulent diffusion—satisfactory perhaps for engineering approximations, but not from a theoretical or experimental point of view. As opposed to molecular diffusion, fluid point velocities are correlated over a significant time compared to the time scale of the motion. They are also finite as compared to molecular velocities which, in the simple random walk, are infinite in the

---

[1] The work reported in this paper was referred to in the review paper by Dr. Corrsin during the Symposium. The author prepared this Addendum at the request of the Editors of the Proceedings.

[2] On leave from Department of Engineering, Swarthmore College.

limit. As an heuristic example, Goldstein (1951) introduced a binary walk in which a particle's velocity has probability p of being the same as that at the preceding step. The sequence of particle velocities thus forms a first order Markov chain. In the limit $dx, dt \to 0$, requiring $dx/dt = V$ and $p = 1 - (dt/2T)$ with V, T constants, the probability distribution satisfies the telegraph equation

$$(2) \quad \frac{\partial^2 P(x, t)}{\partial t^2} + \frac{1}{T}\frac{\partial P(x, t)}{\partial t} = V^2 \frac{\partial^2 P(x, t)}{\partial x^2}$$

Additional memory can be added so that the particle's velocity is conditional on the previous n steps. The case for $n = 2, 3$ has been worked out (Patterson, 1966), but in the limit Eq. (2) is obtained.

To further emulate the problems of turbulent diffusion, S. Corrsin invented a walk in which probabilities were specified not for the walking particle, but for a one-dimensional, unsteady, binary velocity field. Given a particular realization of a velocity field, a trajectory of a fluid point starting from some given position and time could be easily determined. With a number of realizations, statistics of a dispersing particle could be obtained by computing an ensemble average. This is the classic Euler–Lagrange problem of turbulent diffusion. Given statistical information in laboratory coordinates, how does one compute or estimate the statistics of material points? For general classes of random binary fields, this is done by computer experiment (Patterson and Corrsin, 1966), but there is some merit in having a restricted problem that can be solved analytically. Such a problem was proposed by Lumley and Corrsin (1959), and its solution is sketched out here. In more detail, it is found in Patterson (1966).

Let the probability of a velocity being the same as its neighbor spatially be p, and its neighbor temporally be q. Expressed in another way, let the sequence of binary velocities for fixed t be a first order Markov chain with transition probability p, and let the sequence of velocities for fixed x be a first-order Markov chain with transition probability q. As we shall see below, the intuitively attractive idea (Lumley, 1961) that the sequence of particle velocities will form a first-order Markov chain with transition probability $\beta = pq + (1 - p) \times (1 - q)$ is not correct, and the probability that a particle will have the same velocity at time $t + dt$ as time t depends on the entire past history of the particle's trajectory. In the limit as $dx, dt \to 0$ subject to the constraints $dx/dt = V$, $p = 1 - (dx/2L)$, $q = 1 - (dt/2T)$, with V, T, L constants, the probability distribution for the diffusing particles satisfies the following integrodifferential equation:

$$(3) \quad \frac{\partial^2 P}{\partial t^2} + \frac{1}{T}\left(1 + \frac{VT}{L}\right)\frac{\partial P}{\partial t} - V^2 \frac{\partial^2 P}{\partial x^2} \\ = -\frac{1}{4T^2}\left(\frac{VT}{L}\right)\left(1 + \frac{VT}{L}\right)\int_0^t A(t - t')\frac{\partial P(x, t')}{\partial t'}dt'$$

with

$$A(t) = \frac{e^{-t/2T}I_1(t/2T)}{1/2(t/2T)}$$

In the limit $L \to \infty$, this reduces to Goldstein's problem, and we recover the telegraph equation.

The merit in Eq. (3) and its corollaries lies less in the possibility that it might model turbulent diffusion, but more in the heuristic example it gives by illustrating the extreme difficulties in obtaining a solution to an Euler–Lagrange problem and the inefficacy of *ad hoc* type solutions. As regards solutions of the Euler–Lagrange problem where the random velocity fields concerned are solutions to the Navier–Stokes equations, there exists to my knowledge only one analytical solution—that of Kampé de Fériet (1959) for a quite restricted class of one-dimensional shearing motions, and one computer experiment using numerically generated statistically isotropic, but nonstationary fields (Riley and Patterson, 1974).

2. Finite Walks

2.1 Preliminary Mathematics

Binary random walks where a particle moves either one place to the right or one place to the left at each step can be treated mathematically in a very straightforward way by using binary algebra. Let a given scalar function take on values 1 or 0 (true or false, respectively, if one thinks of the function as the occurrence of an event). Define multiplication identically with conventional multiplication and addition as addition modulo 2, so that $1 \oplus 1 = 0$. Define the complement of a scalar A as

(4) $$A' = 1 \oplus A$$

(Note that $A \oplus A = 0$, $A \oplus A' = 1$, $AA' = 0$. Most proofs can be done by exhaustion; i.e., by enumerating all possibilities.)

If A is a random binary event, the ensemble average $\langle A \rangle$ is defined as

(5) $$\langle A \rangle = \lim_{n \to \infty} \frac{1}{n} \sum_{i=1}^{n} A_i$$

where A_i is the value of A in the ith realization of the ensemble. By the strong law of large numbers (Feller, 1957),

(6) $\text{Prob}\{A = 1\}$ = probability that $A = 1$ or probability that the event, occurs

$= \langle A \rangle$

If A and B are two events, which are said to be independent, then

$$\text{Prob}\{AB\} = \text{probability that both } A \text{ and } B \text{ occur}$$
$$= \text{Prob}\{A\} \text{ Prob}\{B\}$$

or

(7) $$\langle AB \rangle = \langle A \rangle \langle B \rangle$$

Also if A and B are mutually exclusive, i.e., $AB = 0$, then

(8) $$\langle A \oplus B \rangle = \langle A \rangle + \langle B \rangle$$

Because of the binary nature of digital computers, binary representation is a very natural way of coding an experiment. It is such an efficient way, moreover, that ensembles of 25,000 realizations can be generated on relatively small machines. This can give remarkably good statistics.

To describe the class of walks on random binary fields, define the following variables:

$$U(m, n) = \begin{cases} 1 & \text{if the velocity at point } m \text{ in space and } n \text{ in time equals } +1, \\ 0 & \text{if the velocity equals } -1, \end{cases}$$

$X(m, n) = 1$ if a particle which left $(0, 0)$ in a particular realization arrives at point m at time n.

$$V(n) = \begin{cases} 1 & \text{if a particle's velocity at time } n = +1 \\ 0 & \text{if it equals } -1. \end{cases}$$

It is important to realize that once $U(m, n)$ is specified for all m, n in a particular realization, then $X(m, n)$, $V(n)$ are immediately deducible, for

(9) $$V(n) = \sum_{m=-n}^{n} U(m, n) X(m, n)$$

and

$$X(m, n) = V(n - 1) X(m - 1, n - 1) \oplus V'(n - 1) X(m + 1, n - 1)$$

with $X(m, 0) = 0$ except for $m = 0$, where $X(0, 0) = 1$.

Statistical quantities of interest include $\langle X(m, n) \rangle$ the probability that a particle reaches m at time n, $E(m, n)$ the statistical correlation between two velocities separated by a distance (m, n) in space–time, and $R(j; n)$ the autocorrelation of a particle's velocity at time n with its velocity at time $n + j$. After some manipulation, we have

(10)
$$E(m, n) = 4\langle U(M, N)U(M + m, N + n)\rangle - 1$$
$$R(j; n) = 4\langle V(n)V(n + j)\rangle - 1.$$

If the velocity field is statistically homogeneous, $E(m, n)$ is independent of M and N. In this case though, $R(j; n)$ will generally not be stationary, i.e., independent of n.

2.2. Markovian Velocity Fields

As outlined in Section 1, and in order to ensure statistical homogeneity, a sequence of velocities is generated for time 0, say $m = -3, -2, -1, 0, 1, 2, \ldots$, with transition probability p, and then this sequence and its *complement* are the elements in another Markov chain with transition probability q for times $1, 2, 3, \ldots$. The result is a velocity field in space–time which has random sized rectangular domains of velocity $+1$ or velocity -1.

More precisely, let Z, $W(m)$, and $Y(n)$ be binary, *independent* random variables such that

(11)
$$\text{Prob}\{Z = 1\} = \tfrac{1}{2}$$
$$\text{Prob}\{W(m) = 1\} = p \quad \text{for every} \quad m$$
$$\text{Prob}\{Y(n) = 1\} = q \quad \text{for every} \quad n.$$

Then the velocity field is generated by

(12)
$$U(m + 1, 0) = W(m)U(m, 0) \oplus W'(m)U'(m, 0)$$
$$U(0, n + 1) = Y(n)U(0, n) \oplus Y'(n)U'(0, n)$$
$$U(0, 0) = Z$$
$$U(m, n) = U(0, 0) \oplus U(0, n) \oplus U(m, 0).$$

The point $(0, 0)$ is arbitrary and could be made (M, N). The Eulerian velocity correlations may be determined by substituting Eq. (12) into Eq. (10), and after some manipulation

(13)
$$E(m, n) = (Zp - 1)^m (Zq - 1)^n.$$

While it is not an easy matter to deduce the Lagrangian statistics, as will be indicated in Section 2.3, some degenerate cases are of interest.

(i) $q = \tfrac{1}{2}$, p arbitrary. This is the classical random walk discussed, for instance, in Feller (1957). The probability distribution $\langle X(m, n)\rangle$ is

$$\binom{n}{(n + m)/2} 2^{-n}$$

and the particle velocities are uncorrelated.

(ii) $p = 1$, q arbitrary. This is the case discussed by Goldstein (1951). $\langle X(m, n) \rangle$ satisfies a difference equation, which in the limit of a continuous distribution becomes the telegraph equation.

(iii) $q = 1$, p arbitrary. After an initial flight, the particle is "trapped" in a region with velocity $+1$ to the left and velocity -1 to the right. $\langle X(m, n) \rangle$ quickly reaches an asymptotic state *independent* of time.

2.3. Particle Statistics for Arbitrary P, Q

The details of solving for $\langle X(m, n) \rangle$ or for $R(j; n)$ are sufficiently complicated so that only the briefest of descriptions are given. A considerable portion of the derivation is given in Patterson (1966), but it is not entirely complete. As far as possible, solutions were checked against computer experiments and found correct.

In a walk of n steps, there are 2^n distinct trajectories. Because of this distinctness (or disjointness), the probability of any particular event equals the sum of the probabilities of every trajectory which "contains" that event. The probability of each trajectory, in turn, equals the sum of the probabilities of all configurations of the field which contain that trajectory. To illustrate schematically where 1 indicates a $+1$ velocity and 0 indicates a -1 velocity:

$$V(2)V(3) = \begin{matrix} & 1 & & & & 1 & & & & 1 & & & & 1 \\ & 1 & & & & 1 & & & & 1 & & & & 1 \\ & & \oplus & & & & \oplus & & & & \oplus & & & \\ & 1 & & & & 0 & & & & 0 & & & & 1 \\ & 0 & & & & 0 & 1 & & & 1 & & & & \end{matrix}$$

and in turn

(14)
$$= \begin{matrix} 1 & 1\,1 & 0\,1 \\ 1 & 1\,1 & 1\,0 \\ & \oplus & \\ 0 & 0\,0 & 1\,0 \\ 1 & 1\,1 & 1\,0 \end{matrix}$$

or

$$\langle V(0)V'(1)V(2)V(3) \rangle = \tfrac{1}{2}p(1-q)^2 q + \tfrac{1}{2}(1-p)q^2(1-q)$$

and so forth. Natural groupings of p and q lead to the definition of the parameters

(15) $$\beta = pq + (1-p)(1-q), \quad \theta = q(1-q)$$

and ultimately

(16) $$\langle V(2)V(3) \rangle = \tfrac{1}{2}\beta\theta + \tfrac{1}{2}\beta\theta + \tfrac{1}{2}(1-\beta)\theta + \tfrac{1}{2}\beta^3.$$

For comparison,
$$V(0)V(1) = \begin{matrix} & 1 & 1 & 1 \\ & & & \\ & 1 & 1 & 1 \end{matrix} = \begin{matrix} 0 & 1 \\ \oplus & \\ 1 & 0, \end{matrix}$$

and

(17) $\langle V(0)V(1)\rangle = \frac{1}{2}pq + \frac{1}{2}(1-p)(1-q)$
$= \frac{1}{2}\beta,$

immediately showing the nonstationarity of $R(j; n)$. By proving such identities as

(18) $\langle X(m, n-2)U'(m, n-2)U(m-1, n-1)U(m, n)\rangle$
$= \theta\langle X(m, n-2)U'(m, n-2)\rangle$

and far more complicated ones, as for

$\langle X(m-2, n-2)U(m-2, n-2)U(m-1, n-1)U'(m, n)\rangle$

a difference equation for $\langle X(m, n)\rangle$ may be derived:

(19) $Y(m, n)$
$= (\delta + \beta^2)Y(m, n-2) - (1-\beta)\delta Z(m, n-2)$
$+ 2\theta^2 \sum_{k=0}^{[(n-6)/2]} A_{2k+4} Y(m, n-4-2k)$
$+ (1-\beta)\theta^2 \sum_{k=0}^{[(n-8)/2]} \sum_{j=0}^{[(n-8-2k)/2]} A_{2k+4} A_{2j+4} \{(1-\beta)Y(m, n-6-2k-2j)$
$+ \delta Z(m, n-6-2k-2j)\}$

where

$Y(m, n) = \langle X(m, n)\rangle - \beta[\langle X(m-1, n-1)\rangle + \langle X(m+1, n-1)\rangle]$
$+ (2\beta - 1)\langle X(m, n-2)\rangle$

$Z(m, n-2) = \frac{1}{2}[\langle X(m-1, n-3)\rangle + \langle X(m+1, n-3)\rangle]$
$- \langle X(m, n-4)\rangle$

$\delta = 2[\beta(1-\beta) - \theta], \quad A_{2k+4} = F(-k, \frac{3}{2}; 3; 4\theta).$

Initially
$Y(0, 2) = Y(-1, 3) = Y(1, 3) = 0,$
$Z(0, 2) = -\frac{1}{2}, \quad Z(-1, 3) = Z(1, 3) = -\beta/2.$

$F(a, b; c; z)$ is the hypergeometric function defined as

(20) $$F(a, b; c; z) = \sum_{n=0}^{\infty} \frac{(a)_n (b)_n}{(c)_n} \frac{z^n}{n!}$$

with

$$(a)_0 = 1$$
$$(a)_n = a(a+1) \cdots (a+n-1).$$

The consequences of Eq. (19) are really only apparent in the limit of a continuous distribution (Section 3), but two interesting points may be extracted from the finite case. First, the probability of a first return to the origin at time $n = 2k + 4$ turns out to be[3]

(21) $$f_{2k+4} = 2\langle V(n)V(n+1)\rangle \theta F(-k, \tfrac{3}{2}; 3; 4\theta)$$

and the probability of a return to origin at some time is

(22) $$f_2 + \sum_{k=0}^{\infty} f_{2k+4} = \begin{cases} 1, & 0 < \theta \leq \tfrac{1}{4} \\ 2\langle V(n)V'(n+1)\rangle, & \theta = 0, \end{cases}$$

an analog to Polya's theorem, except for degenerate case (iii).

Secondly, the slope of the mean square displacement $S(n) = D(n+1) - D(n)$,

(23) $$D(n) = \sum_{m=-n}^{n} m^2 \langle X(m, n) \rangle$$

which is related to the velocity autocorrelation function is found for large n to be

(24) $$S(\infty) = (1 + \beta)\theta / (1 - \beta)(1 - \beta + \theta).$$

An ad hoc estimate of $S(\infty)$ might be gotten by computing

(25) $$I_E = \sum_{k=0}^{\infty} E(k, k)$$

and comparing $2I_E - 1 = \beta/(1 - \beta)$ against $S(\infty)$. It is found to be an upper bound for $S(\infty)$, but is a result opposite to that found for more general binary velocity fields (Patterson and Corrsin, 1966).

---

[3] A conjecture in Patterson (1966) has been replaced by an inductive proof.

3. The Limit of a Continuous Distribution

Equation (19) becomes far more tractable (although an analytical solution has not yet been obtained) if we pass to the limit of a continuous distribution by letting

(26) $$x = m\, dx, \qquad t = n\, dt$$

where dx and dt approach zero as m and n approach infinity in such a way that x and t remain fixed and

(27) $$\lim_{\substack{n\to\infty \\ m\to\infty}} dx/dt = V, \text{ a constant.}$$

Letting

(28) $$p = 1 - (dx/2L), \qquad q = 1 - (dt/2T)$$

where L and T are constants,

(29) $$\lim_{\substack{dx\to 0 \\ dt\to 0}} E(x/dx, t/dt) = \mathscr{E}(x, t) = e^{-|x|/L} e^{-t/T}$$

and

(30) $$\int_0^\infty \mathscr{E}(x, 0)\, dx = L, \qquad \int_0^\infty \mathscr{E}(0, t)\, dt = T,$$

so that L, T are measures of the distance (in space–time) over which the velocity field is correlated.

Noting that only half the points in space–time are touched in the finite walk, define a continuous probability density as

(31) $$P(x, t) = \lim_{\substack{dx\to 0 \\ dt\to 0}} \frac{\langle X(x/dx, t/dt)\rangle}{2\, dx}$$

Similarly, if

(32) $$\mathscr{Y}(x, t) = \lim_{\substack{dx\to 0 \\ dt\to 0}} \frac{Y(x/dx, t/dt)}{dx(dt)^2}$$

$$\mathscr{Z}(x, t) = \lim_{\substack{dx\to 0 \\ dt\to 0}} \frac{Z(x/dx, t/dt)}{dx\, dt}$$

$$A(t) = \lim_{dt\to 0} A_{2k+4}; \qquad k = t/2\, dt,[4]$$

---

[4] A factor of 2 is missing in Patterson (1966) and the expression for $A(t)$ in Eq. (33) is changed accordingly.

then Eq. (19) becomes, after expressing each term by a Taylor series expansion and passing to the limit,

$$(33) \quad \frac{\partial \mathcal{Y}(x,t)}{\partial t} + \frac{1}{2T}\mathcal{Y}(x,t) + \frac{V}{4LT}\left(1 + \frac{VT}{L}\right)\mathcal{Z}(x,t)$$

$$= \frac{1}{8T^2}\int_0^t d\tau \left[A(t-\tau)\mathcal{Y}(x,\tau) + \frac{V}{8LT}\left(1 + \frac{VT}{L}\right)B(t-\tau)\mathcal{Z}(x,\tau)\right]$$

where

$$\mathcal{Y}(x,t) = \frac{\partial^2 P(x,t)}{\partial t^2} + \frac{1}{T}\left(1 + \frac{VT}{L}\right)\frac{\partial P(x,t)}{\partial t} - V^2 \frac{\partial^2 P(x,t)}{\partial x^2}$$

$$\mathcal{Z}(x,t) = \frac{\partial P(x,t)}{\partial t}$$

$$A(t) = \frac{e^{-t/2T}I_1(t/2T)}{\tfrac{1}{2}(t/2T)}$$

$$B(t) = \int_0^t A(\tau)A(t-\tau)\,d\tau.$$

The limit for $m = \pm n$ is singular, and we have

$$(34) \quad \begin{aligned} \text{Prob}\{\xi > Vt\} &= 0 \\ \text{Prob}\{\xi = Vt\} &= \tfrac{1}{2}\exp\{-(1 + (VT/L))(t/2T)\} \\ \text{Prob}\{x < \xi \le x + dx\} &= P(x,t)\,dx \\ \text{Prob}\{\xi = -Vt\} &= \tfrac{1}{2}\exp\{-(1 + (VT/L))(t/2T)\} \\ \text{Prob}\{\xi < Vt\} &= 0. \end{aligned}$$

Alternatively

$$(35) \quad \int_{-Vt}^{Vt} P(x,t)\,dx = 1 - \exp\{-(1 + (VT/L))(t/2T)\}.$$

Equation (33) may be further simplified by applying a Laplace transform with respect to time, and assuming that $\mathcal{Y}(x,t) \to 0$ as $t \to 0$. Then,

$$(36) \quad \mathcal{Y}(x,t) = -\frac{1}{4T^2}\left(\frac{VT}{L}\right)\left(1 + \frac{VT}{L}\right)\int_0^t A(t-\tau)\frac{\partial P(x,\tau)}{\partial \tau}\,d\tau.$$

As $L \to \infty$, we approach Goldstein's case, and $\mathcal{Y}(x,t) = 0$ which is the telegraph equation with solution

$$(37) \quad P(x,t) = \frac{1}{2VT}e^{-t/2T}\left[I_0(Y) + \frac{t}{2T}\frac{I_1(Y)}{Y}\right]$$

$$Y = (V^2 t^2 - x^2)^{1/2}/2VT$$

It is expected that the solution to Eq. (36) should exhibit behavior similar to Eq. (37), although only rough estimates of the effect of integrating $\partial P/\partial t$ over its past history have been made.

ACKNOWLEDGEMENTS

The author wishes to thank S. Corrsin for his encouragement and persistence. This work was supported by the Office of Naval Research, Contract NONR 4010 (5), and a portion by the Sloan Foundation.

REFERENCES

Chandrasekhar, S. (1943). Stochastic problems in physics and astronomy. *Rev. Mod. Phys.* **15**, 1.
Einstein, A. (1905). Über die von der molekularkinetischen Theorie der Wärme geforderte Bewegung von in ruhenden Flüssigkeiten suspendierten Teilchen. *Ann. Phys. (Leipzig)* **17**, 549.
Feller, W. (1957). "An Introduction to Probability Theory and Its Applications," 2nd Ed. Wiley, New York.
Goldstein, S. (1951). On diffusion by discontinuous movements, and on the telegraph equation. *Quart. J. Mech. Appl. Math.* **4**, Part 2, 129.
Kampé de Fériet, J. (1959). Statistical mechanics and theoretical models of diffusion processes. *Advan. Geophys.* **6**, 139.
Lumley, J. L. (1961). Distribution and dispersion in the Euler-Lagrange random walk. *Appl. Sci. Res., Sec. A* **10**, 153.
Lumley, J. L., and Corrsin, S. (1959). A random walk with both Lagrangian and Eulerian statistics. *Advan. Geophys.* **6**, 179.
Patterson, G. S., Jr. (1966). Walks on a random binary velocity field. Ph.D. Thesis, Johns Hopkins Univ., Baltimore, Maryland.
Patterson, G. S., Jr., and Corrsin, S. (1966). Computer experiments on random walks with both Eulerian and Lagrangian statistics. *In* "Dynamics of Fluids and Plasmas" (S. I. Pai, ed.), pp. 275–307. Academic Press, New York.
Rayleigh, Lord (1880). On the resultant of a large number of vibrations of the same pitch and of arbitrary phase. *Phil. Mag.* **10**, 73. [Also *Sci. Pap.* **1**, 491.]
Riley, J. J., and Patterson, G. S., Jr. (1974). Turbulent diffusion on numerically integrated velocity fields. *Phys. Fluids* (to be published).
von Smoluchowski. (1916). Drei Vorträge über Diffusion, Brownsche Bewegung und Koagulation von Kolloidteilchen. *Phys. Z.* **17**, 557.

HEIGHT OF THE MIXED LAYER IN THE STABLY STRATIFIED PLANETARY BOUNDARY LAYER

JOOST A. BUSINGER AND S. P. S. ARYA

Department of Atmospheric Sciences
University of Washington, Seattle, Washington 98195, U.S.A.

1. Introduction

The modeling of the stable boundary layer is difficult because usually a transition from turbulent to laminar flow takes place with increasing height. Consequently, because the processes under laminar conditions are very slow, steady state is usually not reached in the laminar region but may be approached in the turbulent region. Complex transient interactions between the laminar and turbulent regimes may occur. In this paper an attempt is made to formulate a steady-state stable boundary layer which in reality may occur only after a very long time, possibly during the polar night in the arctic. Nevertheless, it is felt that having a reasonable steady-state model may help us in understanding the more complex real case in a qualitative way.

2. Diagnostic Elements to the Problem

The entire boundary layer may be laminar provided the stratification is sufficiently stable. This requires a rather extreme condition if we assume that the Richardson number must be larger than 0.25. To illustrate this point, let us assume a light geostrophic wind $G = 1$ m sec^{-1}. The maximum shear which occurs at the surface is then

(1) $$(\partial \bar{u}/\partial z)_0 = (f/2\nu)^{1/2} G$$

where f is the Coriolis parameter (about 10^{-4} sec^{-1}) and ν is the kinematic viscosity for air (about 0.15 cm^2 sec^{-1}). In order that

(2) $$\text{Ri} = \frac{g}{\theta} \frac{\partial \bar{\theta}/\partial z}{(\partial \bar{u}/\partial z)^2} > 0.25$$

we must have $\partial\bar{\theta}/\partial z > 25°C/m$, which is a very large temperature gradient indeed. The value of Ri_{cr} near the surface is probably larger than the value in the free stream, but the order of magnitude will not change.

The preceding argument is given to illustrate the likelihood of turbulence near the surface in the stable boundary layer under most conditions. Examples of actual observations demonstrating a turbulent layer below the laminar flow are given in Figs. 1 and 2. These were taken during a comparison experiment in the summer of 1970 at Tsimlyansk, USSR. A roving probe sampled the velocity and temperature signals sequentially at five predetermined heights (0.5, 1.0, 2.0, 4.0, and 5.7 m), staying for about 15 sec at each level. In Fig. 1, the records marked as T and T_w are dry- and wet-bulb thermocouple signals and U is the wind speed measured by a cup

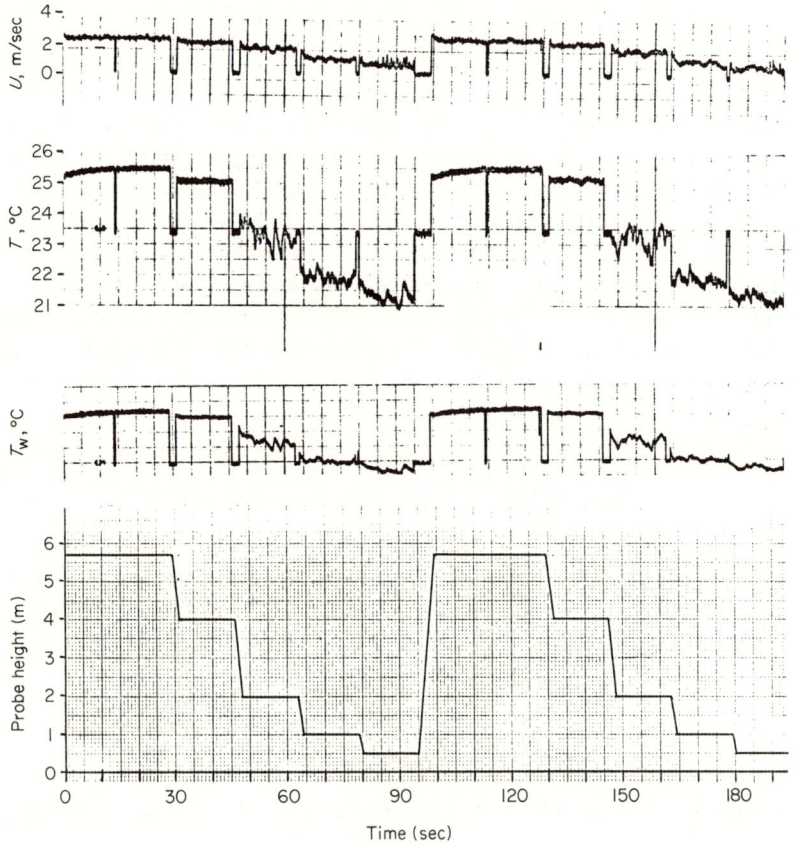

FIG. 1. A sample of velocity and temperature signals recorded at Tsymlyansk, USSR on July 1, 1970 showing the occurrence of a low level transition from turbulent to laminar flow.

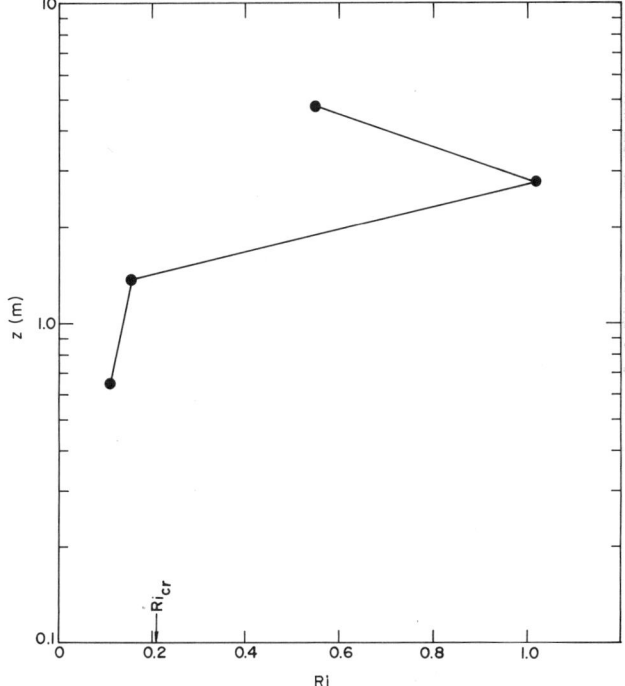

FIG. 2. Variation of Richardson number with height for the run in Fig. 1.

anemometer. The probe height as a function of time is also indicated. It is clearly seen, especially from the fast response temperature signals, that break-down of turbulence and a transition to the laminar state occurred in this particular instance at a low height between 2 and 4 m. The plot of the Richardson number with height (Fig. 2) also indicates that for $z \gtrsim 2m$, $Ri > Ri_{cr} \simeq 0.21$ (Businger et al., 1971; Arya, 1972).

Extensive observations of profiles and fluxes in the stably stratified turbulent surface layer have been reported both in the atmosphere (e.g., Webb, 1970, Businger et al., 1971), and in the wind tunnel (Arya and Plate, 1969). The results can be summarized simply by the following equations for the wind and temperature profile

(3a) $$\phi_m = 1 + \beta \zeta$$
(3b) $$\phi_h = 0.74 + \beta \zeta$$

where

(4) $$\phi_m \equiv \frac{kz}{\bar{u}_*} \frac{\partial \bar{u}}{\partial z}; \quad \phi_h \equiv \frac{kz}{\theta_*} \frac{\partial \bar{\theta}}{\partial z}; \quad \theta_* \equiv -\frac{\overline{(w'\theta')}_0}{u_*}$$

β is a constant about 4.7 ± 0.5; $\zeta \equiv z/L$; $L \equiv -\bar{\theta}u_*^3/kg(\overline{w'\theta'})_0$, the Obukhov length. The subscript zero refers to the value at the surface, and k is von Kármán's constant. When Eqs. (3) and (4) are integrated, the well-known log-linear profile is obtained.

A simple similarity argument may be given which leads to the above equations. Near the surface, the height above the surface is the appropriate scaling length, but when $z > L$, the appropriate scaling length is L because the buoyancy inhibits vertical excursions over distances larger than L. So, for $z > L$, we have $\partial \bar{u}/\partial z \propto u_*/L$, which leads to a linear profile.

It is not a priori obvious that L should be chosen as the scaling length. However, a simple argument to strengthen this notion may be given: A parcel of air with a vertical velocity w' will travel a distance l' in the vertical until its kinetic energy has been converted into potential energy. Thus we have

$$(5) \qquad (g/\theta)(\partial \bar{\theta}/\partial z)\overline{l'^2} \propto \overline{w'^2}$$

In the stable surface layer

$$(6) \qquad \overline{w'^2} \propto u_*^2 \quad \text{and} \quad [(g/\theta)(\partial\bar{\theta}/\partial z)]^{1/2} = f_3,$$

the Brunt–Väissallä frequency. Therefore

$$(7) \qquad l \equiv (\overline{l'^2})^{1/2} \propto u_* f_3^{-1}$$

Furthermore,

$$(8) \qquad \partial\bar{\theta}/\partial z = -\overline{w'\theta'}/K_h$$

and since l has the characteristics of a mixing length, we may assume that the eddy thermal diffusivity

$$(9) \qquad K_h \propto u_* l$$

Now substitute (9), (8), and (6) into (7), we obtain

$$(10) \qquad l \propto -\left(\frac{\theta}{g}\frac{u_*^3}{\overline{w'\theta'}}l\right)^{1/2} \propto (Ll)^{1/2} \quad \text{or} \quad l \propto L$$

A consequence of Eqs. (3a) and (3b) is that for large values of ζ, Ri approaches a constant

$$(11) \qquad \text{Ri} \to 1/\beta$$

This constant is presumably the critical Ri number (see Obukhov, 1946; Webb, 1970; Monin and Yaglom, 1971), which means that the profile behavior in the surface layer predicts an asymptotic approach to Ri_{cr} and not a transition from turbulent to laminar flow. This is a consequence of the concept of the surface layer where it is assumed that the fluxes are independent of height. This approximation is valid only over a limited height interval: especially in the stable boundary layer, it is soon necessary to take the variability of the fluxes with height into account.

A first approximation of the change of momentum flux with height is given by

$$-\partial \overline{u'w'}/\partial z = \rho^{-1}\,\partial p/\partial x \tag{12}$$

where $\partial p/\partial x$ is the pressure gradient in the direction of the mean wind speed in the surface layer. If we assume that the thermal wind is negligible, we may integrate (12) to

$$-\overline{u'w'} = u_*^2 + \rho^{-1}(\partial p/\partial x)z \tag{13}$$

which says that the stress linearly decreases with height and becomes zero at a height h, where

$$h = \frac{-\rho u_*^2}{\partial p/\partial x} \tag{14}$$

Equations (12)–(14) are based on the assumption of no turning of wind with height and, as such, these are more representative of unstable conditions (see, e.g., Deardorff, 1972; Wyngaard et al., this volume) than of neutral and stable ones. For the latter, the stress profile is expected to have some curvature, which increases with stability. Therefore, Eq. (14) would considerably underestimate the height of neutral and stable boundary layers. In the following section we present a simple model for calculating this height as a function of stability and also for the wind and stress profiles.

3. A Model of the Turbulent Stable Boundary Layer

The most sophisticated models of the steady and homogeneous atmospheric boundary layer, which are capable of supplying many fine details of turbulence structure, are those based on the three-dimensional numerical integration of the Navier–Stokes equations (Deardorff, 1970, 1972). The stably stratified boundary layer has remained elusive to this type of modeling because the subgrid-scale motions, which have to be parametrized due to limitations of costs and computer capacity, are considered to dominate the flow.

Next in hierarchy and considerably cheaper are the models based on high order closure techniques (Donaldson, 1973; Wyngaard et al., this volume; Lumley et al., this volume). Although, the lowest order closure assumptions have been found to be satisfactory for neutral and unstable boundary layers, their validity under stable conditions still remains to be demonstrated. There are some problems which have to be resolved before this method can be successfully employed for determining the detailed structure of the stable atmospheric boundary layer. For example, very little is known about the process of transition from turbulent to laminar flow due to increasing stability. The usual closure assumptions are not expected to remain valid near the critical conditions.

The steady-state barotropic boundary layer is described by the following equations of motion:

(15) $$\partial \overline{u'w'}/\partial z = f(\bar{v} - v_g)$$

(16) $$\partial \overline{v'w'}/\partial z = f(u_g - \bar{u})$$

Here, a right-handed coordinate system is used for the Northern Hemisphere with x axis oriented in the direction of surface shear and z axis in the vertical; $\bar{u}$ and $\bar{v}$ are the mean velocity components in x and y directions; $u_g = -(1/\rho f)\, \partial p/\partial y$ and $v_g = (1/\rho f)\, \partial p/\partial x$ are the two components of the geostrophic wind; and $\overline{u'w'}$ and $\overline{v'w'}$ are the components of the momentum flux.

Introducing an eddy viscosity K_m such that

(17) $$\overline{u'w'} = -K_m\, \partial \bar{u}/\partial z$$

(18) $$\overline{v'w'} = -K_m\, \partial \bar{v}/\partial z$$

Eqs. (15) and (16), after differentiation once with respect to z, can be written as

(19) $$K_* \, d^2 T_x/d\xi^2 + T_y = 0$$

(20) $$K_* \, d^2 T_y/d\xi^2 - T_x = 0$$

in which the various dimensionless variables are defined as

(21) $$T_x = -\overline{u'w'}/u_*^2$$

(22) $$T_y = -\overline{v'w'}/u_*^2$$

(23) $$\xi = fz/u_*$$

(24) $$K_* = K_m f/u_*^2$$

The dimensionless eddy viscosity K_* is expected to be a function of ξ as well as of stability. In many eddy viscosity models K_m is prescribed explicitly and somewhat arbitrarily. A more rational approach is to use the turbulent energy equation for deriving an equation for K_m (see, e.g., Monin, 1950; Blackadar, 1962; Bobyleva et al., 1967; Peterson, 1969). This usually requires gross simplification or neglect of the turbulent transport terms and parametrization of the energy dissipation in terms of some length and velocity scales which are then prescribed. Although this has worked well for the neutral case, the validity of the various assumptions involved is highly questionable for stable conditions.

From the observations in the stable atmosphere, there is a strong indication that the log-linear profile extends well into the height range where the flux variation must be considered (see Carl et al., 1973). In neutral conditions also the logarithmic law holds up to 10–15 % of the boundary layer height even though the $\overline{u'w'}$ flux may vary by more than 40 % in this layer (see, e.g., Thuillier and Lappe, 1964; Panofsky, 1973; Wyngaard et al., this volume). Recently, Tennekes (1973) has given an explanation for this.

Based on the above observational evidence an expression for the eddy viscosity function can be derived for the lower part of neutral and stable boundary layers if we can approximate the flux variation in this layer by a simple function. Equation (13) may be considered to be the lowest-order approximation, some of the limitation of which we have pointed out earlier. A better choice seems to be the exponential function

$$T_x = \exp(-|v_g/u_*|\xi) \tag{25}$$

which has the same initial slope as implied by Eq. (13) (this, of course, is required by the equation of motion) and at the same time has some finite curvature. Equation (25) fits the calculated profiles for the neutral case by Deardorff (1972) and Wyngaard et al. (this volume) quite well up to a value of $\xi \approx 0.15$. The eddy viscosity distribution which is consistent with Eqs. (3a) and (25) is given as

$$K_* = \frac{k\xi \exp(-|v_g/u_*|\xi)}{1 + \beta\xi\mu_*} \tag{26}$$

where

$$\mu_* = u_*/|f|L \tag{27}$$

is a stability parameter.

The prescription of K_* through Eq. (26) [a similar form was proposed earlier by Arya (1973), for $\mu_* = 0$] is not explicit, since v_g/u_* is not known until after the solution of Eqs. (19) and (20). To begin, some reasonable value of v_g/u_* is assumed for a given μ_*, and Eqs. (19), (20), and (26) are solved numerically using a version of the "shooting method" (for more details, see Arya, 1973). The new calculated value of v_g/u_* replaces the old one and the equations are solved iteratively until the assumed and calculated values of this parameter agree within a specified tolerance. Although our assumptions or hypotheses implicit in the derivation of Eq. (26) are not valid in the upper layer, the above eddy viscosity distribution will be assumed for the whole boundary layer since the solutions are considered not too sensitive to the details of K_* distribution in the upper layer (see, e.g., Arya, 1973).

The following boundary conditions were used in the solution of Eqs. (19)–(20).

At $\zeta = \zeta_0$,

(28a) $$T_x = \exp(-|v_g/u_*|\zeta_0)$$

(28b) $$T_y = 0$$

At $\zeta = \zeta_t$,

(29a) $$T_x = 0$$

(29b) $$T_y = 0$$

Here, the lower and upper boundaries were specified so that ζ_0 was small enough and ζ_t large enough not to affect the solutions. This occurred when $\zeta_0 \lesssim 0.01 H_*$, and $\zeta_t \gtrsim H_*$, H_* being the dimensionless boundary layer height where the fluxes and their gradients become vanishingly small (this implies that $T_x \to 0$, $T_y \to 0$, $u \to u_g$ and $v \to v_g$, as $\zeta \to H_*$). For a given stability condition (μ_*), the boundary layer would set its own thickness naturally once the above restrictions on ζ_0 and ζ_t were satisfied. This required several trials with different values of these since H_* was not known to begin with.

4. The Model Results and Comparison with Other Studies

The solution to Eqs. (19), (20), and (26), with the boundary condition as specified above yields the profiles of K_*, $\overline{u'w'}/u_*^2 = -T_x$, $\overline{v'w'}/u_*^2 = -T_y$, $(u_g - \bar{u})/u_* = -\partial T_y/\partial \zeta$ and $(v_g - \bar{v})/u_* = \partial T_x/\partial \zeta$, for different values of μ_*. These are shown in Figs. 3–9. For the neutral case ($\mu_* = 0$), our results are compared with those obtained by Wyngaard et al. (this volume) using a higher order closure scheme (see Figs. 3–5). Recognizing that they come from two basically different models, the good agreement between them is very encouraging and lends much credence to the simple model used here.

The model calculations were done for various values of μ_* in the range 0–250, which probably covers all stable conditions of practical interest. The strong effect of stability on eddy viscosity distribution can be seen from Fig. 5. With increasing μ_* the K_* distribution becomes flatter and flatter; both the maximum value and the dimensionless height where it occurs decrease by almost two orders of magnitude. Following Businger et al. (1971) we have used here a value of $k = 0.35$.

The velocity defect profiles are shown in Figs. 6 and 7. Both $\bar{u}$ and $\bar{v}$ components overshoot their respective geostrophic values before coming back to geostrophic equilibrium at a greater height. The maximum overshoot normalized by u_* increases with stability although the change is hardly significant if seen as a percentage of u_g/u_* or v_g/u_*. The stress profiles (Figs. 8 and 9) are similar to those for the neutral Ekman layer except for the increase in their curvature with increasing stability.

The most significant effect of stability brought out by the calculated velocity and stress profiles is the change in the boundary layer height, whichever way one defines it. In Fig. 10 we have represented the dimensionless heights ξ_u and ξ_v, where after the first cross-over and overshoot $\bar{u} = u_g$ and $\bar{v} = v_g$, respectively. The latter is about 25 % greater than the former. The level $\xi_{\tau x}$ where $\overline{u'w'} = 0$ after the first zero cross-over and change in sign falls in between ξ_u and ξ_v. It corresponds to a minimum in the $\bar{u}$ profile. The height corresponding to the maximum of $\bar{u}$ profile is also sometimes used as a measure of the boundary layer height (Clarke, 1970). This, however, is only about half of $\xi_{\tau x}$ and, since there is still considerable amount of the lateral

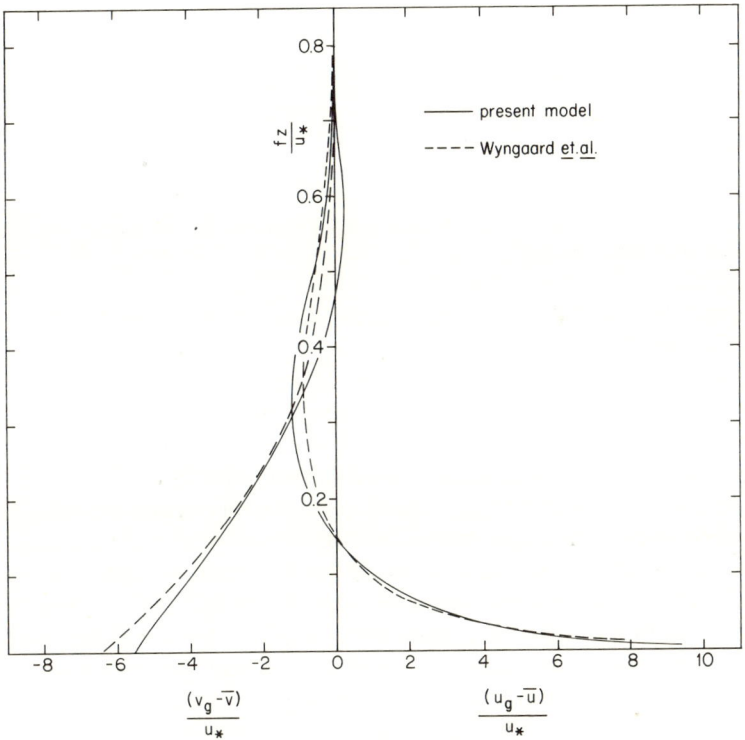

FIG. 3. Comparison of the neutral velocity defect profiles from the present model with those of Wyngaard et al. (this volume).

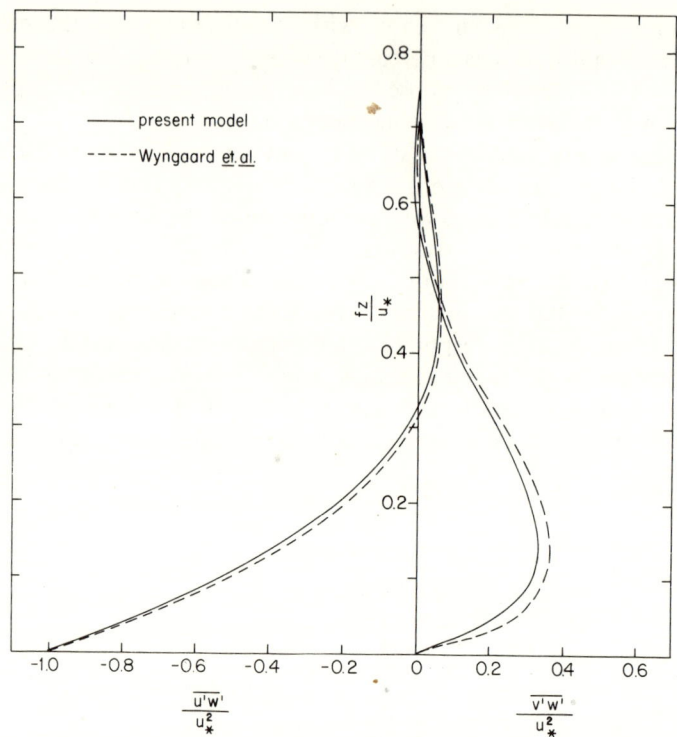

FIG. 4. Comparison of the neutral stress profiles from the present model with those of Wyngaard et al. (this volume).

stress and velocity defect remaining at this level, it should not be considered as the top of the boundary layer.

Alternative ways of defining the boundary layer height have been discussed by Zilitinkevich (1972). The most convenient, especially from a theoretical point of view, is the dimensionless height ξ_τ where the magnitude of the normalized stress reduces to an arbitrarily specified small fraction. If we take this fraction to be 1%, ξ_τ turns out to be very close or identical to ξ_u (represented by open circles in Fig. 10). Based on some rough assumptions about the limiting or critical value of the flux Richardson number and about the relationship between the boundary layer height and an "effective" eddy viscosity, Zilitinkevich (1972) predicted $\xi_\tau \sim \mu_*^{-1/2}$, for the asymptotic case of large stability. This relationship is also confirmed by the results of our model; the dotted line in Fig. 10 representing

$$\xi_\tau = 0.72\mu_*^{-1/2} \tag{30}$$

nicely fits through the computed points, especially for $\mu_* \gtrsim 50$.

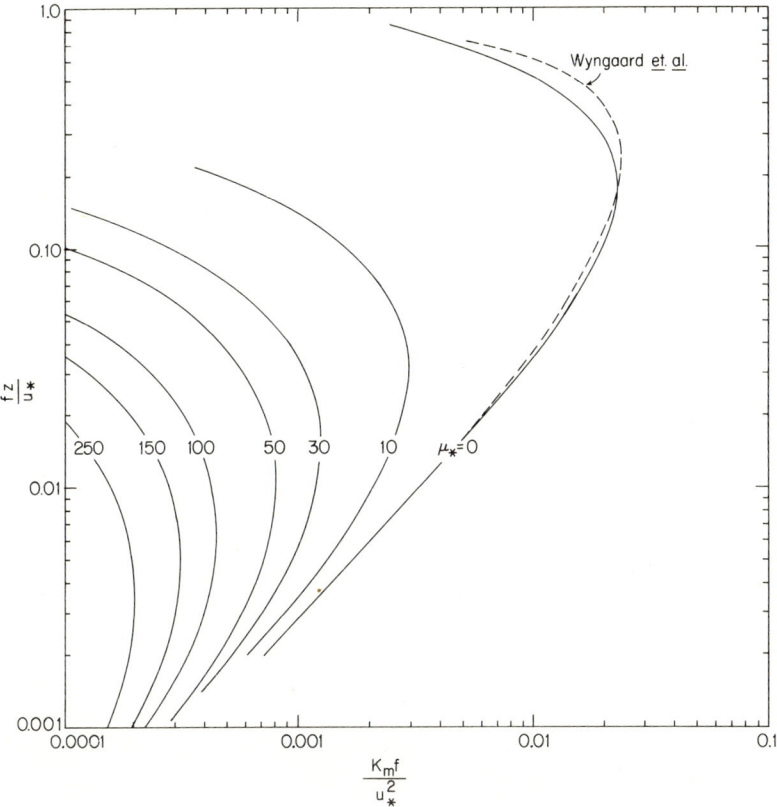

FIG. 5. Eddy-viscosity distributions for different stabilities.

In the classic case of a laminar Ekman layer, the boundary layer height varies as $(v/f)^{1/2}$, where v is the kinematic viscosity of the fluid. An analogous relation is obtained for a turbulent Ekman layer with constant eddy viscosity. The assumption of a constant K_m is, however, unrealistic, the typical K_m distributions being as shown in Fig. 5. These are characterized by a maximum $(K_{*\text{max}})$ occurring somewhere above the surface layer. Figure 11 shows that for all stability conditions the boundary layer height is still uniquely related to this maximum value through a simple power-law relation

$$\xi_\tau = 5.03(K_{*\text{max}})^{0.55}, \tag{31}$$

in which the exponent is not far different from $\tfrac{1}{2}$ for the classical Ekman spiral. This may provide a justification for the crucial assumption made

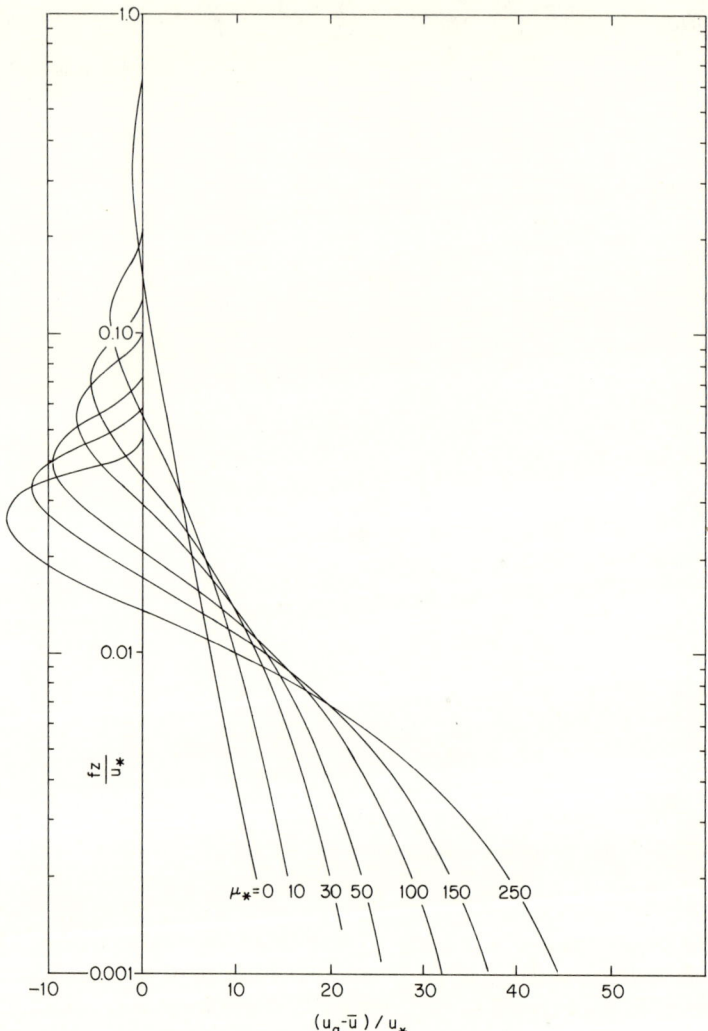

Fig. 6. The longitudinal velocity defect profiles for different stabilities.

earlier by Zilitinkevich (1972) relating the boundary layer thickness to the stability parameter.

The geostrophic drag relations for the stratified boundary layer are usually expressed as (see, e.g., Zilitinkevich and Chalikov, 1968; Clarke, 1970)

(32) $$u_g/u_* = k^{-1}[\ln|u_*/fz_0| - A(\mu_*)]$$

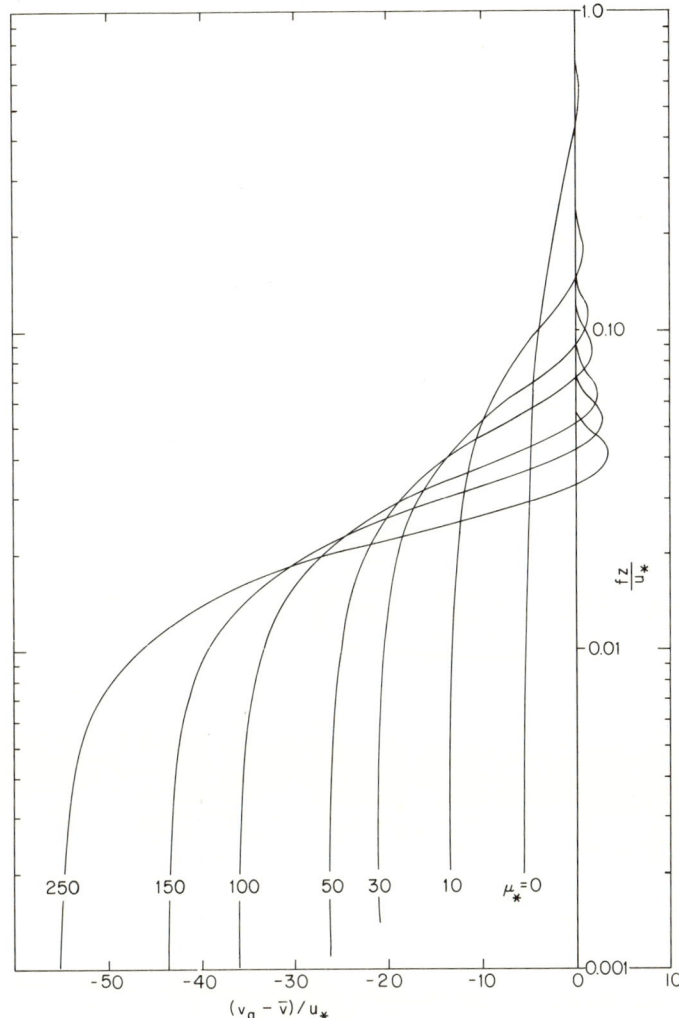

FIG. 7. The lateral velocity defect profiles for different stabilities.

(33) $$v_g/u_* = -(B(\mu_*)/k)\,\text{Sign}(f)$$

in which $A(\mu_*)$ and $B(\mu_*)$ are some universal functions of the stability parameter μ_*. Sometimes A and B appear interchanged and $\mu = k\mu_*$ is used as a stability parameter, although unnecessary use of the Von Karman constant is becoming a problem in view of the recent controversy about its exact value.

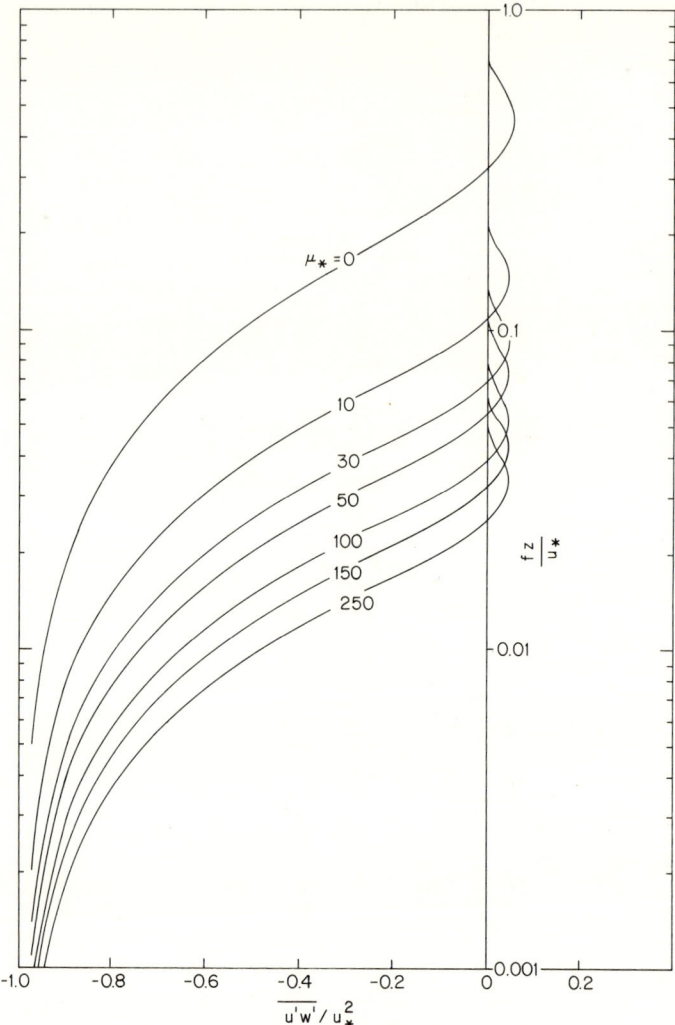

FIG. 8. The longitudinal stress profiles for different stabilities.

By matching the calculated velocity defect profiles to the log-linear profile in the lowest layer, it is easy to show that

(34) $$A(\mu_*) = \lim_{\xi \to \xi_0} [k \, dT_y/d\xi - \ln \xi - \beta \xi \mu_*]$$

(35) $$B(\mu_*) = \lim_{\xi \to \xi_0} [-k \, dT_x/d\xi]$$

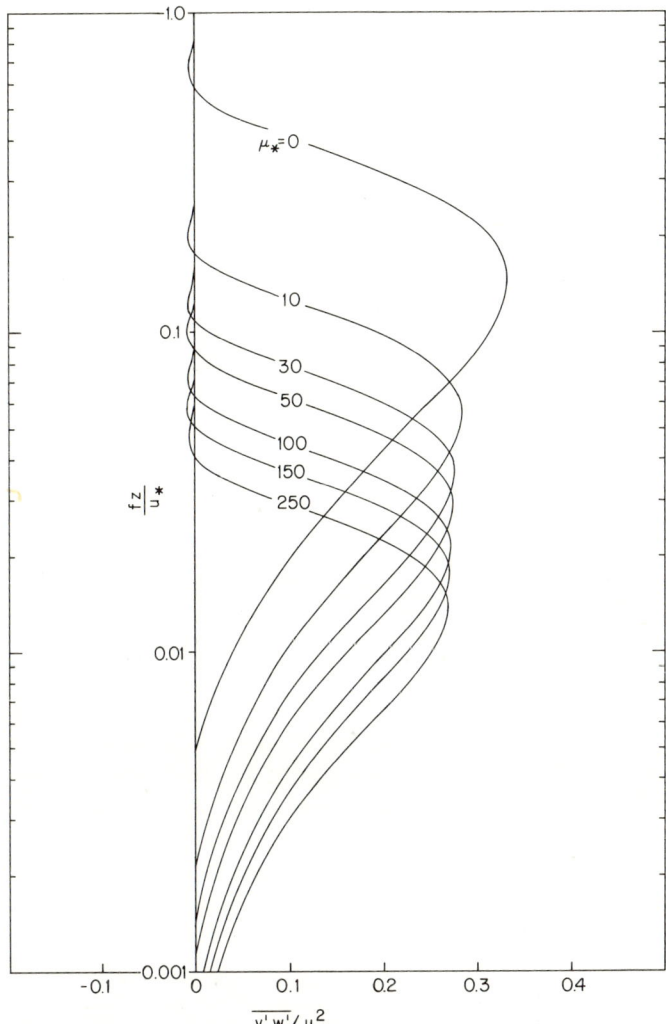

FIG. 9. The lateral stress profiles for different stabilities.

The model results show that for a given μ_*, the right-hand sides of Eqs. (34) and (35) approach some constant values independent of ξ, as $\xi \to \xi_0$, indicating the existence of a matching layer in which both the inner and outer similarity laws apply.

The similarity functions $A(\mu_*)$ and $B(\mu_*)$ are represented in Fig. 12. These have also been determined empirically by Zilitinkevich and Chalikov (1968)

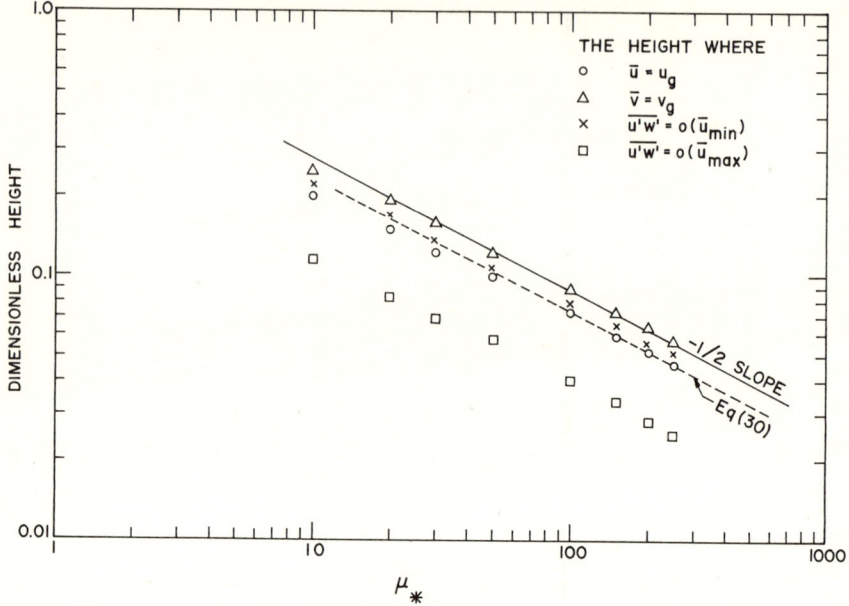

FIG. 10. The boundary layer height defined in various ways as a function of stability.

using the Great Plains data, and by Clarke (1970) using the Australian expedition data. Unfortunately, their results are markedly different with too much scatter of data points to permit any reliable estimate of these supposedly universal functions. In a separate study the junior author has undertaken to resolve these differences and to investigate the possible effects of thermal winds and accelerations. An analysis of the "Wangara" data (Clarke et al., 1971) by him has led to improved estimates of the above similarity functions. There is still considerable scatter of individual data points and only the best fitted third degree polynomials through them are shown in Fig. 12. Note that the agreement with our theoretical curves is good in view of the large scatter in the original data.

The results of Bobyleva et al. (1967) are also shown in Fig. 12 for comparison with the present model. The agreement between the two is very close for $B(\mu_*)$, but very poor for $A(\mu_*)$. Our model is in better agreement with the empirical estimates of the latter. We also point out that the velocity defect profiles calculated by Bobyleva et al. (1967), although quite similar to ours in the lower layer, had very sharp kinks just below the top of the boundary layer. This may be indicative of the failure of some of their model assumptions in stable conditions.

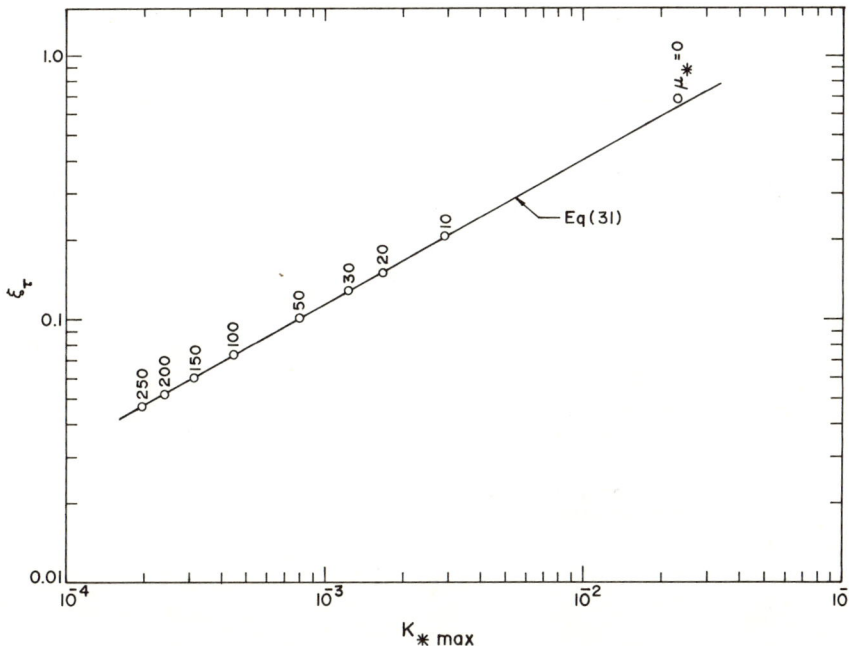

FIG. 11. The boundary layer height as a function of the maximum eddy viscosity.

5. Conclusions

It is argued that under most gravitationally stable conditions in the atmosphere there is always a good likelihood of a turbulent or mixing layer occurring near the surface. Simple reasoning is given in support of the viewpoint that with increasing stability the length scale of the vertical motion becomes independent of the height above the surface, but proportional to the so-called Obukhov length. Under these conditions, the mean velocity and temperature profiles become linear (or rather, log-linear if one considers the flow very close to the surface).

Although the logarithmic law in the neutral case or the log-linear law in stable conditions was originally thought to apply only in the so-called constant flux layer, there is now considerable empirical and theoretical evidence that these simple relations are valid for the x component (in the direction of surface shear) of velocity to much larger heights where the variation of fluxes and the turning of wind due to rotational effects must be considered. Based on this observation and a reasonable approximation of the momentum flux variation in the vertical, an expression is derived for the eddy viscosity

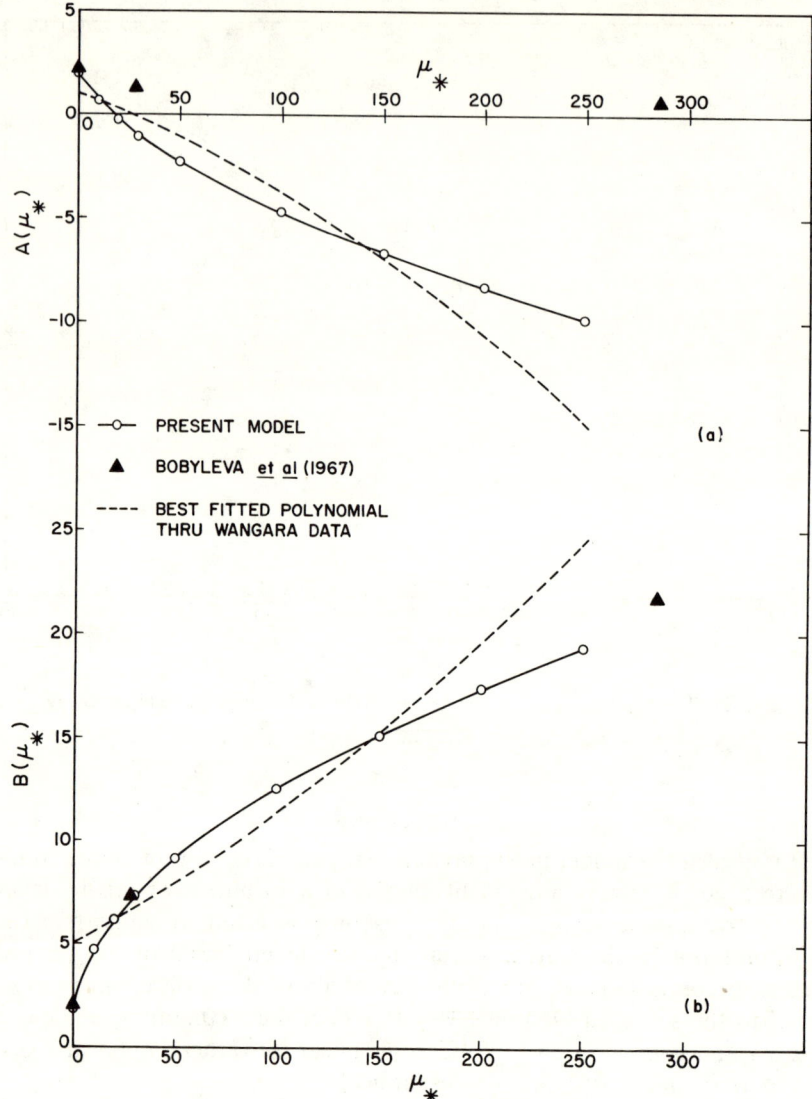

Fig. 12. Comparison of the similarity functions $A(\mu_*)$ and $B(\mu_*)$ calculated from the present model with those from Bobyleva et al. (1967) and from the Wangara data.

distribution. This together with the equations of mean motion and appropriate boundary conditions form the basis of our simple steady-state model of the stable boundary layer. The numerical solution based on a version of the "shooting method" yields velocity defect and stress profiles for different values of the dimensionless stability parameter.

For the neutral case, the results of our model compare quite well with an earlier model by Bobyleva et al. (1967) and also with a more sophisticated numerical model recently proposed by Wyngaard et al. (this volume). The two most significant results for the stable boundary layer are:

(1) The dimensionless boundary layer height varies in inverse proportion to the square-root of the stability parameter; this trend was also predicted by Zilitinkevich (1972) from totally different reasoning.

(2) The calculated forms of the stability dependent functions in the geostrophic drag relations are in fair agreement with their empirical estimates from the "Wangara" data.

Acknowledgments

This research was supported by the National Science Foundation through Grants GA-14680 and GV-32342. The authors wish to thank Mr. Vince Wong for helping in the computer calculations.

References

Arya, S. P. S. (1972). The critical condition for the maintenance of turbulence in stratified flows. *Quart. J. Roy. Meteorol. Soc.* **98**, 264–273.
Arya, S. P. S. (1973). Neutral planetary boundary layer above a nonhomogeneous surface. *Geophys. Fluid Dyn.* **4**, 333–355.
Arya, S. P. S., and Plate, E. J. (1969). Modeling of the stably stratified atmospheric boundary layer. *J. Atmos. Sci.* **26**, 656–665.
Blackadar, A. K. (1962). The vertical distribution of wind and turbulent exchange in a neutral atmosphere. *J. Geophys. Res.* **67**, 3095–3103.
Bobyleva, I. M., Zilitinkevich, S. S., and Laikhtman, D. L. (1967). Turbulent regime in a thermally stratified planetary atmospheric boundary layer. "Atmospheric Turbulence and Radiowave Propagation," pp. 179–190. Nauka, Moscow.
Businger, J. A., Wyngaard, J. C., Izumi, Y., and Bradley, E. F. (1971). Flux-profile relationships in the atmospheric surface layer. *J. Atmos. Sci.* **28**, 181–189.
Carl, D. M., Tarbell, T. C., and Panofsky, H. A. (1973). Profiles of wind and temperature from towers over homogeneous terrain. *J. Atmos. Sci.* **30**, 788–794.
Clarke, R. H. (1970). Observational studies in the atmospheric boundary layer. *Quart. J. Roy. Meteorol. Soc.* **96**, 91–114.
Clarke, R. H., Dyer, A. J., Brook, R. R., Reid, D. G., and Troup, A. J. (1971). "The Wangara Experiment: Boundary Layer Data," Div. Meteorol. Phys. Pap. No. 19. Commonw. Sci. Ind. Res. Organ., East Melbourne.
Deardorff, J. W. (1970). A three-dimensional numerical investigation of the idealized planetary boundary layer. *Geophys. Fluid Dyn.* **1**, 377–410.

Deardorff, J. W. (1972). Numerical investigation of neutral and unstable planetary boundary layers. *J. Appl. Meteorol.* **11**, 91–115.

Donaldson, C. duP. (1973). Construction of a dynamic model of the production of atmospheric turbulence and the dispersal of atmospheric pollutants. *In* "Workshop on Micrometeorology" (D. A. Haugen, ed.), Ch. 8, pp. 313–392.

Monin, A. S. (1950). Dynamic turbulence in the atmosphere. *Izv. Akad. Nauk SSSR, Ser. Geogr. Geofiz.* **14**, No. 3.

Monin, A. S., and Yaglom, A. M. (1971). "Statistical Fluid Mechanics," Vol. 1. MIT Press, Cambridge, Massachusetts.

Obukhov, A. M. (1946). Turbulence in an atmosphere with a non-uniform temperature. [Engl. transl. in *Boundary-Layer Meteorol.* **2**, 7–29 (1971).]

Panofsky, H. A. (1973). The boundary layer above 30 m. *Boundary-Layer Meteorol.* **4**, 251–264.

Peterson, E. W. (1969). Modification of mean flow and turbulent energy by a change in surface roughness under conditions of neutral stability. *Quart. J. Roy. Meteorol. Soc.* **95**, 561–575.

Tennekes, H. (1973). The logarithmic wind profile. *J. Atmos. Sci.* **30**, 234–238.

Thuillier, R. H., and Lappe, U. O. (1964). Wind and temperature profile characteristics from observations on a 1400 ft. tower. *J. Appl. Meteorol.* **3**, 299–306.

Webb, E. K. (1970). Profile relationships: The log-linear range and extension to strong stability. *Quart. J. Roy. Meteorol. Soc.* **96**, 67–90.

Zilitinkevich, S. S. (1972). On the determination of the height of the Ekman Boundary Layer. *Boundary-Layer Meteorol.* **3**, 141–145.

Zilitinkevich, S. S., and Chalikov, D. V. (1968). The laws of resistance and of heat and moisture exchange in the interaction between the atmosphere and an underlying surface. *Izv. Akad. Nauk SSSR, Fiz. Atmos. Okeana* **4**, No. 7, 438–441. (Engl. Transl.)

TRANSPORT OF HEAT ACROSS A PLANE TURBULENT MIXING LAYER[1]

HEINRICH E. FIEDLER

Institut für Thermo- und Fluiddynamik Technische Universität
Berlin, Germany

1. INTRODUCTION

It is a well-known characteristic of turbulent shear flows that the spread of scalar quantities, i.e., heat or matter, is faster than the spread of momentum. The turbulent Prandtl number $Pr_T = \varepsilon_M/\varepsilon_Q$ conventionally used in describing this phenomenon is therefore $0.5 \leq Pr_T \leq 1.0$, with ε_M and ε_Q being the diffusivities of momentum and of scalar quantities, respectively. The upper and lower limits of the Prandtl number are matched by Prandtl's mixing length hypothesis or Taylor's vorticity transport hypothesis, respectively. In reality, however, the Prandtl number has neither the same value for different flows nor for different locations within a single flow. This behaviour is not yet fully understood; it seems, however, to exclude the possibility of identical transport mechanisms for momentum and scalar quantities.

More recent conceptions, originating from Townsend's concepts of bulk convection transfer and of a double structure of turbulence are therefore based on the ideas of different transfer mechanisms and of a strong significance of the intermittent structure of the flow at free boundaries (see Townsend, 1956; Yen, 1967; Mayer and Divoky, 1966; Tyldesley, 1969).

The experimental investigation reported here is aimed at finding a model for the transport mechanism of a scalar quantity in typical turbulent shear flows. As a first configuration the two-dimensional shear layer was chosen for the following reasons:

(a) Its velocity structure is well known (Wygnanski and Fiedler, 1970).
(b) Self-preservation is attained at an early stage.
(c) Heat-transfer measurements have not been reported previously.
(d) This flow has constant boundary values.

In the present state of the investigation measurements of the temperature field alone have been obtained. For more detailed and specific information

---

[1] This work was supported by Deutsche Forschungsgemeinshaft.

and for interpretation of the structure of the temperature field the method of conditional sampling, as introduced by Kibens and Kovasznay (1969) was applied.

2. Basic Equations and Definitions

The temperature field can be mathematically described by four equations; the equation of motion (Reynolds equation), the continuity equation for the mean velocities, the heat-transfer equation, and the equation of the balance of temperature fluctuations (according to Corrsin, 1952). By rendering these equations dimensionless and taking the similarity of the flow into account one obtains, for the Reynolds equation

$$(\eta \bar{u}^* - \bar{v}^*) \, d\bar{u}^*/d\eta = d(\overline{u'v'})^*/d\eta \tag{1}$$

for the equation of continuity,

$$\eta \, d\bar{u}^*/d\eta = d\bar{v}^*/d\eta \tag{2}$$

for the heat-transfer equation,

$$(\eta \bar{u}^* - \bar{v}^*) \, d\Delta \bar{T}^*/d\eta = d(\overline{v'T'})^*/d\eta \tag{3}$$

and for the balance equation of the temperature fluctuations:

$$(\bar{v}^* - \eta \bar{u}^*) \frac{d\overline{(T')^{*2}}}{d\eta} + 2\overline{(v'T')}^* \frac{d\Delta \bar{T}^*}{d\eta} + \frac{d}{d\eta} \overline{(v'T'^2)}^* \tag{4}$$

$$+ \frac{2ax}{u_0 \Delta T_G^2} \left\{ \overline{\left(\frac{\partial T'}{\partial x}\right)^2} + \overline{\left(\frac{\partial T'}{\partial y}\right)^2} + \overline{\left(\frac{\partial T'}{\partial z}\right)^2} \right\} = 0$$

where conventionally

$$u = \bar{u} + u' \qquad u_0 = \text{maximum velocity}$$
$$v = \bar{v} + v' \qquad \Delta T_G = \text{maximum temperature difference}$$
$$\Delta T = \Delta \bar{T} + T'$$

and further

$$\eta = y/x; \qquad \bar{u}^* = \bar{u}/u_0 \text{, etc.}$$
$$\Delta \bar{T}^* = \Delta \bar{T}/\Delta T_G \text{, etc.}$$

From Eq. (4) follows

$$\overline{\left(\frac{\partial T'}{\partial x}\right)^2} \sim \overline{\left(\frac{\partial T'}{\partial y}\right)^2} \sim \overline{\left(\frac{\partial T'}{\partial z}\right)^2} \sim \frac{u_0 \Delta T_G^2}{2ax} \sim \frac{\text{Re}_x \, \text{Pr} \, \Delta T_G^2}{x^2} \tag{5}$$

i.e., the dissipative structure of the temperature field is, in contrast to the large scale structure, a function not only of η but also of x.

The following definitions for the mean values (averaging time T usually 120 s) were used:

(a) For conventional averages (Q being an arbitrary function)

$$\bar{Q} = \lim_{T \to \infty} \frac{1}{T} \int_0^T Q(t)\, dt$$

(b) For conditional averages, turbulent:

$$\bar{Q}_T = \lim_{T \to \infty} \frac{1}{\gamma T} \int_0^T \delta(t) Q(t)\, dt$$

nonturbulent:

$$\bar{Q}_P = \lim_{T \to \infty} \frac{1}{(1-\gamma)T} \int_0^T (1 - \delta(t)) Q(t)\, dt$$

where $\delta(t) = 1$ in turbulent flow and $\delta(t) = 0$ in nonturbulent flow. Further,

$$Q = Q_T \gamma + Q_P(1 - \gamma)$$

or

$$Q = Q_T(\gamma_1 + \gamma_2 - 1) + Q_{P_1}(1 - \gamma_1) + Q_{P_2}(1 - \gamma_2)$$

where

$$\gamma = \overline{\delta(t)}$$

3. Experimental Arrangement

A schematic of the complete experimental setup and of the test section is given in Figs. 1 and 2. The air is electrically heated at the entrance of the radial fan to obtain a uniform heat distribution in the flow. The nozzle contraction ratio is 6 : 1. By insulating the complete plenum chamber and the nozzle, the temperature boundary layer at the nozzle exit was reduced to approximately 5 mm. The thickness of the exit flow boundary layer was only 8mm due to boundary layer bleeding at the nozzle entrance. Over the main region of the nozzle flow the mean temperature and velocity distribution inhomogeneities were less than 3 % of the mean. All measurements reported were obtained with $u_0 \approx 8$ m/s and $\Delta T_G \approx 26°C$.

To measure mean and fluctuating temperature characteristics a resistor probe was used (DISA 55 F 05). By means of electronic compensation circuits the high frequency response of this probe could be improved to better than 2000 Hz. A low frequency response lag due to the influence of the prongs was corrected in the same way. The δ signal necessary for obtaining intermittency distributions and conditional averages was obtained from the

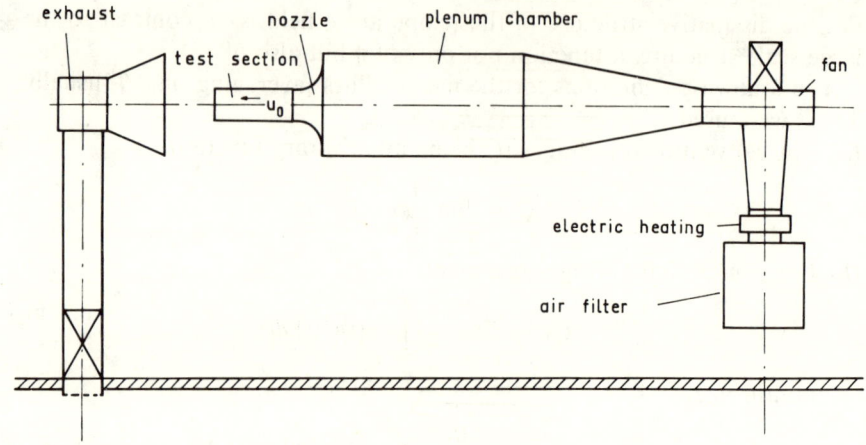

FIG. 1. Test arrangement.

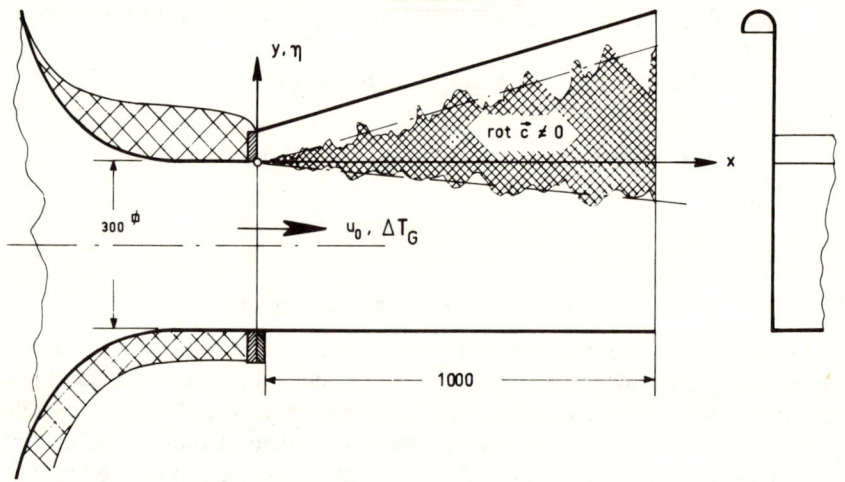

FIG. 2. Schematic of test section.

$\Delta u'/\Delta y \approx \partial u'/\partial y$ signal of an ordinary double (parallel) hotwire probe without compensation for temperature, where identity of the turbulent front for the temperature and velocity fields was assumed (Fiedler and Head, 1966).

4. Self-Preserving Region and Virtual Origin

From a number of preliminary measurements of $\Delta \bar{T}$ and T' it was found that the self-preserving region of the temperature field is at $x \geq 500$ mm with its virtual origin at $y_0 = 0$ and $x_0 = -50$ mm (see Fig. 3).

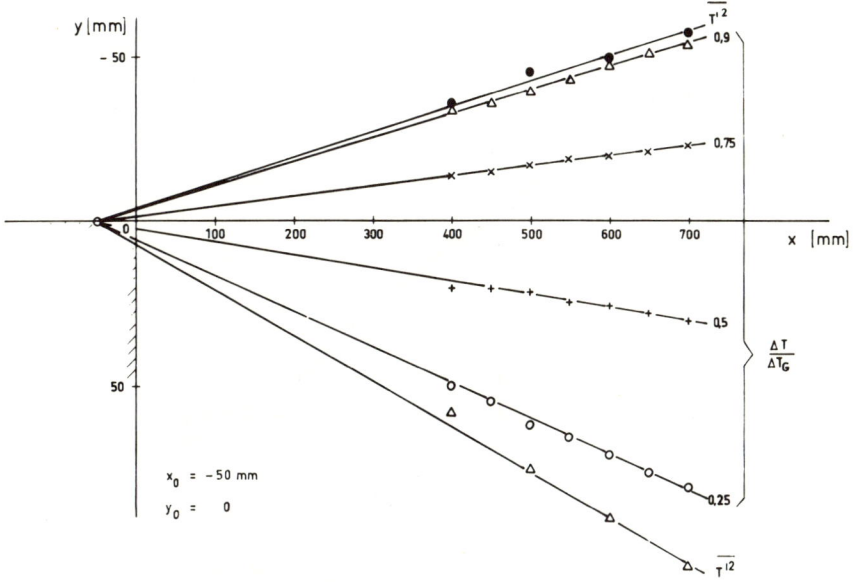

FIG. 3. Self-preserving region and virtual origin.

The measurements which follow herein were restricted to the self-preserving region and plotted versus the non dimensional coordinate $\eta = y/(x - x_0)$.

5. The Distribution of Mean Velocity and of Mean Temperature

The distribution of the mean velocity was measured for $\Delta T_G = 0$ and $\Delta T_G = 26°C$. For the cold flow the spread parameter was $\sigma = 11$. In the heated flow case the spread, especially in the outer (i.e., low velocity) region of the shear layer is somewhat larger. The lateral distribution of the mean temperature, shown in Fig. 4, exhibits a strikingly different character when compared with the mean velocity. As can be seen from Fig. 5, the velocity distribution curve has a single inflexion point indicating an ordinary gradient diffusion mechanism for the momentum. In contrast to this the mean temperature distribution curve has three inflexion points (S-shape) with a somewhat flat region at large values of γ. This peculiar shape of the curve suggests a transport mechanism largely determined by quasi convective action due to large scale vortex motion (bulk convection).

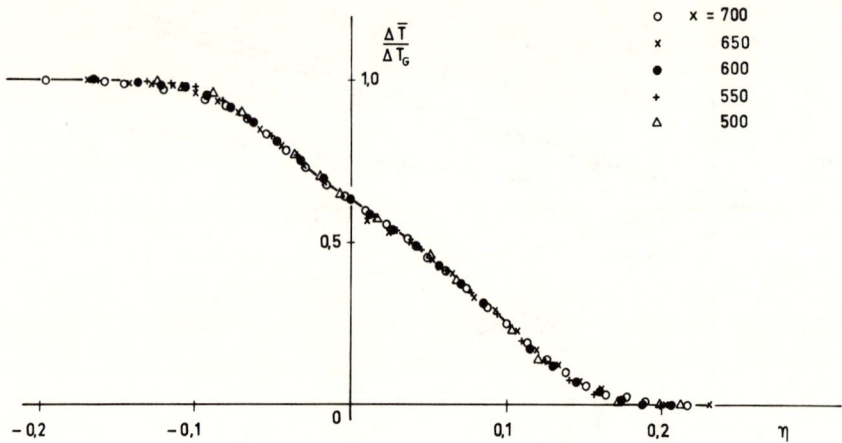

Fig. 4. Distribution of mean temperature.

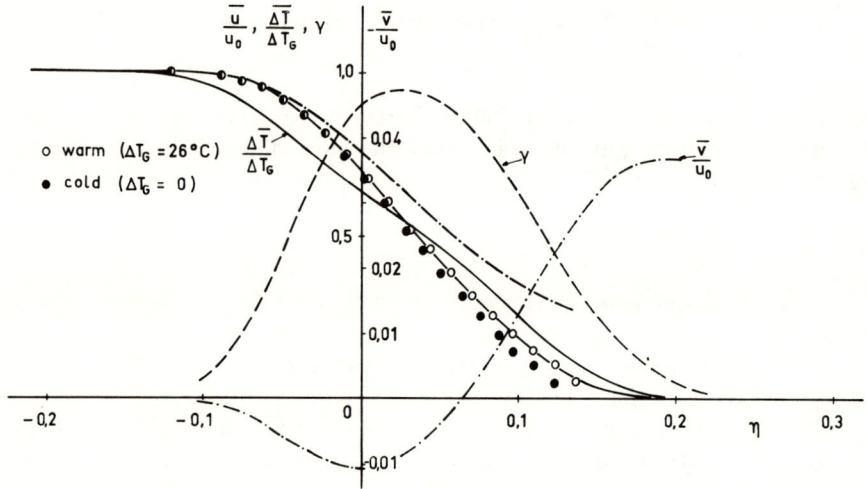

Fig. 5. Velocity distribution in self preserving region with $\Delta T_G = 0$ and $\Delta T_G = 26°C$.

6. Measurements of Fluctuating Quantities

The distribution of the fluctuation intensity across the flow is shown in Fig.6. It has two maxima and one minimum all of which are approximately located at the positions of the inflexion points of the mean temperature curve. The maximum value of the fluctuating variance (at $\eta \approx 0.07$) $\overline{T'^2}/\Delta T_G|_{max} \approx 0.044$, is considerably larger than the maximum velocity fluctuation variance, which in the same kind of flow is $\overline{u'^2}/u_0^2|_{max} \approx 0.030$ (Wygnanski and Fiedler, 1970).

The existence of the three extrema is obvious from the typical shape of the T' fluctuations, which for three different values of η are shown in Fig. 7. For γ slightly smaller than γ_{max}, the sawtooth amplitudes retain their values; the mean temperature, however, shifts away from the mean between inner and outer temperature thus increasing the maximum temperature fluctuation peaks and thereby the variance of ΔT about the mean.

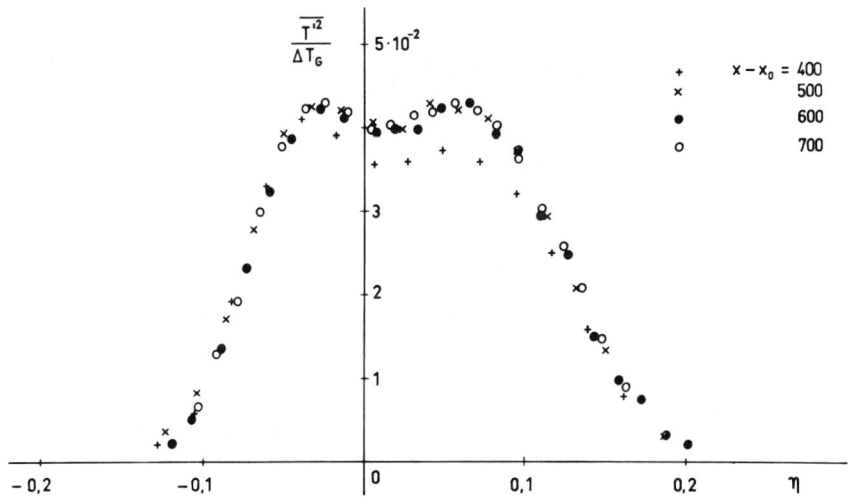

FIG. 6. Distribution of temperature fluctuation variance.

For further characterisation of the fluctuating signals, skewness and flatness distributions were measured (Figs. 8 and 9). Further measurements included spectra of T' at different locations of η (Fig. 10), integral spectra (Fig. 11) showing a clear double structure, and auto-correlations (Fig. 12) showing a strong periodic component of the fluctuating signal. Measurements of intensities of fluctuation time derivatives in the self-preserving regime exhibited a dependence on x in accordance with Eq. (5).

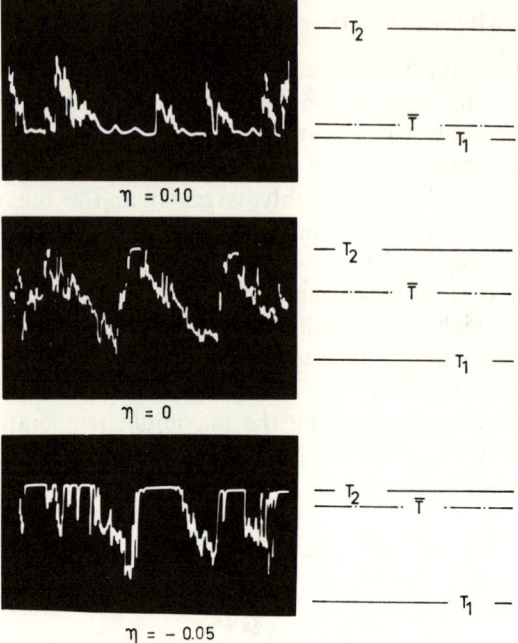

FIG. 7. Typical $\Delta T(t)$ signals at different positions of η.

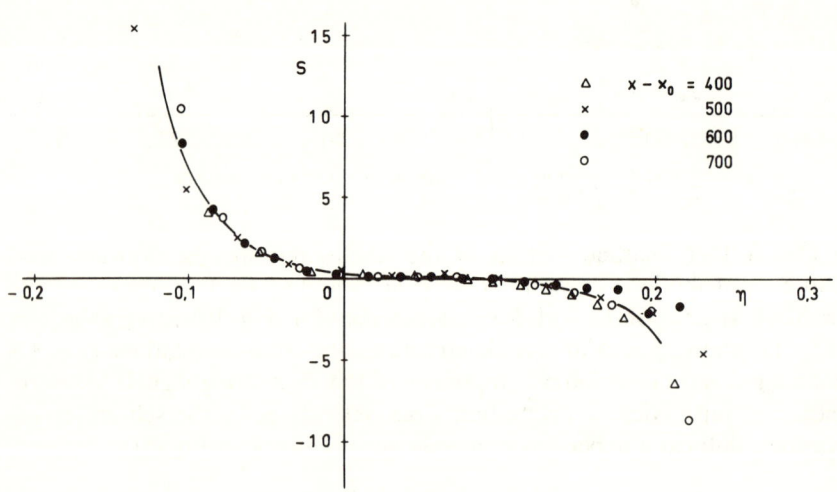

FIG. 8. Skewness distribution of T'.

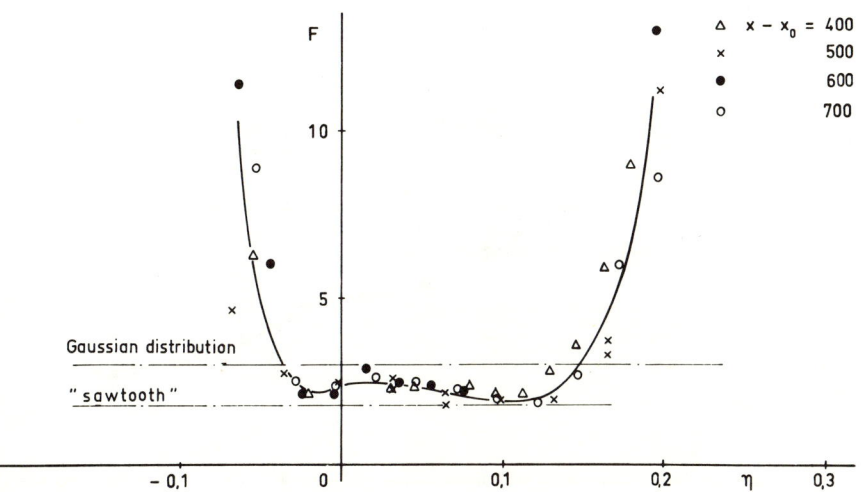

Fig. 9. Flatness distribution of T'.

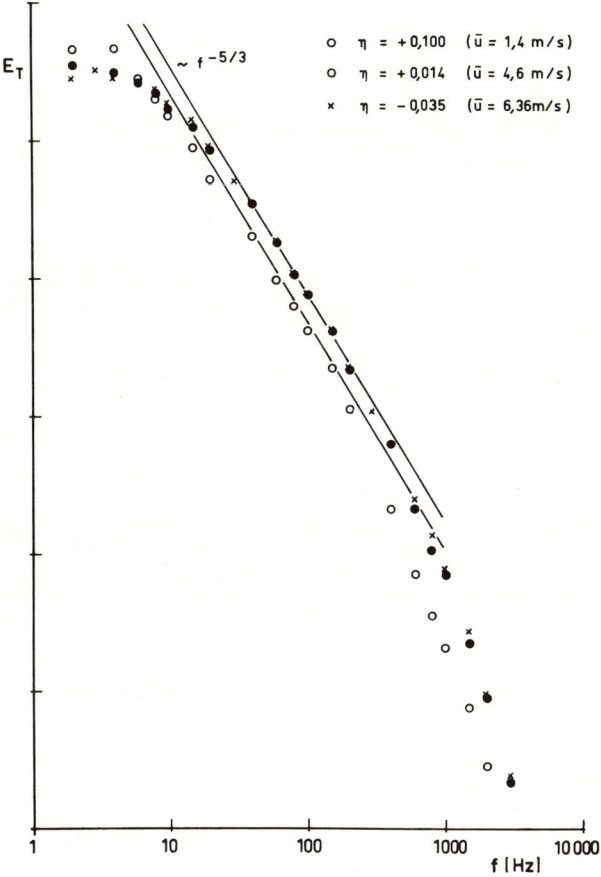

Fig. 10. Spectra of temperature fluctuations in three singular points.

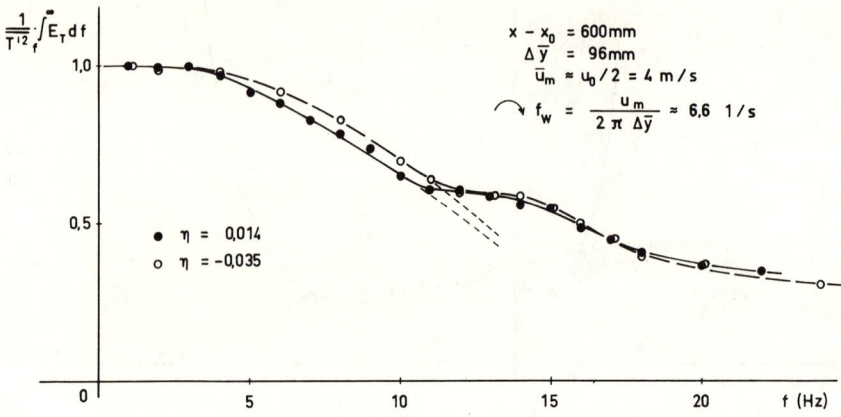

Fig. 11. Integral spectra of T' for two points in the flow.

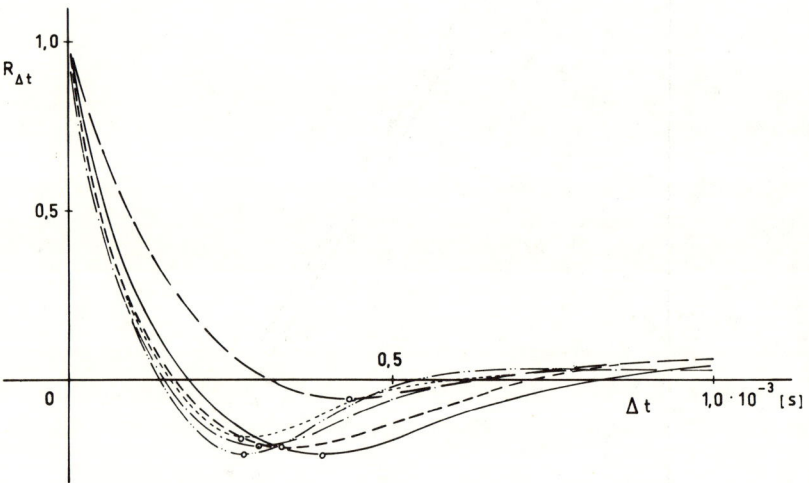

Fig. 12. Auto-correlation of T' at various positions of η.

7. Characteristics of the Temperature Field in the Turbulent Regime

Figure 13 shows a smoke photograph of the flow exhibiting clearly the intermittent front at its low velocity side.

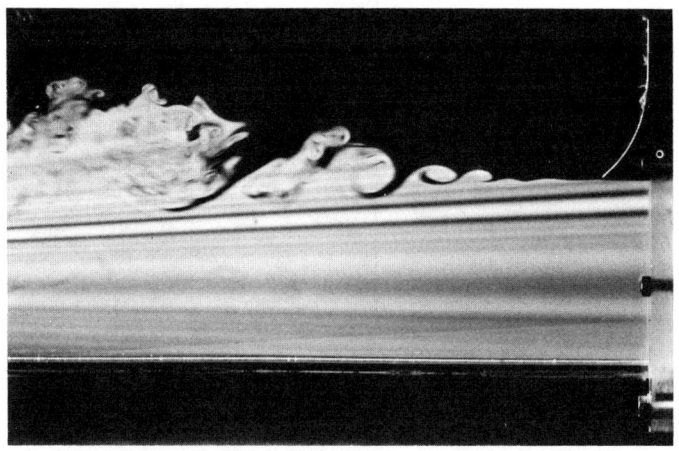

FIG. 13. Smoke photograph of shear layer.

7.1. The Intermittency Distribution

The statistics of this front is described by the intermittency factor γ which is plotted in Fig. 14. Since in these measurements the δ signal was obtained directly, conditional averages of the temperature characteristics could also be obtained.

7.2. The Mean Temperature in the Turbulent and Nonturbulent Regions

These measurements are presented in Fig. 15. The mean temperature distribution in the turbulent zone is rather flat and homogeneous, which is in contrast to the mean velocity characteristic (Wygnanski and Fielder, 1970).

7.3. The Variance of the Temperature Fluctuation

In measuring the mean variance of the temperature fluctuation one has to consider the fact that the mean values of the temperature in the two zones of the flow are different from each other and different from $\Delta \bar{T}$. In defining the variance in the turbulent zone as $\overline{T_T'^2}/\Delta T_G^2$, where $\int_0^T T_T' \delta(t)\, dt = 0$ we have

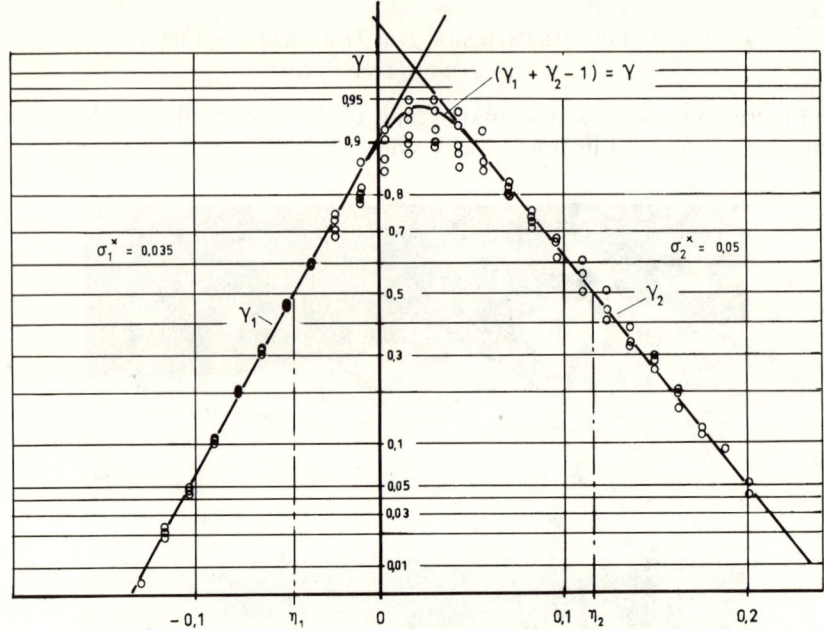

Fig. 14. Distribution of intermittency.

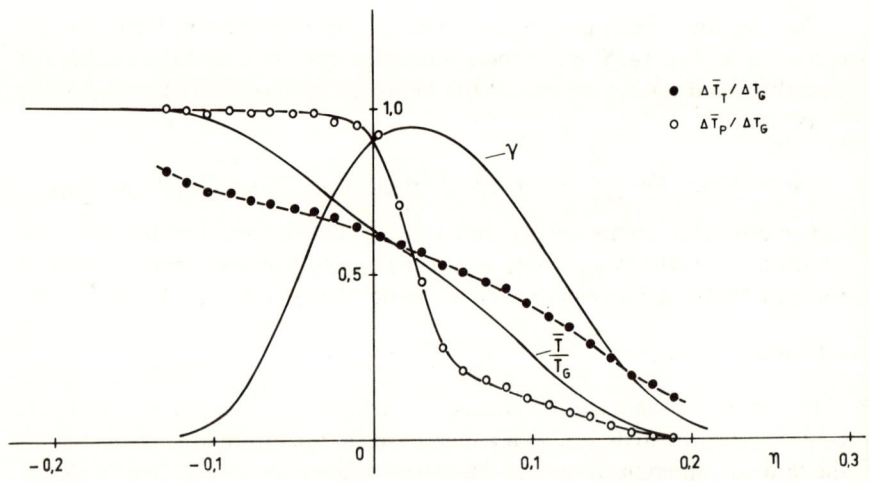

Fig. 15. Zone averages of the mean temperature.

to determine $T'_T = \Delta T_T - \Delta \bar{T}_T$ which has been done directly with a special electronic circuit $(\overline{T'^2_{T,M}})$ and indirectly by measuring

$$\overline{T'^2} = \overline{(T_T - \Delta \bar{T})^2}$$

from which one obtains the desired value of $\overline{T'^2_T}$ using the relation

$$\overline{T'^2_T} = -\overline{(\Delta \bar{T}_T - \Delta \bar{T})^2} + \overline{T'^2} = \overline{T'^2_{T,A}}$$

where the bracketed term was evaluated from Fig. 15. The result is presented in Fig. 16 showing a remarkably flat and homogeneous distribution.

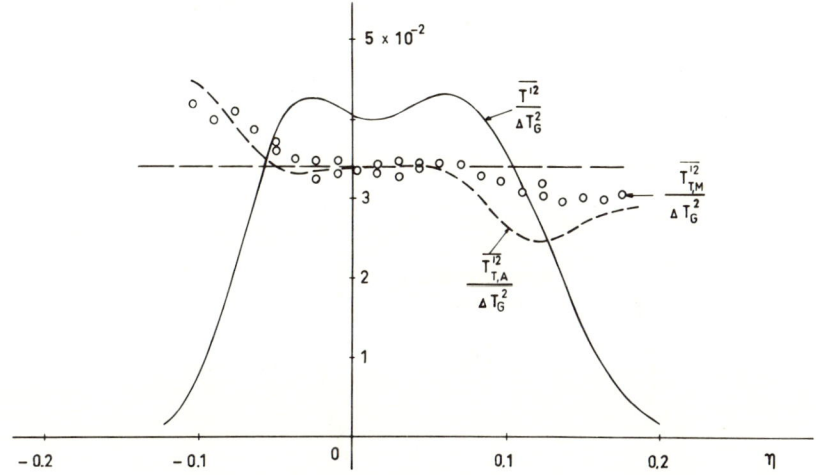

FIG. 16. Temperature fluctuation variance in the turbulent regime.

8. Summary and Discussion

The reported observations may be summarized as follows:

(a) From the shape of the temperature fluctuations as well as from the mean quantity distributions it can be assumed that the transport mechanism is largely due to a large scale vortical motion (see also Brown and Roshko, 1971).

(b) The mean temperature distribution in the turbulent domain is largely homogeneous.

(c) The temperature across a single vortex is linear in the mean.

(d) The temperature fluctuation variance in the turbulent domain is approximately constant.

To discuss the transport mechanism it seems appropriate to start with Townsend's model:

$$\overline{v'T'} \sim \mathscr{V}\Delta T_G - \varepsilon\, d\Delta\bar{T}/d_y$$

($\mathscr{V}$ = bulk convection velocity).

From Fig. 17, which shows a schematic of the heat-transport in a single vortex, the following relations may be derived:

$$\overline{(v'T')}_{T\text{ (vortex)}} = q_K + q_{D\eta} - q_{Dx}$$

where

$$q_K = k_1 u_0 \Delta T_G = \text{lateral convection}$$

$$q_{Dx} = -\varepsilon \frac{d(\Delta\bar{T}_{\text{vortex}})}{dx} = \text{longitudinal diffusion}$$

$$q_{Dy} = -\varepsilon \frac{d(\Delta\bar{T}_{\text{vortex}})}{dy} = \text{lateral diffusion}$$

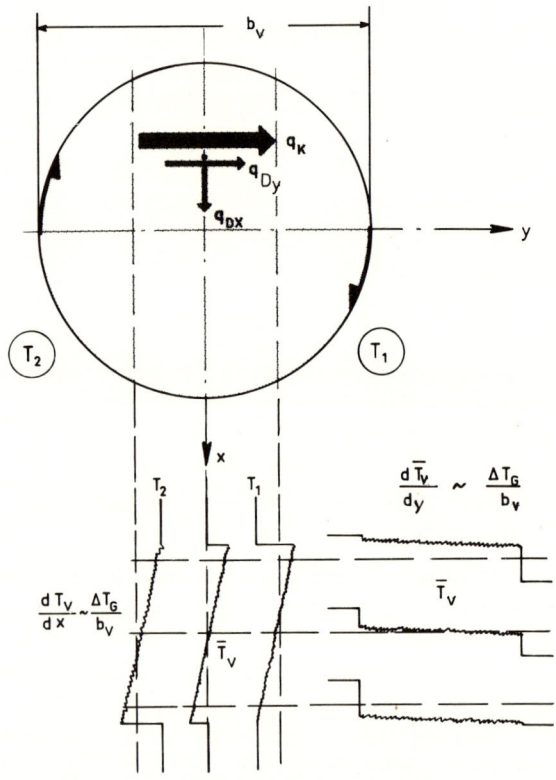

FIG. 17. Model of heat transport in single vortex.

From the observations it appears justified to put

$$\varepsilon \sim u_0 b_v \approx \text{const}$$

and

$$d(\Delta \bar{T}_v)/dx \sim d(\Delta \bar{T}_v)/dy \sim -\Delta T_G/b_v$$

so that one obtains

$$\overline{(v'T')}^*_{T(v)} = \frac{\overline{(v'T')}_{T(v)}}{u_0 \Delta T_G} \approx \text{const} = k$$

and

(6) $$\overline{(v'T')}^* \approx \gamma k$$

with the additional assumption that $\overline{(v'T')}_P \approx 0$ (the subscript index v refers to vortex).

For a verification of this result $\overline{(v'T')}^*$ was computed from Eqs. (2) and (3) using the measured $\Delta \bar{T}$ and $\bar{u}$ distributions and plotted in Fig. 18, where it agrees quite satisfactorily with the γ distribution. Accordingly the mean temperature distribution which was then computed from Eqs. (3) and (6)

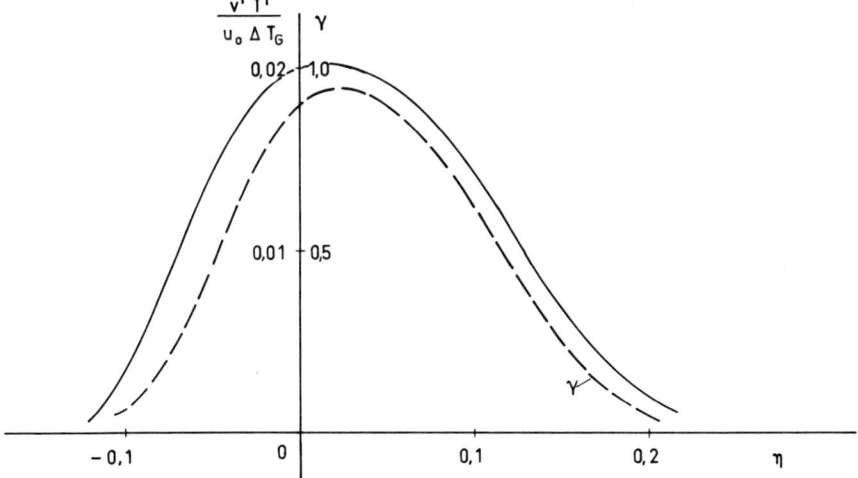

FIG. 18. The turbulent diffusion term, evaluated from $\Delta \bar{T}/\Delta T_G(\eta)$ and $\bar{u}/u_0(\eta)$.

and the measured $\bar{u}$ and γ distribution agrees remarkably well with the measurements of $\Delta \bar{T}$ as can be seen from Fig. 19.

The mean temperature in the large vortices can be computed on the basis of the entrainment balance. It is

$$\frac{\Delta \bar{T}}{\Delta T_G}(\gamma_{max}) = \frac{\dot{V}_{E2}/\dot{V}_{E1}}{1 + (\dot{V}_{E2}/\dot{V}_{E1})}$$

where $\dot{V}_{E1}$ is the entrainment flow rate at low velocity side $(u = 0, T = 0)$ and $\dot{V}_{E2}$ is the entrainment flow rate at high velocity side $(u = u_0, T = \Delta T_G)$. From the mean velocity profile we obtain

$$\dot{V}_{E2}/\dot{V}_{E1} = 1.58;$$

thus

$$\frac{\Delta \bar{T}}{\Delta T_G}(\gamma_{max}) = 0.61$$

which is in excellent agreement with the measured value of

$$\frac{\Delta \bar{T}}{\Delta T_G}(\gamma_{max}) = 0.60.$$

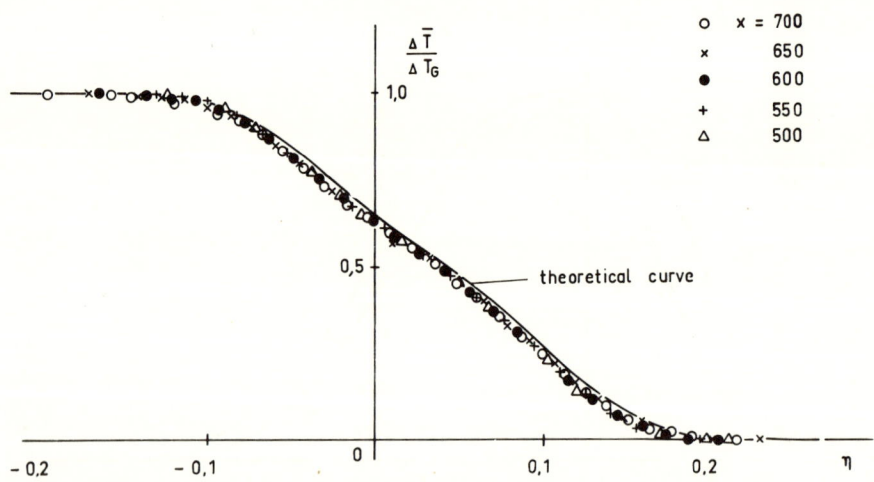

FIG. 19. Computed temperature profile compared with measured distribution.

References

Brown, G., and Roshko, A. (1971). *AGARD Conf. Proc.* No. 93.
Corrsin, S. (1952). *J. Appl. Phys.* **23**, 1.
Fiedler, H., and Head, M. R. (1966). *J. Fluid Mech.* **25**, 4.
Kibens, V., and Kovasznay, L. S. G. (1969). Dep. Mech., Johns Hopkins Univ., Baltimore, Maryland.
Mayer, E., and Divoky, D. (1966). *AIAA J.* **1**, 11.
Townsend, A. A. (1956). "The Structure of Turbulent Shear Flow." Cambridge Univ. Press, London and New York.
Tyldesley, J. R. (1969). *Int. J. Heat Mass Transfer* **12**, 4.
Wygnanski, I., and Fiedler, H. (1970). *J. Fluid Mech.* **41**, 2.
Yen, K. T. (1967). R 67 SD 3. Space Sci. Lab., General Electric Co., Schenectady, New York.

THERMODYNAMIC MODEL FOR THE DEVELOPMENT OF A CONVECTIVELY UNSTABLE BOUNDARY LAYER

D. J. CARSON AND F. B. SMITH

Boundary Layer Research Branch, Meteorological Office, Bracknell, Berkshire, England

1. INTRODUCTION

At the present time there is a great deal of interest being shown in the study of the nonstationary aspects of the atmospheric boundary layer. One reason for this is the desire to incorporate the boundary layer realistically into numerical forecasting and general circulation models. Another is the need to provide reliable estimates of the depth and character of the mixing layer for use in schemes dealing with the dispersion of concentrations of atmospheric pollutants where steady-state theories have proved to be totally inadequate.

The stability of the mixing layer is of primary importance in short-range pollution studies; however its depth h becomes increasingly more important in determining ground-level concentrations when the distance of travel is greater than about $10h$ from the source. The inclusion of such nonsteady features in a practical scheme for estimating the vertical dispersion of pollutants has been outlined by Smith (1972). Also, the ability to specify the diurnal cycle of boundary-layer evolution becomes important when dealing with pollutants tracked for several days on a regional scale, and such effects are discussed elsewhere in this Symposium (Pasquill, Vol. 18B, p.1).

Although intricate parameterizations and numerical models are being developed for the study of evolving boundary layers (Deardorff, 1972a,b, 1973), there remains a need to provide relatively simple parametrizations for general use and a start would be to consider the development of the important dry, convectively unstable layer capped by a stable layer.

Observations of the boundary layer in convective situations show that in general it is a diurnally evolving system with discontinuities around sunrise and sunset. The discontinuities arise because we distinguish between the relatively shallow nocturnal inversion layer in which buoyancy and viscous effects suppress any mechanically generated turbulent motions and the more rapidly evolving daytime convectively unstable layer which is generally capped by a nonturbulent stable layer. Two of the main factors controlling the development of the convective layer are the flux of sensible eddy heat entering the boundary layer at the ground and also a turbulent mixing process which occurs at the interface between the well-mixed boundary layer air and the nonturbulent air in the capping stable layer. The process whereby stable air from above is mixed into the developing convectively unstable boundary layer is called *entrainment*.

The part played by the dynamics in the entrainment process has received little attention and the simplest parameterizations treat it from purely thermal considerations. However, recent observations (Readings et al., 1973) indicate that wind shear at the interface may be of fundamental importance not only to the entrainment of eddy momentum but also of eddy sensible (and latent) heat.

This contribution sets out (i) to draw attention to the general results and limitations of the simple thermal approach, and (ii) to provide a combined, if grossly simplified, dynamical and thermal approach to the parameterization of the entrainment process and the development of the convectively unstable boundary layer.

2. Simple Thermal Models

The history of simple thermal models for parameterizing the development of the convectively unstable boundary layer capped by a stable layer can be traced through the papers of Ball (1960), Lilly (1968), Deardorff et al. (1969), Tennekes (1973) and Carson (1973). We summarize here the method and results of the model discussed in depth by Carson (1973) and independently proposed by Betts (1973).

The potential temperature profile, as illustrated in Fig. 2, is defined by

$$(1) \quad \theta(z, t) = \begin{cases} \theta_c(t), & z < h, \\ \theta_s(z, t) = \theta_0 + \gamma(t)z, & z > h, \end{cases}$$

where h is nominally the depth of the convectively unstable boundary layer, θ_0 is the effective surface temperature obtained by extrapolating the stable lapse rate down to $z = 0$, and $\gamma(t)$ is the vertical gradient of θ in the capping stable layer.

Advection, radiation and evaporation are not considered here although in certain circumstances each or all of these processes can be important. In this case the simple heat balance equation is

$$(2) \quad \frac{\partial H}{\partial z} = -\rho c_p \, d\theta/dt = -\rho c_p [\partial \theta/\partial t + w(z) \, \partial \theta/\partial z],$$

such that

(3) $$H(z, t) = 0, \quad z > h,$$

where $H(z, t)$ is the sensible eddy heat flux, $w(z)$ is the mean, synoptically induced vertical velocity, ρ is the mean air density and c_p is the specific heat of air at constant pressure.

Certain features of the simple model, such as the linearity of $w(z)$ and $H(z, t)$ with height and the exponential increase of $\gamma(t)$ with time are derived from Eqs. (1)–(3), as indeed is the expression for the entrained sensible eddy heat flux

(4) $$H(h, t) = -\rho c_p w_e(t) \Delta\theta(t),$$

where

(5) $$w_e(t) = dh/dt - w(h)$$

is the entrainment rate and

(6) $$\Delta\theta(t) = \theta_s(h, t) - \theta_c(t)$$

is the step discontinuity in θ across the interface at $z = h$.

The system of equations is closed by parameterizing the entrainment process at $z = h$. In this simple model it is postulated that the degree of entrainment is controlled, to a first approximation, by the intensity of the thermal bombardment of the interface which, in turn, is directly proportional to the magnitude of the surface sensible heat flux. Hence the closure equation is

(7) $$H(h, t) = -AH(0, t), \quad 0 \leq A \leq 1$$

Straightforward analysis produces an ordinary differential equation for the development of the convectively unstable layer

(8) $$h \, d(\gamma h)/dt = [H(0, t) - 2H(h, t)]/\rho c_p$$

which although not explicitly dependent on A, does depend on A being constant. Strictly, in keeping with the assumptions, Eq. (8) should be written

(9) $$dh^2/dt + 2\tilde{\beta}h^2 = 2(1 + 2A)H(0, t)/\rho c_p \gamma(t),$$

where

(10) $$\tilde{\beta} = -w(z)/z = \text{constant},$$

and

(11) $$\gamma(t) = \gamma(0) \exp(\tilde{\beta}t).$$

Integration of Eq. (9), with $h(0) = 0$, gives

$$(12) \qquad h^2(t) = \frac{2(1 + 2A)\exp(-2\tilde{\beta}t)}{\rho c_p \gamma(0)} \int_0^t \exp(\tilde{\beta}\tau) H(0, \tau)\, d\tau,$$

and the corresponding evolutionary expressions for $w_e(t)$, $\Delta\theta(t)$ and $\theta_c(t)$ are

$$(13) \qquad w_e(t) = (1 + 2A)H(0, t)/\rho c_p \gamma(t)h(t),$$

$$(14) \qquad \Delta\theta(t) = A\gamma(t)h(t)/(1 + 2A),$$

and

$$(15) \qquad \theta_c(t) = \theta_0 + [(1 + A)/(1 + 2A)]\gamma(t)h(t).$$

The value of A which characterises the degree of interfacial mixing realised in the atmosphere during the typical development of a convectively unstable boundary layer remains to be chosen.

The extreme value $A = 1$ derives from Ball (1960) who, in his consideration of the integrated local turbulent kinetic energy balance equation, assumed that the contribution from molecular dissipation could be neglected. The other extreme, $A = 0$, describes the situation where the boundary layer is growing without entraining heat across the interface, i.e., $\Delta\theta$ in Eq. (4) is zero whereas $w_e(t)$ remains finite. In such circumstances the interface is a passive one with no mixing across it and we shall use the term *encroachment* of the stable layer by the unstable layer to describe this process. Such a state is strictly never realised in the atmosphere but is closely approached in the laboratory studies of penetrative convection by Deardorff et al. (1969) and when weak thermal activity is eroding a strong inversion, such as a nocturnally established inversion (Carson, 1973).

Available evidence favours small values of A; 0.2 is suggested by Deardorff (private communication) and Tennekes (1973), and Betts (1973) quotes evidence for 0.25. It seems unlikely that A remains constant throughout the various phases of boundary layer evolution and Carson (1973) from his analysis of the O'Neill 1953 data has suggested that A varies quite significantly during the day, being very small soon after dawn and reaching a maximum value, as high as 0.5, for a few hours following the time of maximum surface heating.

The uncertainty about A and its likely time dependence may limit the range of applicability of the simple thermal model. Further, entrainment is essentially a dynamical process and therefore it seems inadequate to propose a model which omits the dynamics of the interfacial region. We seek then a simple model which will give the parameterization a combined dynamical and thermal basis and at the same time avoid the restriction that $H(h, t)/H(0, t)$ be constant.

3. A Simple Thermodynamic Model

3.1. The Entrainment Process

Observations such as those of Readings *et al.* (1973) and Browning *et al.* (1973) show that in general there is not only a temperature change across the convoluted interface between the deepening convectively unstable boundary layer and the capping stable layer but also a finite shear in the wind velocity. The general situation in the vicinity of the interface is illustrated schematically in Fig. 1 and we envisage several mechanisms contributing to the

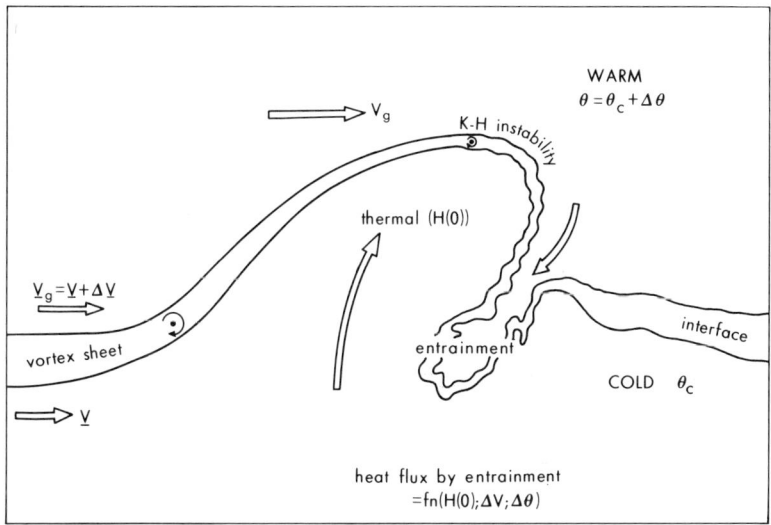

FIG. 1. Schematic representation of the dynamical and thermal effects which control the entrainment process at the interface between a deepening convectively unstable boundary layer and a capping nonturbulent stable layer.

mixing process whereby stable air is entrained into the developing boundary layer across such an interface.

The change in temperature across the interface serves to create a narrow layer or zone which with the accompanying change in wind speed is also a zone of marked vorticity. Bombardment of the interface by thermals causes three-dimensional domes to protrude into the stable layer thereby stretching the vortex sheet and further enhancing the local vorticity. The net effect is a torque which causes a wavelike overturning of the convective dome which enables a tongue of relatively warm air to undercut the dome's colder boundary layer air. At the same time, small-scale interfacial Kelvin–Helmholtz instabilities grow on the crest of the dome, where gradients have been intensified (Readings et al., 1973). These may then be advected into the tongue by the wind shear and there play an important role in enhancing the mixing of the tongue into the convectively unstable boundary layer.

It is therefore postulated that the entrainment process and hence A of Eq. (7) are governed not only by $H(0, t)$, which partly determines the strength of the thermals, but also by $\Delta\theta(t)$ and the magnitude of the wind shear, $\Delta V(t)$, which relate to the dynamical stability of the interface. On dimensional grounds, then, the simplest formulation for A is

$$(16) \qquad H(h, t)/H(0, t) \equiv -A = -\kappa[\rho c_p \, \Delta V \, \Delta\theta/H(0, t)]^a,$$

where a, κ are two "constants" which must be determined from observations.

3.2. Dynamical Considerations

It is necessary in our dynamical formulation to include the parameters needed to determine the degree of entrainment as expressed in Eq. (16) and, as a first attempt, we construct a simple model based on the idealised wind profiles of Fig. 2. A right-handed system of axes is chosen such that the

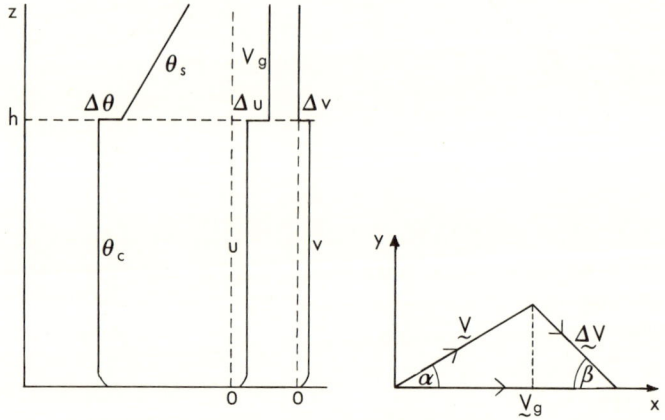

FIG. 2. Schematic representation of the idealised profiles of potential temperature θ and the components u, v of the horizontal wind velocity $\mathbf{V}$ used in the simple thermodynamic model. Also illustrated is the nature of the wind velocity shear, $\Delta\mathbf{V}$, across the interface at $z = h$.

x-axis is directed along the geostrophic wind $\mathbf{V}_g$. For $z < h$, the mean wind components are assumed to be virtually constant with height and, at $z = h$, we include a step discontinuity $\Delta\mathbf{V}$ in the wind velocity which defines the angles α and β, α being the turning of the wind in the boundary layer from the geostrophic direction. For $z > h$, the mean wind is assumed to be $\mathbf{V}_g$.

Assuming quasi-stationarity and horizontal homogeneity for all the relevant variables, including the pressure gradient, the boundary layer momentum equations can be written

(17) $\partial \tau_x/\partial z = -f\rho V \sin \alpha =$ constant

$\partial \tau_y/\partial z = -f\rho(V_g - V \cos \alpha) =$ constant,

implying

(18) $\tau_x(z) = \tau_x(0) - f\rho V \sin \alpha \ z$

$\tau_y(z) = \tau_y(0) - f\rho(V_g - V \cos \alpha)z$

and, in particular,

(19) $\tau_x(0) - \tau_x(h) = f\rho V \sin \alpha \ h$

$\tau_y(0) - \tau_y(h) = f\rho(V_g - V \cos \alpha)h$

where $\tau = (\tau_x, \tau_y)$ is the shearing stress, $V = |\mathbf{V}|$, $V_g = |\mathbf{V_g}|$, and f is the Coriolis parameter.

Entrainment of stable air through the interface not only implies a transfer of heat but also momentum and Deardorff (1973) has recently used this concept to explain the large values of eddy momentum flux in the upper regions of a developing convectively unstable boundary layer, obtained from measurements analysed by Angell (1972).

In a form analogous to that for the heat flux due to entrainment, Eq. (4), the shearing stress components at $z = h$ are expressed as

(20) $\tau_x(h) = \rho(dh/dt) \Delta u = \rho(dh/dt) \Delta V \cos \beta$

and

$\tau_y(h) = \rho(dh/dt) \Delta v = -\rho(dh/dt) \Delta V \sin \beta$

where $\Delta \mathbf{V} = (\Delta u, \Delta v)$, $\Delta V = |\Delta \mathbf{V}|$, $\tau(z) = 0$ for $z > h$ and the mean, synoptically induced, vertical velocity is assumed to be zero.

In the surface layer the shearing stress is specified by means of a drag coefficient C_D in the usual way,

(21) $\tau_x(0) = \rho C_D V^2 \cos \alpha$

$\tau_y(0) = \rho C_D V^2 \sin \alpha.$

Finally, from Fig. 2, we have the kinematical relationships,

(22) $V \sin \alpha = \Delta V \sin \beta$

(23) $V \cos \alpha = V_g - \Delta V \cos \beta$

and

(24) $V^2 = V_g^2 + (\Delta V)^2 - 2V_g \Delta V \cos \beta.$

3.3. Thermal Considerations

In keeping with the simple thermal model of Section 2, the adopted profile of potential temperature in the thermodynamic model is that illustrated in Fig. 2 and defined by Eq. (1) (in this analysis, however, γ is assumed constant with time). This profile and the heat balance equation in the absence of radiational effects, Eqs. (2) and (3), give us three further equations for the thermodynamic model. These are:

Heat input by entrainment:

$$H(h, t) = -\rho c_p (dh/dt)\, \Delta\theta. \tag{25}$$

Heat balance for the whole of the mixing layer:

$$\rho c_p h\, d\theta_c/dt = H(0, t) - H(h, t). \tag{26}$$

The magnitude of the temperature step across the interface:

$$\Delta\theta = \theta_0 + \gamma h - \theta_c. \tag{27}$$

A knowledge of the constant parameters a, κ, C_D, f, ρc_p and the variables $H(0, t)$, γ, V_g and initial values of h, $\Delta\theta$ enable us to use the system of Eqs. (16)–(27) to determine the evolutionary profiles of h, $\Delta\theta$, θ_c, V, ΔV, α, β, τ_x, τ_y, and $H(h)$.

4. Provisional Results and Conclusions

The two constants a and κ await observational determination and so at this stage it is possible only to indicate the nature of the results which can be obtained with the simple thermodynamic model. The full details of the procedure for determining the evolutions of h, $\Delta\theta$, θ_c, V, ΔV, α, β, τ_x, τ_y, and $H(h)$ are given in the Appendix.

The results of a test case with $a = \frac{1}{2}$ and $\kappa = 0.1$ are presented in Figs. 3 and 4. Parameters kept constant throughout the integration are;

$$C_D = 4 \times 10^{-3}, \quad \rho c_p = 10^3 \text{ joule m}^{-3}\,°\text{K}^{-1}, \quad f = 10^{-4} \text{ sec}^{-1},$$

$$V_g = 10 \text{ m sec}^{-1} \quad \text{and} \quad \gamma = 6 \times 10^{-3}\,°\text{K m}^{-1}.$$

We adopted a sinusoidal surface heat flux defined by $H(0, t) = \hat{H} \sin(\Omega t)$,

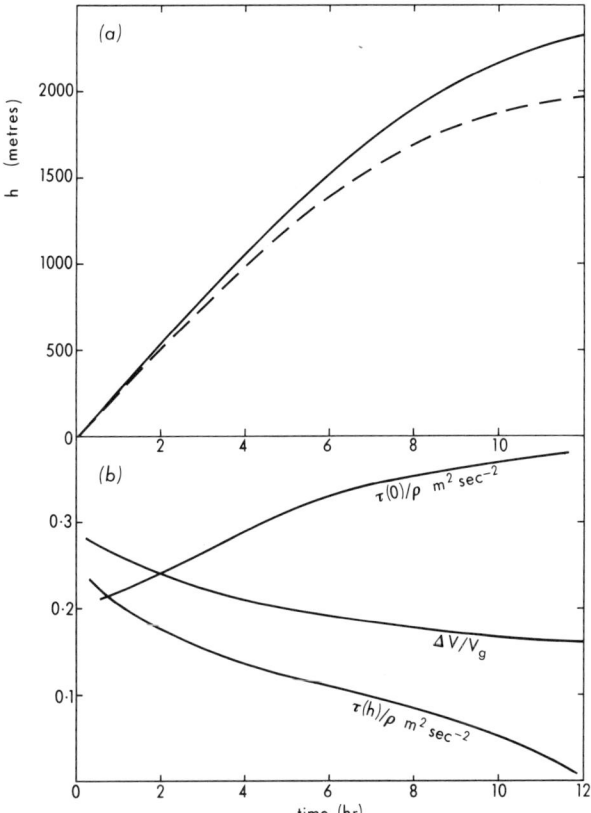

FIG. 3. (a) The full line shows the development of the depth h of the convectively unstable boundary layer obtained from the trial integration using the thermodynamic model. The broken line shows the corresponding development obtained from the simple thermal model with $A = 0.2$, $\tilde{\beta} = 0$, $\gamma(0) = 6 \times 10^{-3}\,°\text{K m}^{-1}$ and a sinusoidal $H(0, t)$ as specified in Fig. 4a. (b) The evolutions of $\Delta V/V_g$, $\tau(0)/\rho$, and $\tau(h)/\rho$ obtained from the trial integration using the thermodynamic model.

where Ω is the Earth's angular rotation and $\hat{H} = 300\,\text{W m}^{-2}$ is the maximum value. The integration was started at $t = 0.4$ hr and the initial values of h and $\Delta\theta$ were obtained from Eqs. (12) and (14) of the thermal model with $\tilde{\beta} = 0$ and $A = 0.2$. The time step for the integration was 0.2 hr.

Figure 3a shows the development of the depth h of the convectively unstable boundary layer. Also shown is the corresponding development derived from the purely thermal model with $\tilde{\beta} = 0$, $A = 0.2$, all other parameters being as stipulated for the thermodynamic model. Because it is not possible to compare critically the two curves, the thermal model is given for

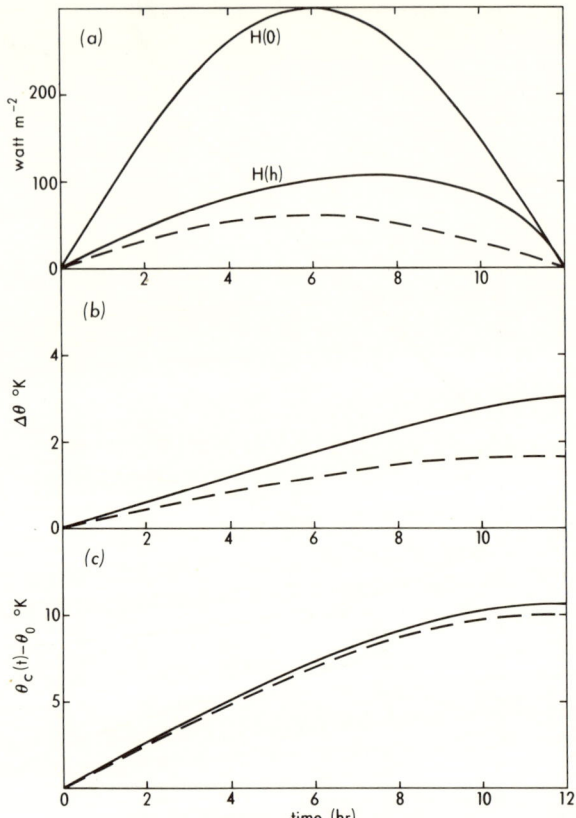

Fig. 4. (a) The sinusoidal surface heat flux $H(0, t)$ adopted for both the thermodynamic and thermal models. $H(h; t)$ (full line) is the entrained heat flux derived from the thermodynamic model and $H(h, t) = 0.2H(0, t)$ (broken line) is the entrained heat flux in the simple thermal model with $A = 0.2$. (b) $\Delta\theta$, the step in the potential temperature across the interface at $z = h$, for the trial thermodynamic model (full line) and the simple thermal model with $A = 0.2$ (broken line). (c) The warming of the convectively unstable boundary layer throughout the day expressed as $[\theta_c(t) - \theta_0]$, for the trial thermodynamic model (full line) and the simple thermal model with $A = 0.2$ (broken line).

information only. Figure 3b shows the variations with time of $\Delta V/V_g$, $\tau(0)/\rho$, and $\tau(h)/\rho$ where $\tau = |\tau|$ and again the trends do not appear to be contrary to expectation. Figure 4a shows the specified surface sensible heat flux $H(0, t)$ and the derived entrained heat flux $H(h, t)$. Also given is $H(h, t) = 0.2H(0, t)$ from the thermal model and here we note the differences in shape of the two curves for $H(h, t)$. In particular $H(h, t)$ from the thermodynamic model is not distributed symmetrically about $t = 6$ hr as is the $H(h, t)$ from the simple thermal model, i.e., A in the thermodynamic model is a function of time.

Figures 4b and 4c show the evolutions of $\Delta\theta$ and $[\theta_c(t) - \theta_0]$ for both models and again they exhibit similar trends.

The nature of these provisional results for the trial values of a and κ provide encouragement for following up these ideas. In particular we await further observational data such that we can determine values of a and κ directly and compare the computed evolutions with the actual ones.

Appendix

Details of the Numerical Integration Procedure

Substitution of Eqs. (20)–(23) into Eq. (19) gives

(A1) $\quad C_D V(V_g - \Delta V \cos \beta) - (dh/dt) \Delta V \cos \beta = fh \Delta V \sin \beta,$

and

(A2) $\quad C_D V \Delta V \sin \beta + (dh/dt) \Delta V \sin \beta = fh \Delta V \cos \beta.$

If we now introduce the nondimensional variables,

(A3) $\quad v = V/V_g,$

(A4) $\quad \varepsilon = \Delta V/V_g,$

(A5) $\quad \eta = fh/C_D V_g,$

and

(A6) $\quad \lambda = (dh/dt)/C_D V_g,$

then Eq. (A2) implies

(A7) $\quad \tan \beta = \eta/(v + \lambda),$

which with Eq. (A1) gives

(A8) $\quad v^2 = \varepsilon^2[\eta^2 + (v + \lambda)^2].$

Further, Eqs. (24), (A7), and (A8) give

(A9) $\quad \varepsilon = \left[\dfrac{v(1 - v^2)}{v + 2\lambda}\right]^{1/2},$

which when substituted in Eq. (A8) gives

(A10) $$\eta = \left[\frac{v^2(v+\lambda)^2 - \lambda^2}{1-v^2}\right]^{1/2}, \qquad \text{valid for} \quad \lambda \leq \frac{v^2}{1-v},$$

i.e.

(A11) $$\lambda = \frac{v^3 \pm [v^4 - \eta^2(1-v^2)^2]^{1/2}}{1-v^2}, \qquad \text{valid for} \quad \eta \leq \frac{v^2}{1-v^2}.$$

Equation (22) provides α from

(A12) $$\sin \alpha = \frac{\varepsilon}{v} \sin \beta,$$

and Eqs. (16) and (25) give

(A13) $$(\kappa/C_D)[H(0,t)/\rho c_p \, \Delta\theta \, V_g]^{1-a} = \lambda \varepsilon^{-a},$$

which couples the dynamical and thermal properties of the system.

The dynamical equations (A9)–(A11) relate the variables $v, \varepsilon, \eta, \lambda$ to each

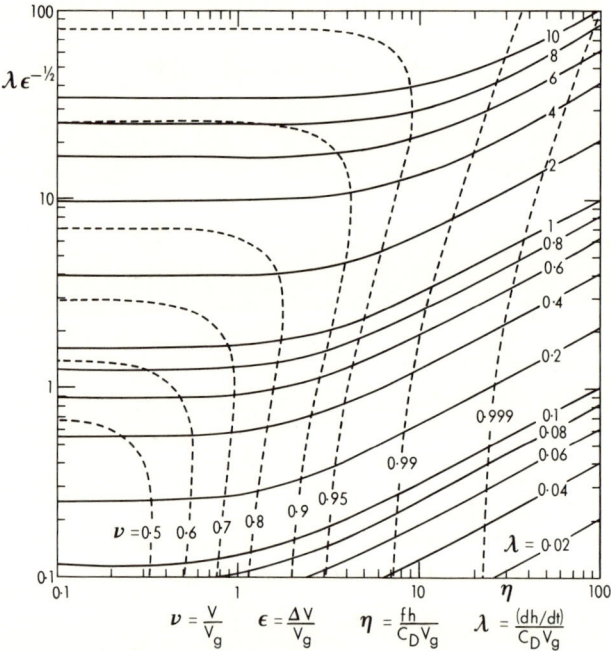

FIG. 5. Contours of the dynamical parameters $\lambda = (dh/dt)/C_D V_g$ (full lines) and $v = V/V_g$ (broken lines) as functions of $\eta = fh/C_D V_g$ and $(\kappa/C_D)[H(0,t)/\rho c_p \Delta\theta V_g]^{1/2} = \lambda\varepsilon^{-1/2}$, as determined by the thermodynamic model with $a = \frac{1}{2}$.

other as shown graphically for $a = \frac{1}{2}$ in Fig. 5 and, finally, the remaining thermal equations are from Eqs. (25)–(27),

(A14) $$\rho c_p h \, d\theta_c/dt = H(0, t) + \rho c_p (dh/dt) \, \Delta\theta$$

and

(A15) $$\Delta\theta = \theta_0 + \gamma h - \theta_c .$$

The steps in the integration procedure are:

(i) *Constant Parameters.* Values are assumed for a, κ, C_D, f, and ρc_p which remain constant throughout the development.

(ii) *Known Time Variations.* The values of $H(0, t)$, γ, and V_g are assumed known at all times. Strictly speaking, the analysis of Section 3 has assumed values of γ and V_g which remain constant throughout the integration.

(iii) *The Entrainment Equation.* At $t = t_0$, the start of the integration, $h(t_0)$ and $\Delta\theta(t_0)$ are given as initial conditions and at any other time $t_i > t_0$, $h(t_i)$ and $\Delta\theta(t_i)$ will have been estimated from the system of equations integrated over the previous time step from t_{i-1}. All the variables and parameters needed to evaluate $\eta(t_i)$ and the L.H.S. of Eq. (A13) are therefore known at any time t_i. The value of the L.H.S. of Eq. (A13) is the value of $[\lambda\varepsilon^{-a}]_i$.

(iv) *The Dynamical Equations.* Knowing $\eta(t_i)$ and $[\lambda\varepsilon^{-a}]_i$, Eqs. (A7)–(A12) are solved (possibly with the help of a graph similar to Fig. 5 corresponding to the chosen value of a) to give v_i, λ_i, ε_i, α_i, β_i and integration over some small time step δt gives,

(A16) $$h(t_{i+1}) \doteq h(t_i) + [\delta h]_i$$

where

(A17) $$[\delta h]_i = \lambda_i C_D V_g \, \delta t.$$

(v) *The Thermal Equations.* Equation (A14) gives

(A18) $$[\delta\theta_c]_i = \left[\frac{H(0, t_i)}{\rho c_p} + \lambda_i C_D V_g \, \Delta\theta(t_i) \right] \frac{\delta t}{h(t_i)},$$

and Eq. (A15) can be expressed as

(A19) $$\Delta\theta(t_{i+1}) \doteq \Delta\theta(t_i) + \gamma(t_i)[\delta h]_i - [\delta\theta_c]_i .$$

(vi) With the new values $h(t_{i+1})$ and $\Delta\theta(t_{i+1})$ we return to (i) and repeat the process for the next step in the integration procedure.

This integration procedure provides us with all the parameters and variables needed to determine the evolutions of, for example, h, $\Delta\theta$, θ_c, V, ΔV, α, β, τ_x, τ_y, and $H(h)$.

REFERENCES

Angell, J. K. (1972). *J. Atmos. Sci.* **29**, 1252–1261.
Ball, F. K. (1960). *Quart. J. Roy. Meteorol. Soc.* **86**, 483–494.
Betts, A. K. (1973). *Quart. J. Roy. Meteorol. Soc.* **99**, 178–196.
Browning, K. A., Starr, J. R., and Whyman, A. J. (1973). *Boundary-Layer Meteorol.* **4**, 91–111.
Carson, D. J. (1973). *Quart. J. Roy. Meteorol. Soc.* **99**, 450–467.
Deardorff, J. W. (1972a). *Mon. Weather Rev.* **100**, 93–106.
Deardorff, J. W. (1972b). *J. Atmos. Sci.* **28**, 91–115.
Deardorff, J. W. (1973). *J. Atmos. Sci.* **30**, 1070–1076.
Deardorff, J. W., Willis, G. E., and Lilly, D. K. (1969). *J. Fluid Mech.* **35**, 7–31.
Lilly, D. K. (1968). *Quart. J. Roy. Meteorol. Soc.* **94**, 292–309.
Readings, C. J., Golton, E., and Browning, K. A. (1973). *Boundary-Layer Meteorol.* **4**, 275–287.
Smith, F. B. (1972). *Proc. Meet. Expert Panel Air Pollut. Modeling, 3rd, NATO/CCMS, Paris* **17**, 1–14.
Tennekes, H. (1973). *J. Atmos. Sci.* **30**, 558–567.

VERTICAL COMPONENT OF TURBULENCE IN CONVECTIVE CONDITIONS

S. J. CAUGHEY AND C. J. READINGS

Meteorological Research Unit, R.A.F. Cardington, Bedford, U.K.

1. INTRODUCTION

The magnitude of the variance of the vertical wind component σ_w^2 is of great importance in the study of the diffusion of contaminants as well as for its bearing on the energy balance in the atmospheric boundary layer. However, although its behavior is quite well understood in the constant-stress layer (e.g., Panofsky and Mazzola, 1971), very little is known higher up in the atmosphere. Panofsky and McCormick (1960) argued that its value is determined by the height above the ground z and by the *local* rate at which turbulent energy is added to the system by the processes represented by the convective and mechanical production terms in the turbulent energy balance equations. Applying dimensional analysis to these assumptions gives

$$\sigma_w = A(z(\varepsilon_1 + \delta\varepsilon_2))^{1/3} \tag{1}$$

where A and δ are empirical constants; ε_1 is the rate at which mechanical energy is being added and ε_2 is the rate at which energy is added by convection. δ is introduced to "allow for the possibility that convection is more efficient than wind shear in producing a vertical component of turbulence." Although there is no *a priori* reason for δ being independent of height and stability, there are no observations that contradict this hypothesis. Of course, higher up in the atmosphere z may not be the appropriate length scale, and at any height σ_w may be affected by other processes such as advection. This approach of defining σ_w in terms of local parameters has recently been developed further by Yokoyama (1971) and Fiedler and Panofsky (1972).

The present paper reconsiders this approach using hitherto unpublished data that have been obtained at heights of up to 1 km, during the past few years at two different sites—one in England (Cardington, Bedfordshire) and the other in the United States (Fort Eglin, Florida). All the measurements were made with the Cardington turbulence probe (see Readings and Butler,

1972) mounted on the flying cable of a tethered balloon or on a fixed support. This instrument measures the instantaneous values of the total wind component (V), the inclination of the wind to the horizontal (ϕ), and the temperature. The outputs were sampled once a second and the various turbulence parameters calculated after the removal of any linear trends. The momentum flux was derived from the equation

$$\overline{u'w'} = \overline{(V\cos\phi - \overline{V\cos\phi})(V\sin\phi - \overline{V\sin\phi})}$$

and the vertical axis was defined either by reference to a sonic anemometer (Florida data) or else by the assumption $\bar{w} = 0$ (Cardington data). Since the equipment was almost always mounted at least 150 m below the balloon the results are considered not to have been significantly affected by balloon movement (see Readings and Butler, 1972). All the runs considered in this paper were of one hour's duration and none of these measurements was made in or above an inversion.

2. The Value of σ_w in the Convective Limit

When the rate of buoyant production is much greater than the rate for mechanical production, Eq. (1) simplifies to

(2) $$\sigma_w = A\delta^{1/3}\varepsilon_2^{1/3}z^{1/3}$$

But $\varepsilon_2 = (g/T)\overline{w'\theta'}$ where g is the acceleration due to gravity, T is the absolute temperature, θ is the potential temperature, and the prime superscript denotes a fluctuation about the mean. Thus Eq. (2) may be rewritten as

(3) $$\sigma_w = A\delta^{1/3}(g/T)^{1/3}(\overline{w'\theta'}\,z)^{1/3}$$

If the assumptions are correct this equation should be applicable at high levels where the mechanical production term will often be negligible or at low levels when the atmosphere is very convective. For a given value of $(z\overline{w'\theta'})^{1/3}$, Eq. (3) should give the *minimum* value of σ_w since any mechanically generated turbulence would increase σ_w. These predictions may be tested by plotting σ_w against $(z\overline{w'\theta'})^{1/3}$ (see Fig. 1). The data have been classified according to the values of z/L_1 where L_1 is the *local* value of the Monin-Obukhov length (assuming von Kármán's constant, $k = 0.4$). It can be seen that

(a) the less unstable points lie above the more unstable ones; and
(b) there is a tendency for the most unstable data to scatter about a limiting line, just as would be expected if the concept of a convective limit is correct.

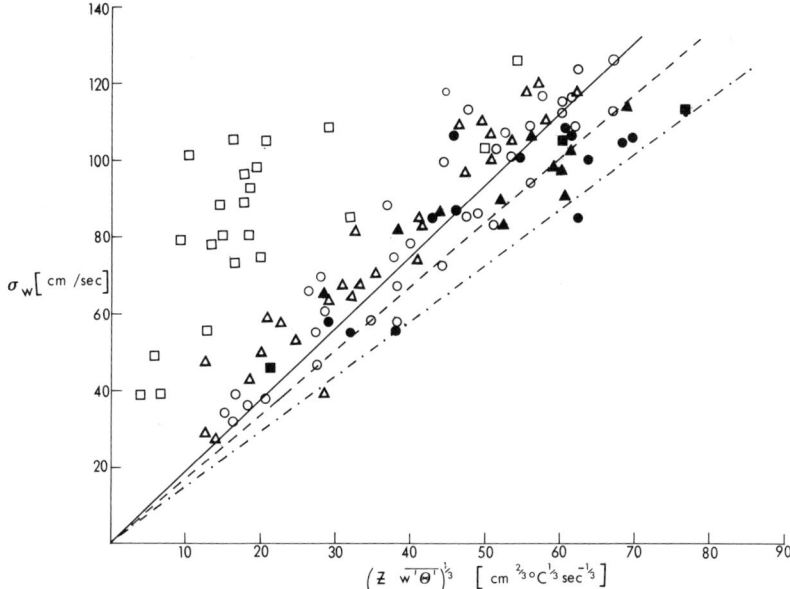

FIG. 1. A graph of σ_w versus $(\overline{zw'\theta'})^{1/3}$ with the data classified according to the local value of z/L_1. Range of z/L_1: □, > -0.5, △, -0.5 to -1.0, ○, -1.0 to -5.0, ●, -5.0 to -10.0, ▲, -10.0 to -100.0, ■, < -100.0.
The following lines are plotted in the figure. Equation of line: $\sigma_w = 1.23(g/T)^{1/3}(\overline{zw'\theta'})^{1/3}$ (———), $\sigma_w = 1.10(g/T)^{1/3}(\overline{zw'\theta'})^{1/3}$ (– – –), $\sigma_w = 0.96(g/T)^{1/3}(\overline{zw'\theta'})^{1/3}$ (— · —).

If it is assumed that the ratio of the two diffusivities (i.e., those for heat and momentum) is unity in near-neutral conditions then Panofsky and McCormick's (1960) work leads to the conclusion

$$A\delta^{1/3} = 1.23$$

but as K_H/K_M increases with instability this is an upper limit. The solid line drawn in Fig. 1 corresponds to this value. Although this line is near the lower boundary of the scatter of observations, quite a few points lie below it and a line of smaller slope would be more appropriate. Yokoyama (1971) has recently reported that $A\delta^{1/3} = 0.96$ and the line corresponding to this value has also been included in Fig. 1. Although this value is based on data for $z < 150$ m, it is obviously a very good lower limit provided the scatter is ignored.

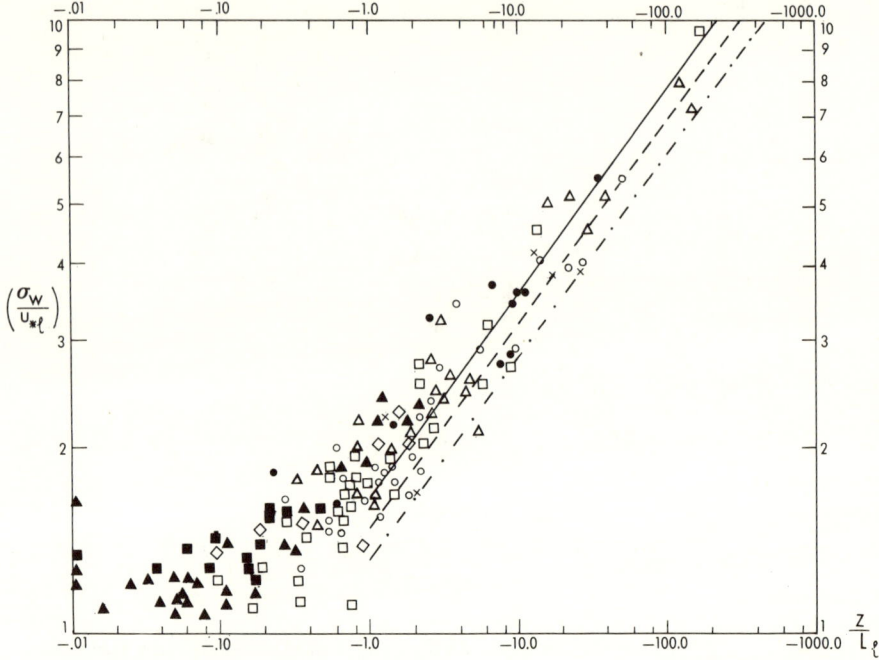

FIG. 2. A logarithmic plot of σ_w/u_{*1} versus z/L_1 with the data classified according to height/site:

Height range (in metres)	Cardington	Site Florida	Kansas
0–16	▲	—	■
16–155	□	○	◇
155–310	●	△	—
> 310	×	—	—

The three lines plotted in Fig. 1 are also plotted using the same notation.

Equation (3) may also be recast in the form

(4) $$\sigma_w/u_{*1} = (A\delta^{1/3}/k^{1/3})(-z/L_1)^{1/3}$$

where u_{*1} is the *local* value of the friction velocity. This equation should also be applicable to the data in the convective limit or at large values of $|-z/L_1|$ if this is a meaningful stability parameter. These conclusions are tested in Fig. 2 which is a logarithmic plot of σ_w/u_{*1} against z/L_1. The data are classified according to height/site, and the measurements made with sonic anemometers by AFCRL (see Haugen *et al.*, 1971) are included for reference. It can be seen that:

(a) there is no separation of the points according to height or site;
(b) there is a clear relation between σ_w/u_{*1} and z/L_1; and
(c) there is an approach to a one-third power law at large values of $|-z/L_1|$.

Those points are very important as they confirm the use of z/L_1 as a stability parameter and point to the relevance of *local* parameterisation even when $z \approx 1$ km. They also tend to confirm the use of z as a scale length up to this altitude. However it is important to bear in mind that none of the measurements used in this paper were made in or above an inversion.

The two lines previously discussed are also included in Fig. 2 using the same notation as in Fig. 1. It can be seen that though they represent the general trend of the data when $|-z/L_1| > 20$, one tends to lie above the points while the other tends to lie below them. This was confirmed by regressing the two variables for various minimum values of $|-z/L_1|$ and it was found that the relation

(5) $$\sigma_w/u_{*1} = (1.10/k^{1/3})|-z/L_1|^{1/3}$$

represented the data quite well when $|-z/L_1| \geq 20$; this line has also been included in Fig. 2. Extending the range of acceptable stabilities down to $|-z/L_1| \geq 1$ or 10 reduces the power law to ~ 0.29 and increases the constant of proportionality. Thus it appears that the convective limit will not be attained unless $|-z/L_1| \geq 20$; for smaller values the slope will decrease and the constant of proportionality increase, as is apparent in Fig. 1 of Wyngaard *et al.* (1971).

The line corresponding to Eq. (5) has also been included in Fig. 1, and it can be seen that the more unstable data scatter quite well about it.

3. Conclusions

The results presented in this paper have clearly established the dependence of σ_w on the local state of the atmosphere and have indicated the existence of a convective limit when $|-z/L_1| \geq 20$. These conclusions appear valid up to heights ~ 1 km provided an inversion is not present. The value of the constant of proportionality in the convective limit $A\delta^{1/3}$ appears to lie in the range 1.10–0.96. If these values are combined with the neutral limit of σ_w/u_{*1} ($= A/k^{1/3}$), the value of δ may be derived. Pasquill (1972) recently tabulated the various estimates of this neutral limit and it appears that a value of 1.3 is reasonable. With $k = 0.4$, this gives

$$1.0 \leq \delta \leq 1.5$$

which in the lower limit may imply that convection and wind shear are equally efficient in producing a vertical component of turbulence. However the present data set includes relatively few points for which $|-z/L_1| \geq 20$, so these conclusions must remain tentative until more measurements are available.

Acknowledgments

The authors wish to express their thanks to the many people involved in the collection and processing of the data.

References

Fieldler, F., and Panofsky, H. A. (1972). The geostrophic drag coefficient and the effective roughness length. *Quart. J. Roy. Meteorol. Soc.* **98**, 213–220.

Haugen, D. A., Kaimal, J. C., and Bradley, E. F. (1971). An experimental study of Reynolds stress and heat flux in the atmospheric surface layer. *Quart. J. Roy. Meteorol. Soc.* **97**, 168–180.

Panofsky, H. A., and McCormick, R. A. (1960). The spectrum of vertical velocity near the surface. *Quart. J. Roy. Meteorol. Soc.* **86**, 495–503.

Panofsky, H. A., and Mazzola, C. (1971). Variances and spectra of vertical velocity just above the surface layer. *Boundary-Layer Meteorol.* **2**, No. 1, 30–37.

Pasquill, F. (1972). Some aspects of boundary layer description. *Quart. J. Roy. Meteorol. Soc.* **98**, 469–494.

Readings, C. J., and Butler, H. E. (1972). The measurement of atmospheric turbulence from a captive balloon. *Meteorol. Mag.* **101**, 286–298.

Wyngaard, J. C., Cote, O. R., and Izumi, Y. (1971). Local free convection, similarity, and the budgets of shear stress and heat flux. *J. Atmos. Sci.* **28**, 1171–1182.

Yokoyama, O. (1971). An experimental study on the structure of turbulence in the lowest 500 metres of the atmosphere and diffusion in it. *Rep. Nat. Inst. Pollut. Resourc.* No. 2.

SURFACE LAYER OF THE ATMOSPHERE IN UNSTABLE CONDITIONS

ROBERT R. LONG

Department of Mechanics and Materials Science
The Johns Hopkins University, Baltimore, Maryland 21218, U.S.A.

1. INTRODUCTION

When the sun heats the ground or when cold air flows over a warm surface, the lower portion of the atmosphere becomes unstable and turbulent convection results in which the potential energy created by the heating is converted into kinetic energy of the turbulence. In addition, the shear provides another source of turbulent kinetic energy. The accompanying energy loss is accomplished by a cascade of energy from the larger eddies to the smaller eddies where the energy is ultimately dissipated. The temperature fluctuations also tend to be reduced by the action of molecular heat conduction in the small eddies.

Many investigators have explored the problem of the surface layer from an observational and theoretical viewpoint. A valuable reference in English is a recent translation of a book by Monin and Yaglom (1971) which emphasizes contributions by scientists in the U.S.S.R. Many experiments have been constructed to study convection between two stationary horizontal surfaces, heated below and cooled above (Malkus, 1954a,b; Thomas and Townsend, 1957; Silveston, 1958; Croft, 1958; Townsend, 1959; Globe and Dropkin, 1959; Somerscales and Dropkin, 1966; Deardorff and Willis, 1967; Somerscales and Gazda, 1969), and one recent experiment also involves shear (Townsend, 1972).

If one divides the scientists interested in turbulent convection into two groups, one finds that the experimentalists all attach major importance to the molecular coefficients of viscosity and conduction, v and K, whereas the atmospheric scientists believe that these quantities are unimportant in the surface layer.

There are two reasons for this. One is that laboratory experiments are on much smaller scales than those characteristic of the atmosphere, so that the relevant Reynolds numbers or Rayleigh numbers in the laboratory are much smaller than those in the atmosphere. The second reason is that all experiments, except the recent one by Townsend, are without shear whereas the atmosphere is usually in appreciable mean motion. We illustrate the importance of the latter remark by considering an idealized experiment involving statistically steady, turbulent flow of a Boussinesq liquid between horizontal smooth plates at $z = 0$ and $z = H$. The lower plate is heated and moves along the x axis at a speed $-\Delta u$; the upper is cooled and moves along the x axis at speed Δu. Thus, the ensemble mean velocity $\bar{u}$ at $z = H/2$ is zero. The temperature difference corresponds to an increment of buoyancy $2\Delta\rho$ where buoyancy is defined as

$$\rho = [(\rho_1 - \rho_0)/\rho_0]g \tag{1}$$

Here ρ_1 is density, ρ_0 is the mean value of the density at $z = H/2$, and g is gravity. Thus, $\bar{\rho} = 0$ at $z = H/2$. Two extreme cases correspond to stationary, heated plates and to moving plates with $\Delta\rho = 0$.

2. Discussion of the Two Cases

If $\Delta u = 0$, $\Delta\rho \neq 0$, dimensional analysis leads to forms of the mean quantities in terms of several nondimensional parameters. The mean buoyancy gradient, for example, is

$$\bar{\rho}_z = (\Delta\rho/z)f_3(P, \xi_0, \xi) \tag{2}$$

where z is the vertical coordinate and

$$P = \nu/K, \quad \xi_0 = H(\Delta\rho)^{1/3}/K^{2/3}, \quad \xi = z(\Delta\rho)^{1/3}/K^{2/3} \tag{3}$$

In Eqs. (3), P is the Prandtl number, ξ_0 is proportional to the cube root of the Rayleigh number,

$$\text{Ra} = H^3 \Delta\rho/\nu K$$

and ξ may be regarded as the ratio of the height above the lower surface to the thickness $\delta_T \sim K^{2/3}/(\Delta\rho)^{1/3}$ of the thermal boundary layer. We confine attention, of course, to situations and regions in which ξ_0 and ξ are large.

Many measurements have been made of the buoyancy flux $q = K\bar{\rho}_z - \overline{w\rho}$, where we now denote fluctuating velocities by u, v, w, and fluctuating buoyancy by ρ. The flux q is constant with height in a steady state and therefore can be expressed as

$$q = (\Delta\rho)^{4/3}K^{1/3}f_1(P, \xi_0) \tag{4}$$

All experiments indicate a very weak dependence on ξ_0 when this number is large (Turner, 1973) and it seems likely that f_1 may be considered independent of ξ_0 at large Rayleigh numbers. It is then inescapable that q is directly influenced by the molecular coefficients even at the very high Rayleigh numbers characteristic of the atmosphere. Since the heat transport is by the eddies, they too are directly influenced by the molecular coefficients.

Let us now consider the case of moving plates but with $\Delta\rho = 0$. It proves to be useful to express mean quantities in terms of the momentum flux $\tau = v\overline{u_z} - \overline{wu}$. We may write, for example,

$$(5) \qquad \overline{u}_z = \frac{\tau^{1/2}}{z} h_4(z_+, R_\tau), \qquad \sigma_u = \tau^{1/2} h_5(z_+, R_\tau)$$

where

$$(6) \qquad z_+ = z\tau^{1/2}/v, \qquad R_\tau = \tau^{1/2}H/v, \qquad \tau/(\Delta u)^2 = h_2(R_\tau)$$

Experiments with flow in channels and pipes indicate that R_τ (or H) is not important in (5) if R_τ is large and z is less than $H/2$, and, furthermore, that viscosity is not important either above a thin layer of about $30v/\tau^{1/2}$ in thickness. Thus

$$(7) \qquad \overline{u}_z = A\tau^{1/2}/z, \qquad \sigma_u = \tau^{1/2}A_1$$

near the lower surface but well above the viscous boundary layer. In (7), A and A_1 are universal constants. The integral for $\overline{u}$ involves the logarithm of z and the existence of this "logarithmic layer" is well established in the laboratory above smooth or rough surfaces and in the lower atmosphere in neutral conditions. Although viscosity, or the size of the roughnesses, may be ignored in obtaining $\overline{u}_z$, these must be taken into account in determining the constant of integration for $\overline{u}$. The picture emerges that the fluid at a given height z "feels" the flux of momentum τ but does not feel the viscosity or the roughnesses except as they serve to fix the level of zero velocity if the logarithmic profile is assumed to hold at all heights. In σ_u, for example, since A_1 is a universal constant, viscosity has no importance and the energy-containing eddies are completely unaffected by molecular properties.

3. Similarity Theories

The success of the above theory of shearing flow of a homogeneous fluid in predicting the properties of the surface layer in neutral conditions has encouraged atmospheric scientists to believe that molecular properties are unimportant when the ground is also heated or cooled despite the clear warning provided by the case of zero shear that a complete neglect of molecular quantities in the presence of heating is not always possible even in the largest systems. This approach is associated with the names of Monin and Oboukhov (Monin and Yaglom, 1971) and regards as fundamental the quantities τ and q which are reasonably constant in the lower 50–100 meters.

Neglecting v and K and assuming H is infinite, we may then write, for example,

(8) $\quad \bar{p}_z = (q/\tau^{1/2}z)\varphi_3(\zeta), \quad \bar{u}_z = (\tau^{1/2}/z)\varphi_4(\zeta), \quad \sigma_u = \tau^{1/2}\varphi_5(\zeta)$

where

(9) $\quad\quad\quad\quad\quad \zeta = z/l_m, \quad l_m = \tau^{3/2}/q$

In Eq. (9), $L = l_m A$ is the traditional expression for the Monin–Obukhov length, where A is the constant in Eq. (7).

Although this theory is simple and appealing, we emphasize that one should properly include two other numbers in the functions of Eqs. (8), namely P and an appropriately defined Reynolds number. Since we are interested only in the atmosphere in this paper, we may set P equal to a constant, say $P = 1$, i.e., $K = v$, and ignore K. The Reynolds number is obviously based on the friction velocity and the length l_m, so that we may write

(10) $\quad\quad \bar{p}_z = (q/\tau^{1/2}z)\varphi_3(\zeta, R_m), \quad R_m = \tau^2/qv$

Although (10) is more accurate than the expression for $\bar{p}_z$ in (8), we acknowledge that our experience with homogeneous fluids suggests that R_m *is negligible when R_m is large*. If R_m is small, we must be cautious. In the first place a small Reynolds number suggests quite naturally that molecular quantities may be important, and in the second place small R_m is associated with the case of zero or weak shear and the above discussion indicates that v and K cannot be neglected in this case.

We illustrate the importance of R_m by considering the arguments of Priestley (1954) used to determine the behavior of $\bar{p}_z$ or of $\varphi_3(\zeta)$ when z is large. The contention is made that large ζ in Eqs. (8) means *either large z or small τ* so that the behavior of $\varphi_3(\zeta)$ as $\zeta \to \infty$ may be obtained by requiring that τ disappear from the analysis. Then, for large z, we have

(11) $\quad\quad\quad\quad \bar{p}_z = \text{const } q^{2/3}z^{-4/3}$

There is controversy concerning this prediction (and predictions of the forms of other mean quantities using the same argument). Soviet scientists believe that the predictions are "well satisfied" (Monin and Yaglom, 1971) while acknowledging some deviations at larger values of ζ. Western scientists have less confidence. There is a general belief that $\bar{p}_z \propto z^{-3/2}$, for example, as indicated by excellent atmospheric data of Dyer (1965) and Businger *et al.* (1971). These data seem to be good enough to distinguish clearly between the observed $z^{-3/2}$ behavior and the similarity theory of Eq. (11).

In view of this disagreement, let us look more closely at the similarity argument. If we use the more accurate Eq. (10) instead of Eq. (8), we have the immediate problem that small τ is not the same as large z because of the presence of R_m. In addition, of course, if we let τ approach zero, the Reynolds number approaches zero. Let us proceed, nevertheless, by allowing τ to approach zero. We may write

(12) $$\bar{\rho}_z = q^{2/3} z^{-4/3} g_3(\zeta, R_m)$$

and the similarity argument is that

(13) $$\lim_{\tau \to 0} g_3(\zeta, R_m) = c$$

where c is finite, nonzero constant. It is easy to investigate the regime of weak shear implied by $\tau \to 0$. In fact, for any value of τ or Δu, we may write

(14) $$\bar{\rho}_z = (\Delta\rho/z) f_3(V, \sigma)$$

(15) $$q = (\Delta\rho)^{4/3} v^{1/3} f_1(V)$$

where

(16) $$V = \Delta u/(v\Delta\rho)^{1/3}, \qquad \sigma = z(\Delta\rho)^{1/3}/v^{2/3}$$

It is easy to decide on physical grounds that the following nonzero limits exist

(17) $$\lim_{V \to 0} f_1(V) = a_1, \qquad \lim_{V \to 0} f_3(V, \sigma) = f_3(0, \sigma)$$

Comparing (12) and (14) and using (15), we get

(18) $$\lim_{\tau \to 0} g_3(R_m, \zeta) = (q^{1/12} z^{1/3}/v^{1/4} a_1^{3/4}) f_3(0, \sigma)$$

The similarity argument is that this is a nonzero constant, i.e.,

(19) $$f_3(0, \sigma) = \text{const } \sigma^{-1/3}$$

for all values of z and this is clearly impossible. On the other hand, (12) shows that large z and small τ are not equivalent limit processes so that we may investigate separately the limit as $z \to \infty$. We get

(20) $$\lim_{z \to \infty} g_3(R_m, \zeta) = \lim_{z \to \infty} \frac{q^{1/12} z^{1/3}}{v^{1/4} f_1^{3/4}(V)} f_3(V, \sigma)$$

To obtain a $z^{-4/3}$ law for $\bar{\rho}_z$, we must have

(21) $$\lim_{z \to \infty} f_3(V, \sigma) = f_3(V) \sigma^{-1/3}$$

This is a conceivable behavior, but the argument is certainly not rigorous or even convincing. It leads precisely to the behavior in (11) which may be obtained quite simply for $V = 0$ by assuming, by analogy to shearing flow, that $\bar{\rho}_z$ should involve only q and z in a region above the thermal boundary layer. We see, therefore, that the two similarity arguments, with and without shear, are the same. If either is to be believed, they should predict the behavior of mean quantities both in the atmosphere and laboratory. We have already indicated one of the major disagreements in atmospheric measurements. In the laboratory there is controversy about the behavior of $\bar{\rho}_z$. Townsend (1959) proposed $\bar{\rho}_z \propto z^{-n}$ where $1.3 < n < 2.5$. Data of Croft (1958) yield $z^{-3/2}$.

The similarity arguments can be used to predict other quantities, for example, rms ρ or $\sigma_\rho \propto z^{-1/3}$ compared with measurements in which exponents range from -0.48 to -0.70 (Rossby, 1969; Somerscales and Gazda, 1969). Finally Townsend (1959) gives an interesting result for the "dissipation function" for buoyancy fluctuations δ. He finds $\delta \propto z^{-3/2}$ compared with $z^{-4/3}$ from similarity theory.

We may conclude that evidence in atmospheric and laboratory investigations does not inspire confidence in the similarity theories when heating effects are strong. We have seen that all theories reduce to the argument that mean quantities should depend only on q and z when the shear is zero and that the argument was inspired by the success of the theory that only τ and z are important in shearing flow above a surface. In the latter case, however, τ is proportional to $(\Delta u)^2$ and is either very weakly dependent on v (smooth wall) or independent of v (rough wall). In the case of heating, however, q and, therefore, either σ_w or σ_ρ directly involve the molecular coefficients.

4. A New Theory

A new theory, differing from the similarity theory, has been proposed by the author (Long, 1974) and involves the behavior of a large number of mean quantities including mean buoyancy gradient, mean density gradient, rms velocities and buoyancies, length and time scales, and eddy viscosity and conductivity. We are concentrating in this paper on the properties of the mean buoyancy gradient, and we will therefore confine attention to a few aspects of the new theory relevant to finding $\bar{\rho}_z$ in a region well above the thermal boundary layer. For simplicity, we will again assume $v = K$ and $H = \infty$. For the time being, we may also consider the shear to be zero.

The similarity theory for $\Delta u = 0$ may be based on simple dimensional analysis once it has been decided that $\bar{\rho}_z$, for example, should depend only on q and z. On the other hand, as shown by Kraichnan (1962), it may also be based on estimates of orders of magnitude of certain quantities as $z \to \infty$ and this approach reveals the essential difference between the present theory and the similarity theory.

The eddy conductivity K_h may be defined by the equation $q = K_h \bar{\rho}_z$ and since q is a constant, the variation of $\bar{\rho}_z$ with height is determined by the variation of K_h. To find this we notice that $q \sim \overline{w\rho}$ implies that $q \sim \sigma_\rho \sigma_w C$ where C is the correlation coefficient. We know, however, that warm parcels rise and cold parcels descend so that C should be of order one. This assumption is strongly supported by observations (Deardorff and Willis, 1967) in which C is 0.5–0.6 over a wide range of Rayleigh numbers. Both theories then should yield $q \sim \sigma_\rho \sigma_w$. The quantity σ_ρ may be estimated by assuming $\sigma_\rho \sim \bar{\rho}_z l$ where l is a length over which the total buoyancy is conserved. It is reasonable that $l \sim z$ as is the case in turbulent shear flow above a surface. Indeed, $l \sim z$ implies $l \sim H$ for the container as a whole; this implies that the energy-containing eddies fill the entire container and this is observed in experiments. The three estimates $q \sim K_h \bar{\rho}_z$, $q \sim \sigma_\rho \sigma_w$, $\sigma_\rho \sim \bar{\rho}_z z$ lead to

$$K_h \sim z\sigma_w$$

and the problem of finding K_h and, therefore, $\bar{\rho}_z$ is reduced to finding σ_w. This is the point of departure for the present theory and the similarity theory. The latter, as shown by Kraichnan (1962) assumes $\sigma_w^2 \sim \sigma_\rho l$ which is a seemingly reasonable assumption that the kinetic and potential energies are of the same order, or that the vertical acceleration of a parcel is of the order of the buoyancy force. This estimate, and the earlier estimates, leads to $\bar{\rho}_z \sim q^{2/3} z^{-4/3}$.

In the present theory the last order of magnitude estimate is not made. Instead experimental observations are used (Malkus, 1954b; Deardorff and Willis, 1967) that the kinetic energy averaged over the entire container is of order $H \Delta\rho$. This result holds with considerable accuracy over a range of Rayleigh numbers from 10^5 to 10^9. This has a physical interpretation that the vertical velocity of a parcel is of an order obtained by imagining that it conserves its density and rises freely in the unstable environment from an origin in the thermal boundary layer. It suggests, therefore, that $\sigma_w^2 \sim z \Delta\rho$ at any given level z. If we adopt this estimate, we obtain

(22) $$\bar{\rho}_z = B(\Delta\rho)^{5/6} v^{1/3} / z^{3/2}$$

where B is a constant for zero shear but in general may be considered a function of V. This contrasts with the $z^{-4/3}$ dependence of the similarity theory. In essence, it is easy to see that the basic difference in the two theories can be reduced to a difference in the estimate of the time scale T of the eddy motion. The present theory assumes T depends on z and $\Delta\rho$ only and the similarity theory assumes that T depends on q and z only. Actually, if one acknowledges that $\Delta\rho$ is a more fundamental parameter than q, the present assumption is preferable if one also feels that vanishingly small molecular coefficients should not directly affect the time scale of the eddies. In any case, experiment and observation will be decisive. We have already stated that experimental observations of $\bar{p}_z$ are inconclusive although it is fair to point out that Croft's observations (Croft, 1958) support Eq. (22). The atmospheric observations, of course, directly support Eq. (22).

The present theory is easily extended to obtain other mean quantities, for example $\sigma_p \propto z^{-1/2}$, and we have already seen that this is closer to observed behavior than the similarity theory. The present theory also yields $\delta \propto z^{-3/2}$ where δ is the "dissipation function" for buoyancy fluctuations and this agrees exactly with Townsend's observations.

5. Conclusions

The surface layer of the atmosphere transmits heat, momentum, and moisture to the free atmosphere and it is obviously of practical importance to understand it thoroughly. In addition, it is a portion of the atmosphere which can be most conveniently examined and measured and then compared with the predictions of theories. The most popular theory involves an approach suggested by the highly successful theory of homogeneous flow over a surface and leads to simple predictions based on dimensional analysis. The above discussion questions the plausibility and the rigor of the similarity arguments and points out a number of disagreements between laboratory and atmospheric observations.

The paper concludes with a development of a new theory which differs fundamentally from the similarity theory in assuming, essentially, that the temperature difference between the heated lower surface and the fluid at infinite heights is a more fundamental quantity than the heat flux. The present theory has a number of predictions that agree exactly with observations or are in better agreement than the similarity theories, but more and better observations are clearly needed.

Acknowledgment

This research was supported by the National Science Foundation under Grant NSF GA-35612.

References

Businger, J. A., Wyngaard, J. D., Izumi, Y., and Bradley, E. F. (1971). Flux-profile relationships in the atmospheric surface layer. *J. Atmos. Sci.* **28**, 181.

Croft, J. F. (1958). The convective regime and temperature distribution above a horizontal heated surface. *Quart. J. Roy. Meteorol. Soc.* **84**, 418.

Deardorff, J. W., and Willis, G. E. (1967). Investigation of turbulent thermal convection between horizontal plates. *J. Fluid Mech.* **28**, 675.

Dyer, A. J. (1965). The flux-gradient relation for turbulent heat transfer in the lower atmosphere. *Quart. J. Roy. Meteorol. Soc.* **91**, 151.

Globe, S., and Dropkin, D. (1959). Natural convection heat transfers in liquids confined by two horizontal plates and heated from below. *J. Heat Transfer* **81**, 24.

Kraichnan, R. H. (1962). Turbulent thermal convection at arbitrary Prandtl numbers. *Phys. Fluids* **5**, 1374.

Long, R. R. (1974). Some properties of turbulent convection with shear. *Geophys. Fluid Dyn.* (in press).

Malkus, W. V. R. (1954a). Discrete transitions in turbulent convection. *Proc. Roy. Soc., Ser. A* **225**, 185.

Malkus, W. V. R. (1954b). The heat transport and spectrum of thermal turbulence. *Proc. Roy. Soc., Ser. A* **225**, 196.

Monin, A. S., and Yaglom, A. M. (1971). "Statistical Fluid Mechanics: Mechanics of Turbulence," Vol. 1. MIT Press, Cambridge, Massachusetts.

Priestley, C. H. B. (1954). Convection from a large horizontal surface. *Aust. J. Phys.* **7**, 176.

Rossby, H. T. (1969). A study of Bénard convection with and without rotation. *J. Fluid Mech.* **36**, 309.

Silveston, P. L. (1958). Wärmedurchgang in Waagrechten. Flüssigkeitsschichten. *Forsch. Ingenieurw.* **24**, 29.

Somerscales, E. F. C., and Dropkin, D. (1966). Experimental investigation of the temperature distribution in a horizontal layer of fluid heated from below. *Int. J. Heat Mass Transfer* **9**, 1189.

Somerscales, E. F. C., and Gazda, I. W. (1969). Thermal convection in high Prandtl number liquids at high Rayleigh numbers. *Int. J. Heat Mass Transfer* **12**, 1491.

Thomas, D. B., and Townsend, A. A. (1957). Turbulent convection over a heated horizontal surface. *J. Fluid Mech.* **2**, 473.

Townsend, A. A. (1959). Temperature fluctuations over a heated horizontal surface. *J. Fluid Mech.* **5**, 209.

Townsend, A. A. (1972). Mixed convection over a heated horizontal plane. *J. Fluid Mech.* **55**, 209.

Turner, J. S. (1973). "Buoyancy Effects in Fluids." Cambridge Univ. Press, London and New York.

NUMERICAL SIMULATION OF LAGRANGIAN TURBULENT QUANTITIES IN TWO AND THREE DIMENSIONS

RICHARD L. PESKIN

Department of Mechanical, Industrial and Aerospace Engineering,
Geophysical Fluid Dynamics Program, Rutgers University
New Brunswick, New Jersey 08903, U.S.A.

1. INTRODUCTION

The diffusive character of fluid dynamic turbulence is one of the least understood, yet most practically important features of such flows. Whether one is concerned with transport of pollutants under actual atmospheric circumstances, or with the fundamental problem of energy cascade in idealized turbulent flows, an understanding of turbulent diffusion is essential; yet such an understanding has remained elusive primarily because turbulent diffusion of single tracer points is controlled by that portion of the energy spectrum, namely, the larger wave numbers, where theoretical analysis is difficult. Furthermore, while diffusion is most naturally described in the Lagrangian system, data acquisition is most practically accomplished using the Eulerian description of the flow. Thus one is faced with the difficult choice of attempting to perform experiments to measure Lagrangian quantities directly, a formidable task, or alternatively using Eulerian measurements and relying on the, as yet, unavailable understanding of the relation between Eulerian and Lagrangian quantities.

Ideally, one would like to be able to perform Lagrangian experiments and simultaneously obtain Eulerian information. While practical problems restrict this possibility in actual physical situations, numerical simulation used as an experimental tool opens up interesting perspectives in this field of study. In principle, both Eulerian and Lagrangian information can be obtained with relative ease from a numerical simulation of turbulent flow, and one can obtain information to establish Eulerian–Lagrangian relations, which are useful in translating Eulerian data into Lagrangian diffusion information. Of course, idealization to this extent is not yet possible in computer simulation. Precise simulation of three-dimensional turbulent flow which would include sufficient resolution in wave number space while at the same time retaining sufficient complexity (i.e., presence of shear, stratification, etc.) to be interesting practically, exceeds the capability of the largest computers now available or projected for the immediate future. Consequently, numerical simulation of three-dimensional turbulent flows for the purpose of obtaining Lagrangian information must, of necessity, rely on the so-called subgrid scale modeling. That is, analytically or heuristically obtained models of small-scale wave number interaction must be incorporated in a deterministic way into the computer program. While such subgrid scale models can cause considerable difficulty in certain aspects of flow simulation, particularly when used in the presence of strong stable stratification, they nevertheless are a necessity, and when using such models to study turbulent diffusion one can rely on the hope that the subgrid scale modeling, a large wave number approximation, will have minimal impact on quantities controlling diffusion since these quantities are determined primarily by the small wave number portion of the spectrum.

In the case of two-dimensional turbulent flows, computer capabilities are such that subgrid scale approximation is not necessary. Of course, two-dimensional turbulent flow is an idealization not yet shown to be a realistic model of even the largest scale motions in the atmosphere. Nevertheless, as a theoretical exercise, and as a possibly practical model for quasi-geostrophic atmospheric turbulence there is some justification for studying the problem. Here it is possible to undertake numerical simulation without use of the subgrid scale model but by dealing with the basic Navier–Stokes equations and truncating these at some sufficiently large wave number to have minimal impact on the large wave number effects controlling diffusion.

In this paper two examples of numerical simulation of Lagrangian turbulence motion are considered. In the first, a two-dimensional simulation is examined using an Eulerian field generation truncated at a reasonably high wave number. In the second case, a three-dimensional simulation is examined, namely, simulated, fully developed turbulent channel flow. Here subgrid scale approximation is essential. In both cases, Lagrangian particles are tracked in order to obtain information about such quantities as Lagrangian autocorrelation, mean-square displacement, relative two-point displacement, and the effects of shear flow on diffusion.

2. SIMULATION OF TURBULENT DIFFUSION FOR TWO-DIMENSIONAL FLOW

Two-dimensional turbulent flow is in a sense a misnomer inasmuch as the two-dimensional restriction prohibits vortex stretching. In any event such an idealization can be considered as an approximation to motions occurring on a large scale in the oceans and atmospheres, i.e., quasi-geostrophic turbulence (Charney, 1971). With the advent of satellites and constant level balloons Lagrangian measurements in the "quasi two-dimensional" region of the atmosphere are a reality. The EOLE experiment tracked 480 balloons at the 200-mb level (Morel and Bandeen, 1973) and forthcoming experiments under GARP promise further expansion of this technique. In the light of these experiments in the stratosphere and in view of the aforementioned ability to treat a two-dimensional simulation without subgrid scale modeling, such simulation seems advisable. In particular, such numerical experiments can eliminate many of the features of real experiments such as anisotropy and nonstationarity. In the present study an Eulerian two-dimensional numerical turbulent field was used to study Lagrangian statistics. Particle velocity correlations, particle mean-square displacement, and Eulerian–Lagrangian relations were examined. Results were limited to single-particle analysis; however, computer-generated movies allowed a qualitative study of cluster dispersion. (Quantitative study of the multi-particle statistics was restricted by grid size limitations.)

The results in general verify the classical theories of diffusion (Taylor, 1921) and the Corrsin hypothesis (Corrsin, 1959) which related the Lagrangian autocorrelation to the Eulerian space–time correlation. Some of the observed behavior of clusters was similar to that observed in the EOLE experiments.

2.1. Eulerian Flow Field Simulation

The technique used in this study was to track particles, that is, fluid points on a two-dimensional numerical simulation of a turbulent flow field. The Eulerian field was that developed by Lilly (1969, 1972). A review and discussion of recent two-dimensional flow fields numerical simulations can be found in the paper by Fox (1972). While the present study used physical space representation, recent work by Orszag (this volume, p. 225) indicates that considerable improvement in accuracy can be obtained by k space techniques, that is Fourier space simulation techniques. However, these do have the disadvantage of imposing considerable programming difficulty, particularly when Lagrangian information is desired.

2.1.1. Two-Dimensional Flow Field Equations. The simulated two-dimensional Eulerian field was generated by solving the basic equations of vorticity and stream function on a 64 × 64 grid and utilizing periodic boundary conditions to maintain isotropy. The equations solved are as follows:

(2.1) $$\partial \zeta / \partial t + \mathbf{V} \cdot \nabla \zeta = F + \nu \nabla^2 \zeta - K \zeta$$
(2.2) $$\mathbf{V} = -\mathbf{i} \, \partial \psi / \partial y + \mathbf{j} \, \partial \psi / \partial x$$
(2.3) $$\zeta = \nabla^2 \psi$$

Because the energy in two-dimensional turbulence cascades toward lower wave numbers, it was essential to introduce a mechanism to remove energy in order to obtain quasi-steady solutions. The energy removal mechanism for this model was a "surface friction coefficient," that is, a vorticity sink linear in vorticity and operative at zero wave number. Physically, such a sink might represent surface friction in the Earth's boundary layer. In addition to energy removal, energy input is required, and this was accomplished by using a constant amplitude random phase forcing function. A normalized random phase Fourier component (at wave number 8) was generated using the following equations:

$$(2.4) \qquad F_{n+1} = R_n F_n + (1 - R_n^2)^{1/2} \hat{F}_{n+1}$$

where subscript n refers to discrete time. Equation (2.4) is an approximation to

$$(2.5) \qquad \frac{\partial F}{\partial t} = -\frac{F}{\tau} + \lim_{\Delta t \to 0} \frac{\hat{F}}{[\tau(\Delta t/2)]^{1/2}}$$

where

$$(2.6) \qquad \tau/\Delta t = \tfrac{1}{2}(1 + R_n)/(1 - R_n)$$

Δt is integration increment, and τ is the correlation time.

$\hat{F}_n$ was obtained by Gaussian selection of wave number amplitudes on a square of width $2k_e$ in k_x, k_y space. This amplitude was controlled normalizing components so that $k_x^2 + k_y^2$ was a constant. Time correlation for the forcing function was provided by a simple Markoff process as indicated by R_n in Eq. (2.6).

All dependent and independent variables were scaled by the amplitude of the forcing function and the input wave number. The remaining parameters, which could be varied, were nondimensional kinematic viscosity, correlation time, surface friction coefficient, and mesh lengths. The last were fixed to a 64 × 64 grid and other results presented in this paper were for a given choice of the parameters.

2.1.2. Eulerian Field Results. Batchelor (1969) and Kraichnan (1967) predicted that energy spectrum for two-dimensional turbulence would divide into two regions either side of the wave number for energy input. The lower wave number portion of the spectrum would behave according to the $-\tfrac{5}{3}$ law, and in this region energy cascades toward lower wave number. At higher wave numbers the spectrum would behave according to the -3 law, and in this region enstrophy, that is mean square vorticity, cascades to the higher wave numbers. In fact, there is some question about the correct description of the higher wave number region and the validity of the -3 law, etc. (Fox, 1972), but these subjects are beyond the scope of this paper. The following equations describe the predicted spectrum according to Batchelor and Kraichnan:

(2.7) $$E(k) = \begin{cases} \alpha(d\bar{E}/dt)^{2/3} k^{-5/3}, & k_c < k < k_e \\ \beta \eta^{2/3} k^{-3}, & k_e < k < k_d \end{cases}$$

$$\bar{E} = \int_0^\infty E(k)\,dk$$

η is the mean square vorticity rate, $k_c = k_c(t)$ is the low wave number limit, k_e is the energy input wave number, and k_d is the dissipation wave number.

Figure 1 is a typical numerically generated vorticity field and Fig. 2 is a typical energy spectrum exhibiting approximate $-\frac{5}{3}$ and approximate -3 regions. (These results should not be taken as necessary evidence for the correctness of the -3 law.) The energy spectrum retains this approximate shape over the total time period used for the diffusion studies and is thus representative of the Eulerian spectrum operative during the time span for the study of Lagrangian statistics.

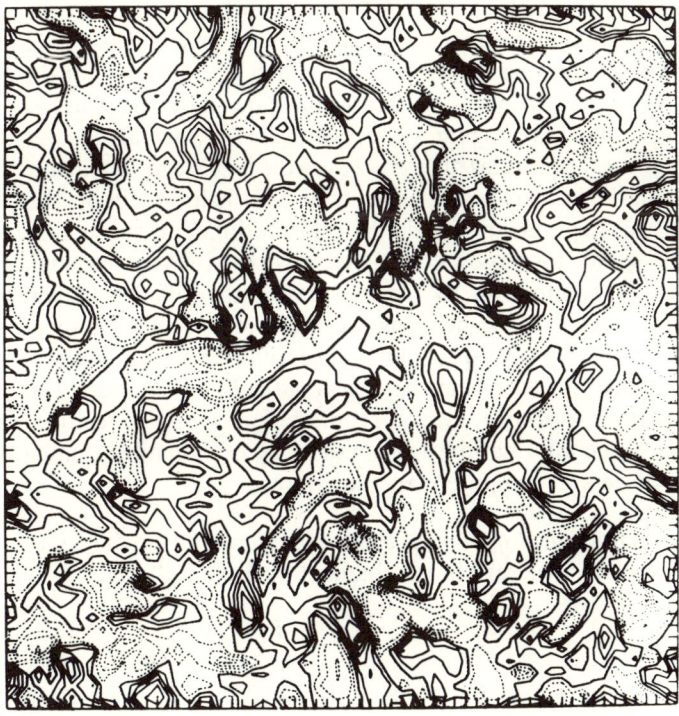

FIG. 1. Typical numerically generated two-dimensional vorticity field.

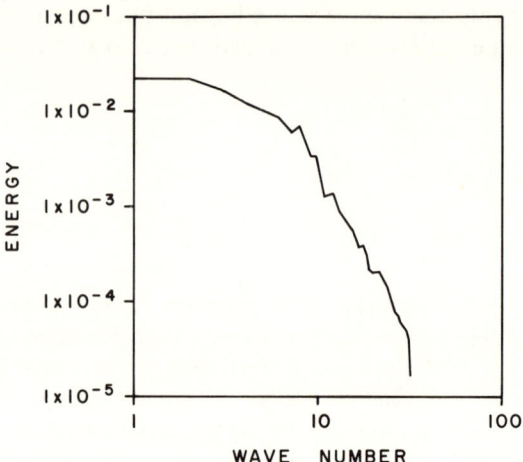

FIG. 2. Two-dimensional energy spectrum generated numerically.

2.2. Generation of the Lagrangian Field

The Lagrangian field was determined by tracking 1024 particles (fluid points) for 800 particle time steps (a nondimensional particle time unit was equal to 50 nondimensional computational time units). This procedure was repeated twice, however, the results presented here are only for the first 800 particle time steps. In view of the approximate stationarity and isotropy of the Eulerian field, averaging was accomplished by averaging over the particles. The basic quantities of interest were the particle mean-square displacement, Lagrangian autocorrelation, and information about the relation between the Lagrangian correlation and the Eulerian space–time correlation. The only information considered was that relevant to single particle statistics. As pointed out by Corrsin (1962), the diffusivity filter function implies that larger scale motions are those responsible for single-particle Lagrangian whereas two-particle relative diffusion is controlled by smaller scale motions. In view of the coarseness of the Eulerian grid, it was felt that the simulation was not totally appropriate for two-point relative statistics. However, some information was obtained by examination, qualitatively, of particle clusters in computer-generated movies.

2.2.1. Lagrangian Equations.
The basic equations for diffusion in a stationary isotropic field are those obtained by Taylor (1921):

$$(2.8) \quad \overline{X^2}(t) = 2\overline{v^2} \int_0^t d\tau \int_0^\tau R_L(T) \, dT$$

$$(2.9) \quad R_L(t) = (1/\overline{v^2})\overline{V(T)v(T+\tau)}$$

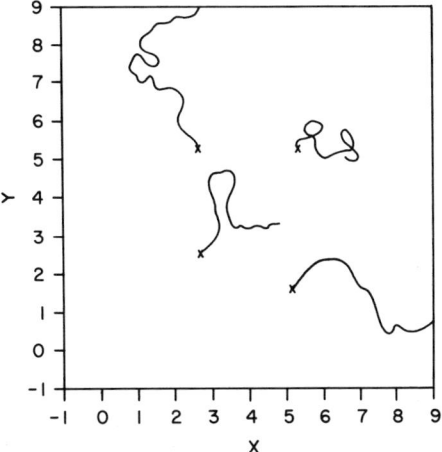

FIG. 3. Sample trajectories of four particles released in the central portion of the two-dimensional field.

It is these two equations which were examined by comparing directly measured mean-square displacement from the computer simulation with computed mean-square displacement using directly measured Lagrangian autocorrelation and the above equations. Figure 3 is a sample of the trajectory for four particles. (Particles that left the field were reintroduced according to the cyclic nature of the boundary conditions and this effect was accounted for in computation of the mean-square displacement.) Figure 4 is

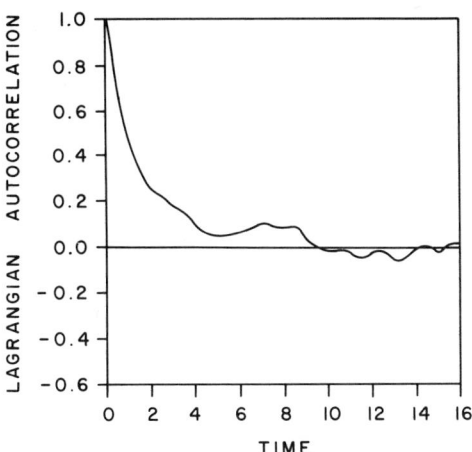

FIG. 4. Lagrangian autocorrelation for two-dimensional flow.

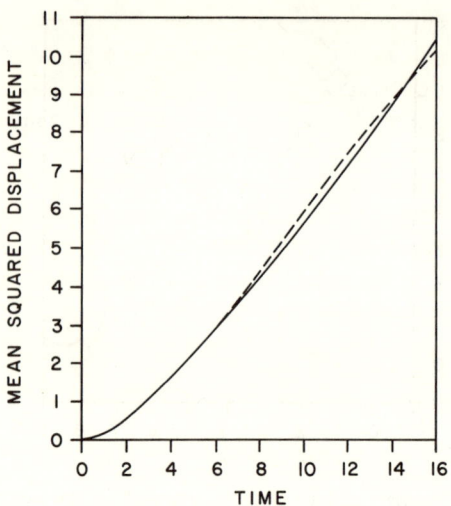

FIG. 5. Numerically calculated mean-square displacement $\overline{X^2}$ and $\overline{X^2}$ computed from Eq. (2.9) (dotted line).

a typical computation for the Lagrangian autocorrelation and Fig. 5 compares the mean-square displacement measured directly with that obtained from Taylor's formula [Eq. (2.8)]. Examination of Fig. 5 indicates that the time length for Lagrangian computation was sufficient inasmuch as the mean-square displacement approaches a region of constant slope, as it must for long times from release. This is further corroborated in Fig. 6 which plots the Lagrangian integral scale as a function of time.

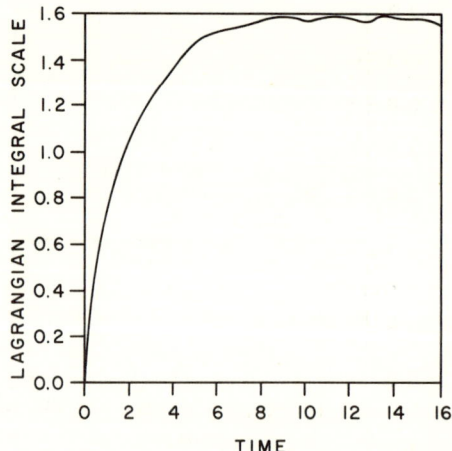

FIG. 6. Numerically calculated Lagrangian integral scale.

In general the results verify the correctness of Taylor's formula for application in this situation and illustrate the appropriateness of this type of technique for Lagrangian statistical studies. Both the Lagrangian autocorrelation and the Lagrangian integral scale exhibit some noise fluctuation at large times, which is indicative of statistical fluctuation caused by use of a finite number of particles which for long times after release result in a reduced number density and consequent increase in statistical noise. This problem was further complicated by a slowly evolving coherent rotation of the whole field. The "Lagrangian Reynolds stress" was also computed to test isotropy. In general this stress was small.

2.2.2. Test of the "Corrsin Hypothesis"

Corrsin (1959) introduced a hypothesis relating Eulerian space–time correlation to Lagrangian autocorrelation. The implications of this hypothesis were examined by Saffman (1963) for long times from release and Peskin (1965) for short time after release. For isotropic stationary fields, Corrsin's relation is

$$(2.10) \qquad R_\text{L}(t) = \tfrac{1}{2} \int_0^\infty [R_\text{N}(r, t) + R_\text{T}(r, t)] P(r, t) \, dr$$

where R_N and R_T are longitudinal and transverse Eulerian space–time correlations and $P(r, t)$ is the probability distribution of particle displacement.

The above result can also be obtained by a "stochastic estimation" model (Peskin, 1971). In general it is difficult to test such hypotheses in actual experiments because of the difficulty in obtaining accurate Eulerian space–time correlations and obtaining accurate Lagrangian autocorrelations. Numerical simulation is an ideal situation for tests of this type of hypothesis. In order to test the hypotheses, the Eulerian field was used to generate the space–time Eulerian correlation in both the longitudinal and lateral directions. The longitudinal space–time correlation is shown in Fig. 7. In addition to the space–time correlation, information is needed about the probability of displacement for the fluid points. As was pointed out by Saffman (1963), it is not unreasonable to assume this displacement to be Gaussian, with dispersion that given by the mean-square displacement as computed from the Taylor formula or directly measured. In a first test of the Corrsin hypothesis, the Lagrangian autocorrelation was compared with a Lagrangian autocorrelation computed using the above equation and employing space–time correlation generated from the Eulerian field and Gaussian particle displacement distribution. The results are shown in Fig. 8. It is interesting to note that while the use of the Gaussian displacement overestimates the magnitude of the Lagrangian correlation the computation correctly reproduced the details including the hump in the correlation around time 8. It

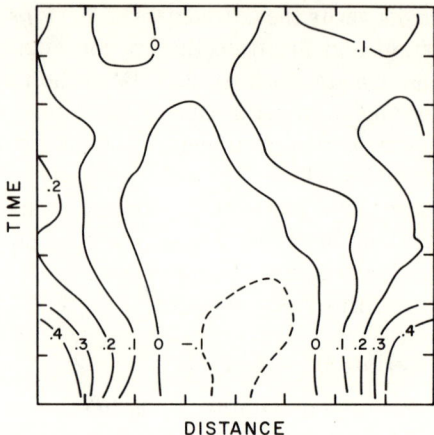

FIG. 7. Longitudinal Eulerian space–time correlation for two-dimensional numerically generated flow.

is apparent that the detailed shape of the Lagrangian autocorrelation is determined in large measure by the Eulerian space–time correlation. The reason for the overestimate was not immediately obvious. Thus it was decided to test the formula using probability displacement functions directly measured from the computer simulation rather than the Gaussian estimate. Figure 9 is an example of such a probability distribution taken at an intermediate time. It can be seen that this probability is clearly not Gaussian. In particular probabilities for large displacements at this time are considerably smaller than those which would be estimated by the Gaussian distribution.

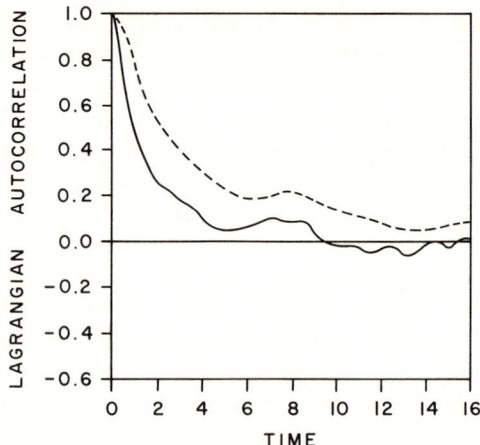

FIG. 8. Lagrangian correlation (Fig. 4) and Corrsin hypothesis estimate (upper curve) using Gaussian probability distribution.

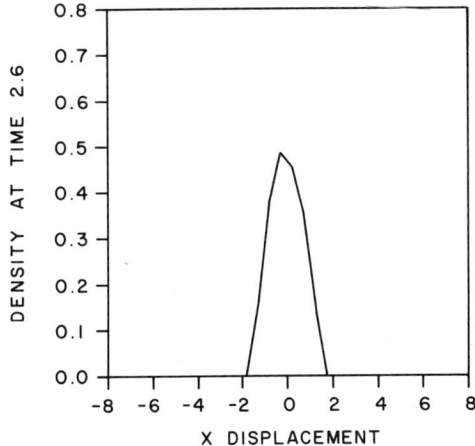

FIG. 9. Sample of particle displacement probability distribution (at time 2.6).

Mechanisms to allow extremely rapid acceleration of particles in short times after release apparently were not present. While this in part may be due to the nature of the simulation, one might also speculate that the absence of vortex stretching in the two-dimensional field implies the absence of mechanisms for rapid particle acceleration. It should be noted that at longer times the displacement distribution did approach a Gaussian form. Figure 10 is a comparison of the Lagrangian autocorrelation with its estimate from the Corrsin equation using the computed rather than the Gaussian probability distribution. It can be seen that the correspondence is quite good. Unfortunately, statistical noise in determination of the probability

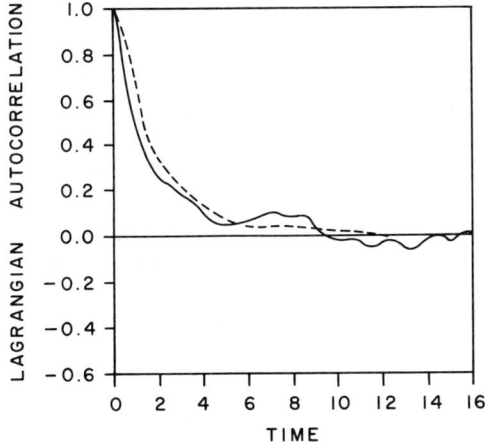

FIG. 10. Comparison of Lagrangian autocorrelation and Corrsin hypothesis estimate (dotted curve) using directly determined probability distribution.

distribution resulted in elimination of some of the fine scale details represented in Fig. 8.

It is evident that numerical simulation of Lagrangian statistics in two-dimensional flow is feasible at least to the extent that single-particle statistics can be studied. The use of physical space rather than k space computation imposes severe limitations on the resolution of the Eulerian field. However, the k space computation would impose serious problems of complexity of computation as far as determination of the Lagrangian field is concerned. In general, for this computation Taylor's formula was verified and the most interesting result was the good correspondence obtained between the Lagrangian autocorrelation as directly determined and that quantity calculated from the Corrsin hypothesis. It is apparent that the probability displacement function for particles is not Gaussian except at long times after release for the two-dimensional case.

2.2.3. Two-Particle Separation. As mentioned above, the filter function arguments are not valid when considering two-particle relative separation. Higher resolution of the Eulerian field is needed because this problem is strongly characterized by initial separation lengths. Computations are planned for this problem using k space rather than physical space representation for the Eulerian field. Lin (1972) estimated the two-point separation law for two-dimensional or quasi-geostrophic turbulence based on the -3 spectrum. His results indicate an exponential separation growth. Peskin (1973) pointed out that the Langevin model theory would predict a t^3 law and at present available data cannot resolve the issue inasmuch as the -3 spectrum may not be correct. An important point is that the lack of energy cascade in the -3 region leaves open the question of physical mechanisms for separation if the -3 for the Eulerian spectrum is correct. It is hoped that the future acquisition of constant entropy level balloon data will resolve this issue. The above-mentioned computer simulations were used to generate computer movies of clusters of particles. Some qualitative information was obtained from these films; in particular, the expected behavior of clusters, namely, that they are persistent for fairly long periods of time and then rapidly accelerate apart, was observed.

3. SIMULATION OF THREE-DIMENSIONAL LAGRANGIAN TURBULENCE IN CHANNEL FLOW

While simulation of two-dimensional Lagrangian turbulence is an interesting exercise and can be effected (for at least single particle statistics) without necessity of ad hoc closure assumptions (other than wave number truncation), the problem lacks certain appeal particularly because of the omission of vortex stretching mechanisms. In addition, the practical importance of shear flow in diffusion dictated that a more complicated three-dimensional study was in order. To this end three-dimensional simulation of turbulent channel flow was used to study the Lagrangian problem. The channel flow problem is relevant because it is a first step toward the important planetary boundary layer simulation problem and it contains regions of shear flow as well as approximate isotropy in the center of the channel. Consequently the channel flow problem contains many features of practical value when one is concerned with turbulent diffusion. Obviously, to deal with such a complicated flow situation, one must give up many of the simplifying features found in the two-dimensional simulation. In particular computation limitations indicate that simulation at reasonable values of the Reynolds number cannot be accomplished with today's computing machinery, without employing some closure model, that is, some model that analytically simulates large wave number or small length scale features of the flow. In effect, the computation reported here employs a subgrid scale model, that is, a model to account for turbulent dissipation on length scales smaller than that represented by the numerical simulation grid. The technique employed is similar to that reported by Deardorff and Peskin (1970), but in the present study, the main emphasis is placed on the Eulerian–Lagrangian problem. In addition the computations were generalized to include the effects of heavy particle (as distinct from fluid point) diffusion in the shear flow field.

3.1. *Techniques for the Generation of the Eulerian Field*

The Eulerian field was generated in the simulated three-dimensional channel flow by a technique similar to that employed by Deardorff (1970). Essentially the computation involved satisfying the Navier–Stokes equations at each point in the internal grid together with application of boundary conditions and a subgrid scale formulation of turbulence to account for motions on a length scale smaller than the grid dimension. More details on this can be found on the aforementioned paper by Deardorff (1970), or the thesis by Kau (1972).

3.1.1. Basic Equations. The equations solved were the Navier–Stokes equations together with the continuity equation:

(3.1) $$\partial \overline{u_i}/\partial t = \overline{R_i} - \partial(\overline{P''} - 2x_1)/\partial x_i$$

(3.2) $$\partial \overline{u_i}/\partial x_i = 0$$

(3.3) $$\overline{P''} = \left(\frac{\overline{P}}{\rho u^{*2}} + \frac{\overline{u'_l u'_l}}{3} + 2x_1\right) - \left\langle \frac{\overline{P}}{\rho u^{*2}} + \frac{\overline{u'_l u'_l}}{3} + 2x_1 \right\rangle$$

(3.4) $$\overline{R_i} = \frac{\partial}{\partial x_i}\left[\left(\overline{u_i}\,\overline{u_j} + \overline{u'_i u'_j} - \delta_{ij}\frac{\overline{u'_l u'_l}}{3}\right) - \delta_{i3}\,\delta_{j3}\left\langle \overline{u'^2_3} + \overline{u'^2_3} - \frac{\overline{u'_l u'_l}}{3}\right\rangle\right]$$

$$\bar{q}(\mathbf{x}) = (1/\Delta x\, \Delta y\, \Delta z) \int_{x-\Delta x/2}^{x+\Delta x/2} \bar{q}(\xi)\, d\xi \tag{3.5}$$

$$\langle \bar{q} \rangle = (1/L_x L_y) \int_0^{L_x} dx \int_0^{L_y} dy\, \bar{q}(x, y, z, t) \tag{3.6}$$

These equations were obtained by employing an averaging technique over the grid volume and collecting the subgrid scale fluctuation terms into a separate stress expression. This expression is represented by a subgrid scale stress model and the model employed here is similar to that first introduced by Smagorinsky (1963):

$$\overline{u_i' u_j'} - \delta_{ij} \frac{\overline{u_l' u_l'}}{3} = -\tilde{K}\left(\frac{\partial \bar{u}_i}{\partial x_j} + \frac{\partial \bar{u}_j}{\partial x_i}\right) \tag{3.7}$$

$$\tilde{K}(x, y, z, t) = (c\Theta)^2 \left[\frac{\partial \bar{u}_i}{\partial x_j}\left(\frac{\partial \bar{u}_i}{\partial x_j} + \frac{\partial \bar{u}_j}{\partial x_i}\right)\right]^{1/2} \tag{3.8}$$

$$\Theta \equiv (\Delta x\, \Delta y\, \Delta z)^{1/3}$$

c is an empirical constant.

Discussion of the choice of the empirical constant in the subgrid scale model is found in the thesis by Kau (1972). These differential equations were solved in a region of channel flow $4h \times 0.8h$ by h in dimension, h being the height of the channel, the longest dimension taken in the axial (mean-flow) direction. (See Fig. 11.) In addition to the basic equations in the interior, boundary layer conditions were imposed which were cyclic in the axial and lateral direction and employed the law-of-the-wall on the upper and lower boundary in the vertical direction. Use of this latter boundary condition effectively implies an infinite Reynolds number calculation. (See Deardorff, 1970, for details of fluctuating velocity boundary conditions.)

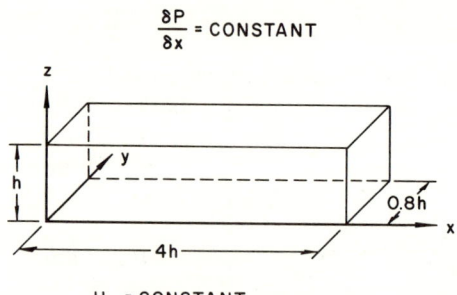

FIG. 11. Coordinate system for the channel flow simulation.

The computation procedure involved the use of a fourth-order Arakawa finite difference scheme and a leap-frog type time differencing scheme. The Poisson equation was solved by use of Fourier transform, in particular, the fast Fourier transform algorithm. Application of this was possible in the axial and transverse directions because of the periodicity of the boundary conditions and effectively implies an exact solution at the mesh points for each vertical level.

3.1.2. Eulerian Results. Computations were performed for all details of the Eulerian field and various output information was obtained by direct graphical plotting techniques and direct generation of microfilm. Figure 12 is a typical velocity field isopleth result. Figure 13 shows the obtained mean velocity profile compared with measurements and Deardorff's computational results. The total Reynolds stress distribution was linear as predicted, and other Eulerian properties of the field were reasonable and compared favorably with experimental and/or computation results obtained in the literature. Figure 14 shows the total fluctuating energy in the field as a function of time and is indicative of the typical problem in computations of this sort employing grid-scale modeling, the problem of reaching steady state. The ability to reach steady state and the time to reach steady state are dependent on the empirical constant and other features of the subgrid scale model. In general, long times are required to reach steady state and this can involve great computational expense. (In the present calculation 3000 time steps are required to reach steady state.) Thus, in performing calculations of

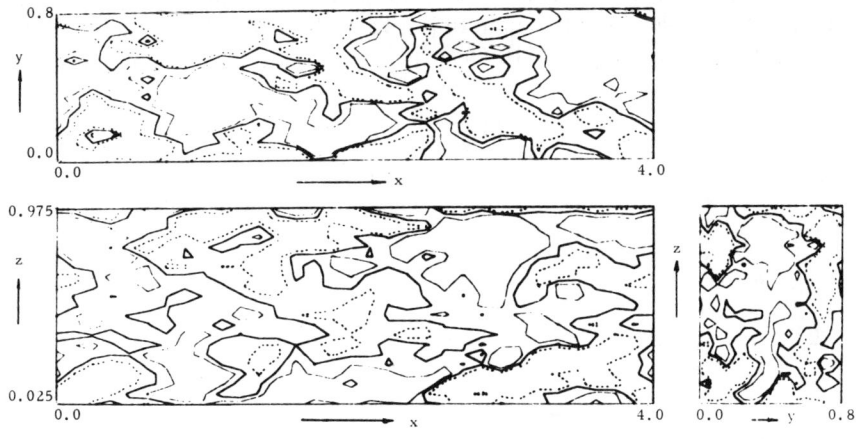

FIG. 12. Typical velocity isopleths in channel flow at $F = 0.175$ (Kau, 1972).

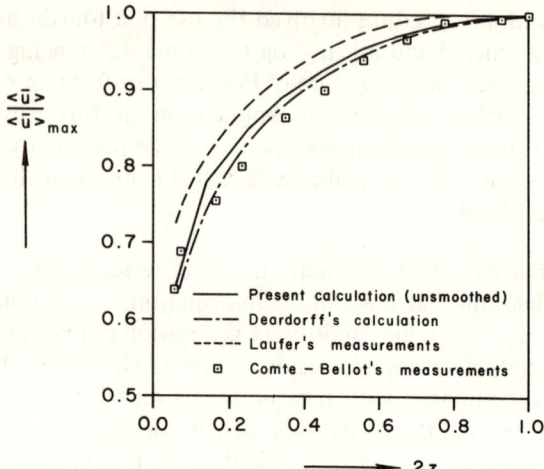

FIG. 13. Calculated mean velocity profiles (Kau, 1972) compared with measurements and other calculations (Deardorff, 1970; Laufer, 1950; Comte-Bellot, 1965).

this type it is important to have them well planned with respect to Eulerian and Lagrangian information desired since any program changes requiring reinitialization of the computation become impractical.

3.2. Lagrangian Information in Three-Dimensional Calculation

While Eulerian information from this type of calculation had been examined previously by Deardorff (1970), detailed information on the Lagrangian problem was the main objective of this study. Some Lagrangian information had been reported by Deardorff and Peskin (1970); however that

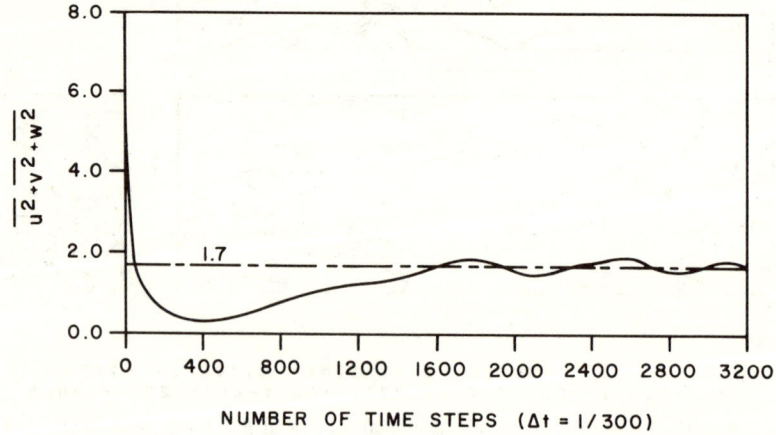

FIG. 14. Fluctuation of total resolvable energy as function of time (Kau, 1972).

paper did not examine space–time Eulerian correlations and their relation to the Lagrangian autocorrelation, nor was there any treatment of heavy particles. The primary purpose of this study was to examine both fluid point and heavy particles in the channel flow with particular emphasis on the effects of the shear flow region, and in addition to examine the space–time correlation and its relation to the Lagrangian autocorrelation, that is, to test the Corrsin hypothesis.

3.2.1. Particle Trajectories. Figure 15 shows the configuration of particles, initially, as viewed from the top of the channel. Particles were released in pairs and tracked with information for both single- and two-point relative diffusion and velocities retained. In the case of heavy particles they were assumed spherical and Stokes drag was employed as a resistance law (see Kau, 1972). Particles were moved from the initial position by invoking linear

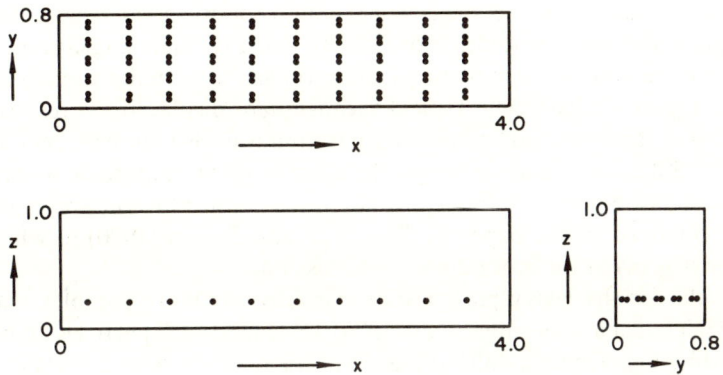

FIG. 15. Position of particles prior to release.

interpolation of the local Eulerian velocity field. Particles leaving the Eulerian field were reintroduced according to the requirements of the cyclic boundary conditions and particles impacting the upper lower walls were reflected into the flow. Effects of reintroduction were included in computation of mean-square displacements.

3.2.2. Lagrangian Results. Figure 16 is a typical set of particle trajectories. Computer motion pictures were generated of this process and the momentum flux to the wall in the case of the fluid points is quite evident from these motion pictures. In general, single fluid point diffusivities behave as expected, that is, they follow Taylor's diffusion law in the case of diffusion in the vertical and cross-stream direction. Diffusion in the down-stream direction is affected by shear. Two-point relative diffusion also shows some expected features, in particular, the initial stage where the particles tend to stay

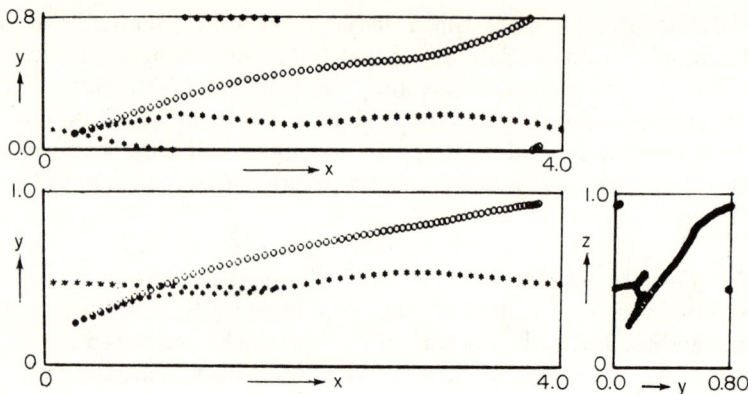

FIG. 16. Typical fluid point and heavy particle trajectories (Kau, 1972). ○: solid particle, *: fluid particle.

together followed by the stage where they rapidly spread apart. Relative two-point diffusion procedes more slowly, initially, than single-point diffusion and then exceeds single-point diffusion as the particles rapidly spread apart. Figure 17 has some typical mean-square displacement results. The effect of shear in the down-stream separation was also quite evident for the relative diffusion. Figure 18 shows the effect of shear on particle separation. The results obtained in this study for fluid point diffusion are similar to those obtained in the paper by Deardorff and Peskin (1970) to which the reader is referred for more detailed discussion.

Results for the heavy particles are considerably more complex and will not be detailed in this paper (Kau, 1972). In general such particles do behave according to a Taylor's diffusion law, but the operative autocorrelation is of course the autocorrelation of the heavy particle velocities which differs from the autocorrelation of fluid point velocities. Heavy particles seem less affected by the presence of shear and the results for heavy particles are strongly dependent on the nature of the vertical boundary conditions. (That is, the effects of slip along the wall which can occur for heavy particles can be important.) Of primary importance in dealing with the heavy particle case is the long time required to achieve steady state. While fluid points are normally tracked by initializing them at local Eulerian velocities, such initialization for heavy particles is in a sense arbitrary and artificial. The time that it can take for the heavy particle Lagrangian field to reach a steady state which is independent of the choice of initial conditions can be quite long, particularly for the larger particles. This effect is quite evident in the numerical computation, and failure to consider this can lead to paradoxical conclusions when one attempts to compare heavy particle to fluid point diffusivities. This has also been observed experimentally (Carlson, 1973). In

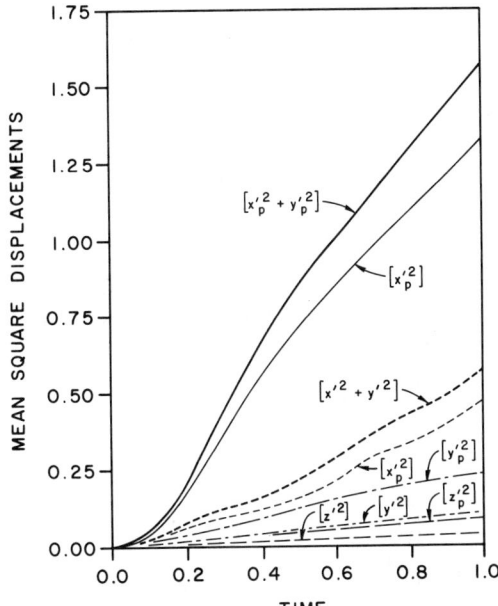

FIG. 17. Mean-square displacements in channel flow (x_p denotes heavy particle, x denotes fluid particle).

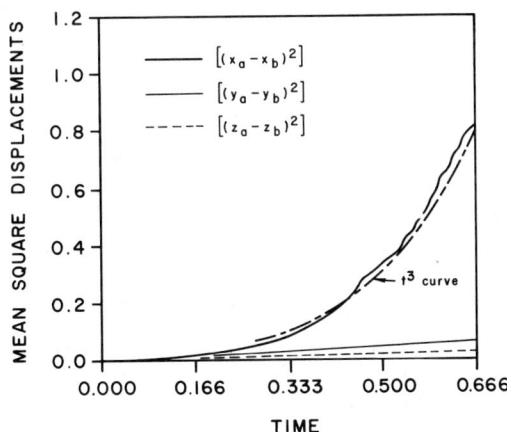

FIG. 18. Mean-square relative separation for two fluid particles (Kau, 1972).

general for isotropic flow the heavy particle diffusivity should not exceed the fluid point diffusivity (Peskin, 1971). However, effects such as wall slip can modify these conclusions (Depew and Kramer, 1973). Any attempt to use numerical simulation to verify theoretical conclusions must consider the problem of achieving steady state. In the presently reported calculations the ratio of particle to fluid point diffusivities approaches the expected value given a long enough computation time.

3.3. The Eulerian–Lagrangian Problem

As previously mentioned one purpose of this calculation was to treat the Eulerian–Lagrangian problem for three-dimensional turbulent calculation. To this end, the Eulerian space–time correlations were computed and the Corrsin hypothesis was invoked to estimate the Lagrangian autocorrelation which compared with directly measured autocorrelations. Figure 19 shows

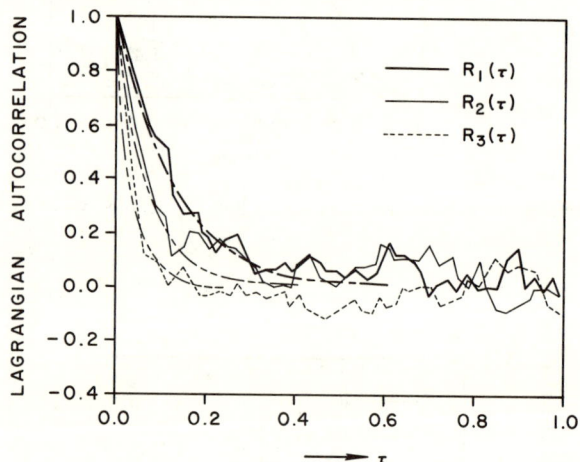

FIG. 19. Fluid particle autocorrelations (dash-dot lines are best fit exponential curves).

the typical fluid point autocorrelation. As was expected the heavy particle autocorrelations exhibited a significantly greater time scale, that is, particle inertia tends to preserve the flow memory of the Lagrangian field. Eulerian space–time correlations were obtained from the calculations. One of the things that is possible to do with a numerical calculation is to examine relevant time scales. In the present case the results were used to compute Lagrangian time scales and compare them with the Eulerian time scale. The calculation implies that the Lagrangian time scale is slightly greater than the Eulerian scale. This conclusion was also reached by Riley (1971), who treated the numerical simulation of Lagrangian diffusion in a homogeneous model shear field. However, the point is as yet unsettled inasmuch as theoretical conclusions (Peskin, 1965; Kraichnan, 1964) indicate that the reverse may be true. This problem needs more detailed study.

Figure 20 shows the results of the test of the Corrsin hypothesis for the three-dimensional calculation. One can see a quite good correspondence in the autocorrelations and in addition it appears that results do not differ greatly depending upon use of the actual or Gaussian probability distribution. This result was obtained in the center of the flow where approximate isotropy holds, and additional calculations are planned considering modified hypothesis for the shear region.

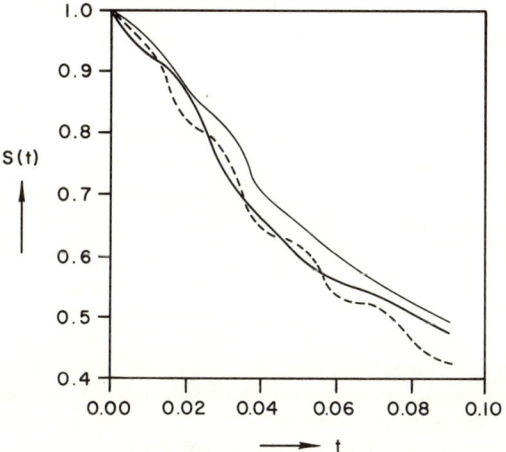

FIG. 20. Comparison of calculated autocorrelation (heavy line) with Corrsin hypothesis (light line) and Corrsin hypothesis using Gaussian probability (dotted line).

4. CONCLUSIONS

For both two- and three-dimensional flow fields, it appears that numerical simulation is a valuable tool for Lagrangian study. While such calculations can be quite costly and are not recommended for application to practical problems involving complex boundaries and sources, these computations can provide a good deal of information about the Lagrangian field and its relation to the Eulerian field. In general, such information is either inaccessible or very difficult to obtain by laboratory measurement techniques.

The fact that single-point diffusion is controlled by larger scales of motion (filter function effect) aids the usefulness of numerical calculation because practical considerations limit the refinement of mesh possible with today's computers. However, adequate study of the two-point relative diffusion problem requires greater computer resolution or, alternatively, an exact or reliable approximate analytical description of the smaller scale motion of the turbulence. This conclusion can be translated to practical situations, that is, fairly coarse mesh descriptions can provide adequate information on single-point diffusion but attempts to use the same mesh sizes to study relative two-point diffusion can lead to serious errors.

Finally, it must be noted that while one can get extensive information about the relation between the Eulerian and Lagrangian field quite easily from a numerical simulation, in a sense, numerical simulations provide almost too much information. Little is known about the Eulerian–Lagrangian relation problem and the mere fact that a good deal of output information is available from a numerical simulation cannot add appreciably to the understanding of this problem unless the information is properly used. That is, numerical simulation, of and by itself, is of little help in solving the Eulerian–Lagrangian problem unless more theoretical background is obtained so that the correct questions can be asked. Essentially, the numerical simulation is a means of providing the data we need to study the Eulerian–Lagrangian problem, but stronger theoretical understanding of this problem is essential.

Acknowledgments

The author wishes to acknowledge Dr. Douglas Lilly, NCAR[1] for his aid and discussion of the two-dimensional diffusion problem. Valuable discussions were also held with Dr. James Deardorff, NCAR, on both the two- and the three-dimensional diffusion study. Two-dimensional diffusion work was done while the author was in residence at NCAR at that time holding a Rutgers University Research Council Faculty Fellowship. Dr. E. J. Kau, Aero-Chem Research Laboratories, contributed greatly to the study of the three-dimensional diffusion problem. Portions of this work were sponsored by the National Science Foundation, Division of Engineering, Grant No. GK-31788X, and Atmospheric Sciences Section, Grant No. GA-33421.

References

Batchelor, G. K. (1969). *Phys. Fluids* **12**, Suppl. II, 233–239.
Carlson, C. R. (1973). Turbulent gas-solid flow measurements utilizing a laser-doppler velocimeter. Ph.D. Thesis, Rutgers Univ., New Brunswick, New Jersey.
Charney, J. G. (1971). *J. Atmos. Sci.* **28**, 287–295.
Comte-Bellot, G. (1965). "Ecoulement turbulent entre deux parois parallèles." Publ. Sci. Tech. Ministère de L'Air, Paris.
Corrsin, S. (1959). *Advan. Geophys.* **6**, 161.

---

[1] The National Center for Atmospheric Research is sponsored by the National Science Foundation.

Corrsin, S. (1962). "Mechanics of Turbulence," pp. 27–52. Gordon & Breach, New York.
Deardorff, J. W. (1970). *J. Fluid Mech.* **41**, 453–480.
Deardorff, J. W., and Peskin, R. L. (1970). *Phys. Fluids* **13**, 584–595.
Depew, C. A., and Kramer, T. J. (1973). *Advan. Heat Transfer* **9**, 113–180.
Fox, D. (1972). Numerical procedures for studying incompressible two-dimensional turbulence. *AIAA Pap.* No. 72-152.
Kau, C. J. (1972). Numerical study of turbulent diffusion in three-dimensional channel flow. Ph.D. Thesis, Rutgers Univ., New Brunswick, New Jersey.
Kraichnan, R. H. (1964). *Phys. Fluids* **7**, 142–143.
Kraichnan, R. H. (1967). *Phys. Fluids* **10**, 1417–1423.
Laufer, J. (1950). "Investigation of turbulent flow in a two-dimensional channel." NACA Tech. Note 2123.
Lilly, D. (1969). *Phys. Fluids* **12**, Suppl., 240–249.
Lilly, D. (1972). *Geophys. Fluid Dyn.* **3**, 289–319.
Lin, J. R. (1972). *J. Atmos. Sci.* **29**, 394–396.
Morel, T., and Bandeen, W. (1973). *Bull. Amer. Meteorol. Soc.* **54**, No. 4, 298–306.
Peskin, R. L. (1965). *Phys. Fluids* **8**, 993–994.
Peskin, R. L. (1971). "Stochastic Hydraulics" (C. L. Chiu, ed.), p. 251. Univ. of Pittsburgh, Sch. of Eng. Press, Pittsburgh, Pennsylvania.
Peskin, R. L. (1973). *J. Atmos. Sci.* **30**, 733–734.
Riley, J. J., Jr. (1971). Computer simulations of turbulent dispersion. Ph.D. Thesis, Johns Hopkins Univ., Baltimore, Maryland.
Saffman, T. G. (1963). *Appl. Sci. Res. Sect. A* **11**, 245–255.
Smagorinsky, J. (1963). *Mon. Weather Rev.* **91**, 99–164.
Taylor, G. I. (1921). *Proc. London Math. Soc., A* **20**, 196–211.

THE EULERIAN–LAGRANGIAN RELATIONSHIP RESULTING FROM A TURBULENCE MODEL

HJALMAR FRANZ

Deutsches Hydrographisches Institut, Hamburg, Germany

In two papers (Franz, 1970, 1973) the author has published a relation between the Lagrangian and the Eulerian correlation functions. Here the consequences will be presented for the power spectra and other functions related to the correlations. For a better understanding of the physics involved, a short review of the main assumptions of the model is given.

(1) The turbulent velocity fluctuations are regarded to be caused by vortices penetrating each other. Each vortex is displaced by the mean current and by all the other vortices.

(2) The time dependency of the vortices is taken into account because otherwise the Lagrangian integral time scale would tend to zero for large times.

(3) The correlations between different vortices are assumed to be small in the mean compared with the autocorrelations.

(4) In each system the mean correlation is found by weighting the autocorrelation with a distribution function common for both systems. That distribution function of unknown shape depends above all on a characteristic frequency and on a characteristic length.

(5) That frequency—characteristic for the Lagrangian autocorrelation—and that length—characteristic for the Eulerian autocorrelation—are related in the mean. As a consequence the distribution function depends further on one scale parameter only.

Especially the last assumption is necessary to follow a relation between the Lagrangian and the Eulerian mean correlations R. The relation is given implicitly by integrals over the distribution function D

(1) $$R_L(t) = \int_0^\infty D(\omega) K_L(\omega t)\, d\omega$$

(2) $$R_E(t) = \int_0^\infty D(\omega) K_E(\omega t; I)\, d\omega$$

with the autocorrelations K as kernels of known shape and the turbulence intensity I. Of course there are two relations like Eqs. (1) and (2), one for the longitudinal and one for the two lateral Eulerian correlations. Finally, the explicit expression can be arrived at by the numerical elimination of the unknown distribution function from the equations above in the general form

(3) $$R_L(t) = \int_0^\infty T(s) R_E(st) \, ds.$$

The transfer function T is determined by the autocorrelations only

(4) $$K_L(z) = \int_0^\infty T(s) K_E(sz; I) \, ds.$$

Because the turbulence intensity is a pure Eulerian quantity, the transfer function T has to depend on I. The structure of the actual turbulence field is contained in the distribution function D.

Some general properties of T are very convenient for the mathematical treatment:

(5) $$T(-s) = T(s)$$

(6) $$\int_0^\infty T(s) s^n \, ds \neq \infty \quad \text{for all} \quad n \geq 0$$

especially

(7) $$\int_0^\infty T(s) \, ds = 1.$$

The following equations hold for every relation between the Lagrangian and Eulerian correlation functions representable in the general form of Eqs. (1) and (2).

Substituting R_E in Eq. (3) by its cosine transform, one gets the Lagrangian correlation expressed by the Eulerian power spectrum P_E:

(8) $$R_L(t) = \int_0^\infty L(vt) P_E(v) \, dv$$

with

(9) $$L(r) = \int_0^\infty T(s) \cos(rs) \, ds.$$

The cosine transform of Eq. (3) yields

(10) $$P_L(v) = \int_0^\infty s^{-1} T(s^{-1}) P_E(sv) \, ds,$$

the relation between the Lagrangian and the Eulerian power spectra. Replacing P_E in Eq. (10) by its cosine transform one obtains the Lagrangian power spectrum expressed by the Eulerian correlation function

$$P_L(v) = (2/\pi) \int_0^\infty M(vt) R_E(t)\, dt \tag{11}$$

with

$$M(r) = \int_0^\infty s^{-1} T(s^{-1}) \cos(rs)\, ds \tag{12}$$

and

$$M(0) = \int_0^\infty s^{-1} T(s^{-1})\, ds = I_L/I_E \tag{13}$$

the ratio of the Lagrangian and the Eulerian integral scale is easy to deduce from Eqs. (10) and (12).

The transfer functions L and M are connected by the self-reciprocal divisor transform (see Oberhettinger, 1972, for pairs of transformation)

$$M(r) = \int_0^\infty L(s)[(2/\pi) K_0(2(rs)^{1/2}) - Y_0(2(rs)^{1/2})]\, ds \tag{14}$$

with the modified and the ordinary Bessel function of the second kind and zero order. Of course $\cos(s)$ is one of the eigenfunctions of Eq. (14).

Starting from the inversion of Eq. (3) with the transfer function T^* as the resolvent of T, Eqs. (6)–(14) are all followed with L^* as the resolvent of M and M^* the resolvent of L.

The relations between the transfer functions T, L, M, and their resolvents are simple only in implicit forms:

$$\int_0^\infty s^{-1} T^*(s) T(rs^{-1})\, ds = (2/\pi) \int_0^\infty L^*(s) M(rs)\, ds \tag{15}$$

$$= \delta(r - 1)$$

with the Dirac delta function, and

$$\int_0^\infty T^*(s) L(rs)\, ds = \int_0^\infty s^{-1} T^*(s) M(rs^{-1})\, ds \tag{16}$$

$$= \cos(r).$$

By the transfer function T there are further given the relations between the exchange coefficient A and the variance σ^2 (both defined in the Lagrangian system) on the one side and the corresponding functions derived formally from the Eulerian correlation on the other side by integration of Eq. (3):

(17) $$A_L(t) = \int_0^\infty s^{-1} T(s) A_E(st)\, ds$$

and

(18) $$\sigma_L^2(t) = \int_0^\infty s^{-2} T(s) \sigma_E^2(st)\, ds.$$

From the known asymptotic expression for the variance

(19) $$\sigma_L^2(t) \Rightarrow I_L(t - t_{0L})$$

with constant I_L and t_{0L} and from Eqs. (10) and (18) it follows at last

(20) $$\int_0^\infty s^{-2} T(s)\, ds = (I_L t_{0L})/(I_E t_{0E}).$$

Together with Eq. (13) the asymptotic behavior of σ^2 can be expressed very simply by Eulerian quantities

(21) $$\sigma_L^2(t) \Rightarrow I_E \left[t \int_0^\infty s^{-1} T(s)\, ds - t_{0E} \int_0^\infty s^{-2} T(s)\, ds \right].$$

Finally, there are two matters to be mentioned. First the kernels K in Eqs. (1), (2), and (4) and therefore the transfer functions also contain two further parameters indeterminable by means of the model described above. If and only if both parameters were proved to be constant or almost constant, the transfer functions would not depend on the structure of turbulence but only on the turbulence intensity. This matter remains to be decided by experimental research.

Second, neither the Lagrangian nor the Eulerian correlation contains complete information on turbulence but each a certain part only, and these parts will never be the same. Only the part common to both can be comprised by a relation between the correlations. Therefore such a relation has to be regarded as an estimation and not as a function. And due to the statistical nature of the relation the two parameters determined by experiments will vary within a certain range in any case.

References

Franz, H. (1970). *Beitr. Phys. Freien Atmos.* **43**, 93–118.
Franz, H. (1973). *Proc. Phys. Processes Responsible Dispersal Pollutants Sea, Aarhus, 1972*
Oberhettinger, F. (1972). "Table of Bessel Transforms," Springer-Verlag, Berlin and New York.

COMPUTATIONAL MODELING OF TURBULENT TRANSPORT[1]

JOHN L. LUMLEY AND BEJAN KHAJEH-NOURI

The Pennsylvania State University, University Park, Pennsylvania 16802, U.S.A.

1. INTRODUCTION

Inexpensive semiquantitative numerical simulation of turbulent transport would have many applications. In particular, when applied to pollution dispersal in an urban atmospheric environment, or in an estuary, it would permit rational decision making by the government bodies involved. There are, in addition, problems in oceanography and meteorology (such as thermocline formation) on which a reasonably realistic, though semiempirical, computational technique could shed light.

Before attempting to construct such a method, we should consider two basic, related questions. Is the method within our grasp conceptually and computationally? That is, do we understand turbulence well enough to model it with acceptable accuracy, and can the model provide simulation at acceptable cost, with available computational facilities? These are very real questions for, although we understand a great deal about turbulence, there is even more that we do not know; also, even techniques that do not exceed the capacity of the most recent generation of computers can easily exceed the ability (or willingness) of reasonable men to pay, if computer time must be accounted for.

A great deal of success has been had with direct numerical simulation of turbulence (Orszag and Patterson, 1972). This involves no dynamical modeling. However, Fox and Lilly (1972) have shown that such an approach rapidly recedes beyond economic reach as the Reynolds number increases.

---

[1] This work supported in part by the Environmental Protection Agency, through its Select Research Group in Meteorology, and in part by the Atmospheric Sciences Section of the National Science Foundation under Grant No. GA-35422X. Prepared for presentation at the Second IUGG-IUTAM Symposium on Atmospheric Diffusion and Environmental Pollution, Charlottesville, Virginia, 1973; a preliminary version was presented by proxy at the IAHR-AIRH International Symposium on Stratified Flows, Novosibirsk, 1972.

The next most successful approach is direct simulation with so-called subgrid scale modeling (e.g., Deardorff, 1974). This recognizes that it is economically impossible to carry in the computation the smallest scales at high Reynolds numbers; the grid scale is made as small as is economically feasible, and the motion on scales below the grid scale is modeled, making use of the well-supported property of turbulence dynamics (Tennekes and Lumley, 1972) that the precise nature of the dissipative mechanism does not influence the large scales of the motion, if the Reynolds number is high enough.

The success of this method depends not only on having a sufficiently high Reynolds number to have an inertial subrange (so that energy-containing and dissipative scales are, to a first approximation, dynamically related only by the value of the spectral energy flux) but also on being able to use a grid scale that is small enough to lie in the inertial subrange. It suffers from the disadvantage that one calculation is not sufficient: each calculation is a realization in an ensemble, and a sufficient number of independent runs must be made to obtain stable statistics; for example, of the order of 200 runs to obtain 10% accuracy in second order quantities. To date, this technique has been used primarily for flows having a homogeneous direction; in such flows, the number of runs necessary to obtain stable statistics can be substantially reduced by spacial averaging in the homogeneous direction.

If one is attempting to model the flow in a fully three-dimensional region such as an urban environment, two facts rapidly become clear: the number of points required to make even a crude model of the region precludes the use of a grid scale lying in the inertial subrange, and a single calculation is so expensive as to preclude the possibility of doing statistics on an ensemble of them (since there is no homogeneous direction for averaging). We must then use a grid scale lying in the energy-containing range and compute only the statistical properties of the turbulence. The so-called subgrid scale motions now contain virtually all the turbulence, and the dynamical modeling becomes much more critical. In fact, since the entire influence of the turbulence is being computed through the moments, it is no longer correct to think of turbulence quantities as being subgrid scale; for example, the scale of the turbulence is not now related to the grid scale, but must be obtained from dynamical considerations.

No good direct model of second order turbulence quantities exists. The only practical model is the so-called eddy diffusivity, or "K-theory" model, which has been used with some success in simple situations (Tennekes and Lumley, 1972) to predict first order quantities. This model is known to fail, however, in situations which are rapidly changing in space or time. A number of authors, realizing that a good model is desirable but that good second order models are not available, have decided to carry the equations for second moments exactly, and model third order terms (Donaldson, 1972a,b; Mellor, 1974; Daly and Harlow, 1970; Jones and Launder, 1972; Ng and Spalding, 1972). There is some justification for this approach (which we will follow); as we shall show later, while it is not possible to construct a rational model at second order, it is at third order. However, the model constructed still rests on a fallacy: kinetic theory concepts are embodied, implying that length and time scales of the transporting mechanism (the turbulence) are small relative to length and time scales of the mean motion. This is, of course, known not to be the case for turbulence. There is thus an article of faith involved: if a crude assumption for second moments predicts first moments adequately, perhaps a crude assumption for third moments will predict second moments adequately.

Some of these third order closure schemes are incomplete in the sense that they provide no prediction for one of the scales (which may be taken to be equivalent to a length scale) (Donaldson, 1972a,b; Mellor, 1974). Others (Daly and Harlow, 1970; Ng and Spalding, 1972; Jones and Launder, 1972) do provide a supplementary equation equivalent to one for a length scale. All of them, however, suffer from a basic flaw: they do not present any method for generating the models used for the third order terms. Since the models are constructed on an ad hoc basis, usually being required to have only the same general tensor character as the terms modeled, models are occasionally constructed that behave incorrectly with Reynolds number (Corrsin, 1972), or as we shall see, the right term is included for the wrong reason, or important terms are omitted.

We will present here two related techniques which make it possible to generate, in a consistent and straightforward manner, models of all orders of the third moments, and of all order in Reynolds number. The technique is equally applicable to stratification, to pollution dispersal, to chemical reactions, etc. Many of the terms generated are essentially those suggested by other authors on an ad hoc basis. However, in the case of the third order transport terms, we will find that it is inconsistent within the model not to allow the flux of one second order quantity to be produced by gradients of *all* the others, much as a molecular flux of salt can be produced in a liquid by a temperature gradient, and vice versa. This opens the possibility of up-gradient diffusion, an important process in atmospheric modeling. Unlike the situation in kinetic theory, where the cross-diffusion coefficients are ordinarily small, the turbulent cross-diffusion coefficients may be substantial. However, in the (artificial) situation of constant eddy viscosity and constant structure, the forms obtained reduce to the classical forms assumed by other authors on an ad hoc basis.

It is possible that we will conclude ultimately that these third order closures, though within our reach computationally, are, for some purposes, inadequate models of the second moments (although preliminary results, as we shall see later, appear quite favorable). It should not be necessary to point out, however, that we cannot reach a rational conclusion on this question unless we are sure that the closure used is based on a small number of explicitly stated, readily grasped principles, and that all terms, and only those terms, generated by these principles are used. Otherwise, we will not know if an unsatisfactory result can be attributed to the omission of a vital term, the inclusion of an extraneous one, or the use of an incorrect basic principle. The model that we will present, in common with the other third order closures, contains many undetermined constants. It has long been part of the folk wisdom in turbulence that a model can be made to fit a flow, given sufficiently many constants.[2] While there is some justice to this, it is not quite true. A model may be incapable of reproducing a certain qualitative behavior, regardless of the values assigned to the constants. Models with many constants can still be intellectually satisfactory, so long as the constants are not optimized for each flow, or group of flows, and so long as the physical interpretation of the constant is clear. It is important that the values of the constants governing each physically distinct effect be determined by computation in a situation in which that effect is not influenced by others. Otherwise, one is in danger of adjusting the wrong constant for the right reason. As an aside, if this principle is applied conscientiously, it quickly becomes clear that despite the wealth of experimental data collected over the past few decades, there is a remarkable dearth of well-documented *elementary* turbulent flows, in which one effect at a time is carefully studied.

2. THE HIGH REYNOLDS NUMBER APPROXIMATION AND THE DISSIPATION EQUATIONS

In Tennekes and Lumley (1972), it is shown how orders of magnitude may be assigned to various correlations appearing in the dynamical equations. Roughly, instantaneous quantities appearing in the correlations are of two types, belonging either to the energy containing range of eddies or to the dissipation range. The former has characteristic frequency u'/l (where $\bar{\varepsilon} = u'^3/l$, $3u'^2$ is twice the mean fluctuating energy q^2, and $\bar{\varepsilon}$ is the mean dissipation of energy per unit mass), while the latter has characteristic frequency u'/λ, where $\lambda \sim 4lR_l^{-1/2}$, $R_l = u'l/\nu$. The correlation coefficient between two quantities from the same range may usually be taken as unity, but the coefficient between two quantities, each from a different range, is of the order of the time scale ratio $\lambda/l \sim 4R_l^{-1/2}$.

---

[2] Bradshaw is credited with the remark, at the Stanford Conference on Computation of Turbulent Boundary Layers, that with six constants he could create an elephant.

In addition, of course, we may make use of the more familiar fact that derivatives which are external to correlations correspond to scales in the energy containing range, while derivatives within the correlation correspond to dissipation scales. We wish to apply this sort of reasoning to every term appearing in the equations, but particularly to the equations for the dissipation of energy (and of temperature or concentration variance); applied to these equations, it is particularly productive because the dynamics of these quantities is dominated by the small scales, and interacts only weakly with the energy containing eddies. Proceeding in this way, the equation for the mean dissipation of energy may be reduced [as is done with the (equivalent) vorticity equation in Tennekes and Lumley, 1972] to the form

$$\dot{\bar{\varepsilon}} + \bar{\varepsilon}_{,j} U_j + \overline{(\varepsilon u_j)}_{,j} = -2\nu \overline{u_{i,\kappa} u_{i,j} u_{j,\kappa}} - 2\nu^2 \overline{u_{i,\kappa j} u_{i,\kappa j}} \qquad (1)$$

As is discussed in Tennekes and Lumley (1972), the two terms on the right are of order one but differ by order $R_l^{-1/2}$. The remaining terms are of order $R_l^{-1/2}$. Other terms (many of which appear in the full equations) are of higher order. The first term on the right represents the production of velocity gradients by stretching by fluctuating strain rate, while the second represents the destruction of these gradients by viscosity.

Several authors (Daly and Harlow, 1970; Jones and Launder, 1972; Reynolds, 1970; Ng and Spalding, 1972) have retained the terms on the left-hand side

$$2 \overline{u_{i,\kappa} u_{i,j}} U_{j,\kappa} \qquad (2)$$

and another term of similar form, correctly feeling that there must be some source of dissipation. However, these terms are of (relative order R_l^{-1} since (as is shown in Lumley, 1970))

$$\nu \overline{u_{i,\kappa} u_{i,j}} = \varepsilon(\delta_{\kappa j}/3 + O(S_{\kappa j} \lambda/u')) \qquad (3)$$

where $S_{\kappa j}$ is the mean strain rate; since U_i is incompressible, only the second term contributes. Since a term like (3), namely $\nu \overline{u_{i,\kappa} u_{j,\kappa}}$ appears in the Reynolds stress equation, we should mention here that this term also has the same behavior as (3) (cf. Corrsin, 1972). It has been modeled by some authors (Daly and Harlow, 1970; Donaldson, 1972a,b) as proportional to $\overline{u_i u_j}$, whereas the ratio of off-diagonal to diagonal terms must vanish as $R_l^{-1/2}$, as shown by (3).

The proper source of the production of dissipation is in the first term in the right-hand side of (1). In the following section, we will apply a formal procedure to obtain an unambiguous expression for the right-hand side of (1), and we will find that we obtain a term of form similar to that retained by the authors mentioned in connection with (2), so that in a sense they have been using the right term for the wrong reason. Before going to the formal procedure, however, it will be instructive to carry out a physical analysis of the right-hand side of (1), to see how a production term can be retained at infinite Reynolds number.

The right-hand side of (1) represents a balance between stretching and dissipation, and it must be possible for the (relatively small) mismatch to be of either sign. That is, consider the stretching of a single vortex to equilibrium: if the stretching is suddenly increased, momentarily the first term will dominate the second, setting up more vorticity until equilibrium is again attained, at a higher level; if the stretching is reduced, the process is reversed. Put in statistical terms, if the spectral energy flux increases, the first term should dominate the second until the dissipation has been increased to match the flux, and vice versa. There is, in addition an unsteady effect; in a fluctuating turbulent velocity field, equilibrium is never attained since the strain rate changes before it can be achieved. Hence, there is always a fluctuating mismatch; although to first order, we would expect this to average to zero, we would expect nonlinear effects to produce a small net loss.

The response to these effects should depend on the time scale ratio. The time scale of the dissipative eddies is $(v/\bar{\varepsilon})^{1/2}$; if the turbulence were isotropic and decaying, a time scale descriptive of the unsteady stretching would be $q^2/2\bar{\varepsilon}$. If there is an input to the spectral flux, there will be another time scale associated with this input. In a homogeneous flow, the input is characterized by P, the production, and the time scale will be $q^2/2P$. In an inhomogeneous situation, it is more difficult to find a simple way of characterizing the input, since part of P is transported. In the next section we will obtain by formal means an appropriate expression in the inhomogeneous situation; here we will retain P, which may be thought of as an approximation for small inhomogeneity. This is effectively what was done by Daly and Harlow (1970), Jones and Launder (1972) and Ng and Spalding (1972). The inverse of the time scale thus may be written as $(2\bar{\varepsilon}/q^2)F(P/\bar{\varepsilon})$, where F is an unknown function. If the production (and hence the anisotropy) is small, we may expand to obtain $(2\bar{\varepsilon}/q^2)(1 - aP/\bar{\varepsilon})$. It is, of course, not legitimate to use such an expression for values of $P/\bar{\varepsilon} \sim O(1)$; the expression will serve at least to predict qualitative behavior, however, since it reverses sign as we have reasoned it must.

The difference on the right-hand side of (1) might consequently be modeled as

(4) $\qquad -\varepsilon(\bar{\varepsilon}/v)^{1/2}\{0 + b(2\bar{\varepsilon}/q^2)[1 - a(p/\bar{\varepsilon})](v/\bar{\varepsilon})^{1/2} + O(1/R_l)\}$

where the 0 symbolizes the equality of the terms at infinite Reynolds number. To second order in time, the direct response to the mean flow distortions will appear through S_{ij}, etc.; these cannot appear to first order because they are of the wrong tensor rank; a scalar term is needed and can be made from S_{ij} only by a quadratic form. Note that the terms are the same order as the others retained in the equations because the time scale $(v/\bar{\varepsilon})^{1/2}$ cancels one of the factors in the magnitude $(\bar{\varepsilon}/v)^{1/2}$; i.e., as the Reynolds number increases, the magnitude of each term grows, but the difference shrinks at the same rate.

In Lumley (1970), a similar analysis was carried out, though less physical and more formal; consequently, although the order of the term obtained there was correct, the fact that it should be reversible was missed. The general conclusion obtained there, regarding the continual growth of the length scale in a homogeneous flow is correct, however, so long as $a \neq 1$, as may be easily verified using (4) in the analysis there.

As in Lumley (1970), one of the coefficients may be identified by reference to homogeneous decay; we find that

(5) $$b = 2$$

Thus, (1) becomes

(6) $$\dot{\bar{\varepsilon}} + \bar{\varepsilon}_{,j} U_j + \overline{(\varepsilon u_j)}_{,j} = -4(\bar{\varepsilon}^2/q^2) + 4a(\bar{\varepsilon}P/q^2)$$

This is essentially the form used by the author referred to above.

The temperature (or contaminant) dissipation equation may be attacked in exactly the same way. Presuming that the Prandtl number is of order one (a very large or very small value can lead to the retention or discard of different terms) we obtain

(7) $$\dot{\bar{\varepsilon}}_\theta + \bar{\varepsilon}_{\theta,j} U_j + \overline{(\varepsilon_\theta u_j)}_{,j} = -2\kappa \overline{\theta_{,j}\theta_{,i} u_{i,j}} - 2\kappa^2 \overline{\theta_{,ij}\theta_{,ij}}$$

The interpretation of the terms is exactly the same, and the dynamical reasoning is the same. The time scale characterizing the small scales (again presuming the Prandtl number to be of order unity) is $(v/\bar{\varepsilon})^{1/2}$; the time scale characterizing the fluctuating stretching is $q^2/2\bar{\varepsilon}$, and that characterizing the input to the spectral flux is $q^2/2P$ (again, we will obtain a better expression than P for the input to the spectral flux later). Hence, the form is very similar to (4), and we obtain

(8) $$-\varepsilon_\theta(\bar{\varepsilon}/v)^{1/2}\left\{0 + c\frac{2\bar{\varepsilon}}{q^2}[1 - d(P/\bar{\varepsilon})](v/\bar{\varepsilon})^{1/2} + O(1/R_l)\right\}$$

giving

(9) $$\dot{\bar{\varepsilon}}_\theta + \bar{\varepsilon}_{\theta,j} U_j + \overline{(\varepsilon_\theta u_j)}_{,j} = -5(\overline{\varepsilon\varepsilon_\theta}/q^2) + 5d(\overline{\varepsilon_\theta}/q^2)P$$

Again, the relation between the coefficients is obtained by reference to homogeneous decay data, specifically Gibson and Schwarz (1963). In the following section, we will obtain by formal methods improved approximations to (6) and (9).

3. Modeling the Third Moments

If we consistently apply the Reynolds/Peclet number order of magnitude analysis given here, we obtain the set of equations given below. [In addition to (1) and (7), we are considering here only velocity and temperature; the treatment of a passive contaminant, or of an active contaminant other than temperature, can be handled in exactly the same way. We write the equations in the Boussinesq approximation—Phillips, 1966.]

(10) $\dot{U}_i + U_{i,j}U_j + (\overline{u_i u_j})_{,j} = -p_{,i}/\rho_0 + \delta_{3i}g\Theta/T_0$, $\quad U_{i,i} = 0$

(11) $\dot{\Theta} + \Theta_{,i}U_i + (\overline{\theta u_i})_{,i} = 0$

(12) $\dot{\overline{u_i u_\kappa}} + U_{i,j}\overline{u_j u_\kappa} + U_{\kappa,j}\overline{u_j u_i} + (\overline{u_i u_\kappa})_{,j}U_j + (\overline{u_i u_\kappa u_j})_{,j}$
$\qquad = -(\overline{u_\kappa p_{,i}} + \overline{u_i p_{,\kappa}})/\rho_0 + (\overline{u_\kappa \theta}\delta_{3i} + \overline{u_i \theta}\delta_{3\kappa})g/T_0 + -2\bar{\varepsilon}\delta_{i\kappa}/3$

(13) $\dot{\overline{\theta^2}} + 2\Theta_{,j}\overline{u_j\theta} + \overline{\theta^2}_{,j}U_j + (\overline{\theta^2 u_j})_{,j} = -2\varepsilon_\theta$

(14) $\dot{\overline{\theta u_i}} + U_{i,j}\overline{\theta u_j} + \Theta_{,j}\overline{u_i u_j} + (\overline{\theta u_i})_{,j}U_j + (\overline{\theta u_i u_j})_{,j}$
$\qquad = -\overline{\theta p_{,i}}/\rho_0 + \delta_{3i}\overline{\theta^2}g/T_0$

where Θ and θ are, respectively, the mean and fluctuating parts of the temperature. Equations (10)–(14), plus (1) and (7) constitute the set we will consider.

The averages are to be understood as ensemble averages, so that the equations can accommodate evolution of the turbulent field, or changing mean or boundary conditions. In addition, such phenomena as internal waves can be accommodated so long as the period is long compared to any characteristic time of the turbulence, so that there is no direct coupling. We have neglected $\kappa\Theta_{,jj}$, $\nu U_{i,jj}$, $\nu(\overline{u_i u_\kappa})_{,jj}$, $\kappa\overline{\theta^2}_{,jj}$, $\nu(\overline{\theta u_{i,j}})_{,j}$ and $\kappa(\overline{u_i \theta_{,j}})_{,j}$ which are of order R_l^{-1} in their respective equations, and $\nu\overline{\theta_{,j}u_{i,j}}$ and $\kappa\overline{u_{i,j}\theta_{,j}}$ which are of order $R_l^{-1/2}$. In the atmosphere, typically $R_l^{1/2} \sim 10^3$, so that the neglect of these terms may be expected to be an excellent approximation. We are also neglecting off-diagonal components of the last term in (12), in accord with (3).

Let us consider first the term

(15) $\qquad -(\overline{u_\kappa p_{,i}} + \overline{u_i p_{,\kappa}})/\rho_0$

in Eq. (12). Part of this term is a transport term; let us subtract the trace, and consider

(16) $\qquad -(\overline{u_\kappa p_{,i}} + \overline{u_i p_{,\kappa}})/\rho_0 + 2(\overline{u_j p})_{,j}\delta_{i\kappa}/3\rho_0 = A_{i\kappa}$

say.

Now, from Eqs. (10)–(12), if we knew the distributions of $\overline{u_i u_j}$, $\overline{\theta u_i}$ and the value of g/T_0 for all time and space, we would know[3] U_i and Θ; having U_i and Θ, together with $\overline{u_i u_j}$, $\overline{u_i \theta g/T_0}$, and $\bar{\varepsilon}$, Eq. (12) would give us the transport term plus $A_{i\kappa}$. Thus it must be possible to write the sum of $A_{i\kappa}$ and the transport term as a functional of $\overline{u_i u_j}$, $\overline{\theta u_i}$, $\bar{\varepsilon}$, g/T_0 (the latter occurring with and without $\overline{\theta u_i}$). The trace of the transport term itself may be so written [by applying the same reasoning to the trace of Eq. (12), since $A_{ii} = 0$]. It is surely a small step to assume that the *whole* transport term may be so written, and hence that $A_{i\kappa}$ may be so written:

$$(17) \qquad A_{i\kappa} = \mathscr{F}_{i\kappa}\{\overline{u_i u_\kappa}, \overline{\theta u_i}, \bar{\varepsilon}, g/T_0\}$$

where the functional extends over all space, and all earlier time. We have not included the direction of the gravity vector since this is part of the structure of the equations; information about the direction of gravity must be forced to appear in $\overline{u_i u_j}$ and $\overline{\theta u_i}$.

This far, no approximation is involved [except the assumption that either of the third order terms in (12) may be written as (17) if both may be]. Now, we wish to introduce the approximation of *weak anisotropy* (which implies *weak inhomogeneity*) and quasi-steadiness. Both these concepts are kinetic theory concepts: that time and length scales of inhomogeneity are large relative to time and length scales of the turbulence, and that the turbulence is nearly in equilibrium (isotropic). These are known to be poor descriptors of turbulence, equivalent to gradient transport concepts. However, we are applying them here to third-order quantities, handling the second-order ones exactly; it is hoped that the predictions will be less sensitive to assumptions made at this level. The assumption at least provides an exact model of a physically realizable process (something like a very rarified gas), obtainable from turbulence in a conceptually (though not physically) possible way (by letting the turbulent length and time scales become short), so that one may hope that the predictions will not be unphysical (i.e., producing negative energy, etc.) and should bear some qualitative resemblance to turbulence.

To implement the approximation of weak anisotropy, we will expand the right-hand side of (17) in a functional power series (assuming fading memory and limited awareness, as in Lumley, 1967); in keeping with quasi-steadiness, we will neglect time derivatives. Carrying out the expansion requires the following steps: express $\overline{u_i u_j}$ as $\overline{u_i u_j} - q^2 \delta_{ij}/3 = a_{ij}$, say, and q^2, writing $\overline{\theta u_i} = b_i$. The functional (17) is now a function of a symmetric tensor

---

[3] The question of when and where we may expect to find a unique functional relation (a "constitutive relation") between the turbulent fluxes and the mean field gradients is extensively discussed in Lumley (1970); the answer generally is, away from the boundaries in space and time. We can expect to determine in this way, of course, only mean velocity *gradients*, since the equations are invariant under rigid translation.

a_{ij}, a vector b_i, and three scalars, $\bar{\varepsilon}$, q^2, and g/T_0. In the isotropic limit a_{ij}, b_i, and g/T_0 all vanish, as do all gradients. Thus $a_{ij,\kappa}$ would be a second order term, and $q^2_{,i}$ a first order term. First, form the functional Taylor series in the gradients of the arguments. Second, the tensor coefficients are now functions of the local values of a_{ij}, b_i, etc.; express them in invariant form, arranging them according to order (note that the invariants of a_{ij} are of second and third order, etc.). Expand in powers of g/T_0; finally, apply dimensional analysis to the coefficients.

The net result consists of terms of two types: those that would be present in a homogeneous flow, and corrections for inhomogeneity. Through third order, the homogeneous terms are

$$(18) \qquad A_{ij} = -\{(1 + B_1 \mathrm{II}/q)a_{ij} + B_2(a^2_{ij} - \mathrm{II}\delta_{ij}/3)/q^2\}/\mathcal{T}$$

where $a^2_{ij} = a_{i\kappa}a_{\kappa j}$, and $\mathrm{II} = a_{ij}a_{ji}$, the second invariant. $\mathcal{T} = cq^2/\bar{\varepsilon}$, where c may be evaluated from the initial rate of return to isotropy (see, e.g., Tucker and Reynolds, 1968). Later, we will identify $\mathcal{T}$ as the Lagrangian integral time scale, which permits approximate evaluation of $c \sim \tfrac{1}{8}$.

The lowest order term $a_{ij}/\mathcal{T}$ may be identified as that suggested by Rotta (1951). The form (18) is consistent with the observations of Champagne et al. (1970) that the principal axes of $A_{i\kappa}$ and $a_{i\kappa}$ were the same, in a homogeneous flow. Initially, it had been hoped that first order terms would provide an adequate model; however, the second and third order terms prove to be necessary, at least in the wake. The third order term speeds the return to isotropy for large anisotropy, while the second provides some redistribution in the presence of shear. In the wake, the peak of $\overline{w^2}$ is directly attributable to the second order term only, while the peak in $\overline{u^2}$ cannot be reduced to reasonable proportions without the third order term.

Through third order, the terms involving first derivatives are buoyancy corrections:

$$(19) \qquad \beta_1(g/T_0)(b_i\delta_{j\kappa} + b_\kappa \delta_{ij} - b_j\delta_{i\kappa}2/3)q^2_{,j}/q^2$$

$$\beta_2(g/T_0)(b_i\delta_{j\kappa} + b_\kappa \delta_{ij} - b_j\delta_{i\kappa}2/3)\bar{\varepsilon}_{,j}/\bar{\varepsilon}$$

where use has been made of the fact that $A_{i\kappa} = A_{\kappa i}$, $A_{ii} = 0$. There are also tensorially appropriate terms in $q^2_{,ij}$, $\bar{\varepsilon}_{,ij}$, $q^2_{,i}\bar{\varepsilon}_{,j}$, $q^2_{,i}q^2_{,j}$, $\bar{\varepsilon}_{,i}\bar{\varepsilon}_{,j}$, and $gb_{i,j}/T_0$, the coefficients of the first five being first order functions of $\bar{\varepsilon}$, q^2, and a_{ij}; the sixth term, being a third order buoyancy correction, has a coefficient which is a function only of $\bar{\varepsilon}$, q^2. Finally, there are third order terms in $a_{i\kappa,jl}$, $a_{i\kappa,j}q^2_{,l}$, and $a_{i\kappa,j}\bar{\varepsilon}_{,l}$, with coefficients in $\bar{\varepsilon}$, q^2. It is not expected that all these terms will be equally important. In the (isothermal) wake, we found that only the terms in $q^2_{,ij}$, $\bar{\varepsilon}_{,ij}$, and $a_{i\kappa,jl}$ were dynamically important.

Note that the type of second order terms suggested by Rotta (1951) and

Reynolds (1970) and others do not appear; our formalism excludes *explicit* dependence on the mean velocity profile, and forces implicit dependence through second order quantities.[4]

Now, we may apply the same reasoning to the term in (14). Examining the equations, we find that

(20) $\qquad -\overline{\theta p_{,i}}/\rho_0 = \mathscr{F}_i\{\overline{\theta u_i}, \overline{u_i u_j}, g\overline{\theta^2}/T_0, g/T_0, \bar{\varepsilon}\}$

The dissipation $\bar{\varepsilon}$ does not appear directly, since (assuming high Reynolds and Peclet numbers) this equation has no dissipation terms. However, in the homogeneous case, $\overline{\theta u_i}$ and $\overline{u_i u_j}$ (and g/T_0) are not sufficient to determine U_i and Θ; some other quantity must be included to fix a length scale for the turbulence. This is reflected in the group in (19) being dimensionally incomplete without $\bar{\varepsilon}$ (in the homogeneous case); that is, a nontrivial form can be found only at an order higher than the first. We will include $\bar{\varepsilon}$, noting that including redundant quantities can do no harm.

Applying our expansion procedure, we find for the homogeneous part (to third order)

(21) $\qquad -\overline{\theta p_{,i}}/\rho_0 = -\{(17/16 + a_1 \mathrm{II}/q^4 + a_2 \mathscr{T}^2 g^2 \overline{\theta^2}/q^2 T_0^2)\delta_{ij}$
$\qquad \qquad + a_3 a_{ij}/q^2 + a_4 a_{ij}^2/q^4\} b_j/\mathscr{T}$

where $\mathscr{T}$ is the same as in (18), and the coefficient has been evaluated by reference to homogeneous decay data again. That is, the criterion was applied that, in the limit of small anisotropy, i.e., late in the decay, $\overline{\theta u_i}(q^2/\overline{\theta^2})^{1/2}$ should decay at the same rate as a_{ij}. The lowest order part of (21), $b_i/\mathscr{T}$, has been suggested on an ad hoc basis by Donaldson (1972a,b).

The inhomogeneous terms may easily be generated; through second order they are (with numerical coefficients)

(22) $\qquad g\overline{\theta^2}_{,i}/T_0; \qquad (q\mathscr{T}g\overline{\theta^2}/T_0)q_{,i}^2; \qquad (\mathscr{T}^2/q)(g\overline{\theta^2}/T_0)\bar{\varepsilon}_{,i}$

These are all buoyancy corrections; third order terms bring in

(23) $\qquad b_{i,j\kappa}; \qquad b_{i,j}\bar{\varepsilon}_{,\kappa}; \qquad b_{i,j}q_{,\kappa}^2; \qquad g\overline{\theta^2}_{,i}\bar{\varepsilon}_{,j}/T_0; \qquad g\overline{\theta^2}_{,i}q_{,j}^2/T_0$

plus anisotropy corrections to the first and second order terms. It will be interesting to see by calculation whether it is necessary to retain terms of this order.

---

[4] Two situations do require explicit dependence on the mean velocity profile: *rapid* application of mean distortion to an existing flow, when the mean fields and the fluxes are not uniquely related, and higher order corrections for rotation (due to the mean velocity field); i.e., lack of material indifference, as discussed in Lumley (1970). The resulting changes in the form of A_{ik} are discussed in detail in Lumley and Khajeh-Nouri (1974).

4. The Transport Terms

The various transport terms may be attacked in the same way. Adopting the forms (18) and (20), we find

(24) $$\overline{\varepsilon_\theta u_i} = \mathscr{F}_i\{\overline{u_i u_j}, \overline{\theta u_i}, \bar{\varepsilon}, \bar{\varepsilon}_\theta, g/T_0\}$$

Following exactly the same procedure, we obtain to first order

(25) $$\overline{\varepsilon_\theta u_i} = -A_{41}\bar{\varepsilon}_\theta \mathscr{T}(q^2/2)_{,i} - A_{42}\bar{\varepsilon}_\theta \mathscr{T}^2 \bar{\varepsilon}_{,i} - A_{44}(q^2/3)\mathscr{T}\bar{\varepsilon}_{\theta,i}$$
$$+ A_{45}(\bar{\varepsilon}_\theta/\mathscr{T})^{1/2}\overline{\theta u_i}$$

so that gradients of several quantities, and another flux, can produce a flux of $\bar{\varepsilon}_\theta$.

The same reasoning will produce

(26) $$\overline{\theta^2 u_i} = \mathscr{F}_i\{\overline{u_i u_j}, \overline{\theta u_i}, \overline{\theta^2}, \bar{\varepsilon}_\theta, g/T_0\}$$

so that

(27) $$\overline{\theta^2 u_i} = -A_{31}\overline{\theta^2}\mathscr{T}_\theta(q^2/2)_{,i} - A_{33}(q^2/3)\mathscr{T}_\theta\overline{\theta^2}_{,i}$$
$$- A_{34}\mathscr{T}_\theta^2(q/3)\bar{\varepsilon}_{\theta,i} + A_{35}\overline{\theta^2}^{1/2}\overline{\theta u_i}$$

where $\mathscr{T}_\theta = \overline{\theta^2}/8\bar{\varepsilon}_\theta$.

The number of additional constants introduced in (25) and (27) is somewhat startling. It is encouraging that we find a simpler result when we examine $\overline{\theta u_i u_j}$. Thus

(28) $$\overline{\theta u_i u_j} = \mathscr{F}_{ij}\{\overline{\theta u_i}, \overline{u_i u_j}, \overline{\theta^2}g/T_0, \bar{\varepsilon}, g/T_0\}$$

which leads (to first order) to

(29) $$\overline{\theta u_i u_j} = A_{53}\mathscr{T}q\overline{\theta^2}\,\delta_{ij}g/T_0$$

Whether or not it is necessary to carry second order terms in any particular case is a point that can be settled in general only by computation. We would normally expect to need only first order terms in the fluxes, since taking the divergence raises the order by one. One situation, however, provides a clear need for second order terms: if the first order terms vanish identically in a particular physical problem. In (28), if gravity is not important, the transport of thermal flux vanishes. Since this is surely not true, second order terms are necessary to provide an adequate model when stratification is weak:

(30) $$\overline{\theta u_i u_j} = \delta_{ij}(A_{53}\mathcal{T}q\overline{\theta^2}g/T_0 + A_{54}q^2\mathcal{T}\overline{(\theta u_p)}_{,p})$$
$$+ A_{55}\mathcal{T}q(3\overline{u_i u_j}/q^2 - \delta_{ij})\overline{\theta^2}g/T_0 + A_{56}q^2\mathcal{T}[\overline{(\theta u_i)}_{,j} + \overline{(\theta u_j)}_{,i}]$$

Only the terms in A_{54} and A_{56} will remain if gravitational effects are relatively weak. The term in A_{55} simply corrects the term in A_{53} for anisotropy. We may give a simple physical interpretation to the terms in A_{53} and A_{55}: $\theta g/T_0$ is the buoyant acceleration, and $\mathcal{T}q$ the turbulent length scale; hence, $\mathcal{T}q\theta g/T_0$ is that part of the turbulent energy that is correlated with the temperature fluctuation. The terms in A_{54} and A_{56} are simply gradient transport.

The purely mechanical transport terms are also relatively simple. For the mechanical dissipation we have

(31) $$\overline{\varepsilon u_i} = \mathcal{F}_i\{\overline{u_i u_j}, \overline{\theta u_i}, \bar{\varepsilon}, g/T_0\}$$

leading to

(32) $$\overline{\varepsilon u_i} = -A_{21}(q^2/3)(q^2/2)_{,i} - A_{22}\mathcal{T}(q^2/3)\bar{\varepsilon}_{,i}$$

In an exactly similar way, we find

(33) $$\overline{(u_i u_\kappa + 2\delta_{i\kappa} p/3\rho_0)u_j} = \mathcal{F}_{i\kappa j}\{\overline{u_i u_j}, \overline{u_i \theta}, \bar{\varepsilon}, g/T_0\}$$

which gives (to first order)

(34) $$\overline{(u_i u_\kappa + 2\delta_{i\kappa} p/3\rho_0)u_j} = -A_{11}\mathcal{T}(q^2/3)[\delta_{i\kappa}\delta_{jl} + a(\delta_{ij}\delta_{\kappa l} + \delta_{il}\delta_{\kappa j})]q^2_{,l}$$
$$- A_{12}\mathcal{T}^2(q^2/3)[\delta_{i\kappa}\delta_{jl} + b(\delta_{ij}\delta_{\kappa l} + \delta_{il}\delta_{\kappa j})]\bar{\varepsilon}_{,l}$$

where a and b are constants. In each case we have tried to choose signs and numerical coefficients which will result in the constants A_{pq} being positive and of order one.

Just as we found for the transport of thermal flux, there is a situation in which the first order transport of Reynolds stress vanishes, and second order terms are consequently necessary. Specifically, in a simple shear $U_1(x_2)$, in which only two gradients are nonzero, $\overline{u_1 u_2 u_2} = 0$ from (33). Since this is a dynamically important quantity, we will need second order terms (in the wake calculations to be described later, the omission of this transport results in undesirable, and unrealistic, hyperbolic behavior, with steepening fronts, etc.). The second order terms are

(35)
$$\mathcal{T}(q^2/3)[\alpha_1 a_{i\kappa,j} + \alpha_2(a_{ij,\kappa} + a_{\kappa j,i}) + \alpha_3 \delta_{i\kappa} a_{jp,p} + \alpha_4(\delta_{ij} a_{\kappa p,p} + \delta_{\kappa j} a_{ip,p})]$$
$$+ \mathcal{T}[\beta_1 \delta_{i\kappa} a_{jl} + \beta_2 a_{i\kappa}\delta_{jl} + \beta_3(\delta_{ij}a_{\kappa l} + \delta_{\kappa j}a_{il}) + \beta_4(\delta_{il}a_{\kappa j} + \delta_{\kappa l}a_{ij})]q^2_{,l}$$
$$+ \mathcal{T}^2[\gamma_1 \delta_{i\kappa} a_{jl} + \gamma_2 a_{i\kappa}\delta_{jl} + \gamma_3(\delta_{ij}a_{\kappa l} + \delta_{\kappa j}a_{il}) + \gamma_4 \delta_{il}a_{\kappa j}$$
$$+ \delta_{\kappa l}a_{ij})]\bar{\varepsilon}_{,l}$$

5. The Dissipation Equations

Finally, we may apply the same procedure to obtain the forms for the right-hand sides of (1) and (7). The right-hand side of (1) must be a scalar functional of the same variables as (31). Complete to terms of third order, the right-hand side of (1) is

$$\text{RHS}(1) = -4\bar{\varepsilon}^2/q^2 + \bar{\varepsilon}\{c_1 \text{II}/q^4 + c_2 \text{III}/q^6\}/T \tag{36}$$

and for (7) we obtain [with the same dependency as (24)]

$$\begin{aligned}\text{RHS}(7) = &-5\bar{\varepsilon}\bar{\varepsilon}_\theta/q^2 + \bar{\varepsilon}_\theta\{d_1 \text{II}/q^4 + d_2 \text{III}/q^6 \\ &+ d_3 b_i b_i/\bar{\varepsilon}_\theta \mathcal{T} q^2 + d_4 a_{ij} b_i b_j/\bar{\varepsilon}_\theta \mathcal{T} q^4 \\ &+ d_5 g \bar{\varepsilon}_\theta^{1/2} \mathcal{T}^{3/2}/T_0 q + d_6 g \bar{\varepsilon}_\theta^{1/2} \text{II} \mathcal{T}^{3/2}/T_0 q^5 \\ &+ d_7 b_i b_i \mathcal{T}^{1/2} g/\bar{\varepsilon}_\theta^{1/2} q^3 T_0\}\end{aligned} \tag{37}$$

We have accepted the coefficients determined from the homogeneous decay data. III is the third invariant of a_{ij}, $\text{III} = a_{ij} a_{j\kappa} a_{\kappa i}$.

If we look at (36), keeping only second-order terms, we see that what we should have used in (6) and (9) in place of P is $\text{II} = a_{ij} a_{ji}$ (with a suitable coefficient to make the dimensions correct). It is reasonable to associate II with the spectral flux since the inequality of components indicates that straining of the turbulence is taking place. II is something like P; if we made the simplistic K-theory approximation that $a_{ij} \propto S_{ij}$ where S_{ij} is the mean strain rate, then $P \propto \text{II}$. In reality, however, $\text{II} \neq 0$ in regions of most flows where P vanishes. Hence, we will still have production of $\bar{\varepsilon}$ there. The same reasoning applies to (37), where it is also evident that, even to second order, our physical reasoning resulted in the neglect of a number of important terms.

6. Evaluation of Coefficients

The evaluation (or elimination) of most of these coefficients will have to await detailed calculation of flows in which many measurements exist. A few statements can be made, however, by considering simple flows.

First, consider a steady, homogeneous, isotropic turbulence with a linear temperature gradient and no gravity or mean velocity. Then (14) reduces to [using (21)]

$$\theta_{,j} \overline{u_i u_j} 16 \mathcal{T}/17 = -\overline{\theta u_i} \tag{38}$$

This is just the Lagrangian transport form for a passive scalar. Hence, evidently $16\mathcal{T}/17$ may be identified with the Lagrangian integral time scale. This scale has been estimated from first principles (by a very crude technique) as $l/3u'$, and from wake decay data as $l/2.8u'$ (see Tennekes and Lumley, 1972, p. 229), where $\bar{\varepsilon} = u'^3/l$, and $3u'^2 = q^2$. This gives roughly $\mathcal{T} = q^2/8\bar{\varepsilon}$.

If we consider a steady, homogeneous flow with $U_i = (U_{1,3}x_3, 0, 0)$, $U_{1,3} = \text{const}$, and $\theta = \theta_{,3}x_3$, and no gravity, we do obtain "K-theory" forms

(39) $\overline{\theta u_3} \propto -\mathcal{T}\overline{u_3^2}\theta_{,3}\ 16/17;\ \overline{u_1 u_3} \propto -\mathcal{T}\overline{u_3^2}U_{1,3}$

To first order, the coefficients are unity. However, if we keep our higher order terms, the coefficients are complicated functions of $U'\mathcal{T}$ and the anisotropy. Thus, although the ratio K_M/K_H begins at 17/16 for very weak shear, it rapidly changes as the shear becomes more intense (measured in terms of $U'\mathcal{T}$).

If the first order model is applied to the constant stress layer of a neutral turbulent boundary layer, we obtain (using $\mathcal{T} = q^2/8\bar{\varepsilon}$)

(40) $\overline{u_1^2} = q^2/2,\quad \overline{u_2^2} = \overline{u_3^2} = q^2/4,$

$-\overline{u_1 u_2}/(\overline{u_1^2}\ \overline{u_2^2})^{1/2} = \tfrac{1}{2}\overline{u_2^2}/u_*^2 = \overline{u_3^2}/u_*^2 = 2^{1/2},\quad \overline{u_1^2}/u_*^2 = 2^{3/2}$

Including higher order terms precludes algebraic evaluation. In addition, one may obtain a relation between A_{22} and the c_1 of Eq. (36). In a similar way, if a constant heat flux is added to the neutral constant stress layer, a relation may be obtained among the constants A_{42}, A_{44}, and A_{45}.

Further evaluation of the constants will have to await further computations. One general guideline has suggested itself, however, which provides at least an estimate of the magnitudes of some of the constants, and eliminates others: in a region of constant eddy viscosity, and constant structure (which exists only conceptually), the transport terms should reduce to the K-theory forms.[5] That is (considering the purely mechanical case), in such a region, writing $Q_{ij} = \overline{u_i u_j}$,

(41) $\varepsilon \propto q^4,\quad Q_{ij}/q^2 = \text{const}$

which leads to

(42) $2q_{,i}^2/q^2 = \varepsilon_{,i}/\varepsilon,\quad Q_{ij,\kappa} = (Q_{ij}/q^2)q_{,\kappa}^2$

---

[5] Because such a region is characterized by a single length and velocity scale at each point, and only in such a region would K-theory forms be expected to hold (cf. Tennekes and Lumley, 1972).

If, for example, we are considering $q_{,\kappa}^2$, then equally good expressions are

(43) $$q_{,\kappa}^2; \quad (q^2/2\varepsilon)\varepsilon_{,\kappa}; \quad q^2 Q^{ji} Q_{ij,\kappa}/3$$

where Q^{ji} is the inverse of Q_{ij}, $Q^{pi}Q_{ij} = \delta_j^p$. We should thus expect the transport of q^2 to be a form like

(44) $$\mathcal{T} Q_{ij}\{A q_{,j}^2 + B(q^2/2\varepsilon)\varepsilon_{,j} + C q^2 Q^{ji} Q_{ij,\kappa}/3\}$$

where $A + B + C \sim 1$. This expression may be expanded, keeping only second order terms, to eliminate many of the unknown constants.

7. Computation in the Two-Dimensional Isothermal Wake

For a first computation, we should begin with an isothermal, mechanical flow; when the various constants have been evaluated, if the results are satisfactory, we can proceed to a flow with temperature fluctuations, but without stratification, and finally to flows with stratification.

We have selected for our first computation the two-dimensional wake.[6] We feel that a flow without boundaries is more sensitive to the values of the constants, and our calculations have borne out that feeling. In a flow with boundary conditions, interior points can never depart very far from the boundary values. In a flow without boundaries, however, the variables must develop on their own. Other reasons for selecting the wake are the existence of up-gradient energy transport near the center line, which provides a demanding test of the model, the existence of a similarity solution, and the fact that it is one dimensional.

The equations programmed neglect streamwise transport; they are thus the equations describing the lateral development with *time* of a linear wake of constant cross section, created instantaneously. Making use of symmetry, only one-half the wake was programmed.

A modified leap-frog method was used: all terms other than transport terms were evaluated at level κ, and the time difference was centered at level κ. Centered space differences were used, evaluated at level $\kappa - 1$. Since this produces a second order difference equation in time (modeling a first order differential equation), an extra initial condition must be supplied; hence, the first two time steps are set equal. The second solution is oscillatory, and corresponds to amplification of the error (from the true solution) made in

---

[6] Properly speaking, we should begin with a homogeneous flow, like that of Champagne *et al.* (1970), to evaluate the coefficients in the homogeneous forms. When we began the computation, however, we did not realize that second and third order terms would be necessary, and hence were under the impression that the only such first order coefficient (c in $\mathcal{T}$) had been evaluated in Lumley (1970). It will now, of course, be necessary to redo this calculation. (See Lumley and Khajeh-Nouri, 1974.)

this assignment of the value at the second time step. This manifests itself by a gradual separation of the values of the solution at alternate time steps. This is a well understood phenomenon, and we used the classical cure: the solutions at κ and $\kappa + 1$ were averaged and the process restarted every 10 time steps.

This differencing scheme is only conditionally stable, so that a limitation must be placed on the size of the time step. The classical limitation is $\Delta\tau \leq \frac{1}{4}\Delta y^2$, in dimensionless variables in which the diffusion coefficients are unity. In our variables, normalized by the velocity defect on the centerline and the standard deviation of the defect profile, the diffusion coefficient is 12.5 (roughly) so that $\Delta t \leq 3.1\,\Delta y^2$ is the appropriate restriction. We used $\Delta t = 3\,\Delta y^2$, and $\Delta y = 0.1$.

Because the wake width grows continually, and the centerline defect shrinks, development is slower and slower, and larger and larger time and space steps can be used as time progresses, both from the viewpoint of truncation error and from the viewpoint of stability. In addition, the size of the computational mesh required constantly increases. Consequently, as the computation progressed, $\overline{v^2}$ was continually monitored at the last mesh point; when it had grown to a value about 1% of the peak, a new, empty mesh point was added. When the number of meshpoints had doubled, the computation was stopped, every other mesh point was discarded, the values renormalized by the new velocity and length scales, and the computation restarted, effectively increasing Δy by 2 and Δt by 4. Each of these doublings corresponded to a dimensionless time lapse of about 5 (defined by $\int dt/\mathcal{T}$, where $\mathcal{T} = q^2/8\varepsilon$ on the centerline). About three doublings are required to get within 10% of the final values.

If the initial wake width is taken as θ, the momentum thickness, then three doublings corresponds to $x/d = 500$ (cf. Tennekes and Lumley, 1970). Each successive doubling corresponds to a fourfold increase in streamwise distance. To get within 1% of the final values, roughly two more doublings are required, that is, an x/d of roughly 8000 (see Fig. 1).

Coefficients were approximated by the following techniques: in the wake, outboard of the zero of the advection (cf. Tennekes and Lumley, 1972) there is a region in which $-\overline{uv} \sim q^2 \sim \overline{v^2}$, production $\sim$ dissipation, advection $\sim$ transport, and hence $\overline{v^2}\mathcal{T} \sim$ const. In this region, the equations may be solved exactly; compatability among the equations requires that the coefficient in the net q^2 transport be unity, that in the $\overline{\varepsilon}$ transport be $\frac{1}{2}$ and that in the $-\overline{uv}$ transport be $\overline{v^2}/q^2$. Using this, plus the concepts described in Eqs. (41)–(44), and in addition neglecting pressure transport [i.e., requiring threefold symmetry of (33)] reduced the transport coefficients to five, two of known magnitude.

We found generally that the values of the various transport coefficients are

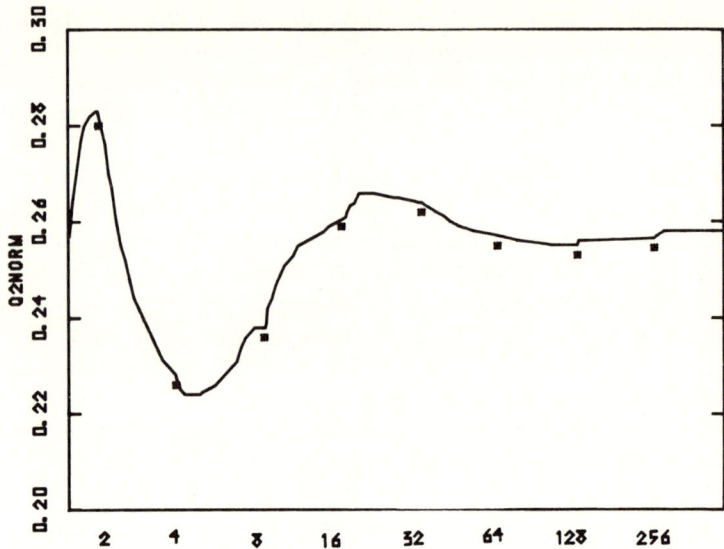

FIG. 1. Convergence of the normalized mean square total velocity (twice the energy). The squares indicate points at which the wake was renormalized. Thus, between such points, the scale is linear, but increases by a factor of four at each point (hence the change in slope). Each point indicates a doubling of the width. The abscissa values hence are multiples of the initial wake width. The corresponding values of x/d can be obtained as roughly eight times the square of the abscissa. Hence, the third doubling corresponds roughly to $x/d = 512$.

not critical, so long as the overall magnitude of the transport is correct. The result is most sensitive to the coefficients in the pressure gradient–velocity correlation, and in the production of dissipation. The latter controls the overall energy level. As already mentioned, the peak in $\overline{w^2}$ is directly attributable to the second order term in the pressure gradient–velocity correlation, and the proportion of the peak in $\overline{u^2}$ to the third order term.

As can be seen from Figs. 2–8 (where the calculations are compared with the data of Townsend, 1956), the axial values of $\overline{u^2}$ and $\overline{w^2}$ are higher, while those of $\overline{v^2}$ are lower, than Townsend's measurements. Use of the second order inhomogeneity correction to the pressure gradient–velocity correlation, proportional to $q^2_{,ii}$, and $\overline{\varepsilon}_{,ii}$ has the effect of increasing the relative intensity of the component normal to an energy or dissipation trough (which would help fill-in the trough). This term would transfer energy from $\overline{u^2}$ and $\overline{w^2}$ to $\overline{v^2}$ on the axis, and from $\overline{v^2}$ to $\overline{u^2}$ and $\overline{w^2}$ near the energy peak. This would push the ratios in the right direction on the axis. At the peak, the anisotropy is already so great that little more would be produced due to the counter balancing third order terms.

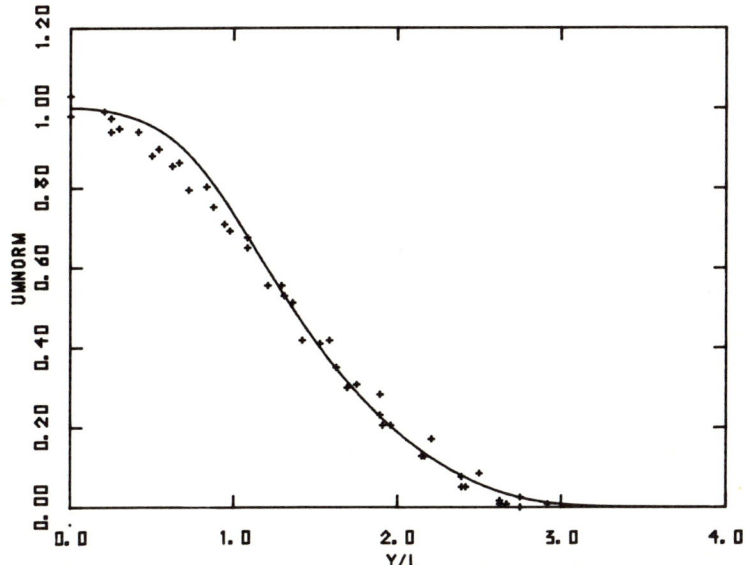

FIG. 2. In Figs. 2–8, the experimental points are those of Townsend (1956) scaled from the published graphs, renormalized and machine plotted. The solid line is machine plotted from the calculated points, approximately 80 per figure, with linear interpolation. The mean velocity-defect profile.

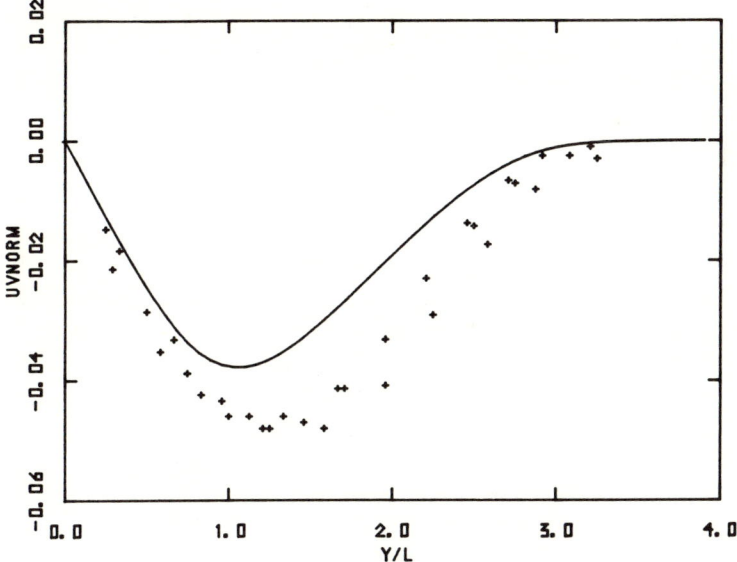

FIG. 3. The Reynolds stress. In the self-preserving state, this can be obtained directly from the first moment of the mean velocity-defect profile. Townsend's values are not quite self-preserving. The constant of proportionality determines the growth; our wake is growing somewhat slower than Townsend's. All turbulent quantities normalized by the mean velocity defect at the origin.

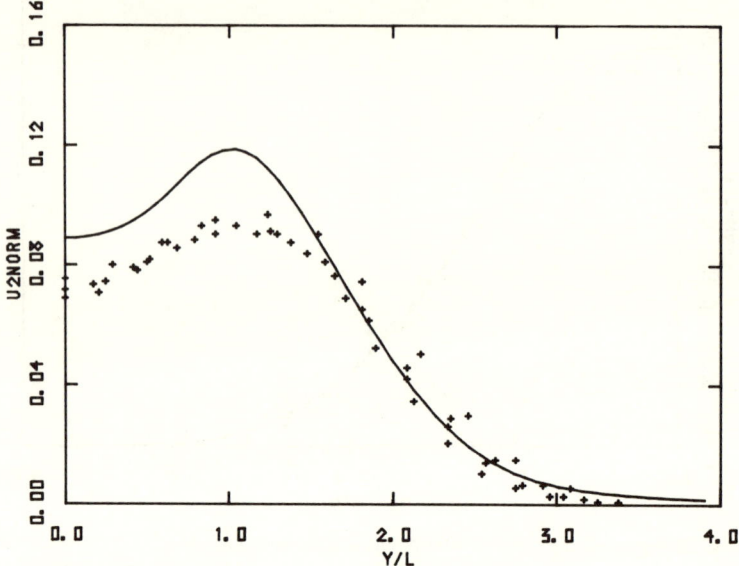

Fig. 4. The normalized mean square value of the streamwise fluctuating velocity.

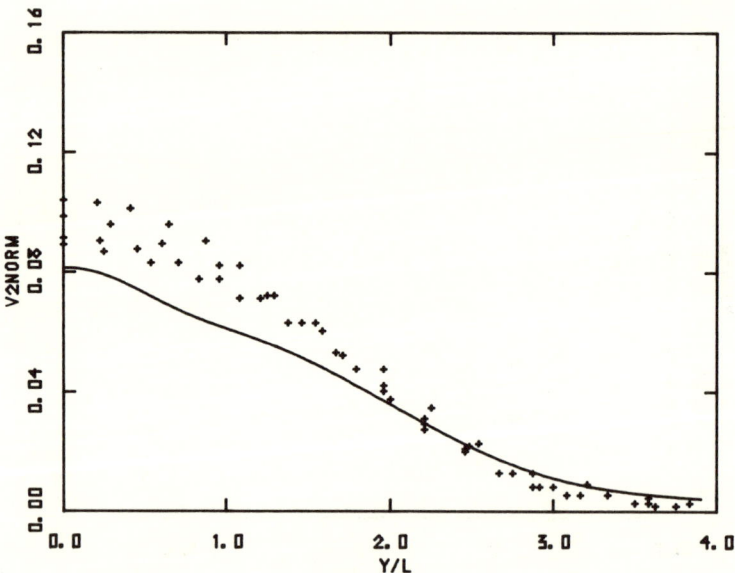

Fig. 5. The normalized mean square value of the fluctuating velocity normal to the plane of the wake.

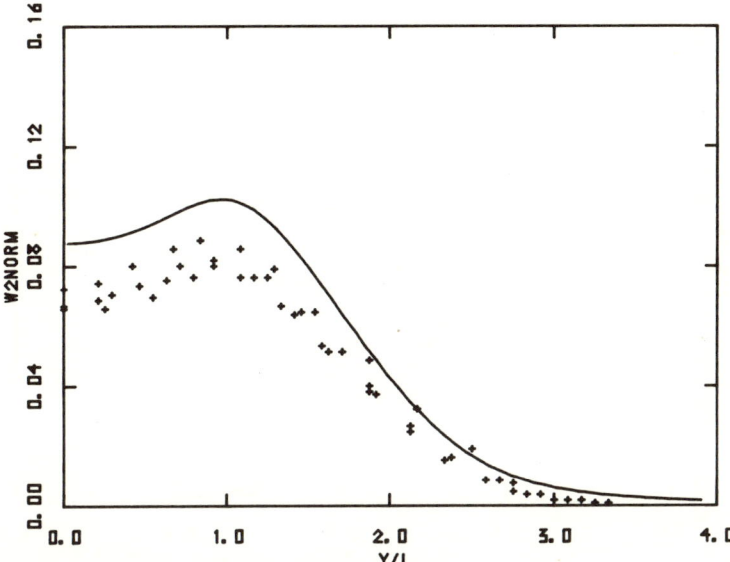

FIG. 6. The normalized mean square value of the cross-stream fluctuating velocity in the plane of the wake.

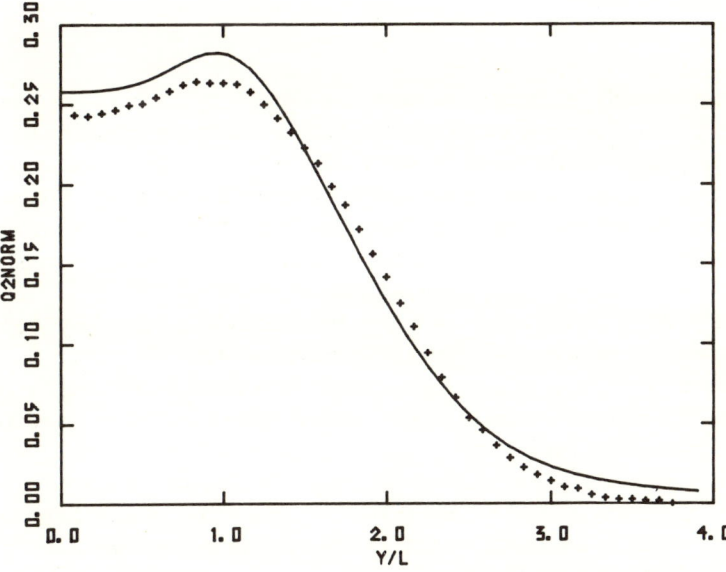

FIG. 7. The normalized mean square value of the total fluctuating velocity (twice the energy). The experimental points are the sum of the three component intensities read from Townsend's faired curves.

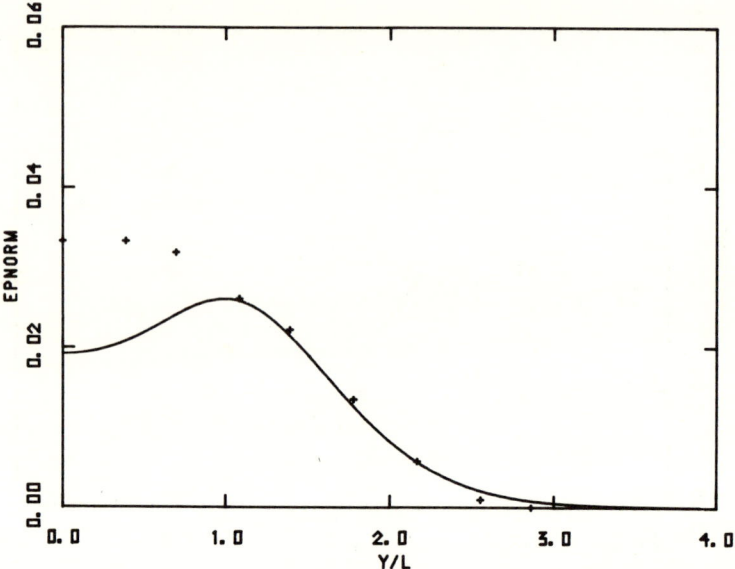

Fig. 8. The normalized dissipation. Townsend's values of the abscissa are clearly incorrect since they result in a displacement of the peak of the production, which can be obtained without assumption from the mean velocity curve. The abscissa has been renormalized so that the peak of the measured production occurs in the same place as that obtained from the mean velocity profile.

Unfortunately, we were not able to use significant values of this term due to a second order nonlinear computational instability, triggered by the addition of points at the edge of the mesh; even very small values of this coefficient required drastic decreases in the size of the time step. This is not a fundamental difficulty, however; we must simply find a differencing scheme for these terms which is more stable, or find a way of adding points to the mesh which will not excite the instability.

The peak in the curve of $\bar{\varepsilon}$ will also be essentially removed by the use of this term; at present, the flow is too isotropic near the center line, and consequently there is not enough $\bar{\varepsilon}$ production there. Increasing the anisotropy there will make the $\bar{\varepsilon}$ production more uniform and permit a reduction in the overall level of $\bar{\varepsilon}$.

It is worth mentioning that in the range of values of x/d in which Townsend's measurements were made (500–1000), our values changed only a total of 4%, which is about the variation observed in Townsend's data; it thus appears perfectly possible that Townsend's measurements are 6–10% from the true self-preserving values (but would appear to be self-preserving). This is borne out by the fact that his measured Reynolds stress is not self-preserving (i.e., is not proportional to yU, where U is the mean velocity defect, an exact result of self-preservation) by about this percentage. The direction of the error and trend (if it exists) in each component depends on the initial values, and would not necessarily be the same in our computation and in the real experiment.

Although the number of terms involved boggles the imagination, the cost of such a calculation is not prohibitive. One run of the wake calculation, to six doublings, programmed on the IBM 370/165, costs about $15.00. It must be remembered that this is an experimental calculation, and the program was written so that the various constants could be adjusted at will; no attempt was made to minimize cost. In the final form, the cost should be substantially lower perhaps of the order of $10.00. If we envision a three-dimensional calculation of a stratified flow with boundaries (which converges some five times as fast as a boundary-free flow) involving some 10^4 points and 17 equations instead of six, we would have an estimated cost of perhaps $3000. This is not a particularly accurate way to make such an estimate, because the larger calculation cannot be done in core, and requires reading out to disk storage. In addition, the pressure must be computed in a three-dimensional calculation. A detailed, direct estimate of the number of computations involved, and storage requirements, suggests a cost of the order of $1500. The difference between the estimates probably represents the different proportion of I/O, CPU, and auxiliary storage access time, and reflects the inefficiency of our present program. It is nevertheless clear that such a calculation is within our grasp for a few thousand dollars.

Acknowledgments

We have profited from many discussions with J. W. Deardorff, R. Owen, W. C. Reynolds, H. Tennekes and J. Wyngaard. A remark of Deardorff led to the understanding of the dissipation production; a remark of Owen suggested the importance of the second and third order terms in the pressure gradient–velocity correlation; Reynolds carried out many calculations in an early stage of the investigation, and contributed much analysis and comment which led to a better understanding of the structure of the equations, and an appreciation of the importance of the second order transport terms; Tennekes of course shares responsibility for the Reynolds number order-of-magnitude analysis, and has made extensive contributions by applying the model to simple situations, bringing flaws to light; Wyngaard has made independent calculations of the stratified atmospheric boundary layer, and his findings influenced the development of the model.

We are particularly grateful to our assembly language programmer, M. Hurvitz, who has been absolutely indispensable.

Generally, the development of the theory has been the responsibility of JLL and the execution of the calculations the responsibility of BK-N, although there has been intensive and extensive interactions between the two, so that they cannot be separated so simply. These calculations will form part of the doctoral dissertation of BK-N.

We wish to acknowledge the assistance of the Applied Research Laboratory in the preparation of the final figures.

References

Champagne, F. H., Harris, V. G., and Corrsin, S. (1970). Experiments on nearly homogeneous turbulent shear flow. *J. Fluid Mech.* **41**, 81–139.

Corrsin, S. (1972). Comment on "Transport equations in turbulence." *Phys. Fluids* **16**, 157–158.

Daly, B. J., and Harlow, F. H. (1970). Transport equations in turbulence. *Phys. Fluids* **13**, 2634–2649.

Deardorff, J. W. (1974). The use of sub-grid transport equations in a three-dimensional model of atmospheric turbulence. *Trans. ASME* (submitted for publication).

Donaldson, C. DuP. (1972a). Calculation of turbulent shear flows for atmospheric and vortex motions. *AIAA J.* **10**, 4–12.

Donaldson, C. DuP. (1972b). Construction of a dynamic model of the production of atmospheric turbulence and the dispersal of atmospheric pollutants. *Amer. Meteorol. Soc. Workshop Micrometeorol.*, Boston, Mass.

Fox, D. G., and Lilly, D. K. (1972). Numerical simulation of turbulent flows. *Rev. Geophys. Space Phys.* **10**, 51–72.

Gibson, C. H., and Schwarz, W. H. (1963). The universal equilibrium spectra of turbulent velocity and scalar fields. *J. Fluid Mech.* **16**, 365–384.

Jones, W. P., and Launder, B. E. (1972). The prediction of laminarization with a two equation model of turbulence. *Int. J. Heat Mass Transfer* **5**, 301–314.

Lumley, J. L. (1967). Rational approach to relations between motions of differing scales in turbulent flows. *Phys. Fluids* **10**, 1405–1408.

Lumley, J. L. (1970). Toward a turbulent constitutive relation. *J. Fluid Mech.* **41**, 413–434.

Lumley, J. L., and Khajeh-Nouri, B. (1974). Modeling homogeneous deformation of turbulence. *Phys. Fluids* (submitted for publication).

Mellor, G. L. (1974). Analytic prediction of the properties of stratified planetary surface layers. *J. Fluid Mech.* (submitted for publication).

Mulhearn, P. J. (1970). "On the Theory of Uniform Shear Flows," Tech. Note F-15. Charles Kolling Res. Lab., Dep. Mech. Eng., Univ. of Sydney, Sydney.

Ng, K. H., and Spalding, D. B. (1972). Turbulence model for boundary layers near walls. *Phys. Fluids* **15**, 20–30.

Orszag, S. A., and Patterson, G. S., Jr. (1972). Numerical simulation of turbulence. In "Statistical Models and Turbulence." Lecture Notes in Physics, Vol. 12, pp. 127–147. Springer-Verlag, Berlin and New York.

Phillips, O. M. (1966). "The Dynamics of the Upper Ocean." Cambridge Univ. Press, London and New York.

Reynolds, W. C. (1970). "Computation of Turbulent Flows—State of the Art, 1970," Rep. MD-27, Thermosci. Div., Dep. Mech. Eng., Stanford Univ., Stanford, California.

Rotta, J. (1951). Statistische Theorie Nichthomogener Turbulenz. *Z. Phys.* **129**, 547–572.

Tennekes, H., and Lumley, J. L. (1972). "A First Course in Turbulence." MIT Press, Cambridge, Massachusetts.

Townsend, A. A. (1956). "The Structure of Turbulent Shear Flow." Cambridge Univ. Press, London and New York.

Tucker, H. J., and Reynolds, A. J. (1968). The distortion of turbulence by irrotational plane strain. *J. Fluid Mech.* **32**, 657–673.

MODELING THE ATMOSPHERIC BOUNDARY LAYER

J. C. WYNGAARD, O. R. COTÉ, AND K. S. RAO

Air Force Cambridge Research Laboratories, Bedford, Massachusetts 01730, U.S.A.

1. INTRODUCTION

Higher order closure models, which use exact equations for the mean field and approximate ones for the turbulence, can reproduce in remarkable detail the structure of turbulent shear flows. Several models of this type are discussed by Reynolds (1970), Bradshaw (1972), Mellor and Herring (1973), and Donaldson (1973). The main differences between models are the closure assumptions used for the turbulence equations.

Developing this type of model for the atmospheric boundary layer is in some ways more difficult. The effects of buoyancy and rotation modify the structural relationships found in shear flows, and presumably buoyant and rotation terms should appear in at least some of the closure assumptions. Because the details of closure are currently rather controversial, even in shear flows, going to the atmosphere only complicates matters. A second difficulty is the lack of atmospheric turbulence data outside the surface layer. Testing and refining models, checking closure assumptions, and establishing model constants will be more difficult than with shear flows.

Lumley (1967, 1970) has suggested how closure can be done rationally, and in Lumley and Khajeh-Nouri (this volume, p. 169), the method is demonstrated. In contrast to other models, this approach minimizes the need for ad hoc closure assumptions and thus represents a considerable advance in the state of the art. Unfortunately, the model is considerably more complex than any proposed to date and it seems that its application to the atmosphere will first require establishing some of the model constants through calculation of simpler flows. Our calculations here, which are based on a simplified version of his model, are therefore only exploratory. Because of the lack of atmospheric data, it is difficult to set model constants, and some of them have arbitrarily been taken as zero; it is also difficult to check predictions with observations, so we have simply compared our steady-state neutral and unstable results with those from the three-dimensional numerical model of Deardorff (1972). The stably stratified case can be done with the same model, and this work will be covered in a future paper.

2. The Problem

We have limited our study to a horizontally homogeneous planetary boundary layer with the x direction (the direction of the surface-layer wind) positive to the east, y positive to the north, and z positive upward. In this coordinate system, the unit vector n_i along the earth's rotation axis is $(0, \cos \phi, \sin \phi)$, where ϕ is latitude. Denoting mean and fluctuating wind vectors by $U_i = (U, V, 0)$ and $u_i = (u, v, w)$, mean and fluctuating temperatures by Θ and θ, the mean field equations, for a dry atmosphere without radiative flux divergence, are

$$\begin{aligned} \partial U/\partial t + \partial \overline{uw}/\partial z &= f(V - V_g) \\ \partial V/\partial t + \partial \overline{vw}/\partial z &= f(U_g - U) \\ \partial \Theta/\partial t + \partial \overline{\theta w}/\partial z &= 0 \end{aligned} \tag{1}$$

Here U_g and V_g are the geostrophic wind components defined by

$$U_g = -(1/f)\,\partial P/\partial y; \qquad V_g = (1/f)\,\partial P/\partial x \tag{2}$$

where P is the mean kinematic pressure and f, the Coriolis parameter, is $2\omega \sin \phi$ with ω the earth's rotation rate. The turbulence covariance equations are

$$\begin{aligned} \dot{\overline{u_i u_k}} &+ U_{i,j}\overline{u_j u_k} + U_{k,j}\overline{u_j u_i} + (\overline{u_i u_j u_k})_{,j} + (\overline{u_i u_k})_{,j} U_j \\ &= -(\overline{u_k p_{,i}} + \overline{u_i p_{,k}}) + (\overline{u_k \theta}\delta_{3i} + \overline{u_i \theta}\delta_{3k})\frac{g}{T_0} - 2\bar{\varepsilon}\frac{\delta_{ik}}{3} \\ &\quad - 2\omega\varepsilon_{ijl}n_j\overline{u_l u_k} - 2\omega\varepsilon_{klm}n_l\overline{u_m u_i} \end{aligned} \tag{3}$$

$$\dot{\overline{\theta^2}} + 2\Theta_{,j}\overline{u_j \theta} + \overline{\theta^2}_{,j}U_j + (\overline{\theta^2 u_j})_{,j} = -2\bar{\varepsilon}_\theta \tag{4}$$

$$\begin{aligned} \dot{\overline{\theta u_i}} &+ U_{i,j}\overline{\theta u_j} + \Theta_{,j}\overline{u_i u_j} + (\overline{\theta u_i})_{,j}U_j + (\overline{\theta u_i u_j})_{,j} \\ &= -\overline{\theta p_{,i}} + \delta_{3i}\overline{\theta^2}\frac{g}{T_0} - 2\omega\varepsilon_{ijk}n_j\overline{u_k \theta} \end{aligned} \tag{5}$$

where we have indicated differentiation by a comma and repeated indices are summed; the other notation is standard. In writing Eqs. (3)–(5), we assume the turbulence Reynolds number is so large that molecular diffusion can be neglected and locally isotropic forms can be used for the molecular destruction terms.

With the scaling arguments presented in Tennekes and Lumley (1972; see also Lumley and Khajeh-Nouri, this volume), a dynamical equation for the turbulent energy dissipation ($\bar{\varepsilon}$) can be written and reduced to

(6) $$\dot{\bar{\varepsilon}} + U_j \bar{\varepsilon}_{,j} + \overline{(u_j \varepsilon)}_{,j} = -2\nu \overline{u_{i,k} u_{i,j} u_{j,k}} - 2\nu^2 \overline{u_{i,kj} u_{i,kj}}$$

3. Closure

Equations (1), (3)–(5), and (6) are our basic set. Since there are more unknowns than equations, we need closure approximations for the flux divergence and pressure covariance terms in Eqs. (3)–(5) and the flux divergence and right-hand side terms in Eq. (6). The object is to express these in terms of other dependent variables for which we have equations. While in many cases the flux divergence terms are not overly important, the pressure terms always are; they cause intercomponent energy transfer and maintain stresses and heat fluxes steady by balancing the production rates. Some inferences on the magnitudes of both types of terms in the surface layer are given by Wyngaard et al. (1971).

The basic closure philosophy (see Lumley and Khajeh-Nouri, this volume) is to replace each of these terms by the sum of (a) its value in an isotropic turbulence field, and (b) a series of terms representing corrections for anisotropy. Consider first the pressure term in the $\overline{u_i u_k}$ equation. Part of this is flux divergence, so the above authors consider

(7) $$-(\overline{u_k p_{,i}} + \overline{u_i p_{,k}}) + 2\overline{(u_j p)}_{,j} \frac{\delta_{ik}}{3} = A_{ik}$$

We write A_{ik} as the sum of its value if the turbulent velocity and temperature fields were isotropic plus a correction for anisotropy:

(8) $$A_{ik} = A_{ik}^I + A_{ik}^A$$

Through the p-equation, obtained by taking the divergence of the u_i equation, we can identify the mechanisms which maintain pressure covariances:

(9) $$\nabla^2 p = \underbrace{-2U_{m,n} u_{n,m}}_{\text{(mean strain)}} - \underbrace{(u_{m,n} u_{n,m} - \overline{u_{m,n} u_{n,m}})}_{\text{(turbulence)}}$$

$$+ \underbrace{\frac{g}{T_0} \theta_{,m} \delta_{3m}}_{\text{(buoyancy)}} - \underbrace{2\omega \epsilon_{mnl} n_n u_{l,m}}_{\text{(rotation)}}$$

One can write an expression for A_{ik} by using the formal solution to Eq. (9). The isotropic part A_{ik}^I can be evaluated exactly; only the mean strain term contributes, and Crow (1968) shows that if the mean strain field changes insignificantly in a distance over which the two-point velocity correlation tensor falls to zero, it is

(10) $$A^I_{ik} = 0.2\overline{u_m u_m}(U_{i,k} + U_{k,i})$$

In principle, A^A_{ik} can be approximated by Lumley's functional expansion, with each of the four terms contributed by Eq. (9) being treated separately. We have used simply the leading term in the total expansion,

(11) $$A^A_{ik} = -\left(\overline{u_i u_k} - q^2 \frac{\delta_{ik}}{3}\right)\frac{c_{ik}}{\tau} \quad \text{(no sum)}$$

adding the constants c_{ik} to allow our simplified model to reproduce in detail the structure of the surface layer. Here $q^2 = \overline{u_i u_i}$ and $\tau = q^2/\bar{\varepsilon}$ is a turbulence relaxation time; since equations for both q^2 and $\bar{\varepsilon}$ are carried, τ is set by the model, not specified as input information. Invariance requirements are not used to constrain c_{ik}; as shown later, reproducing surface-layer structure requires different values for diagonal and off-diagonal components.

Similarly, the pressure term in the heat flux equation (5) is broken into two parts:

(12) $$-\overline{\theta p_{,i}} = B^I_i + B^A_i$$

Only the buoyant term in Eq. (9) contributes to B^I_i, giving

(13) $$B^I_i = -(g/T_0)(\overline{\theta^2}/3)\delta_{3i}$$

Again, the anisotropic part can be approximated through functional expansion. We have kept simply the leading term which emerges, namely,

(14) $$B^A_i = -d_i\overline{\theta u_i}/\tau \quad \text{(no sum)}$$

and have again introduced free constants (the d_i).

We became aware of A^I_{ik} and B^I_i only after we were well into this study. At that point, we added them to the model and found that A^I_{ik} [Eq. (10)] changes the structure slightly but in an unphysical way; it makes some of the neutral turbulence profiles nonmonotone with z, for example. Evidently, if this term is kept, it should be accompanied by other terms representing the anisotropic contribution. These are higher order terms which we dropped and we dropped A^I_{ik} as well. The B^I_i term enters only in the convective case where we found it caused insignificant changes.

The right-hand side of the $\bar{\varepsilon}$ equation (6) is the imbalance between production by vortex stretching and destruction by viscosity. For this, we used the simplified functional expansion result of Lumley and Khajeh-Nouri:

(15) $$\text{Imbalance} = -4\bar{\varepsilon}^2/q^2 + 4a\bar{\varepsilon}P/q^2$$

Here a is a free constant and P is the production rate of turbulent kinetic energy. We used $a = \frac{3}{4}$, but found the results generally indistinguishable from those for $a = 0$.

One can also carry a dynamical equation for $\bar{\varepsilon}_\theta$, the molecular destruction rate of $\overline{\theta^2}$; its structure and modeling are analogous to the $\bar{\varepsilon}$ equation, as discussed by Lumley and Khajeh-Nouri. We have simply used instead

$$\bar{\varepsilon}_\theta = d_4 \overline{\theta^2}/\tau \tag{16}$$

where d_4 is a free constant.

We have not used the functional expansion results for turbulent transport modeling, but a rather simpler, ad hoc gradient diffusion model

$$\overline{f u_i} = -a_t \overline{f_{,j} u_j u_i} \tau \tag{17}$$

where f is stress, heat flux, or dissipation rate, and a_t is a transport constant. Since pressure also contributes to the flux divergence in the stress equations, there we used

$$\overline{(u_i u_k + \tfrac{2}{3} p \delta_{ik}) u_j} = a_t \overline{(-u_i u_k)_{,l} u_l u_j} \tau \tag{18}$$

in keeping with Eq. (7).

Many of the closure constants can be set by requiring the model to reproduce the observed structure of the surface layer. For z approaching, but larger than, the roughness length z_0, we require

$$-\overline{uw}/u_*^2 = 1.0, \quad \overline{vw}/u_*^2 = 0, \quad \overline{uv}/u_*^2 = 0$$
$$\overline{u^2}/u_*^2 = 4.0, \quad \overline{v^2}/u_*^2 = \overline{w^2}/u_*^2 = 1.75$$
$$(kz/u_*) \partial U/\partial z = 1.0, \quad (kz/u_*) \partial V/\partial z = 0, \quad (kz/T_*) \partial \Theta/\partial z = 0.74 \tag{19}$$
$$(kz/u_*^3)\bar{\varepsilon} = 1.0, \quad \overline{\theta^2}/T_*^2 = 4.0$$
$$\overline{\theta u}/u_* T_* = 3.0, \quad \overline{\theta v}/u_* T_* = 0, \quad \overline{w\theta} = -u_* T_* = Q_0$$

The turbulent transport terms in the model equations vanish in this limit, consistent with observations (Wyngaard et al., 1971). If this model is to satisfy the conditions (19), the constants must be

$$c_{11} = c_{22} = c_{33} = 6.7, \quad c_{13} = 13.2 \tag{20}$$
$$d_1 = 4.4, \quad d_3 = 9.7, \quad d_4 = 1.4$$

Note that in the evaluation of c_{13} in Eq. (20), we have dropped the isotropic mean strain term A_{13}^I as discussed earlier. Because all terms in the $\overline{uv}$, $\overline{vw}$, and $\overline{v\theta}$ equations vanish in the surface-layer limit, this leaves c_{12}, c_{23}, and d_2 unspecified. We arbitrarily set the first two equal to c_{13}, and set d_2 midway between d_1 and d_3.

This leaves only the transport coefficients a_t unspecified; there is one in each turbulence equation, a total of 11. Only in the dissipation equation is transport nonzero in the surface layer, however, so we can assign a value to only that constant. Using the $\bar{\varepsilon}$ equation (6) together with the imbalance and transport models (15) and (17), and the surface-layer constraints (19), we get for $\bar{\varepsilon}$:

(21) $$a_t = (4 - 4a)/12.1 = 0.083$$

The remaining a_t values were set at 0.15. The neutral calculations shown later reveal that the structure is not very sensitive to a_t.

4. Computational Techniques

The model set was solved numerically with boundary conditions at $z = h$ and $z = H$. Those at h are given in Eq. (19), and it is most convenient to regard u_* and Q_0 (the surface heat flux) as inputs and the mean wind and temperature at H as dependent variables. It is then also convenient to replace the mean equations (1) with their z derivatives. At $z = H$ we set the turbulence moments, mean shears, and mean temperature gradient to zero.

For some problems, it is appropriate to treat wind and temperature at H as inputs. In this case, the mean set (1) is carried as written, U_g and V_g are specified, and the lower boundary conditions (19) are adjusted to account for the angle between the surface and geostrophic winds. Here u_* and Q_0 are dependent variables, and u_*, for example, can be related to the mean speed $S = (U^2 + V^2)^{1/2}$ at h by

(22) $$S = (u_*/k) \ln(h/z_0)$$

Since the equations are tailored to reproduce observed surface-layer structure, the group hf/u_* need only be chosen small enough that it no longer affects the solutions. This occurred for $hf/u_* \leq 0.002$, the value used.

In the unstable case, the boundary at H is interpreted as an inversion lid, as in Deardorff's (1972) study, and in those runs we maintained $Hf/u_* = 1.0$. The neutral boundary layer sets its own thickness (Blackadar and Tennekes, 1968) to be a multiple of u_*/f, so H should be so large compared to this "natural" thickness that it does not interfere. We chose our value of $Hf/u_* = 1.85$ by increasing H until the neutral structure below no longer changed. This is four times the height used in most of Deardorff's neutral calculations.

We solved the equation set numerically using the DuFort–Frankel (DuFort and Frankel, 1953) method on a logarithmic height grid with 88 intervals; since the method is "leapfrog," only half the grid points are used in each time step. Experiments with coarser and finer grids showed this grid to be a good compromise between accuracy and computation time.

The steady-state solutions shown here were found by letting them evolve from assumed initial states, but they did not depend on the initial state.

Unless stated otherwise, the calculations assumed $k = 0.35$ (Businger et al., 1971) and 45°N latitude.

5. Steady State, Neutral Structure

5.1. Geostrophic Drag Law

Asymptotic similarity arguments (Blackadar and Tennekes, 1968; Tennekes, 1973) show that the "geostrophic drag law" is

$$(23) \qquad \ln\frac{G}{fz_0} = B + \ln\frac{G}{u_*} + \left(\frac{k^2 G^2}{u_*^2} - A^2\right)^{1/2}$$

where A and B are constants defined by

$$(24) \qquad A = -kV_g/u_*, \qquad B = \ln(u_*/fz_0) - kU_g/u_*$$

Some use a different convention in which A and B are reversed.

Our calculations are consistent with this drag law and for the chosen model constants gave $A = 2.3$, $B = 1.8$. They are somewhat sensitive to von Kármán's constant; for $k = 0.40$ the values were $A = 2.5$, $B = 1.1$. Similarity theory ignores any possible latitude dependence of A and B, but some would be expected because both f and $f \cot \phi$ appear in the stress equations (3); hence, f alone cannot completely account for latitude effects. Between 15 and 90°, the calculated variations in A and B were only about 5%, however.

A and B are notoriously difficult to determine experimentally, and the scatter in observations precludes any detailed testing of model calculations. Caldwell et al. (1972) have tabulated A and B values from several sources, both experimental and model calculations. A values (in our convention) range from 1.5 to about 5, and B from -1.6 to 2.5.

5.2. Turbulence Distributions

Our calculated turbulence profiles are nondimensionalized with u_* and f in keeping with the asymptotic similarity theory prediction (Blackadar and Tennekes, 1968) that these are the appropriate scales outside of the surface layer.

Figure 1 shows calculated turbulent energy distributions for a range of values of the transport coefficient a_t. Below $zf/u_* = 0.1$ (which corresponds to $z = 300$ m if $f = 10^{-4}$ sec^{-1} and $u_* = 0.3$ m sec^{-1}), varying a_t by a factor of two from our normal value of 0.15 has negligible effect. Figure 1 also shows good agreement between our curve for $a_t = 0.15$ and Deardorff's result. Figure 2 shows the distributions of energy components.

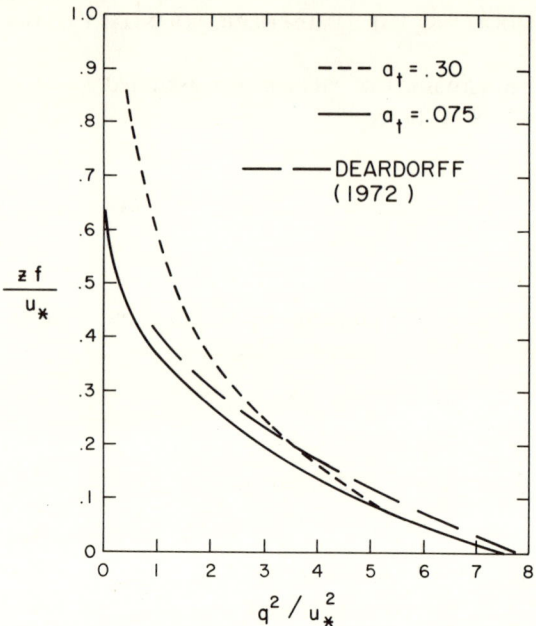

FIG. 1. The vertical profile of turbulent kinetic energy, neutral case, and its sensitivity to the transport coefficient. Also shown is the result from Deardorff's three-dimensional numerical study.

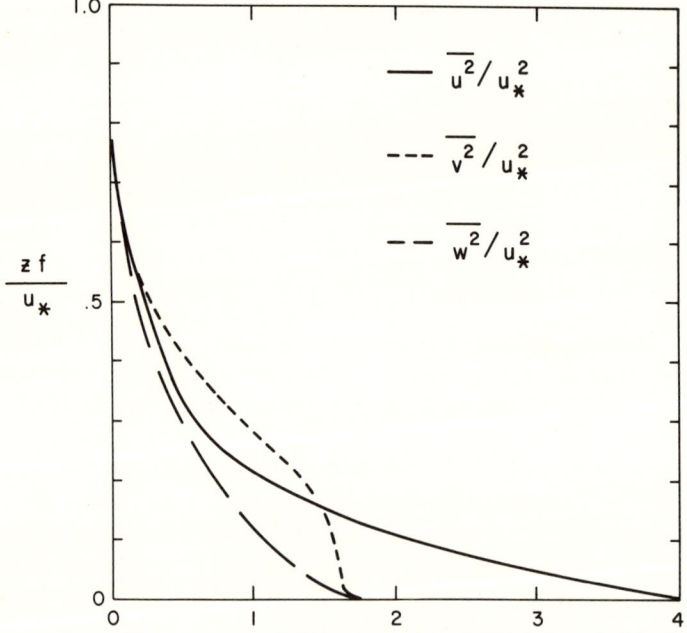

FIG. 2. The vertical distributions of the energy components, neutral case.

Other results of interest are the stress profiles (Figs. 3 and 4). The $\overline{uw}$ profile aloft differs from Deardorff's partly because his own results indicate that his upper boundary conditions should have been applied higher. The differences below are of the order that can be induced by adjusting model constants. Note that all three stress profiles depend very weakly on latitude.

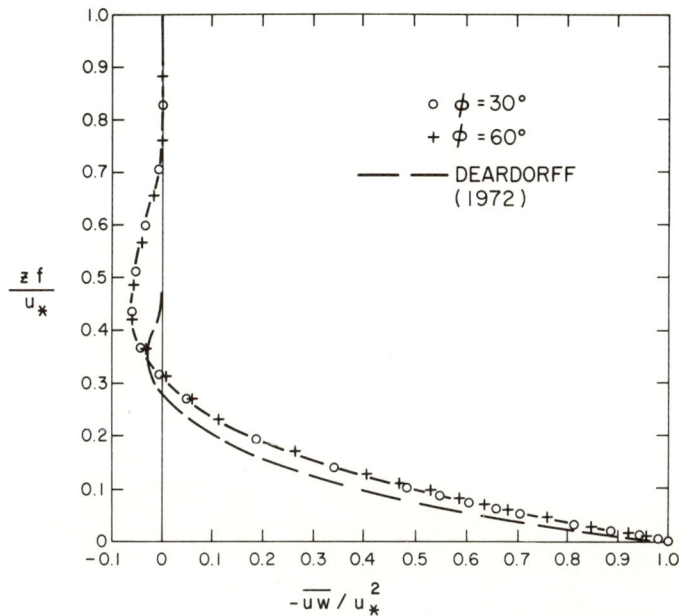

FIG. 3. The neutral $\overline{uw}$ profile for two latitudes and Deardorff's (45°) result.

The horizontal equations of motion (1) show that the stress gradients are *largest* near the surface where it is traditional to assume a "constant stress" layer exists. This paradox can be reconciled by nondimensionalizing Eq. (1) with the surface-layer length and velocity scales u_* and l:

(25) $$l\,\partial(\overline{uw}/u_*^2)/\partial z \sim f V_g l/u_*^2$$

$$l\,\partial(\overline{vw}/u_*^2)/\partial z = f(U_g - U)l/u_*^2$$

Since l scales with z, the dimensionless stress gradients do vanish as $z \to 0$. In surface-layer scales, there is a "constant-stress" layer near the surface; Eq. (25) suggests it be viewed in a logarithmic height plot, as in Fig. 5. Only a slight departure from the surface values is noticeable at $zf/u_* = 0.01$

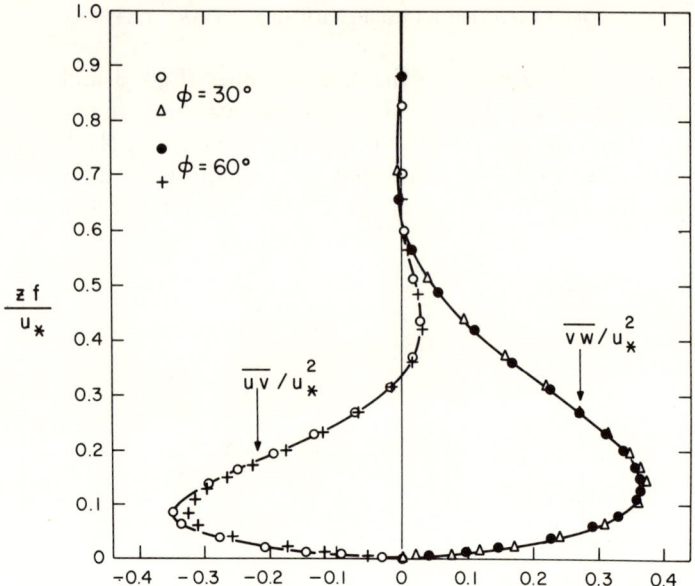

FIG. 4. The neutral $\overline{uv}$ and $\overline{vw}$ profiles for two latitudes.

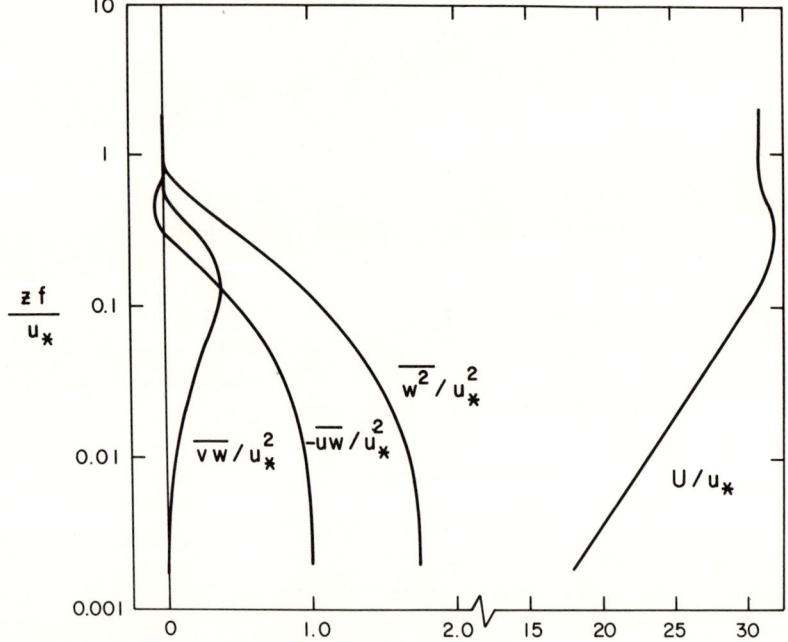

FIG. 5. Mean wind and turbulence profiles, neutral case, plotted against a logarithmic height scale to show the constant stress layer behavior.

($z = 30$ m for our standard conditions). Note that the U-profile is logarithmic until it first reaches U_g. The integral constraint

$$\text{(26)} \qquad \int_0^\infty (U_g - U)\,dz = \frac{1}{f}\int_0^\infty \frac{\partial}{\partial z}\overline{vw}\,dz = 0$$

shows U must overshoot U_g.

Most previous models of the boundary layer have used eddy-diffusivity models (see Estoque, 1973, for a review) in which stresses in the U and V equations, for example, are replaced by

$$\text{(27)} \qquad \overline{uw} = -K_x\,\partial U/\partial z, \qquad \overline{vw} = -K_y\,\partial V/\partial z$$

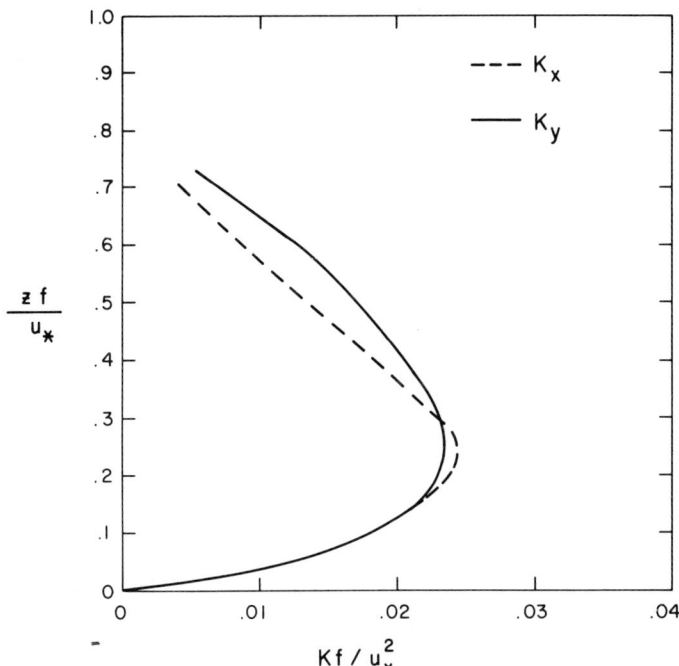

FIG. 6. Vertical distributions of eddy viscosity, neutral case.

and some assumption is made about the behavior of the K's. Our results imply K distributions shown in Fig. 6. Since both wind shear and stress vanish near the top of the layer, the calculated K values are rather uncertain there; in fact, there seems to be no requirement that K remain finite. It is often assumed that $K_x = K_y$, which is consistent with Fig. 6.

5.3. Some Effects of Geostrophic Wind Shear

To investigate the effect of z-dependent geostrophic wind, we used simple rectangular profiles:

$$(28) \quad \partial U_g/\partial z, \partial V_g/\partial z = \begin{cases} \pm 10f, & 0 \leq z \leq 0.5 u_*/f \\ 0, & z \geq 0.5 u_*/f \end{cases}$$

Confining the geostrophic shear to a finite layer assures that the flow above will approach a uniform free stream so the usual upper boundary conditions can be used. Figure 7 shows calculated $\overline{uw}$ distributions for the four combinations in (28). Although the imposed shear is not large (it corresponds to 1 m/sec change in U_g and V_g over 1 km depth at 45° latitude), the stress profiles are severely distorted.

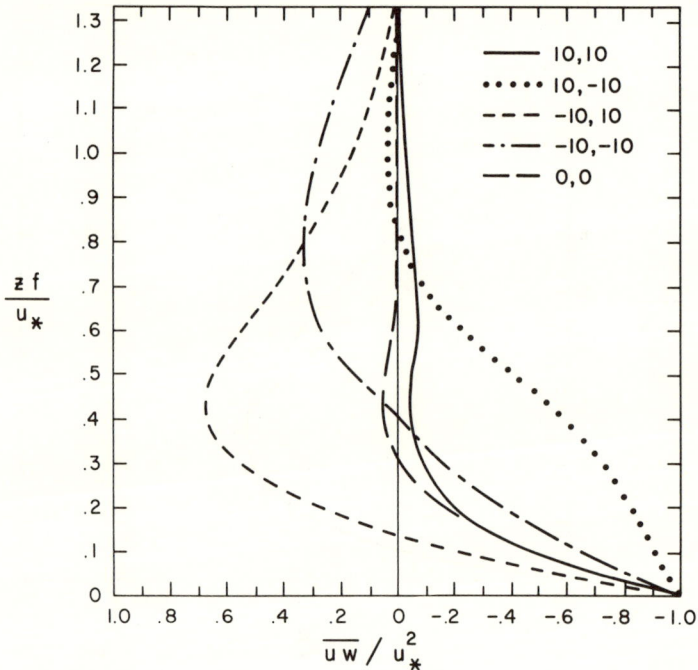

FIG. 7. The effect of geostrophic wind shear on the neutral $\overline{uw}$ profile, for four combinations of $(f^{-1} \partial U_g/\partial z, f^{-1} \partial V_g/\partial z)$.

6. STEADY-STATE UNSTABLE STRUCTURE

By letting the surface heat flux Q_0 be positive, we generate, with the same model set, an unstable planetary boundary layer. We found, as did Deardorff, that for all but very weakly unstable conditions the convective layer depth z_i was the relevant length for scaling vertical distributions. A stability index is z_i/L, where $L = -u_*^3 T_0/gkQ_0$, the Monin-Obukhov length. The flat wind profiles, Fig. 8, are a distinctive feature of the convective layer. The wind direction shift with height found in the neutral case $(z_i/L = 0)$ tends to be wiped out by convection, and at $z_i/L = -10$, Fig. 8 shows the wind shear to be essentially zero over the entire layer. Further increases in $-z_i/L$ serve mainly to diminish the small angle between surface and geostrophic winds and to make still thinner the region of significant wind shear near the surface.

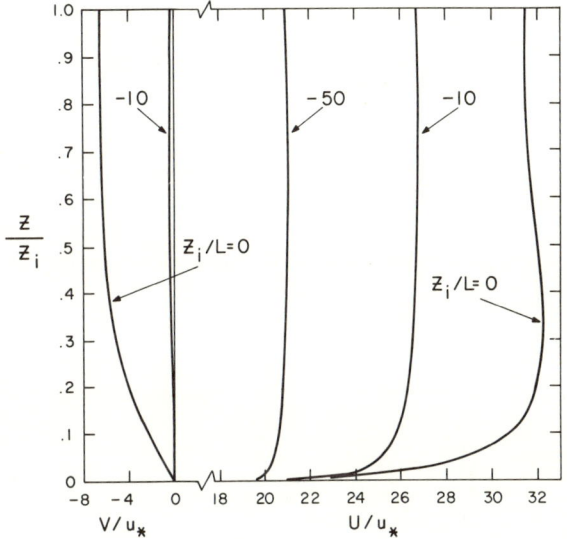

FIG. 8. The evolution of the mean wind profiles with increasing instability.

The differentiated forms of the U and V equations,

(29) $\quad \partial^2 \overline{uw}/\partial z^2 = f \partial V/\partial z; \quad \partial^2 \overline{vw}/\partial z^2 = -f \partial U/\partial z$

require the reduced wind shear in the convective layer to be accompanied by reduced stress profile curvature. The $\overline{uw}$ profile (Fig. 9) goes from its strongly curved neutral shape to an essentially linear profile at $-z_i/L = 50$, in agreement with Deardorff's results (the departure from the correct stress value at the surface in his experiments is due to averaging-time problems). The $\overline{vw}$ profile, on the other hand, approaches its curvature-free limit by approaching zero over the bulk of the layer (Fig. 10). The slight differences between the neutral stress profiles in Figs. 9 and 10 and those in Figs. 3 and 4 are due to changes in the constant a in the $\bar{\varepsilon}$ equation modeling shown in

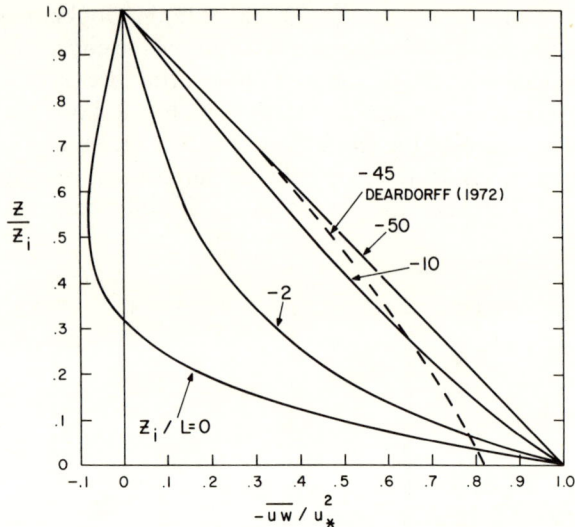

FIG. 9. The approach toward a linear $\overline{uw}$ profile with increasing instability.

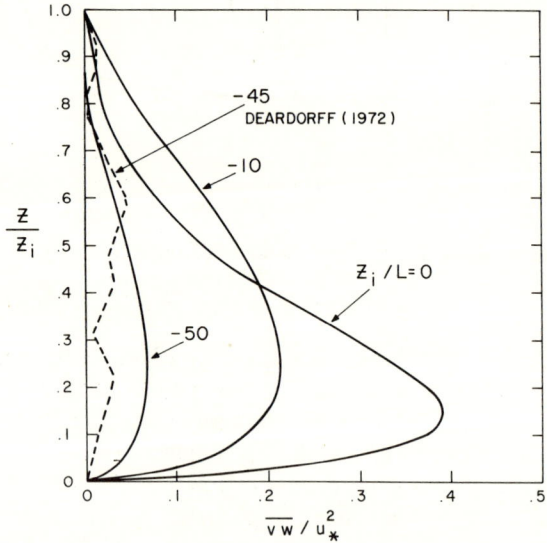

FIG. 10. The shrinking of the $\overline{vw}$ profile with increasing instability.

Eq. (15); a was zero in our first calculations, and later we realized this was inappropriate and used $a = \frac{3}{4}$.

At $z_i/L = -50$, these results show a simple state of affairs. From Eq. (1) we have

(30) $\qquad \overline{\partial uw}/\partial z \simeq u_*^2/z_i \sim -fV_g, \qquad V_g/u_* \simeq -u_*/fz_i$

Presumably the very slight shift in wind direction, corresponding to V changing from 0 to V_g, occurs within the inversion. Equation (1) also shows that $U \simeq U_g$ over most of the convective layer.

Another strong effect of upward heat transfer is the increase in turbulence energy levels over their neutral values. The vertical component (Fig. 11) is

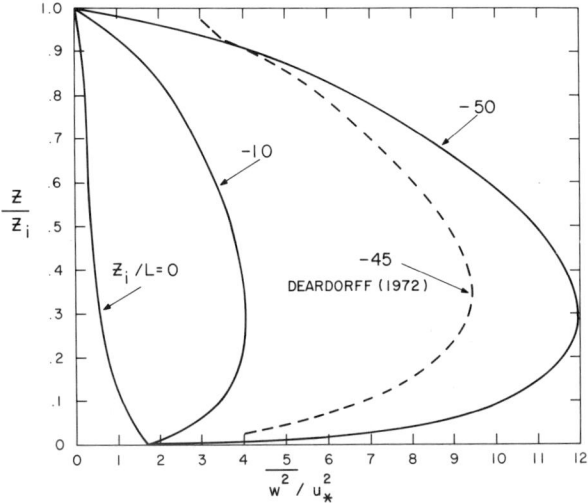

FIG. 11. The evolution of the $\overline{w^2}$ profile with increasing instability.

most sensitive because it receives the buoyant input directly. Fluctuating pressure forces transfer some of this $\overline{w^2}$ energy to the horizontal components, and they grow as well. Figure 12 shows the fluctuating streamwise component energy. The agreement with Deardorff's model is good in midregions, but the two models go opposite ways at the boundaries. A more realistic treatment of the upper boundary would probably reconcile the models at z_i, and the surface-layer differences are probably due to the failure of the assumed Monin–Obukhov similarity for u and v in our model and the increasing importance of sub-grid-scale events in Deardorff's model.

At very large $-z_i/L$, we expect the velocity variances to scale with w_*, defined by

(31) $\qquad\qquad\qquad w_* = [Q_0(g/T_0)z_i]^{1/3}$

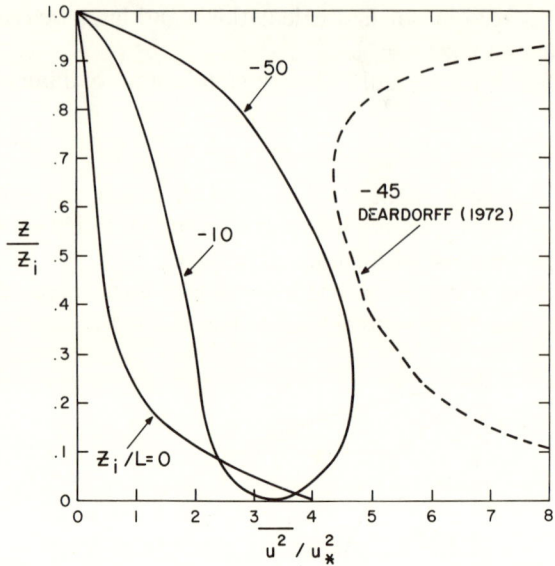

FIG. 12. The evolution of the $\overline{u^2}$ profile with increasing instability.

The $\overline{w^2}$ results (Fig. 13) do scale in a universal way with w_* for $-z_i/L = 10$ and greater. They also agree well with Deardorff's results.

The "local free convection" curve in Fig. 13 is an extrapolation of very unstable surface layer observations (Wyngaard et al., 1971). There is a suggestion in Fig. 13 that the model overestimates $\overline{w^2}$ near the lower boundary. This is probably because the higher order flux-gradient relations are not always as assumed in Eqs. (17) and (18) in convective turbulence. It is difficult to check Eq. (18) experimentally for $i = k = 3$ (for $\overline{w^2}$) because of the pressure flux term, but it is clear that in the unstable surface layer (Wyngaard, 1973) both $\overline{w^3}$ and $\partial \overline{w^2}/\partial z$ are positive.

The horizontal wind components also approach universal profiles when scaled on w_*. Figure 14 shows the lateral energy component $\overline{v^2}$ and suggests, as do Deardorff's results, that the approach to universality is slower than for $\overline{w^2}$.

Similarly, at very large $-z_i/L$ we expect temperature fluctuations to scale with θ_*, defined by

(32) $$\theta_* = (Q_0^2 T_0/gz_i)^{1/3}$$

Our results (Fig. 15) suggest this for $-z_i/L$ as small as -10. Our results indicate significantly larger $\overline{\theta^2}$ aloft than Deardorff found, and seem to be in better agreement than his model with the upper-level data he cites. Note that the surface-layer data, indicated by the "local free convection" curve, agree

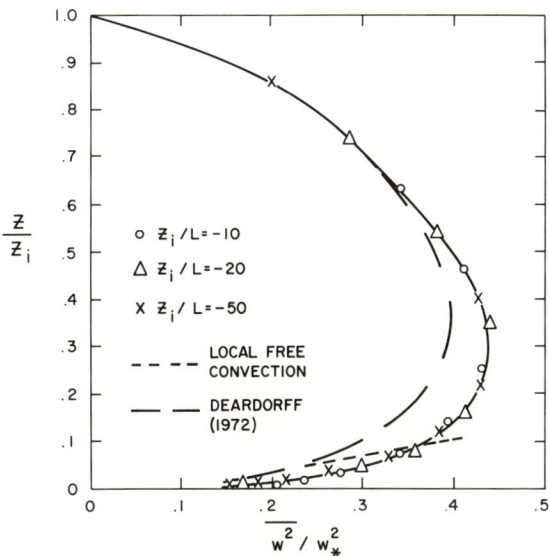

FIG. 13. The $\overline{w^2}$ profile scaled with the convective velocity scale w_*.

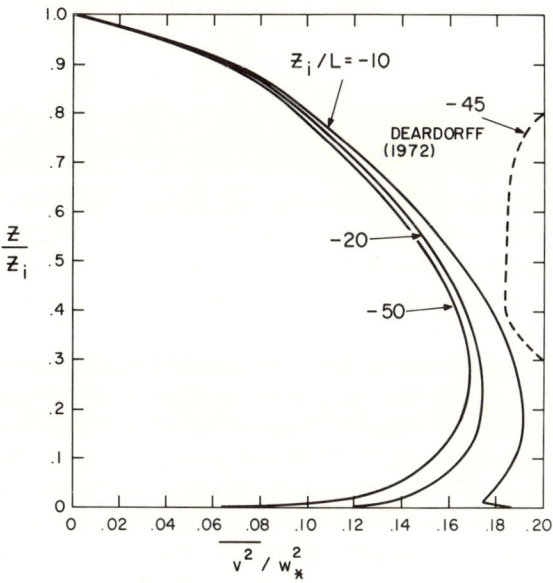

FIG. 14. The $\overline{v^2}$ profiles, scaled with w_*.

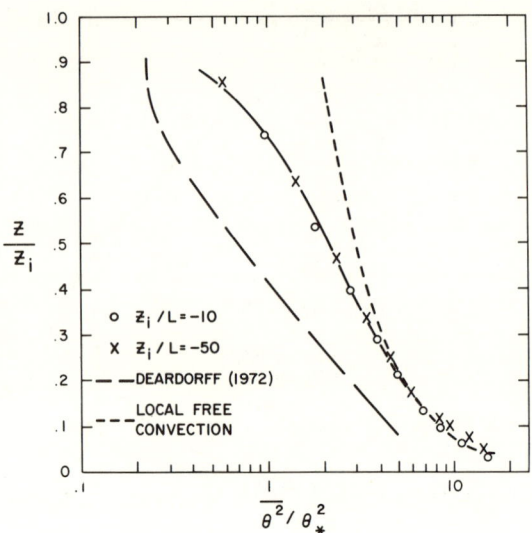

FIG. 15. The $\overline{\theta^2}$ profiles, scaled with the convective temperature scale θ_*.

well with our model calculations. Perhaps this is because our gradient-transport assumption (17) for the vertical flux of $\overline{\theta^2}$ is observed to be qualitatively correct in the unstable surface layer (Wyngaard, 1973).

7. Summary and Conclusions

Our results suggest this is an attractive approach to the calculation of atmospheric boundary layer structure. Broadly speaking, the results look much like those from Deardorff's (1972) three-dimensional model but take relatively little computer time; our neutral case has essentially reached steady state in 5 min computer time (CDC-6600), and the unstable case is faster. Simulating a 24-hr boundary layer cycle takes about 10 mins. Sophisticated programming techniques could considerably improve these times.

Because the model is in an embryonic stage, we have concentrated on the broad features of the results—not their details. Certainly the inversion at z_i, here treated as a rigid lid, needs more realistic treatment; and the model itself needs more attention now, particularly with respect to the higher-order terms in the pressure covariance expansions. The details of the results can be expected to change to some extent as the model is refined.

As we have discussed, there is a rational technique for refining the model, but because the data required for model testing are so difficult to obtain in the atmosphere, it seems we will have to rely heavily on the simulation of various laboratory flows. Fortunately, much of the required data are available from extensive laboratory measurements of both shear and buoyancy-driven turbulence.

ACKNOWLEDGMENTS

We are grateful to Prof. J. L. Lumley for many useful discussions on his functional expansion closure technique and also to our colleagues at AFCRL for helpful suggestions in the course of this work. Thanks are also due to Mr. R. Sizer for preparing the figures. One of us (K.S.R.) is an NRC Resident Research Associate at AFCRL and would like to express his sincere appreciation to the National Research Council for support.

REFERENCES

Blackadar, A. K., and Tennekes, H. (1968). *J. Atmos. Sci.* **25**, 1015–1020.
Bradshaw, P. (1972). *Aeronaut. J.* **76**, 403–418.
Businger, J. A., Wyngaard, J. C., Izumi, Y., and Bradley, E. F. (1971). *J. Atmos. Sci.* **28**, 181–189.
Caldwell, D. R., Van Atta, C. W., and Helland, K. N. (1972). *Geophys. Fluid Dyn.* **3**, 125–160.
Crow, S. C. (1968). *J. Fluid Mech.* **33**, 1–20.
Deardorff, J. W. (1972). *J. Atmos. Sci.* **29**, 91–115.
Donaldson, C. duP. (1973). In "Workshop on Micrometeorology" (D. A. Haugen, ed.), pp. 313–392. Amer. Meteorol. Soc., Boston, Massachusetts.
DuFort, E. C., and Frankel, S. P. (1953). *Math. Tables Aid Comput.* **7**, 135–152.
Estoque, M. A. (1973). In "Workshop on Micrometeorology" (D. A. Haugen, ed.), pp. 217–270. Amer. Meteorol. Soc., Boston, Massachusetts.
Lumley, J. L. (1967). *Phys. Fluids* **10**, 1405–1408.
Lumley, J. L. (1970). *J. Fluid Mech.* **41**, 413–434.
Mellor, G. L., and Herring, H. J. (1973). *AIAA J.* **21**, 590–599.
Reynolds, W. C. (1970). Rep. MD-27. Mech. Eng. Dep., Stanford Univ., Stanford, California.
Tennekes, H. (1973). In "Workshop on Micrometeorology" (D. A. Haugen, ed.), pp. 177–216. Amer. Meteorol. Soc., Boston, Massachusetts.
Tennekes, H., and Lumley, J. L. (1972). "A First Course in Turbulence." MIT Press, Cambridge, Massachusetts.
Wyngaard, J. C. (1973). In "Workshop on Micrometeorology" (D. A. Haugen, ed.), pp. 101–149. Amer. Meteorol. Soc., Boston, Massachusetts.
Wyngaard, J. C., Coté, O. R., and Izumi, Y. (1971). *J. Atmos. Sci.* **28**, 1171–1182.

INSTANTANEOUS VELOCITY AND LENGTH SCALES IN A TURBULENT SHEAR FLOW

P. J. SULLIVAN

Department of Applied Mathematics
The University of Western Ontario, London, Ontario, Canada

1. INTRODUCTION

In an earlier paper (Sullivan, 1971b), the prospect of describing the longitudinal dispersion of marked fluid in an open channel flow using a numerical simulation was explored. This approach consisted of recording the paths of many particles as these traversed the simulating random field that was programmed on a digital computer. The resulting concentration patterns at prescribed time intervals were compared with experimental measurements. Very crude estimates based upon Eulerian measurements were used to construct the simulating random field. Although a quite reasonable qualitative agreement between simulated and experimental results was achieved, it is clear that progress with such a simulation depends critically upon the acquisition of pertinent Lagrangian data. This is particularly true when the dispersion of particles with other than unit specific gravity is to be considered. The particular statistics considered here (e.g. projection of motion on a plane perpendicular to the mean flow vector) are those most relevant to such a simulation model of dispersion in a turbulent shear flow.

Neutrally buoyant particles of approximately $\frac{1}{2}$ mm diameter were injected, using an eyedropper, through the free surface of an open channel flow. The 8.95-cm-deep flow within the glass walled flume was determined to be homogeneous in the lateral direction y and streamwise direction x within a working section that was 0.46 m wide and 2.45 m long. When a particle was injected, approximately mid-way across the working section, it was allowed to travel approximately 10 depths before its motion was recorded. A photographic system, suspended above the flow, travelled at the flow discharge velocity and recorded the three-dimensional coordinate positions of the particle at discrete time intervals of 0.07 sec. The Reynolds number of the flow was $R = u_* d/v = 620$ where u_* is the friction velocity and d is the flow depth. Approximately 150 particle trajectories over the channel length were recorded and contained approximately 200 coordinate positions each.

[A more complete description of the experimental technique is given in Sullivan (1968) and Sullivan (1971b).] The reconstructed particle trajectories and in particular the projection of these paths on the z-y plane as shown in

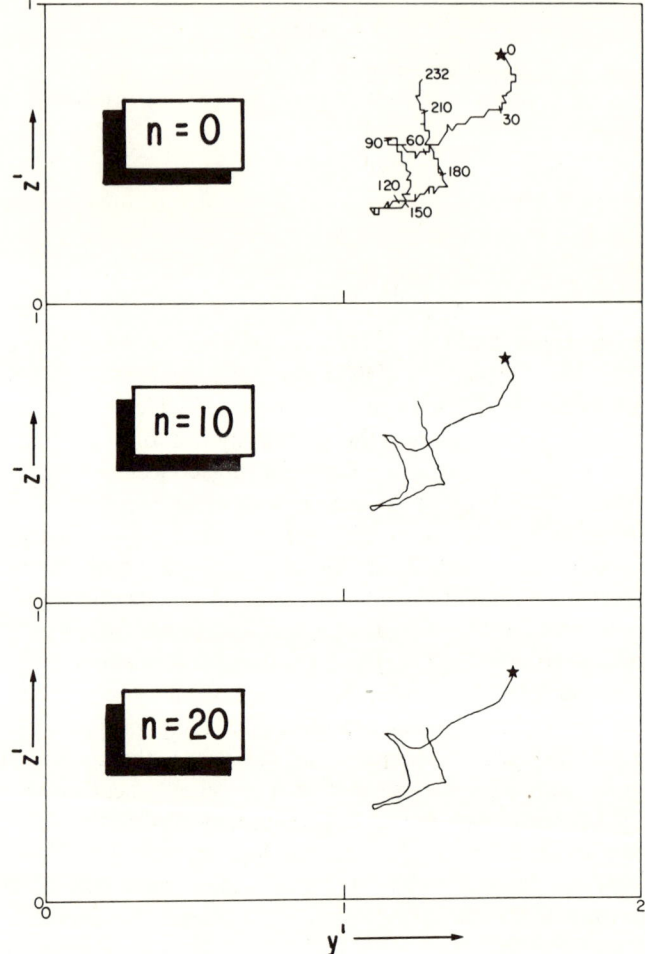

FIG. 1. Projection of a typical particle trajectory onto the z-y plane. n refers to an averaging procedure (see Section 2). The numbers on the curve refer to the number of time increments from the start of recorded motion.

Fig.1 suggest that those paths could be resolved into a series of circular arcs. That is three successive coordinate positions could be used to define a plane upon which a circular arc could be drawn through the three positions. The radius of this arc would then define an instantaneous length scale, and the swept angle θ divided by two time intervals serves to define an instantaneous angular velocity.

It is known (see Aris, 1962) that three points, sufficiently close together along a smooth curve such that $\delta s/r = 2\omega \delta t$ is small with respect to unity, uniquely define a local radius of curvature r. The smallest range of eddy sizes one would wish to represent occur in the dissipation range of the energy spectrum and an estimate of their size is given by Townsend (1956) as

$$R_d = v^{3/4} \left(\frac{3/2(\overline{u^2})^{3/2}}{L_u} \right)^{-1/4} \tag{1}$$

where $(\overline{u^2})^{1/2}$ is a root mean square value of velocity fluctuations and L_u represents an Eulerian integral length scale of the turbulence. An estimate within the open channel flow using $(\overline{u^2})^{1/2} \simeq u_*$ and $L_u \simeq d$ is

$$R_d/d \simeq (2/3R^3)^{1/4}. \tag{2}$$

For this value of r, the requirement of small $\delta s/r$ is approximately $\delta s/r \simeq 2u_* \delta t/R_d < 1$ or

$$\delta t < (d/2u_*)(2/3R^3)^{1/4}. \tag{3}$$

For r values comparable with d, the requirement is that $\delta t < d/2u_*$.

The time interval δt must be significantly less than the reaction time of turbulence $T_R = \lambda^2/10v \simeq \frac{3}{2}(\overline{u})^2/\varepsilon$ where λ is the microscale of turbulence and ε the rate of energy dissipation. With the approximation

$$\varepsilon \simeq \frac{3/2(\overline{u^2})^{3/2}}{L_u} \simeq \frac{3/2(u_*)^3}{d} \tag{4}$$

then

$$T_R \simeq d/u_*. \tag{5}$$

Comparing the nondimensional time interval available in the experiment $\delta t' = \delta t u_*/d = 0.00614$ with the equivalent nondimensional criteria from Eq. (3) that $\delta t' < \frac{1}{2}(2/3R^3)^{1/4} = 0.00727$ and $T'_R = 1$ it is apparent that the criteria of Eq. (3) will not be met for the smallest scales of turbulence, but that the time interval is far below the reaction time of turbulence and the criterion for r comparable with the depth, i.e., $\delta t' < \frac{1}{2}$. The nondimensional size of the particle used in the experiment is 0.00559. This is comparable with the value of $R'_d = R_d/d = 0.00727$ and would preclude an accurate representation of the motion at the dissipation range of eddy sizes.

Direct relationships between the values of ω' and r' and those of the more usual Eulerian measurements are not available. However, if individual radii of curvature r' are found to apply over a large number of time intervals δt such that, for example, arcs that are comparable with $\pi/4$ are formed, then one may expect the average value of r and the Eulerian integral length scale of turbulence in the cross stream direction to be comparable. The numbers along the typical trajectory shown in Fig. 1 refer to the number of time intervals from the start of the record. These figures suggest that one radius of curvature may well describe the trajectory over quite a large number of time intervals.

2. Length Scales

The particle trajectories are smoothed to reduce the small scale erratic behaviour resulting from experimental error in spatial and temporal resolution and perhaps resulting from scales of turbulence far into dissipation range of sizes. In each trajectory the values of adjacent coordinate positions are averaged, i.e., $z_j = (z_i + z_{i+1})/2$ and $y_j = (y_i + y_{i+1})/2$, and this is repeated 20 times for each record. Figure 1 with $n = 10$ and $n = 20$ shows the change in the trajectory for 10 and 20 repetitions of the averaging procedure. The smoothed trajectories with $n = 20$ are used for further statistical information.

The channel depth is subdivided into 10 equal segments on z. Using experimental values that are found in each of these 10 horizontal slabs an experimental probability density function $p(r')$ is compiled. Figures 2a and 2b show the experimental data and the solid curve on these figures is calculated from

(6) $$p(r') = [(\alpha r')^\eta / r' \Gamma(\eta)] \exp(-\alpha r').$$

The values of α and η are determined using an iterative procedure. First the values of α and η are estimated from the first and second moments μ and σ of the experimental data in the range $0.01 < r' < 0.5$. That is

$$\mu = \sum r' p(r') \Delta r' \quad \text{and} \quad \sigma = \sum r'^2 p(r') \Delta r'$$

These values of α_1 and η_1 are then used to generate corrective factors c and g from the gamma distribution to be applied to the first estimates of μ and σ. This procedure is repeated as

(7) $$P_i(r') = \frac{(\alpha_i)^{\eta_i} (r')^{\eta_i - 1}}{\Gamma(\eta_i)} \exp(-\alpha_i r')$$

(8) $$\eta_i = \alpha_i \mu_i, \quad \alpha_i = \frac{\mu_i}{\sigma_i - (\mu_i)^2},$$

(9) $$\mu_{i+1} = \frac{\mu_i}{c_i}, \quad c_i = \frac{\int_a^b r' P_i(r') \, dr'}{\int_0^\infty r' P_i(r') \, dr'},$$

(10) $$\sigma_{i+1} = \frac{\sigma_i}{g}, \quad g = \frac{\int_a^b r'^2 P_i(r') \, dr'}{\int_0^\infty r'^2 P_i(r') \, dr'},$$

$a = 0.01$, $b = 0.5$, until the value of α is changed by less than 1%.

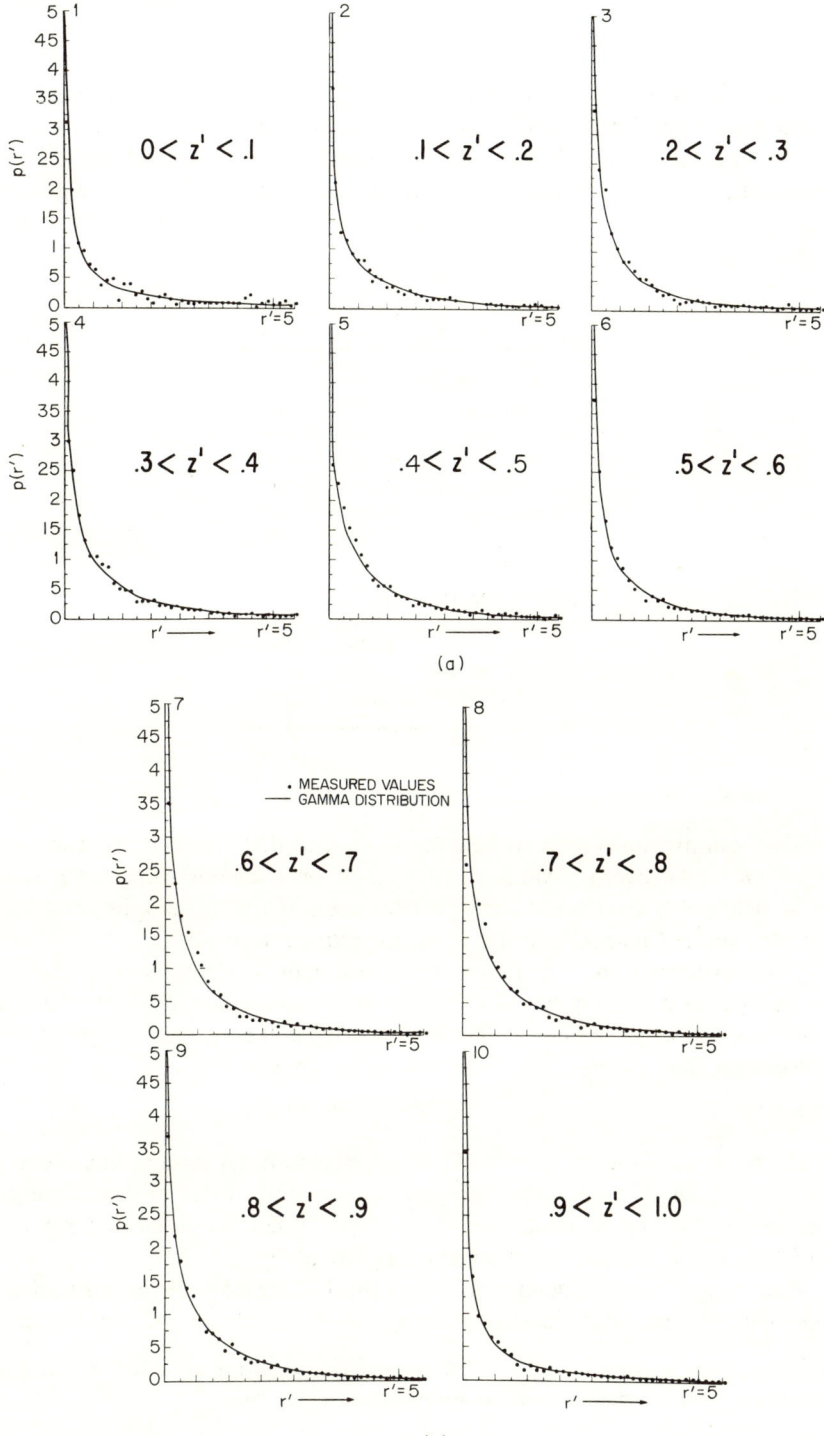

FIG. 2. A comparison at each of 10 sampling intervals on the depth of experimental values with the gamma distribution.

The solid curves on Figs. 2a and 2b appear to be reasonable representations of the experimental data. Table I presents the values of α and η for each of the sampling intervals on z. There appears to be a dependence on z of these values in that the α and η values are as much as a factor of 3 greater near $z' = 0.5$, where $z' = z/d$, than values near the extremities $z' \simeq 0$ and $z' \simeq 1$.

TABLE I

Sampling interval	α	η
$0 < z' \leq 0.1$	1.988	0.248
$0.1 < z' \leq 0.2$	3.629	0.421
$0.2 < z' \leq 0.3$	5.041	0.470
$0.3 < z' \leq 0.4$	5.280	0.582
$0.4 < z' \leq 0.5$	5.294	0.604
$0.5 < z' \leq 0.6$	4.745	0.413
$0.6 < z' \leq 0.7$	5.556	0.502
$0.7 < z' \leq 0.8$	5.642	0.606
$0.8 < z' \leq 0.9$	5.646	0.568
$0.9 < z' \leq 1.0$	2.780	0.198

3. Angular Velocity

The instantaneous angular velocity ω' is recorded for each experimental value of r'. An average value of ω', i.e., $\overline{\omega}'$ on the condition that r' is in $[r' - 0.005, r' + 0.005]$ is determined for each of the 10 sampling sections selected on z'. Figures 3a–c show this experimental result.

There appears to be a definite dependence of $\overline{\omega}'$ on r', and this appears to be independent of z'. Figure 4 shows $\overline{\omega}'$ to have an exponential dependence on r' when all experimental values are included in the average irrespective of z' location. The experimental relationship so determined

$$\overline{\omega}' = 2.2 r'^{-0.698} \tag{11}$$

is shown as a solid curve in Fig. 3 and appears to fit the data points at every sampling interval on z'. The average of the absolute value of the difference between individual ω' values and values given by Eq. (11) when corresponding values of r' are used is approximately 20–30 %.

Equation (11) is independent of z and this feature may suggest an association with the universal inertial subrange of the spectrum.[1] If it is assumed

---

[1] The author is indebted to L. S. G. Kovasznay who suggested, at the IUTAM–IUGG Symposium, that a connection with the inertial subrange may exist.

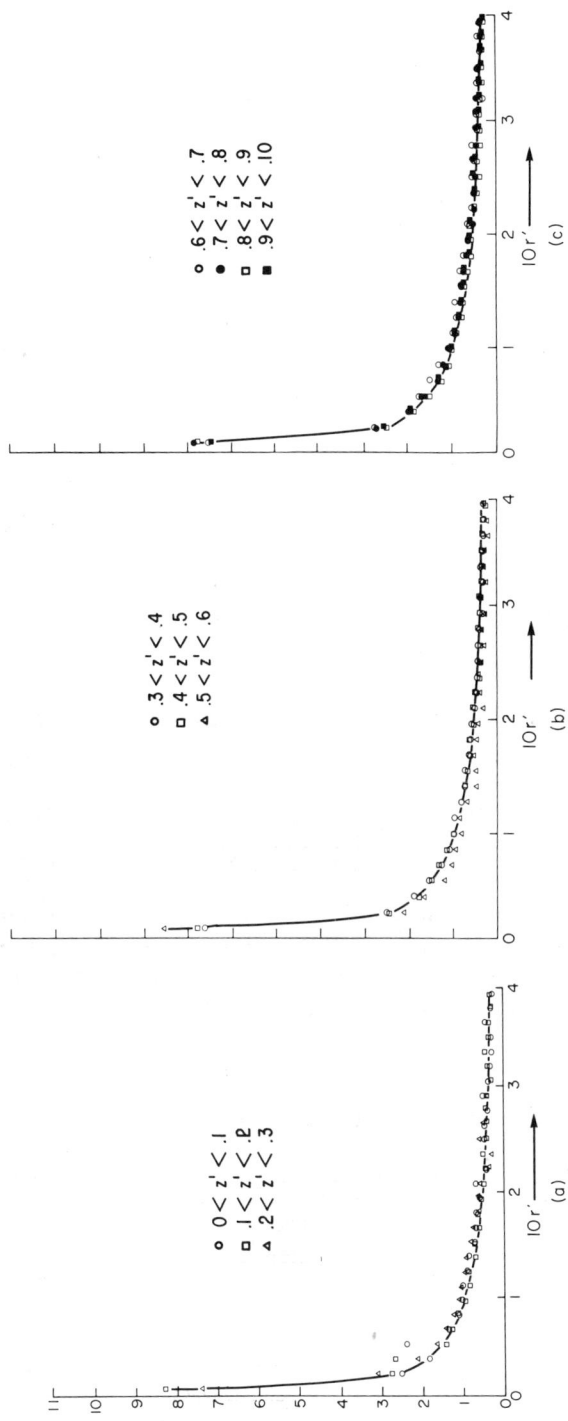

FIG. 3. A comparison of Eq. (11) with experimental values of ω' at each of ten sampling intervals on the depth.

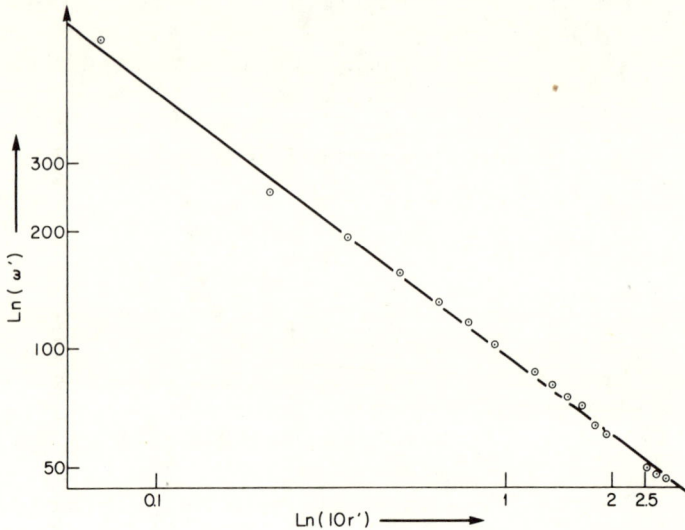

FIG. 4. Experimental values of $\overline{\omega}'$ when all experimental data is used. Solid line given by Eq. (11).

that r is in the inertial subrange and ω depends only on r and ε, then

(12) $$\omega = a\varepsilon^{1/3}r^{-2/3}$$

where a is a universal constant. With the estimate for ε given in Eq. (4), then

(13) $$\omega' \simeq a(\tfrac{3}{2})^{1/3}r'^{-2/3}$$

which by comparison with Eq. (11) suggest that a is approximately 2. The comparison is quite favourable between Eqs. (13) and (11), however it is unlikely that an inertial subrange structure would be present with a value of $R = 620$ and Eq. (11) appears to describe the experimental data over even the largest length scales measured.

4. Time Scales

One would like an estimate of the average persistence time of a particular value of r' along the trajectory. Starting with a value of r' in $[r' - 0.028, r' + 0.028]$ one can construct a "mean history" in terms of r' values a particle encounters following this particular range of initial values. A normalized expression of such a mean history is

(14) $$B(t'; r'_1) = [\overline{r'(t')} - A]/[\overline{r'(0)} - A]$$

where r'_1 refers to the initial value on $[r'_1 - 0.028, r'_1 + 0.028]$ at $t' = 0$ and A is the average value of r' for all realizations recorded; $A = 0.097$. The overbar denotes an ensemble average. It is expected that $B(t', r'_1)$ should have positive values and decrease in a regular way between the initial value of 1 and final value of 0. If for example the value of r' found after one time interval δt was entirely unrelated to the initial value of r'_1, then $B(\delta t; r'_1) \equiv 0$ as $\overline{r'(\delta t)} \equiv A$. A persistence time scale T' is defined using the function $B(t'; r'_1)$ as

$$(15) \qquad T = \int_0^\infty |B| \, dt'.$$

The value of T' is interpreted as the equivalent time interval before which $r'(t') = r'(0)$ and after which $r'(t') = A$.

Figure 5 shows the way in which the persistence time scales T' depend upon r'_1. There appears to be a pronounced increase in values of T' for initial values of r'_1 close to A. Figure 6 shows typical forms of $B(t'; r'_1)$ as well as those for $r'_1 \simeq A$. When $r'_1 \gtrsim A$ then for small t', $B(t', r'_1) > 1$; and for $r'_1 \lesssim A$, there occurs a rapid reduction of $B(t'; r'_1)$ for small t' and negative values occur. It appears that in the neighbourhood of A, values of r'_1 tend to increase on average through the value of A.

One possible interpretation of the tendency of values of r'_1 to increase in the neighbourhood of A may be found in the production and convection of turbulent energy from the mean flow. The generation of turbulence within the wall layer and the projection of this to the rest of the flow are clearly demonstrated by the measurements of Kline et al. (1967). When the flow depth was considered in two halves, the tendency of r' values to increase was somewhat more pronounced for experimental values recorded in the lower

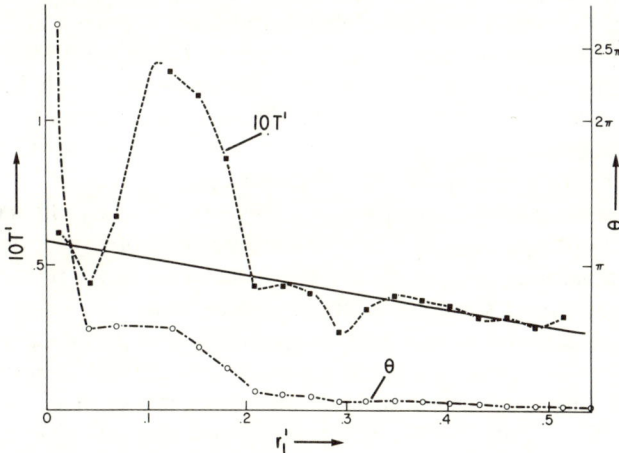

FIG. 5. The values of average residence time T' and swept angle θ for various initial values of r'_1.

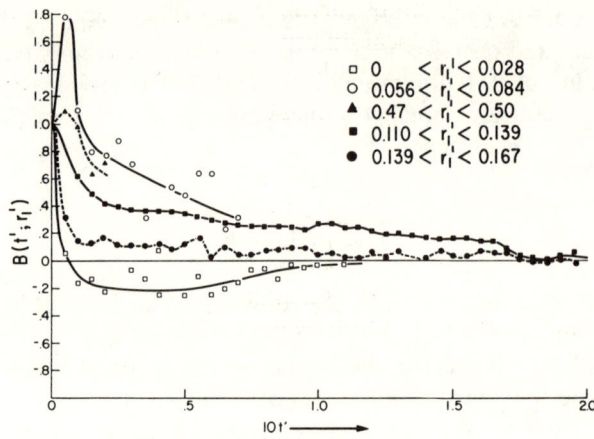

Fig. 6. Some representative values of $B(t'; r'_1)$.

half. The amount of experimental data would not allow any further subdivision. The measurements of r' which in this study occur on a y-z plane may possibly reflect the effects of the transfer of energy in the x-z plane demonstrated by Kline et al.

5. Concluding Remarks

The spatial and temporal resolution of the experimental data used is such that the smallest scales of turbulence are not likely to be well represented and the number of events recorded are far fewer than desired. However within these limitations it would appear that instantaneous length and velocity scales can be simultaneously defined, and there appears to be a definite relationship between the conditional average of the velocity scale and their corresponding length scale. The length scales appear to be well described by a gamma distribution with parameters that depend upon the vertical distance z in an open channel flow. The mean value of this length scale is not inconsistent with an Eulerian integral scale as given by Laufer (1951) for the cross-stream direction in a duct flow, i.e., $2\bar{r}' \simeq 0.2$. Where radii of curvature persist for a sufficient length of time to have portions of trajectories describing arcs comparable with say $\pi/2$ and in particular the larger radii measured, then one may expect some such comparison to exist.

Where this criteria of large arcs is applicable and particularly over the smaller values of r', one may expect a loose application to relative dispersion. When one considers a meandering-plume model of dispersion (see Batchelor, 1952; Sullivan, 1971a), it is expected that velocities associated with length scales of the order of the plume diameter will dominate relative dispersion. One can, for radii of curvature which persist over a reasonable spatial extent, construct a local diffusivity dependent on the width of the plume and velocities associated with this dimension, i.e.,

(16) $$d\bar{r}^2 \propto (\bar{\omega}'r')r'/dt.$$

When Eq. (11) is used for $\bar{\omega}'$,

(17) $$d\bar{r}^2/dt \propto r^{1.302}.$$

This result is very close to the "$\frac{4}{3}$ power law" of relative dispersion. It is interesting that the $\frac{4}{3}$ dependence of a relative diffusivity has been observed in very diverse circumstances which do not always have an inertial subrange in evidence.

ACKNOWLEDGMENTS

The experimental work was undertaken at the Department of Applied Mathematics and Theoretical Physics, University of Cambridge, England in 1967. Subsequent analysis and computation was supported from NRC Grant N62. The author wishes to express his gratitude to the referee for some very useful comments particularly in relation to Eq. (12).

REFERENCES

Aris, R. (1962). "Vectors, Tensors, and the Basic Equations of Fluid Mechanics." Prentice-Hall, Englewood Cliffs, New Jersey.
Batchelor, G. K. (1952). *Proc. Cambridge Phil. Soc.* **48**, 345–362.
Kline, S. J., Reynolds, W. C., Schraub, F. A., and Rundstadler, P. (1967). *J. Fluid Mech.* **30**, 741–773.
Laufer, J. (1951). *NACA (Nat. Adv. Comm. Aeronaut.) Rep. No. 1033*.
Sullivan, P. J. (1968). Ph.D. Thesis, Univ. of Cambridge, Cambridge, England.
Sullivan, P. J. (1971a). *J. Fluid Mech.* **47**, 601–607.
Sullivan, P. J. (1971b). *J. Fluid Mech.* **49**, 551–576.
Townsend, A. A. (1956). "The Structure of Turbulent Shear Flow." Cambridge Univ. Press, London and New York.

NUMERICAL COMPUTATION OF TURBULENT SHEAR FLOWS[1]

STEVEN A. ORSZAG AND YIH-HO PAO

Flow Research, Inc. Kent, Washington 98031, U.S.A.

1. INTRODUCTION

In this paper, we report simulations of turbulent shear flows by direct numerical solution of the three-dimensional Navier–Stokes equations. This approach provides several advantages over more conventional approaches. First, the complete flow field is obtained at all times so that detailed flow characteristics may be obtained that would be difficult to measure in the laboratory. Second, initial conditions can be accurately controlled so that their effect may be determined. Third, numerical simulations are convenient experiments to assess the effect of various physical processes, like chemical reactions, on turbulence and vice versa. Fourth, the simulation results can be used to test and suggest various statistical hypotheses involved in turbulence models and theories.

The present work is an extension to inhomogeneous shear flows of the three-dimensional, homogeneous, isotropic turbulence simulations reported by Orszag and Patterson (1972). In this earlier work, simulations using up to $(32)^3$ Fourier modes to represent each velocity component were performed at microscale Reynolds numbers 20–50 (grid Reynolds numbers 5000–30,000), corresponding to moderate Reynolds number grid turbulence. The results of these homogeneous turbulence simulations agreed well with both turbulence theory (e.g., the direct interaction approximation) and laboratory experiments. Simulations have also been made of two-dimensional "turbulence" using up to $(128)^2$ Fourier modes (Herring *et al.*, 1974) with successful comparisons with the results of turbulence theories. These successful simulations gave impetus to the present extension to shear flows.

In the homogeneous turbulence simulations, it was observed that some important features of the flows were Reynolds number independent (Herring *et al.*, 1974), so that there is basis for assuming that the large-scale features of

---

[1] This work is supported by Fluid Dynamics Branch, Office of Naval Research under Navy Contract No. N00014-72-C-0355, ONR Task No. NR 062-464 and the Climatic Impact Assessment Program, Department of Transportation under Contract DOT-A5-30041.

flows simulated at moderate Reynolds numbers are not very different from those at huge Reynolds numbers. It is also reasonable to assume Reynolds-number-independence of large scales in shear flows *provided* initial and boundary conditions are Reynolds number independent. It seems that Reynolds number dependencies observed in laboratory flows may be ascribed to variations in initial or boundary conditions; for example, Reynolds number variations in turbulent jets seem to be mainly due to variations in inlet conditions (like boundary layer thickness) with Reynolds number.

As a first application of our shear flow turbulence codes, we have simulated the momentumless wake of a self-propelled body. As pointed out by Naudascher (1965), the momentumless wake bears close relationship to grid (homogeneous) turbulence, since both are characterized by a very limited region in which there is significant energy transformation from mean (shear) flow to turbulence (Reynolds stresses).

In Section 2, we summarize the dynamical equations and boundary conditions employed with particular emphasis on the momentumless wake model. In Section 3, some novel aspects of the numerical approximation are discussed, while in Section 4, some results are presented for wake turbulence. Finally, in Section 5, we summarize our results and the future outlook.

2. Equations of Motion

The Navier–Stokes equations of motion for an incompressible fluid are

$$\partial \mathbf{v}(\mathbf{x}, t)/\partial t = \mathbf{v}(\mathbf{x}, t) \times \boldsymbol{\omega}(\mathbf{x}, t) - \nabla \Pi(\mathbf{x}, t) + \nu \nabla^2 \mathbf{v}(\mathbf{x}, t) \qquad (1)$$

$$\nabla \cdot \mathbf{v}(\mathbf{x}, t) = 0 \qquad (2)$$

where $\mathbf{v}(\mathbf{x}, t)$ is the three-dimensional velocity field, $\boldsymbol{\omega}(\mathbf{x}, t) = \nabla \times \mathbf{v}(\mathbf{x}, t)$ is the vorticity, $\Pi(\mathbf{x}, t) = p + \frac{1}{2}v^2$ is the pressure head, $p(x, t)$ is the pressure, and ν is the kinematic viscosity. Equation (1) is written in rotation form to facilitate numerical solution (Section 3).

Boundary conditions require more discussion. In order to simulate the momentumless wake of a self-propelled body the computational domain should include a sizeable region of potential flow both upstream and downstream of the body as well as the body itself. However, this would be very wasteful since most of the computational degrees of freedom would be involved in resolving the flow outside the turbulent wake. Even within the wake, the first several body diameters downstream are a means of adjustment to more self-similar conditions downstream. The effect of these resolution problems is that with presently practicable numerical simulations that involve at most an order of 10^5 degrees of freedom to describe the velocity field, there would remain little more than 10^3 degrees of freedom to determine the flow in the turbulent wake. Clearly, it is not possible to simulate details of a turbulent flow with so little resolution.

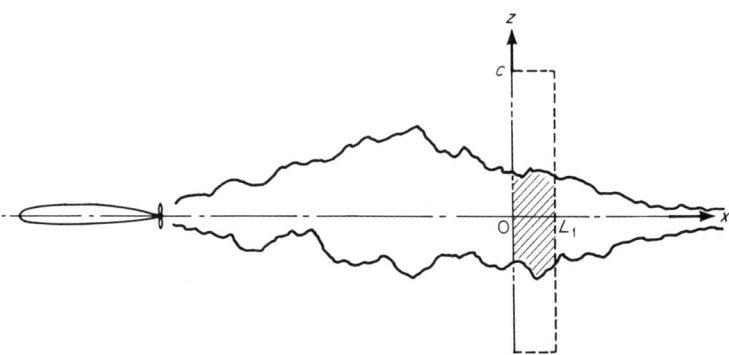

FIG. 1. Cross section of a turbulent wake with the section simulated in the present computations.

In order to avoid the resolution problem just described, we simulate wake flow in the following way. We isolate a slab in the wake region, like the section $[0, L_1]$ shown in Fig. 1, and follow its time evolution by considering it enclosed in a three-dimensional box as shown in Fig. 2. The wake axis is assumed to be along the x_1 axis and periodic boundary conditions are applied at $x_1 = 0, L_1$ and $x_2 = 0, L_2$. On the other hand, rigid free-slip (no-stress) or rigid no-slip boundary conditions are applied at $x_3 = 0, L_3$.

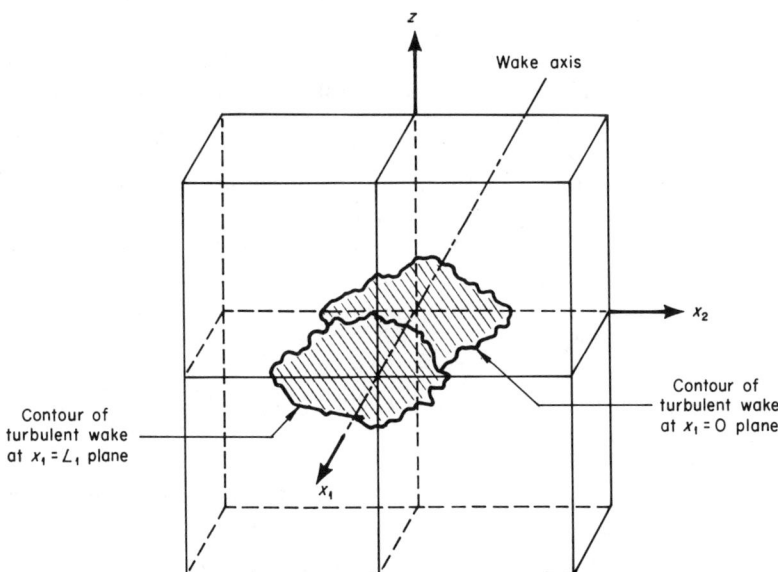

FIG. 2. Spatial box enclosing the section of wake $[0, L_1]$ is as in Fig. 1. Periodic boundary conditions are applied at the sidewalls and either free-slip or no-slip conditions are applied at the top and bottom.

This choice of boundary conditions in the x_3 direction is necessitated by the fact that our computer codes are written to solve also the Boussinesq equations of motion of a stratified fluid for which the x_3 direction is singled as the direction of gravity and the equations of shear flows in a channel with rigid top and bottom boundaries.

Initial conditions are chosen so that the slab of turbulence is a realization of a section of a fully developed turbulent wake, i.e.,

$$\mathbf{v}(\mathbf{x}, 0) = \bar{\mathbf{v}}(\mathbf{x}, 0) + \mathbf{v}'(\mathbf{x}, 0) \tag{3}$$

Here the mean defect velocity $\bar{\mathbf{v}}(\mathbf{x}, 0)$ is chosen to be

$$\bar{\mathbf{v}}(\mathbf{x}, 0) = v_d(r)\hat{x}_1 \tag{4}$$

where $\hat{x}_1$ is the unit vector in the x_1 direction, $r^2 = (x_2 - \tfrac{1}{2}L_2)^2 + (x_3 - \tfrac{1}{2}L_3)^2$ and

$$\int_0^\infty v_d(r) r \, dr = 0 \tag{5}$$

in order to simulate the momentumless wake. The initial fluctuating velocity $\mathbf{v}'(\mathbf{x}, 0)$ is chosen as a realization of an incompressible random velocity field with specified local energy spectrum and turbulence intensity. Some of the details involved in the construction of $\mathbf{v}'$ are explained in Section 3.

Numerical solution of the Navier–Stokes equations from the imposed initial conditions gives the time evolution of the simulated cylindrical section of turbulence. After evolution time t, the results are interpreted as a realization of a section of the wake flow at a distance $x_1 = U_0 t$ downstream from the location of the initial wake section, where U_0 is the body velocity. In other words, the initial flow is chosen to have the same (or rather similar) statistical properties as a section of a turbulent wake and the time evolution of the flow is interpreted as the downstream variation of the wake. The model we solve numerically is statistically homogeneous along the wake axis x_1 but nonstationary in time; the wake in the frame of a uniformly moving body is statistically stationary in time but is inhomogeneous in x_1. It is asserted that the Galilean transformation $x_1 = U_0 t$ relates the numerical and physical experiments.

3. Numerical Methods

3.1. Free-Slip Boundaries

If the boundary conditions at $x_3 = 0, L_3$ are no-stress (free-slip), while those at $x_1 = 0, L_1$ and $x_2 = 0, L_2$ are periodic, then

$$\partial v_1/\partial x_3 = \partial v_2/\partial x_3 = v_3 = 0 \quad \text{on} \quad x_3 = 0, L_3, \tag{6}$$

$$\mathbf{v}(x_1 + mL_1, x_2 + nL_2, x_3) = \mathbf{v}(\mathbf{x})$$

In this case, the velocity field can be expanded in the Fourier series

$$
(7) \quad v_\alpha(\mathbf{x}, t) = \sum_{|k_1|<K_1} \sum_{|k_2|<K_2} \sum_{0\leq k_3<K_3} u_\alpha(\mathbf{k}, t)
$$

$$
\exp[2\pi i(k_1 x_1/L_1 + k_2 x_2/L_2)]
$$

$$
\times \begin{cases} \cos \pi k_3 x_3/L_3, & \alpha = 1, 2 \\ \sin \pi k_3 x_3/L_3, & \alpha = 3 \end{cases}
$$

where the indicated summations are over integers k_1, k_2, k_3. For the simulations reported below, the cutoffs are chosen as $K_1 = K_2 = 16$, $K_3 = 32$, so that the spectral representations (7) each involve about $(32)^3$ independent degrees of freedom (Fourier amplitudes).

We have used the expansions (7) in two kinds of numerical approximation to the Navier–Stokes equations. In the spectral method (Orszag, 1971a), equations for $\mathbf{u}(\mathbf{k})$ are derived by substituting (7) into (1), multiplying the result by $\exp[-2\pi i(k_1 x_1/L_1 + k_2 x_2/L_2)] \cos[\pi k_3 x_3/L_3]$ for $\alpha = 1, 2$ (and by the same expression with the cosine replaced by sine if $\alpha = 3$), and finally integrating the result over the box $0 \leq x_\alpha < L_\alpha$. The resulting equations are, after elimination of the pressure by means of the incompressibility constraint (2),

$$
(8) \quad \frac{\partial \bar{u}_\alpha(\mathbf{k}, t)}{\partial t} + v\bar{k}^2 \bar{u}_\alpha(\mathbf{k}, t) = -i\bar{k}_\beta(\delta_{\alpha\gamma} - \bar{k}_\alpha \bar{k}_\gamma/\bar{k}^2) \sum_{\substack{\mathbf{p}+\mathbf{q}=\mathbf{k} \\ -K_\rho < p_\rho, q_\rho < K_\rho}} \bar{u}_\beta(\mathbf{p}, t)\bar{u}_\gamma(\mathbf{q}, t)
$$

where $\bar{k}_\alpha = 2\pi k_\alpha/L_\alpha$ for $\alpha = 1, 2$ and $\bar{k}_3 = \pi k_3/L_3$, and

$$
(9) \quad \bar{u}_\alpha(\mathbf{k}, t) = \begin{cases} \tfrac{1}{2} u_\alpha(k_1, k_2, |k_3|, t), & \alpha = 1, 2 \\ (1/2i)\,\text{sgn}\,k_3\,u_3(k_1, k_2, |k_3|, t), & \alpha = 3 \end{cases}
$$

Numerical solution of (8) is accomplished using fast Fourier transform methods to evaluate the convolution sums, leapfrog time differencing on the nonlinear terms, and Crank–Nicolson (implicit) time differencing on the viscous terms. Overall, 18 Fourier transforms on $(32)^3$ points must be performed each time step (Orszag, 1971a).

In the pseudospectral method (Fox and Orszag, 1973), the expansions (7) are used as an interpolatory tool to evaluate the derivatives appearing in (1). "Grid" points $x_\alpha = L_\alpha j_\alpha/2K_\alpha$, $j_\alpha = 0, \ldots, 2K_\alpha - 1$ for $\alpha = 1, 2$, $x_3 = L_3 j_3/K_3$, $j_3 = 0, \ldots, K_3$ are introduced and the series (7) are used to evaluate derivatives as, for example,

$$
(10) \quad \omega_3(\mathbf{x}, t) = \sum_{|k_1|<K_1} \sum_{|k_2|<K_2} \sum_{0\leq k_3<K_3} [i\bar{k}_1 u_2(\mathbf{k}, t) - i\bar{k}_2 u_1(\mathbf{k}, t)]
$$

$$
\times \exp[i\bar{k}_1 x_1 + i\bar{k}_2 x_2] \cos \bar{k}_3 x_3
$$

The final result of this procedure is to give a set of equations for the spectral amplitudes $\bar{u}(k)$, defined by (9), that are identical to (8) except for the replacement of the convolution sums by similarly truncated sums with $p_\alpha + q_\alpha = k_\alpha \pm 2K$ or $p_\alpha + q_\alpha = k_\alpha$ (Orszag, 1971a). The additional terms entering the sums in (8) are usually called "aliasing" terms. It has been shown (Fox and Orszag, 1973) that the differences in the results obtained by the present pseudospectral method and the spectral method are generally negligible so long as either method gives an accurate solution of the equations of motion. Since the pseudospectral method may be implemented in only nine Fourier transforms over $(32)^3$ points per time step, it is roughly a factor 2 more efficient than the spectral method and so has been used for most of the simulations reported below.

Both the spectral and pseudospectral methods have been programmed for a CDC-7600. The programs involve double buffering of data between small core memory, large core memory, and disks. The spectral method requires about 6 s per time step on the CDC-7600, while the pseudospectral code requires only 3.2 s per time step, both for the cutoffs $K_1 = K_2 = 16$, $K_3 = 32$. Many of the critical internal loops of the program are written in assembly language to avoid inefficiencies attributable to the Fortran compiler, although further improvements in the code should permit a speedup of nearly 50 %.

Besides the speed advantage of the pseudospectral method over the spectral method, the former has the great advantage that it applies with but the most minor of modifications to problems involving more complicated physics, like chemical reactions or radiation. Since the expansions (7) are used merely as an interpolatory tool in the evaluation of derivatives, they may be similarly used in the evaluation of these more complicated effects. Nevertheless, the pseudospectral method shares all the advantages of the spectral method with regard to accuracy and, especially, efficiency improvement over finite-difference schemes (Orszag and Israeli, 1974). The expression of (1) in rotation form is useful since it gives pointwise energy conservation.

3.2. No-Slip Boundaries

With no-slip boundary conditions applied at $x_3 = 0, L_3$, the velocity satisfies

(11) $\quad \mathbf{v} = 0 \quad$ on $\quad x_3 = 0, L_3, \quad \mathbf{v}(x_1 + mL_1, x_2 + nL_2, x_3) = \mathbf{v}(\mathbf{x})$

instead of (6). It is no longer appropriate to use the Fourier expansions (7), not just because $v_1 = v_2 = 0$ at $x_3 = 0, L_3$, but rather because imposition of a Fourier series representation of the x_3 dependence would induce Gibbs phenomena at the boundaries and result in slow convergence of the Fourier series (Orszag, 1971a).

No-slip boundaries are best treated by Chebyshev expansion in x_3 (Orszag, 1971b). The velocity field is represented as

$$\mathbf{v}(\mathbf{x}, t) = \sum_{|k_1|<K_1} \sum_{|k_2|<K_2} \sum_{0 \le k_3 < K_3} \mathbf{u}(\mathbf{k}, t) \tag{12}$$

$$\times \exp[i\bar{k}_1 x_1 + i\bar{k}_2 x_2] T_{k_3}([2x_3 - L_3]/L_3)$$

where the nth degree Chebyshev polynomial $T_n(x)$ is defined by $T_n(\cos \theta) = \cos n\theta$. It may be shown that, if $\mathbf{v}(\mathbf{x})$ is smooth, the series (12) or any of its termwise derivates do not exhibit Gibbs phenomena at the boundaries. Equivalently, the series (12) converges faster than algebraically as $K_\alpha \to \infty$. Notice that the boundary conditions (11) must be imposed as constraints on (12).

One advantage of (12) is that it leads to pseudospectral approximations to (1), (2) that are very similar in form to that following from the Fourier series (7). In particular, the pseudospectral equations with (12) may be implemented in nine Fourier transforms per time step, and are but slightly less efficient than for free-slip boundaries. With cutoffs $K_1 = K_2 = 16$ and $K_3 = 32$, our code requires 4 s per time step. In this latter code, time differencing is done by Adams–Bashforth differencing (Lilly, 1965) on the nonlinear terms (to avoid instability that would result from use of leapfrog because of the stability induced by the boundary conditions) and Crank–Nicolson on the viscous terms. The pressure computation is done by using (2) to get an equation tridiagonal in the Chebyshev index k_3 for the pressure and diagonal in k_1 and k_2. Solution of the resulting tridiagonal system accounts for most of the additional time required by the rigid boundary code over the free-slip boundary code.

Methods based on the spectral expansion (12) have an important advantage over difference methods, in addition to the advantages they enjoy when periodic or free-slip boundary conditions are applied. The "grid" points for the pseudospectral method based on (12) are $x_3 = \tfrac{1}{2} L_3 (1 + \cos \pi j_3 / K_3)$ for $j_3 = 0, \ldots, K_3$, so that the effective resolution near the walls $x_3 = 0$, L_3 is $\Delta x_3 \sim L_3/K_3^2$. In fact, if there is a boundary layer of thickness δ along either $x_3 = 0$ or $x_3 = L_3$, it may be shown (Orszag and Israeli 1974) that it is sufficient to take $K_3 \sim 3(L_3/\delta)^{1/2}$ to achieve better than 1% accuracy in the boundary layer. In effect, the Chebyshev polynomial expansion gives a highly nonuniform grid near the boundaries. This behavior is particularly appropriate for the study of channel flows, etc., where thin boundary layers are apt to develop.

3.3. Initial Conditions

In the wake model introduced in Section 2, initial conditions of the form of a "cylinder" of turbulence are required. The mean defect velocity (4) may be imposed arbitrarily, but the fluctuating component $\mathbf{v}'(\mathbf{x})$ must satisfy the incompressibility constraint. This is done by writing

(13) $$\mathbf{v}'(\mathbf{x}) = \nabla \times \mathbf{A}'(\mathbf{x})$$

where the vector potential $\mathbf{A}'(\mathbf{x})$ is chosen to achieve the desired local energy spectrum and turbulence intensity. It suffices to choose $\mathbf{A}'(\mathbf{x})$ to be of the form

(14) $$\mathbf{A}'(\mathbf{x}) = [I(r)]^{1/2}\mathbf{B}(\mathbf{x})$$

where the turbulence intensity function $I(r)$ is a nonrandom function of only the distance from the wake axis x_1, and the fluctuation component $\mathbf{B}(\mathbf{x})$ is chosen as a realization of a homogeneous, isotropic random field with specified isotropic energy spectrum, as done by Orszag and Patterson (1972). If the intensity function is chosen to vanish outside a radius r_0 from the wake axis, the resulting turbulent velocity field is, by (13), nonturbulent outside the cylinder of radius r_0 centered on the wake axis.

In summary, our technique of imposing the initial conditions allows arbitrary mean velocity profile, turbulence intensity profile, and local turbulence energy spectrum.

4. Momentumless Wake

The wake model of Section 2 and the numerical methods of Section 3 permit simulation of the turbulent wake of a self-propelled body. Free-slip boundary conditions are applied at $x_3 = 0$, L_3 and the spectral cutoffs $K_1 = K_2 = 16$, $K_3 = 32$ are used. The following choice of initial parameters was made: $L_1 = L_2 = L_3 = 2$, $v_d(r) = v_0 \sin(r/r_1)/(r/r_1)$ where $r_1 = \frac{1}{4}$, $I(r) = \max[1 - r^2/4.5, 0]$, and turbulence energy spectrum $E(k) = Ak^4 \exp(-Bk^2)$ (cf. Orszag and Patterson, 1972]. The initial conditions are chosen to match as closely as possible the results of Naudascher (1965) and Wang (1965) at a location four body diameters behind a self-propelled disk in a wind tunnel. In the numerical simulations, the viscosity is chosen to be $v = 0.005$, so that the Reynolds number of the simulation is roughly 13,000 (run 2a), and $v = 0.003$ with Reynolds number roughly 22,000 (run 3). In comparison, the laboratory experiments of Naudascher and Wang were run at Reynolds numbers of roughly 55,000.

Some measure of the degree of complication of the flow that is simulated is given by Figs. 3 and 4. In Fig. 3, contours of the run 3 axial mean velocity are plotted at $t = 0.8$, corresponding to about seven diameters downstream from the body upon making identification of body velocity by the ratio $\max[v_d(r)]/U_0$ at the station at $x/D = 4$ with the experimental results. The axial mean is determined by averaging over the x_1 direction. In Fig. 4, contours of the run 3 x_1 component of vorticity are plotted at $t = 0.8$.

Fig. 3. Axial mean velocity contours for run 3 at $t = 0.8$.

In Fig. 5, we compare the axial variation of maximum turbulent intensity with the results of Wang and Naudascher. Here U_0 and t_0 are determined as explained above by correspondence with the data at the station at $x = 4D$, while $x_0 = 2D$ according to Naudascher and Wang. It is apparent from these results that the present simulations are in substantial agreement with the laboratory results of Naudascher and Wang, at least over the limited downstream range of the present experiments.

In Fig. 6, we compare the radial variation of axial mean-square turbulent intensity in the laboratory experiments and the numerical calculations. The curve labeled $t = 0$ shows the initial distribution, while that labeled $t = 0.522$ shows the resulting distribution for run 2a at about $6D$ downstream from the body. Again, the agreement with the experimental results is satisfactory.

It is apparent from the simulation results of this section that numerical simulation of turbulent shear flows is well within present computational capabilities. However, the process of extracting useful information about shear flows from numerical simulations is still in its infancy. The most

Fig. 4. Axial component of vorticity contours for run 3 at $t = 0.8$.

significant problem with the present simulations is the scarcity of points in the axial direction with which to compute averages. With just 32 points in the longitudinal direction, statistical errors are large and either multi-time step information or ensembles must be employed to improve the statistics. This situation should be contrasted with the case for homogeneous, isotropic turbulence where space averages suffice to give statistical results generally to within 5 % (Orszag and Patterson, 1972).

5. Conclusion

The present simulations of turbulent shear flows are but a first step toward proper understanding of the basic mechanisms and dynamics of these flows. Detailed comparisons and tests are presently being made between various turbulence modelling hypotheses, laboratory experiments, and the present simulations. As time and the art of numerical simulation progress, simulations like the present ones should be expected to fulfill more and more the need of a laboratory workhorse. Our present simulations

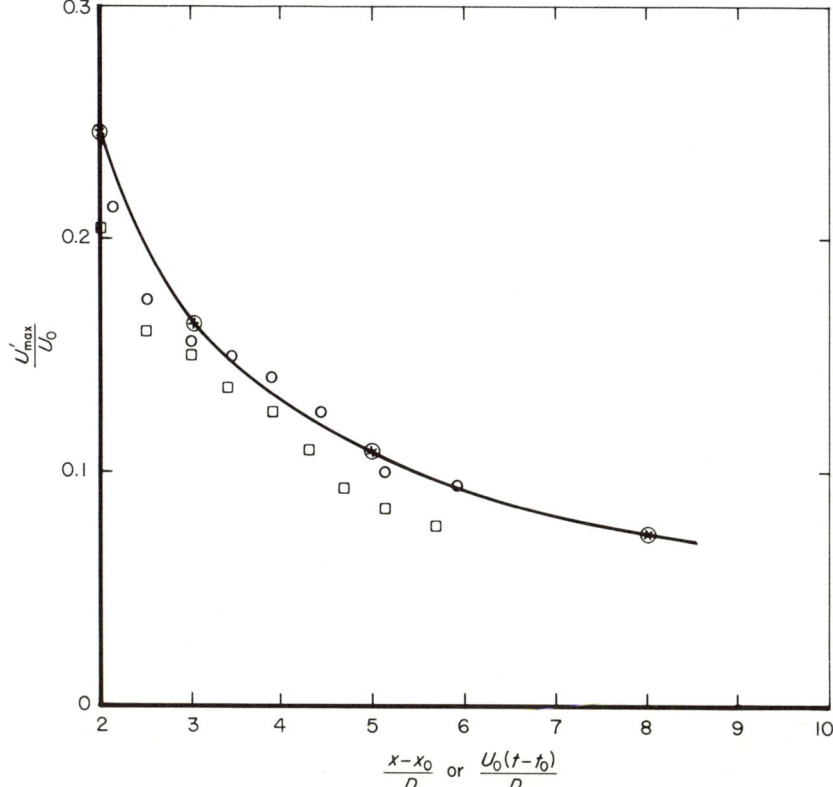

Fig. 5. Axial variation of maximum turbulent intensity. Crossed circles: results of Wang and Naudascher; squares: Run 2; circles: run 3.

provide ample evidence for this. The computer codes we have written are sufficiently general to study channel flows like plane Poiseuille and plane Couette flows, as well as momentumless and momentumfull wakes, both in homogeneous and stratified fluids. Experimental set-up of this variety of shear flows, notwithstanding specifying the form of the shear and turbulence profiles that may be imposed, would be a herculean task. On the other hand, the computer simulations handle all these cases with ease.

In future work on the momentumless wake, we shall report on longer simulation runs now underway and on techniques to improve the statistics of the results. In order to improve the choice of initial conditions, simulations runs are made in which the pseudorandom initial conditions imposed as described in Section 3.3 are let to evolve for about 10 body diameters downstream and then are reapplied, amplified in excitation, at a virtual upstream point in order to begin the calculation anew. This approach avoids

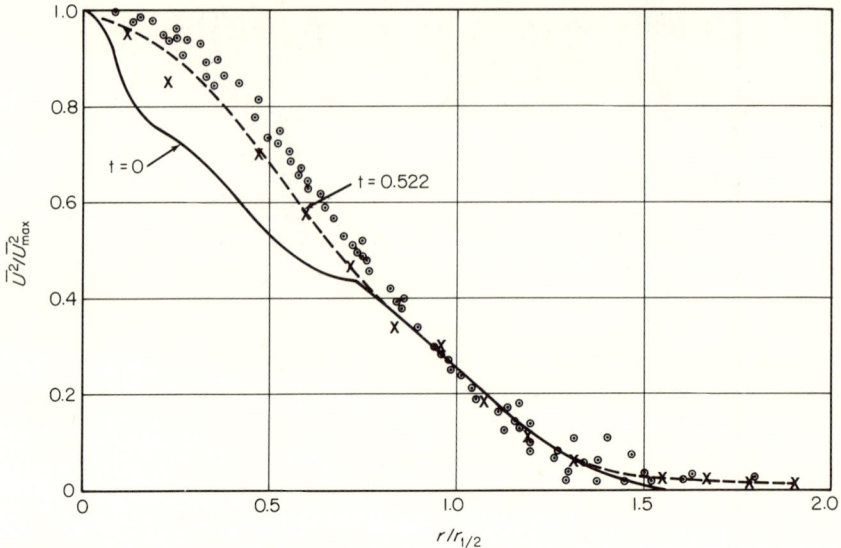

FIG. 6. Radial variation of axial mean-square turbulent intensity at $t = 0.522$ for run 2a. Crosses: run 2a; circles: results of Wang and Naudascher.

the difficulty of modelling imprecisely known laboratory experiments. With the initial conditions imposed as in Section 3.3, the wake is very sensitive to the initial turbulence level relative to the mean shear, while much of this dependence is avoided by the present technique.

References

Fox, D. G., and Orszag, S. A. (1973). *J. Comput. Phys.* **11**, 612–619.
Herring, J. R., Orszag, S. A., Kraichnan, R. H., and Fox, D. G. (1974). *J. Fluid Mech.* (in press).
Lilly, D. K. (1965). *Mon. Weather Rev.* **93**, 11–26.
Naudascher, E. (1965). *J. Fluid Mech.* **22**, 625–656.
Orszag, S. A. (1971a). *Stud. Appl. Math.* **50**, 293–327.
Orszag, S. A. (1971b). *Phys. Rev. Lett.* **26**, 1100–1103.
Orszag, S. A., and Israeli, M. (1974). *Annu. Rev. Fluid Mech.* **6**, 281–318.
Orszag, S. A., and Patterson, G. S. (1972). *Phys. Rev. Lett.* **28**, 76–79.
Wang, H. (1965). Ph.D. Thesis, Univ. of Iowa, Iowa City, Iowa.

ENERGY CASCADE IN LARGE-EDDY SIMULATIONS OF TURBULENT FLUID FLOWS

A. LEONARD[1]

*Department of Mechanical Engineering, Stanford University
Stanford, California 94305, U.S.A.*

1. INTRODUCTION

Computer simulations of three-dimensional turbulent flows which explicitly account for the motions of eddies ranging in size down to the inertial subrange are now possible. In most cases of interest, motions on the order of the dissipation length scale cannot be treated explicitly. Modifications of the Navier–Stokes equations must then be introduced to simulate properly the energy cascade. Considerable "damming up" of the turbulence energy in the large scales would occur, for example, if the unmodified equations were used with an energy-conserving finite-difference scheme on the advective term.

One approach to the problem is to use an eddy viscosity to account for the influence of the subgrid-scale motions on the large-scale fluctuations (Lilly, 1967). In this model, the energy cascade is then viewed solely as an energy loss of the large-scales due to an artificial viscosity arising from subgrid-scale motions. The advective term for the large-scale motions is unmodified.

In this paper, the derivation of smoothed or filtered momentum and continuity equations for the large-scale, energy-containing eddies is reexamined. Noting that the large-scale motions vary in a nonnegligible way over an averaging volume, we investigate a more accurate, modified advective term in the momentum equations for these motions. This term is nonconservative and is shown to lead to significant energy extraction from the large scales due to triple correlations of these motions. The subgrid-scale Reynolds stress term is still present but plays a reduced role as far as the energy cascade is concerned. Similar arguments are applied to the analysis of large-scale fluctuations of a passive scalar.

---

[1] *Present address:* NASA Ames Research Center, Moffett Field, California 94035, U.S.A.

2. Problem of Numerical Simulation

We consider an incompressible flow whose time evolution is given by the Navier–Stokes and continuity equations for the velocity components $u_i(\mathbf{x}, t)$, $i = 1, 2, 3$ and the pressure $p(\mathbf{x}, t)$:

$$\text{(2.1)} \qquad \frac{\partial u_i}{\partial t} + \frac{\partial}{\partial x_j}(u_i u_j) = -\frac{1}{\rho}\frac{\partial p}{\partial x_i} + \nu \nabla^2 u_i,$$

$$\text{(2.2)} \qquad \partial u_i/\partial x_i = 0.$$

These equations, along with appropriate initial and boundary conditions, will yield the flow field for all later times (although for turbulent flows this field is likely to be unstable with respect to small perturbations in the initial or boundary conditions). Due to the wide range of length scales present in real turbulent flows, however, the full numerical simulation of such flows is not yet possible, in general. The required number of mesh points on a three-dimensional grid is proportional to $\text{Re}^{9/4}$ where Re is the Reynolds number (Hirt, 1969). For Re = 50,000 about 10^9 mesh points would be required to simulate all the turbulent eddies down to and including those with a dissipation length scale. On the other hand, the present capability of one of the largest available machines (ILLIAC IV) is about 10^6 mesh points. The most ambitious simulation reported to date in terms of total number of mesh points is a study by Orszag and Patterson (1972) of three-dimensional homogeneous isotropic turbulence using approximately $(32)^3$ mesh points in Fourier space. The Reynolds number based on Taylor microscale was $R_\lambda = 35$, within the range of wind tunnel experiments.

In most situations, however, full simulations are not practical, or even possible. On the other hand, most of the momentum transport and turbulent diffusion is carried out by the large-scale energy-containing eddies. Therefore, simulation of these large-scale fluctuations is often of great interest. Hence, we turn to the problem of deriving momentum and continuity equations for these large-scale turbulent fluctuations.

3. Filtered Momentum and Continuity Equations

If $f(\mathbf{x})$ is a function containing all the scales we define, quite generally, the large-scale or resolvable-scale component of f to be denoted $\bar{f}$ and given by a convolution of f with a filter function $G(\mathbf{x})$,

$$\text{(3.1)} \qquad \bar{f}(\mathbf{x}) = \int G(\mathbf{x} - \mathbf{x}') f(\mathbf{x}') \, d\mathbf{x}'.$$

Integration is over the flow volume. Some examples of filters are shown in Fig.1. The one shown in Fig. 1c corresponds to a truncated Fourier expansion with $|k_i| < \pi/\Delta$. The other two are more localized in the spatial variables and are representative of finite-difference schemes based, for example, on expansions in terms of piecewise continuous polynomials. The filter shown in Fig. 1a was used by Lilly (1967).

Note that by integration by parts we find that

(3.2) $$\overline{\partial f/\partial x_i} = \partial(\bar{f})/\partial x_i,$$

if f vanishes on the boundaries. Filtering Eqs. (2.1) and (2.2) therefore gives

(3.3) $$\frac{\partial \bar{u}_i}{\partial t} + \frac{\partial}{\partial x_j}\overline{(u_i u_j)} = -\frac{1}{\rho}\frac{\partial \bar{p}}{\partial x_i} + \nu \nabla^2 \bar{u}_i$$

(3.4) $$\partial \bar{u}_i/\partial x_i = 0.$$

To avoid writing dynamical equations for $\overline{u_i u_j}$, we must approximate it in terms of combinations of the $\bar{u}_k$ and their derivatives.

If we decompose u_i into its resolvable-scale and subgrid-scale components, $u_i = \bar{u}_i + u'_i$, then

(3.5) $$\overline{u_i u_j} = \overline{\bar{u}_i \bar{u}_j} - \tau_{ij} + \tfrac{1}{3}\eta_{kk}\delta_{ij}$$

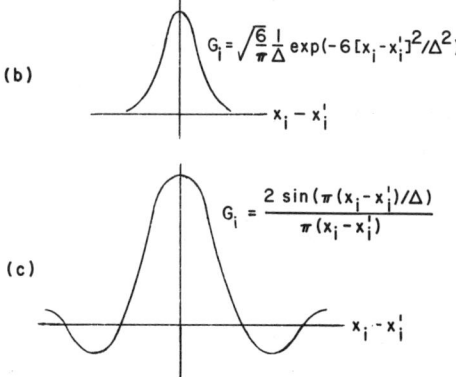

FIG. 1. Possible spatial filters defining large-scale quantities with $G = G_1 G_2 G_3$. The filters of (a) and (b) have identical second moments. The filter of (c) is equivalent to the finite Fourier expansion method.

where

(3.6) $$\tau_{ij} = -(\eta_{ij} - \tfrac{1}{3}\eta_{kk}\delta_{ij})$$

(3.7) $$\eta_{ij} = \overline{u_i'\bar{u}_j} + \overline{\bar{u}_i u_j'} + \overline{u_i' u_j'}.$$

The averaged momentum and continuity equations become

(3.8) $$\frac{\partial \bar{u}_i}{\partial t} + \frac{\partial}{\partial x_j}(\overline{\bar{u}_i \bar{u}_j}) = -\frac{\partial}{\partial x_i}\left(\frac{\bar{p}}{\rho} + \frac{1}{3}\eta_{kk}\right) + \frac{\partial \tau_{ij}}{\partial x_j} + \nu\,\nabla^2 \bar{u}_i$$

(3.9) $$\partial \bar{u}_i/\partial x_i = 0.$$

To proceed one must model τ_{ij} in terms of the $\bar{u}_k$. The function η_{kk} appearing in Eq. (3.8) may be combined with $\bar{p}$ and therefore need not be calculated explicitly.

The usual approach (Lilly, 1967) is to approximate

(3.10) $$\overline{\bar{u}_i \bar{u}_j} \simeq \bar{u}_i \bar{u}_j$$

(or to lump the difference into the definition of η_{ij}) and model τ_{ij} by an eddy viscosity hypothesis,

(3.11) $$\tau_{ij} = K\left(\frac{\partial \bar{u}_i}{\partial x_j} + \frac{\partial \bar{u}_j}{\partial x_i}\right),$$

where K is an eddy viscosity coefficient, variable in space and time. Lilly (1967) has shown that if K is taken to be similar to an expression used by Smagorinsky (1963),

(3.12) $$K = (c\Delta)^2 \left[\frac{\partial \bar{u}_j}{\partial x_i}\left(\frac{\partial \bar{u}_i}{\partial x_j} + \frac{\partial \bar{u}_j}{\partial x_i}\right)\right]^{1/2}$$

where Δ is the mesh spacing (or width of the G function), then the resultant energy dissipation of the large scales is consistent with the Kolmogorov power spectrum. Furthermore, the constant c is dependent only on Kolmogorov's universal constant α. Deardorff (1970) has used this approach to simulate turbulent channel flow with some success but found that the eddy viscosity constant c had to be chosen somewhat lower than that calculated by Lilly, otherwise the turbulence was excessively damped out (see also Deardorff, 1971).

In a recent simulation of an atmospheric boundary layer Deardorff (1973) abandoned the above eddy viscosity model and resorted to developing dynamical equations for the subgrid Reynolds' stresses and other relevant subgrid fluxes. The presence of a stably stratified layer apparently could not be accommodated with the use of an eddy coefficient. Perhaps the use of the modified or filtered advective term $\partial(\overline{\bar{u}_i \bar{u}_j})/\partial x_j$, i.e., avoiding the use of the approximation (3.10), would have remedied the situation. It is shown below that the filtered term plays an important role in the energy extraction from the large scales whereas the unfiltered term, $\partial(\bar{u}_i \bar{u}_j)/\partial x_j$ is energy-conserving up to finite-differencing errors.

The implications of the assumption $\overline{\bar{u}_i \bar{u}_j} = \bar{u}_i \bar{u}_j$ are illustrated in Fig. 2. This assumption is satisfied if the $\bar{u}_k$ remain constant over an averaging volume (Fig. 2a). One might compensate by dealing with a subgrid component u'_k which is effectively larger than that obtained when the variation of $\bar{u}_k$ over an averaging volume is explicitly accounted for (Fig. 2b). In the former case the modeling of the subgrid terms is clearly more critical. An exceptional case is the truncated Fourier expansion (filter of Fig. 1c) where the difference between $\overline{\bar{u}_i \bar{u}_j}$ and $\bar{u}_i \bar{u}_j$ is identically zero in the dynamical equations for the large-scale flow. We comment further on this case in the next section.

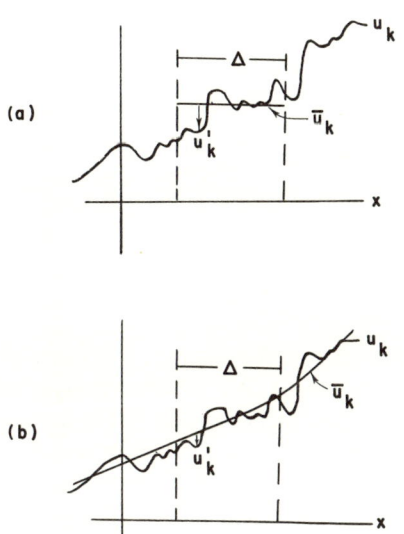

FIG. 2. Two possible definitions of the subgrid-scale component u'_k.

4. Energy Loss of the Large-Scale Turbulence

In the above model, all the energy dissipation of the large scales is viewed as a result of Reynolds stress of the subgrid-scale turbulence and modeled by an eddy viscosity times the squared deformation tensor of the large-scale flow. However, a different mechanism appears to be responsible for a substantial portion of the large-scale dissipation arising from the fact that $\overline{\bar{u}_i \bar{u}_j} - \bar{u}_i \bar{u}_j$ is not generally negligible as discussed above.

Assuming that the influence of the molecular viscosity term is negligible, energy loss of the large scales will occur through the action of the nonlinear, resolvable-scale and subgrid scale terms $\partial(\bar{u}_i\bar{u}_j)/\partial x_j$ and $\partial\tau_{ij}/\partial x_j$, respectively. The rate of energy cascade due to the former we denote ε_{RS} and to the latter, ε_{SGS}. If ε is the total loss rate, then

(4.1) $$\varepsilon = \varepsilon_{RS} + \varepsilon_{SGS}$$

Multiplying Eq. (3.8) by $\bar{u}_i$ and volume averaging we find that

(4.2) $$\varepsilon_{RS} = \langle \bar{u}_i\, \partial(\bar{u}_i\bar{u}_j)/\partial x_j \rangle$$

(4.3) $$\varepsilon_{SGS} = \langle \bar{u}_i\, \partial\tau_{ij}/\partial x_j \rangle$$

where $\langle\ \rangle$ denotes volume averaging

(4.4) $$\langle h \rangle = V^{-1} \int_V h(\mathbf{x})\, d\mathbf{x}$$

We concentrate on the evaluation of ε_{RS}. In terms of the triple correlation tensor, defined by

(4.5) $$\bar{S}_{ik,j}(\boldsymbol{\xi}) = \langle \bar{u}_i(\mathbf{x})\bar{u}_k(\mathbf{x})\bar{u}_j(\mathbf{x} + \boldsymbol{\xi}) \rangle,$$

(an overbar denotes a quantity corresponding to the filtered flow field $\bar{u}_i$) ε_{RS} can be written as

(4.6) $$\varepsilon_{RS} = -\int G(\boldsymbol{\xi}) \frac{\partial}{\partial \xi_j} \bar{S}_{ij,i}(\boldsymbol{\xi})\, d\boldsymbol{\xi}$$

Because of the presence of the filter G, the behavior of $\bar{S}_{ij,i}$ only in the domain $|\boldsymbol{\xi}| \lesssim \Delta$ is important. We assume that the velocity fluctuations in this range of scales are homogeneous and isotropic in which case $\bar{S}_{ij,k}(\boldsymbol{\xi})$ can be written in terms of a single scalar function $\bar{k}(r)$ ($r = |\boldsymbol{\xi}|$) (Hinze, 1959).

(4.7) $$\bar{S}_{ik,j}(\boldsymbol{\xi}) = \bar{q}^3 \left[\left(\bar{k} - r\frac{d\bar{k}}{dr} \right) \frac{\xi_i \xi_k \xi_j}{2r^3} - \frac{\bar{k}}{2} \delta_{ij} \frac{\xi_j}{r} \right. \\ \left. + \frac{1}{4r} \frac{d(r^2 \bar{k})}{dr} \left(\delta_{ij} \frac{\xi_k}{r} + \delta_{kj} \frac{\xi_i}{r} \right) \right],$$

where $\bar{q}^2$ is the filtered turbulence intensity

(4.8) $$\bar{q}^2 = \langle \bar{u}_i \bar{u}_i \rangle.$$

Performing the differentiations and contractions required by (4.6), we find that

(4.9) $$\varepsilon_{RS} = -\bar{q}^3 \int G(\boldsymbol{\xi}) \bar{f}(|\boldsymbol{\xi}|)\, d\boldsymbol{\xi},$$

where

(4.10) $$\bar{f}(r) = \frac{1}{2r^2}\frac{d}{dr}\left(r^3\frac{d\bar{k}}{dr} + 4r^2\bar{k}\right).$$

For a spherically symmetric filter, (4.9) reduces to

(4.11) $$\varepsilon_{RS} = -4\pi\bar{q}^3 \int_0^\infty G(r)\bar{f}(r)r^2\,dr.$$

To proceed further we must obtain at least an approximate form for $\bar{k}(r)$. This function is a scalar triple correlation having the basic definition

(4.12) $$\bar{k}(r) = \langle \bar{u}_1^2(\mathbf{x})\bar{u}_1(\mathbf{x} + \hat{e}_1 r)\rangle/\bar{q}^3.$$

From symmetry requirements and the fact that

$$\langle \bar{u}_1^2\, \partial \bar{u}_1/\partial x\rangle = \tfrac{1}{3}\langle \partial \bar{u}_1^3/\partial x_1\rangle = 0,$$

one can show that $\bar{k}$ has the small r expansion (Hinze, 1959),

(4.13) $$\bar{k}(r) = \frac{\bar{k}_0''' r^3}{3!} + \frac{\bar{k}_0^{(v)} r^5}{5!} + \cdots;$$

this yields for $\bar{f}(r)$ the expansion

(4.14) $$\bar{f}(r) = \tfrac{35}{12}\bar{k}_0''' r^2 + \tfrac{21}{80}\bar{k}_0^{(v)} r^4 + \cdots.$$

Assuming that the $\bar{u}_i$ fluctuations include a portion of the inertial subrange and the smallest scale in the subrange is $\approx \Delta$, then $\bar{k}(r)$ is linear in that subrange with a known coefficient (Kolmogorov, 1941),

(4.15) $$\bar{k}(r) = -2\varepsilon r/15\bar{q}^3 \qquad (\Delta \ll r \ll L_0)$$

where L_0 is the scale of the energy-containing eddies. The resultant behavior of $\bar{f}(r)$ is

(4.16) $$\bar{f}(r) = -\varepsilon/\bar{q}^3 \qquad (\Delta \ll r \ll L_0).$$

Note that using the inertial subrange form of $\bar{f}(r)$ in (4.9) yields $\varepsilon_{RS} = \varepsilon$. This is no accident. In fact $2\bar{q}^3 \bar{f}(r)$ represents the advective term in the Kármán–Howarth equation (von Kármán and Howarth, 1938), a dynamical equation for the correlation $\bar{Q}_{ii}(r) = \langle \bar{u}_i(x)\bar{u}_i(x + \hat{e}_1 r)\rangle$, and in the inertial subrange of the filtered flow this term is solely responsible for the energy loss of $\bar{Q}_{ii}(r)$. Thus $2\bar{q}^3\bar{f}(r) = -2\varepsilon$ for $\Delta \ll r \ll L_0$. Equation (4.15) then follows by using (4.10).

The corrections to (4.16) for small r however will give $\varepsilon_{RS} < \varepsilon$. This is shown qualitatively in Fig. 3. The behavior indicated by curve A assumes that the quadratic term $\tfrac{35}{12}\bar{k}_0''' r^2$ dominates the small r behavior up to the

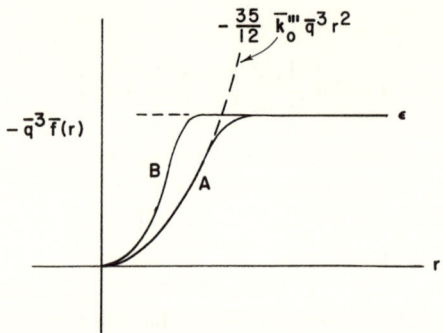

FIG. 3. Two possible interpolations of $\bar{k}(r)$ between known asymptotic forms.

inertial subrange asymptote. Unpublished calculations based on turbulence as given by Burgers' equation (Burgers, 1940, or see Burgers, 1948) ("Burgerlence") suggest that $\bar{k}(r)$ will approach its asymptotic form more quickly as shown by curve B.

Furthermore, turbulence measurements in an atmospheric boundary layer (McConnell, 1973) and an air jet (Clay, 1973) have shown that the inertial subrange behavior of $k(r)$ given by (4.15) persists down to about a Kolmogorov length η whereas curve A assumes a lower bound on the linear range of $k(r)$ at $\simeq 5\eta$.

Nevertheless, as a tentative lower bound to the amount of energy loss attributable to ε_{RS} we will use curve A for $\bar{k}(r)$. With the filter

(4.17) $$G(\xi) = [\Delta^{-1}(6/\pi)^{1/2}]^3 \exp(-6|\xi|^2/\Delta^2),$$

(4.9) gives

(4.18) $$\varepsilon_{RS} = -\tfrac{35}{48}\bar{q}^3 \bar{k}_0''' \Delta^2$$

plus a negligible contribution from the asymptotic form of $\bar{f}(r)$.

We relate $\bar{k}_0'''$ to skewness by noting that

(4.19) $$\langle (\partial \bar{u}_1/\partial x_1)^3 \rangle = \langle \bar{u}_1 \bar{u}_1 \, \partial^3 \bar{u}_1/\partial x_1^3 \rangle$$
$$= \frac{\partial^3}{\partial \xi_1^3} S_{11,1}(\xi)\bigg|_{\xi=0}$$
$$= \bar{q}^3 \bar{k}_0''',$$

and therefore

(4.20) $$\varepsilon_{RS} = -\tfrac{35}{48}\langle (\partial \bar{u}_1/\partial x_1)^3 \rangle \Delta^2$$
$$= \tfrac{35}{48} S \langle (\partial \bar{u}_1/\partial x_1)^2 \rangle^{3/2} \Delta^2,$$

where $\bar{S}$ is the skewness defined by

(4.21) $$\bar{S} = \left\langle \left(\frac{\partial \bar{u}_1}{\partial x_1}\right)^3 \right\rangle \bigg/ \left\langle \left(\frac{\partial \bar{u}_1}{\partial x_1}\right)^2 \right\rangle^{3/2}.$$

Following Lilly (1967) we complete the calculation by relating $\langle (\partial u_1/\partial x_1)^2 \rangle$ to the Kolmogorov spectrum. In the initial range the energy spectrum of turbulence is

(4.22) $$E(k) = \alpha \varepsilon^{2/3} k^{-5/3}.$$

However, the filtering process will truncate the high wave number end of this spectrum so that the spectrum of the filtered turbulence is

(4.23) $$\bar{E}(\mathbf{k}) = \alpha \varepsilon^{2/3} k^{-5/3} |\hat{G}(\mathbf{k})|^2,$$

where $\hat{G}(\mathbf{k})$ is the Fourier transform of the filter function $G(\mathbf{x})$,

(4.24) $$\hat{G}(\mathbf{k}) = \int e^{i\mathbf{k} \cdot \mathbf{x}} G(\mathbf{x})\, d\mathbf{x}.$$

Of particular interest is the integral

(4.25) $$\int_0^\infty k^2 \bar{E}(k)\, dk \int |\hat{G}(\mathbf{k})|^2\, d\Omega_k/4\pi = -\tfrac{1}{2}\langle \bar{u}_i \nabla^2 \bar{u}_i \rangle$$
$$= \tfrac{15}{2} \langle (\partial \bar{u}_1/\partial x_1)^2 \rangle,$$

where again we have used isotropy in the final step. Combining (4.20), (4.23), and (4.25), ε_{RS} takes the form

(4.26) $$\varepsilon_{RS} = \frac{35}{48} \bar{S} \left[\frac{2\alpha}{15} \int_0^\infty k^{1/3}\, dk \int |\hat{G}(\mathbf{k})|^2 \frac{d\Omega_k}{4\pi} \right]^{3/2} \Delta^2 \varepsilon.$$

For the Gaussian filter (4.17) we have

(4.27) $$\hat{G}(\mathbf{k}) = \exp(-\Delta^2 k^2/24)$$

Using $\alpha = 1.62$ (Wyngaard and Pao, 1972), we obtain

(4.28) $$\varepsilon_{RS} = 0.49 \bar{S} \varepsilon.$$

Experimentally determined values of skewness vary slightly, depending on Reynolds number, from $S \approx 0.40$ for wind-tunnel grid turbulence ($R_\lambda = 50\text{--}100$) to $S = 0.60\text{--}0.85$ for atmospheric turbulence ($R_\lambda \approx 10^3\text{--}10^4$) (Wyngaard and Pao, 1972). Thus, without more information on the behavior of $\bar{k}(r)$ we obtain the tentative lower bound for ε_{RS}

(4.29) $$\varepsilon_{RS} \gtrsim 0.3\varepsilon \pm 0.1\varepsilon.$$

The SGS term $\partial \tau_{ij}/\partial x_j$ must account for the remainder of the losses.

As mentioned earlier, with the use of the truncated Fourier representation $\bar{u}_i \bar{u}_j$ has the same large-scale Fourier components as does $\overline{\bar{u}_i \bar{u}_j}$. Therefore, for this case,

$$\varepsilon_{RS} = \langle \bar{u}_i \, \partial(\overline{\bar{u}_i \bar{u}_j})/\partial x_j \rangle \tag{4.30}$$
$$= \langle \bar{u}_i \, \partial(\bar{u}_i \bar{u}_j)/\partial x_j \rangle = 0.$$

The interpretation is related to the fact that large-wave-number Fourier modes need the assistance of small wave-number modes to transfer energy from large scales to small scales. In the Fourier method the sharp cutoff in wave-number space precludes such a transfer whereas the localized spatial filters of Figs. 2a and 2b produce smooth, gradual filtering in wave-number space which evidently allows for energy transfer to the subgrid scales.

5. Turbulent Diffusion of a Passive Scalar

If the numerical simulation is to include the large-scale fluctuations of a passive scalar field $\psi(\mathbf{r}, t)$, one must consider the filtered equation of motion

$$\partial \bar{\psi}/\partial t + \overline{u_j \, \partial \psi/\partial x_j} = \kappa \nabla^2 \bar{\psi} \tag{5.1}$$

where κ is the diffusivity. Decomposing u_j and ψ into large-scale and subgrid-scale components gives for the convection term,

$$\overline{u_j \, \partial \psi/\partial x_j} = \overline{\bar{u}_j \, \partial \bar{\psi}/\partial x_j} + \text{subgrid contributions} \tag{5.2}$$

Analogous to the preceding results, the filtered large-scale convection term on the right-hand side of (5.2) will produce a loss in scalar variance χ due to the mixed triple correlation $\langle \bar{\psi}(\mathbf{x}) \bar{u}_1(\mathbf{x}) \bar{\psi}(\mathbf{x} + \hat{e}_1 r) \rangle$. This correlation is cubic for small r (Corrsin, 1951) and linear ($\simeq \chi r/3$) in the resolvable-scale portion of the convective subrange. Again, the amount of scalar variance dissipation due to the filtered large-scale convection term depends on the extent to which this linearity penetrates into the small r regime.

6. Conclusions

Numerical simulation of all the scales of a turbulent flow, even at modest Reynolds numbers, is generally not practical. However, most information of interest can be obtained by simulating the motion of the large-scale, energy-containing eddies. The large-scale fluctuations satisfy filtered or averaged momentum and continuity equations. Averaging the nonlinear advection term yields two terms. One is the Reynolds stress contribution from the subgrid-scale turbulence and the other is the filtered advection term for the large scales,

$$\partial(\overline{\bar{u}_i \bar{u}_j})/\partial x_j.$$

A significant portion of the large-scale dissipation is provided by appropriate treatment of this term. In evaluating this term, the variations of $\partial(\bar{u}_i\bar{u}_j)/\partial x_j$ within an averaging volume defined by $G(\mathbf{x})$ should be explicitly accounted for. One obvious possibility is to represent $\partial(\bar{u}_i\bar{u}_j)/\partial x_j$ as a weighted average of the values of $\partial(\bar{u}_i\bar{u}_j)/\partial x_j$ at neighboring grid points. Another possibility is to use the Taylor expansion

(6.1) $$\bar{u}_k(\mathbf{x}') = \bar{u}_k(\mathbf{x}) + (\mathbf{x}' - \mathbf{x}) \cdot \nabla \bar{u}_k(\mathbf{x}) + O(|\mathbf{x}' - \mathbf{x}|^2)$$

in the definition

(6.2) $$\overline{\bar{u}_i\bar{u}_j} = \int G(\mathbf{x} - \mathbf{x}')\bar{u}_i(\mathbf{x}')\bar{u}_j(\mathbf{x}')\, d\mathbf{x}',$$

with the result

(6.3) $$\frac{\partial(\overline{\bar{u}_i\bar{u}_j})}{\partial x_j} = \frac{\partial}{\partial x_j}\left(\bar{u}_i\bar{u}_j + \gamma \frac{\partial \bar{u}_i}{\partial x_l}\frac{\partial \bar{u}_j}{\partial x_l}\right)$$

where γ is the one-dimensional second moment of G,

(6.4) $$\gamma = \int_{-\infty}^{\infty} x^2\, dx \int_{-\infty}^{\infty} G(x, y, z)\, dy\, dz.$$

Finally, an expansion of $\bar{u}_i\bar{u}_j(\mathbf{x}')$ including $O(|\mathbf{x} - \mathbf{x}'|)^2$ terms, gives

(6.5) $$\frac{\partial(\overline{\bar{u}_i\bar{u}_j})}{\partial x_j} \simeq \frac{\partial}{\partial x_j}\left(\bar{u}_i\bar{u}_j + \frac{\gamma}{2}\frac{\partial^2}{\partial x_l\, \partial x_l}(\bar{u}_i\bar{u}_j)\right).$$

Using the methods of the previous section, one can show that the approximation (6.3) would yield twice the ε_{RS} determined for curve A, while (6.4) gives the same value of ε_{RS}.

The SGS term $\partial \tau_{ij}/\partial x_j$ must produce the remaining dissipation and probably can be modeled by the eddy viscosity model of (3.11) and (3.12). The value of the eddy coefficient in (3.12) will be somewhat smaller than that calculated by Lilly (1967) ($c = 0.17$). Numerical experiments will probably be required to obtain a satisfactory value.

Similar considerations apply to the simulation of large-scale fluctuations of a passive scalar.

Acknowledgments

The author gratefully acknowledges the important comments of Drs. J. W. Deardorff, D. K. Lilly, and W. C. Reynolds who read a draft of the manuscript. Contributions by Drs. J. H. Ferziger, C. H. Gibson, and S. Corrsin were also most helpful. Messrs. John Clay and Steve McConnell kindly made available their experimental measurements prior to publication. This work was supported by a grant from NASA Ames Research Center.

References

Burgers, J. M. (1940). *Proc. Kon. Ned. Akad. Wetensch* **43**, 2–12.
Burgers, J. M. (1948). *Advan. Appl. Mech.* **1**, 171–199.
Clay, J. (1973). Private communication.
Corrsin, S. (1951). *J. Aeronaut. Sci.* **18**, 417–423.
Deardorff, J. W. (1970). *J. Fluid Mech.* **41**, 453–480.
Deardorff, J. W. (1971). *J. Comput. Phys.* **7**, 120–133.
Deardorff, J. W. (1973). The use of subgrid transport equations in a three-dimensional model of atmospheric turbulence. *Joint Appl. Mech. Fluids Eng. Conf., Ga. Inst. Technol.* ASME Prepr. 73-FE-21.
Hinze, J. O. (1959). "Turbulence," Ch. 3. McGraw-Hill, New York.
Hirt, C. W. (1969). *Phys. Fluids, Suppl.* **II**, 219–227.
Kolmogorov, A. N. (1941). *C. R. Acad. Sci. URSS* **31**, 538–540.
Lilly, D. K. (1967). The representation of small-scale turbulence in numerical simulation experiments. *Proc. IBM Sci. Comput. Symp. Environ. Sci., IBM Data Process. Div., White Plains, N.Y.* pp. 195–210.
McConnell, S. (1973). Ph.D. Thesis, Univ. of California at San Diego.
Orszag, S. A., and Patterson, G. S., Jr. (1972). *Phys. Rev. Lett.* **28**, 76–79.
Smagorinsky, J. (1963). *Mon. Weather Rev.* **91**, 99–164.
von Kármán, T., and Howarth, L. (1938). *Proc. Roy. Soc., Ser. A* **169**, 192–215.
Wyngaard, J. C., and Pao, Y. H. (1972). Some measurements of the fine structure of large Reynolds number turbulence. *In* "Statistical Models and Turbulence" (M. Rosenblatt and C. Van Atta, eds.). Lecture Notes in Physics, Vol. 12, pp. 384–401. Springer-Verlag, Berlin and New York.

STATISTICAL ANALYSIS OF WALL TURBULENCE PHENOMENA

Z. Zarić

Department of Thermal Physics
Boris Kidrič Institute, University of Belgrade, Yugoslavia

1. Introduction

Turbulence is a phenomenon best known for its complexity. Physically, it represents a mechanical system with an extremely large number of active degrees of freedom. The only correct approach to the problem of turbulence is, therefore, a statistical one. Statistical fluid mechanics, however, is still not nearly in a position to provide a theoretical solution to the general problem of turbulence.

From the point of view of engineering predictions, the most successful were the so-called phenomenological theories of turbulence, based on the long-time averaging of the Navier–Stokes equations, introduced by Reynolds. The resulting system of equations is an unclosed one, as the equations for the statistical moments of order n contain unknowns in the form of statistical moments of order $n + 1$. Phenomenological theories provide semiempirical relations between the moments of order $n + 1$ and the moments of order n. Despite their success in a great number of flow configurations, they all fail to be general enough.

Therefore, a physical insight into turbulence and especially into the mechanism of turbulence production is greatly needed. Until recently, experimental research has been handicapped by not being appropriate enough in dealing with a basically statistical phenomenon. The works of Frenkiel and Klebanoff (1967) have marked the widespread use of digital computers, indispensable in statistical turbulence research. It became, however, apparent that the conventional statistical analysis, adequate for Gaussian processes, is insufficient in revealing the details of the turbulence transport mechanism. The departures from Gaussianity, however small, appear to be the real feature of turbulence. They are detected by an averaging process over a long period of time, but in order to find their causes a different type of statistical analysis is needed.

Recent visualization studies by the group in Stanford (Kline et al., 1967;

Kim *et al.*, 1968), as well as by Corino and Brodkey (1969), Grass (1971), and Shlanchiauskas (1972) have been most helpful in revealing extremely interesting structural features in the wall layers. The "bursts" of Kline, which were found to be responsible for the largest part of turbulent energy generation, were later interpreted by other investigators as containing not only ejections of low momentum fluid from the layers closest to the wall, but also inrushes of the high momentum fluid from the outer layers all the way into the viscous sublayer. These two highly intermittent phases have been found to play a leading role in the turbulence production mechanism. Similar conclusions could be drawn from the studies of fluctuating turbulence parameters at the wall, such as velocity gradient (Popovich, 1969), shear stress (Duhamel and Py, 1972), temperature (Meek and Baer, 1970), and heat flux (Armistead and Keyes, 1968).

Intermittency is a phenomenon known to provoke departures from Gaussianity. It is also a phenomenon beginning to grow in importance in turbulence research since the time of the studies of Townsend (1949) and Corrsin and Kistler (1955), who investigated turbulent–nonturbulent interfaces at the edge of a wake, a round jet, and a boundary layer. Kovasznay *et al.* (1970) made an important contribution to the study of intermittency of the free boundaries by introducing conditional sampling and the averaging technique which later was employed by many other investigators. This technique enabled the detection of the apparent interaction between interface intermittency and large-scale fluctuations inside the inner layers, and again this interaction is found to play a dominant role in turbulent energy production. Quite recently, Blackwelder and Kovasznay (1972) found non-zero correlations between the interface intermittency and the turbulent "bursts" in the buffer layer.

Conditional averaging has proven to be a very efficient technique in analyzing intermittent, non-Gaussian statistical phenomena. It is therefore natural to also employ it in the analysis of the inner layer intermittency. In fact, Grass (1971) has employed conditional averaging in revealing the importance of the inrush phases in the wall vicinity. Wallace *et al.* (1972) and Willmarth and Lu (1972) have used the technique in determining the principal contributions of the intermittent phases to the Reynolds stresses. Gupta *et al.* (1971) and Blackwelder and Kaplan (1972) have combined the technique with the use of digital analysis in investigating "bursting" phenomena in the wall layers.

In a previous study we also employed conditional sampling techniques in analyzing wall turbulence phenomena (Zarić, 1972b,d). This paper presents the results of further development of the conditional sampling and averaging analysis as applied to the wall layers, including the viscous sublayer, in an attempt to separate intermittent phases from the overall signal.

2. Experimental Technique

The flow configuration is a simple one. Measurements have been made in air flow at the straight outlet section of a channel used previously for other experiments and described elsewhere (Zarić, 1972a). A fairly low Reynolds number of Re = 39,000 was chosen in order to thicken the viscous sublayer. Maximum velocity in the 40 by 300 mm cross section was 10.2 m/s, with the friction velocity amounting to $u_* = 0.48$ m/s.

Hot wire anemometry is employed as the most suitable existing technique for statistical analysis, in spite of a number of disadvantages when used in the wall vicinity. The pronounced nonlinearity of the signal at very low velocities appearing in the viscous sublayer requires very careful calibration and digital linearization. The wall effect requires special corrections which are still mainly empirical. At high turbulence intensities in the viscous sublayer the fact that the wire is equally sensitive to the two velocity vector components normal to the wire has to be taken into account.

A single, 5-μm diameter, tungsten wire is employed in order to approach the wall as close as possible. The wire could be switched either to a DISA anemometer, or to a Mueller bridge so that the flow temperature could also be detected. More details on the technique are given elsewhere (Zarić, 1972a). Signals from the anemometer are registered on an analogue tape running at 1.524 m/s, using an AMPEX FR-1300 tape recorder operated in the FM mode. The tape is replayed with a 1 : 32 speed reduction and the signal is fed into the digital computer with a frequency of 250 Hz, via an analogue-to-digital conversion system. The real time sampling frequency being 8000 Hz, 15-s long signals are registered on a digital tape in the form of a succession of 120,000 instantaneous values. The statistical analysis is performed on a CDC-3600 digital computer with a 64K memory.

3. Conventional Statistical Analysis

Feeling that amplitude statistical analysis was somehow neglected in wall turbulence research, we have emphasized probability density distribution analysis. Velocity probability densities are determined on the basis of 120,000 instantaneous values for 20 different nondimensional distances from the wall, ranging from $Y^+ = 1.6$ to $Y^+ = 211$. Complete probability density distributions have been presented elsewhere (Zarić, 1972c). Probability density distributions in the buffer layer only are presented in Fig. 1. In Fig. 2 distributions of the turbulence intensities K/U_1, skewness factors S, and flatness factors F are presented as functions of the distance from the wall.

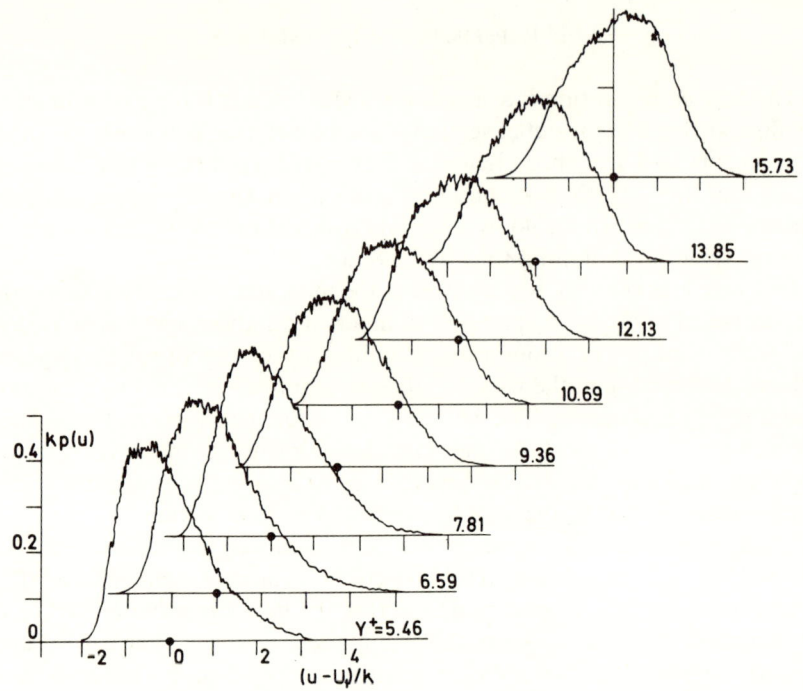

Fig. 1. Velocity probability density distributions in the buffer layer.

Velocity probability density distributions in the wall layers have been presented by only a few authors (Klebanoff, 1955; Marcillat, 1964; Eckelman, 1970), with Comte-Bellot (1963) presenting only measurements of the skewness and the flatness factors. In all these studies the analogue technique was employed. Our data on the distribution of the first four statistical moments agree reasonably well with the results of Comte-Bellot (1963) and Marcillat (1964). What is more important, the general shape of the complete probability density distributions, as a function of the distance from the wall, agrees well with the shape of the distributions presented by Eckelman (1970). It appears, therefore, that some features of these distributions are general enough for wall turbulence. These include:

(1) Distributions are highly non-Gaussian throughout the inner layers. In fact, some of the distributions presented in Fig. 1 are of a shape unlike any known theoretical probability density distribution.

(2) Skewness and flatness factors have very high values in the viscous sublayer. Asymmetry is such that an appreciable difference exists between the average and the most probable velocity.

(3) At about $Y^+ = 15$ the distributions become symmetrical, with the odd order moments being equal to zero. Approximately at the same distance

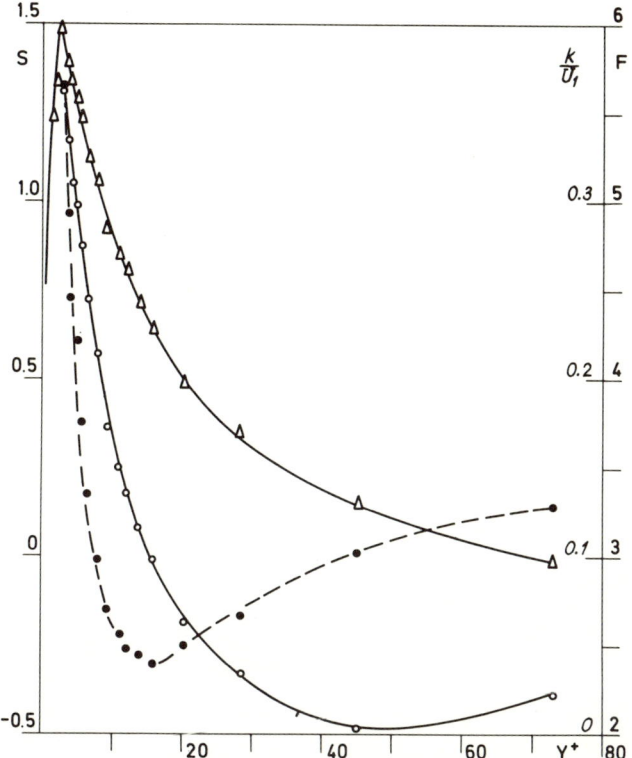

FIG. 2. Distributions of the turbulence intensity K/U_1, the skewness factor S, and the flatness factor F, in the wall layers. △, K/U_1; ○, S; ●, F.

from the wall, even order moments have maximum absolute value. However, relative values, such as flatness and superflatness factors, are at a minimum, which is appreciably lower than the corresponding Gaussian values.

(4) Further from the wall the distributions are becoming asymmetrical in the opposite sense—toward low velocities, so that the odd order moments are negative.

One might argue that the probability density distributions near the wall at high turbulence intensity, being distributions of the velocity vector, could not be Gaussian (Zarić, 1969, 1972a). However, a velocity vector distribution could never be negatively skewed. Thus, other causes have to be found.

On the basis of the mentioned structural studies made by visual methods, we would like to suggest that the principal cause for the non-Gaussian behavior of the probability distributions in the wall layers comes from the apparent intermittency of these layers. The intermittent inrush phase, in which high momentum fluid is being brought into the sublayer (Grass, 1971)

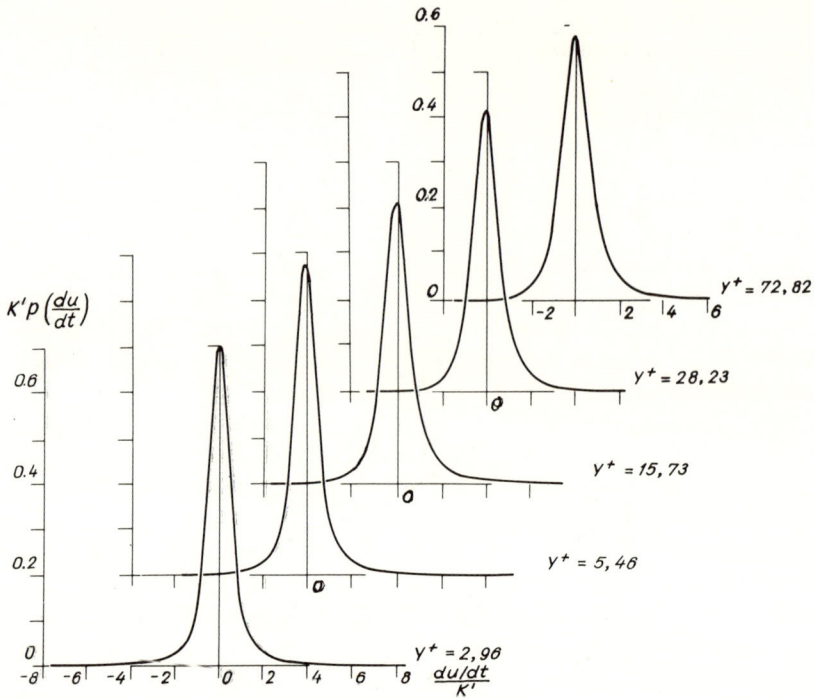

FIG. 3. Probability density distributions of the velocity derivative du/dt in the wall layers.

could well give rise to high positive skewness factors in these layers. On the other hand, intermittent ejections of the low momentum fluid are becoming pronouncedly intensive at distances further from the wall and might well provoke negative skewness in this region.

These suggestions are further confirmed by analysis of the probability density distributions of the velocity time derivative (du/dt) presented in Fig.3. It is seen that all these distributions have very high flatness factors, with high amplitude, low probability excursions, indicating the presence of intermittency. Skewness factors calculated from these distributions are also high and positive throughout the wall layer and reach a maximum at the edge of the viscous sublayer.

4. Conditional Averaging Analysis

Conditional sampling and averaging analysis is applied to the signals from the wall layers with the objective of separating both inrush and ejection intermittent phases from the prevailing quiescent part of the signal. From investigations of free boundaries, for which the technique was developed, it is apparent that in defining the intermittency function

$$I(t) = \begin{cases} 1, & \text{flow turbulent} \\ 0, & \text{flow nonturbulent} \end{cases}$$

the most delicate matter is the choice of criteria by which to judge when the flow is turbulent and when it is not.

We have performed preliminary conditional averaging by employing digital filtering (Zarić, 1972d) in a manner similar to that described by Blackwelder and Kaplan (1972). While a certain degree of intermittent phase separation was obtained, it turned out that the digital filtering technique is not altogether satisfactory.

Aiming toward a more satisfactory separation of the intermittent phases, we decided to establish the criteria by which the intermittency function $I(t)$ could be directly determined. Analysis of the probability density distributions of the velocity derivative (du/dt), shown in Fig. 3, indicated that the du/dt values could be used as an efficient indicator of intermittency. Analysis of the simultaneous traces of shear stress, the longitudinal and the normal velocity component, presented by Wallace et al. (1972), as well as traces of the velocity signal recorded by us, indicates that whenever an intermittent phase occurs, both velocity derivative *and* longitudinal velocity have fluctuations of high amplitude. For these reasons, as the criterion for switching on the intermittency function $[I(t) = 1]$ we have employed the product:

$$Z_{udu} = (u - U_{mp})|du/dt|$$

where U_{mp} is the most probable velocity, determined from the probability distributions of the whole signal. The reason for using U_{mp} instead of the average velocity U_1 is that we expect that the signal from which the intermittent phases are separated will prevail, so that the U_{mp} should be close to the mean velocity of this noncontaminated signal. The sign of $u - U_{mp}$ indicates the intermittent phase in question. Negative values of Z_{udu} correspond to ejections and positive values to inrush phases.

The next step is to define the critical level of the criterion Z_{udu} beyond which the intermittency function is to be switched on. To facilitate this, we have performed preliminary conditional averaging analysis, the condition being simply the sign of Z_{udu}. The rms values calculated from the probability distributions of the product Z_{udu} for positive values of $u - U_{mp}$, $\langle K_{udu} \rangle^+$, and for negative values of $u - U_{mp}$, $\langle K_{udu} \rangle^-$, are plotted in Fig. 4 in function of Y^+. Also plotted in Fig. 4 are the rms values of the velocity derivative du/dt conditionally averaged for positive $(\langle K' \rangle^+)$ and negative $(\langle K' \rangle^-)$ values of $u - U_{mp}$. It follows from Fig. 4 that the characteristics of the two

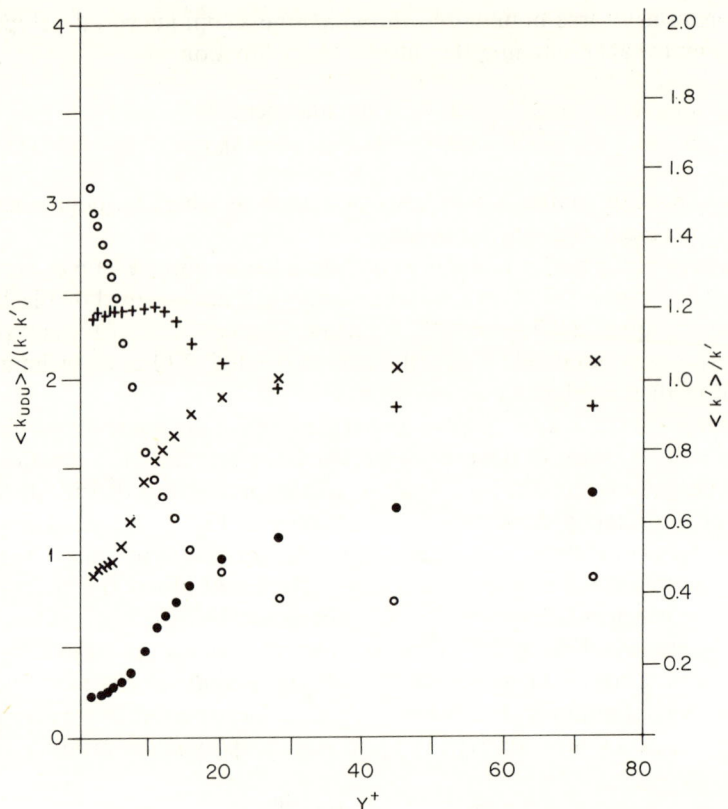

FIG. 4. Distributions of the rms values of the Z_{udu}, $\langle K_{udu}\rangle$, and the velocity derivative $\langle K'\rangle$, conditionally averaged for the positive and the negative values of $u - U_{mp}$. $\bigcirc, \langle K_{udu}\rangle^+$; $\bullet, \langle K_{udu}\rangle^-$; $+, \langle K'\rangle^+$; $\times, \langle K'\rangle^-$.

intermittent phases are different from each other and also that they are dependent on the distance from the wall Y^+. After some preliminary analysis we set the critical level of Z_{udu} which switches on the intermittency function at an rms value corresponding to a given intermittent phase $-\langle K_{udu}\rangle^+$ for the ejections and $\langle K_{udu}\rangle^-$ for the inrush phases. Another delicate matter is when the intermittency function is switched off $[I(t) = 0]$. After preliminary analysis we programmed the intermittency function to remain switched on until

$$|u - U_{mp}| < 2K$$

where K is the rms value of the whole signal, *and*

$$|Z_{udu}| < 0.3\langle K_{udu}\rangle$$

where the $\langle K_{udu}\rangle$ value corresponds to the given intermittent phase.

The results of the analysis of the signal at $Y^+ = 4.88$ are presented in Fig.5. The curves 1, 2, 3, and 4 represent probability distributions of the whole signal, the intermittent inrush phase, the intermittent ejection phase, and the noncontaminated part of the signal, respectively. All these probability distributions are normalized by the rms value of the whole signal (K) in order to obtain an impression of the relative importance of different phases. It is seen that the inrush phase (curve 2) is relatively more pronounced than the ejection phase (curve 3) and that the probability distribution of the noncontaminated signal is closer to a Gaussian distribution.

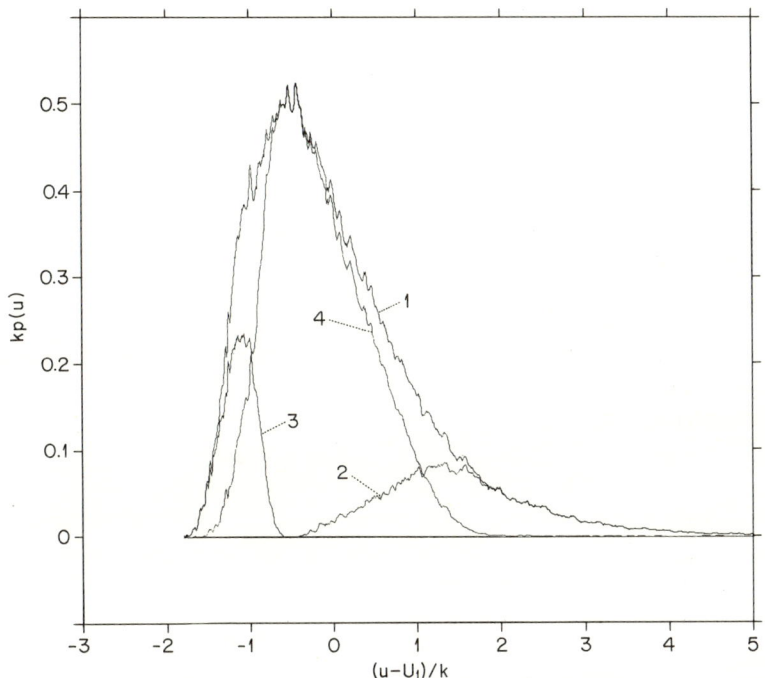

FIG. 5. Probability distributions of the whole signal (1), the inrush phase (2), the ejection phase (3), and the noncontaminated part of the signal (4), at $Y^+ = 4.88$.

The results for $Y^+ = 13.85$ are presented in Fig. 6 with the same notations of the curves. At this Y^+ the intermittent ejection phase (curve 3) has gained much in importance over the ejection phase (curve 2). The probability distribution of the noncontaminated signal is very close to a Gaussian one. The very peculiar shape of the whole signal (curve 1), mentioned earlier as characteristic for these nondimensional distances from the wall, is easily explained by the superposition of the distributions 2, 3, and 4. As suggested

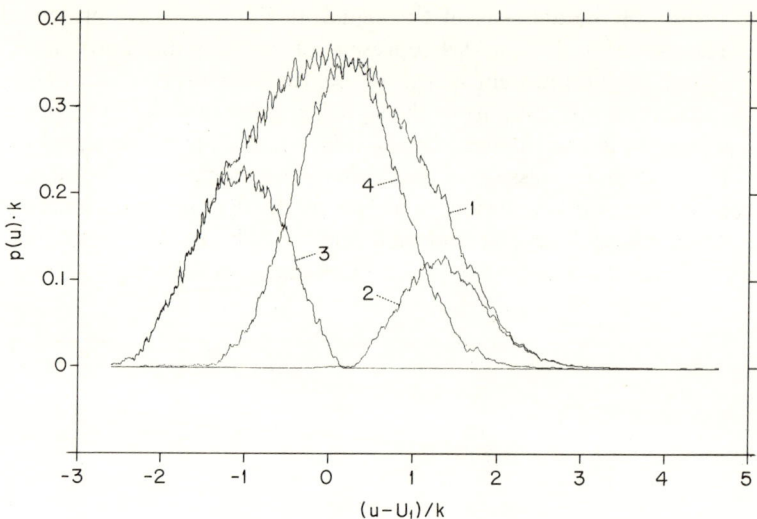

FIG. 6. Probability distributions of the whole signal (1), the inrush phase (2), the ejection phase (3), and the noncontaminated part of the signal (4), at $Y^+ = 13.85$.

earlier, the increased importance of the ejection phase leads to negative skewness further from the wall.

The intermittency factors

$$\gamma = \langle I(t) \rangle$$

corresponding to the ejections (γ_B), the inrush phases (γ_I) and the noncontaminated signal $(1 - \gamma_B - \gamma_I)$ are presented in Fig. 7 as a function of Y^+. The variance (K^2) of the whole signal could be given by

$$K^2 = \langle e_k \rangle_B + \langle e_k \rangle_I + \langle e_k \rangle_0$$

where $\langle e_k \rangle_B = \gamma_B \langle K^2 \rangle_B$ for ejections; $\langle e_k \rangle_I = \gamma_I \langle K^2 \rangle_I$ for the inrush phases, and $\langle e_k \rangle_0 = (1 - \gamma_B - \gamma_I)\langle K^2 \rangle_0$ for the noncontaminated signal. The $\langle e_k \rangle$ values corresponding to different phases are presented in Fig. 8 as a function of Y^+ as an indication of the relative contribution of different phases to the turbulent energy at a given Y^+. It is seen that the contribution of the noncontaminated signal is approximately 20 % throughout the inner layers. Very close to the wall, at $Y^+ = 2$, the contribution of the inrush phases is about 70 % while that of the ejection phase is but a few percent. At about $Y^+ = 12$, the contributions of both intermittent phases are equal. Further on, the contribution of the inrush phases continues to diminish while that of the ejections continues to rise.

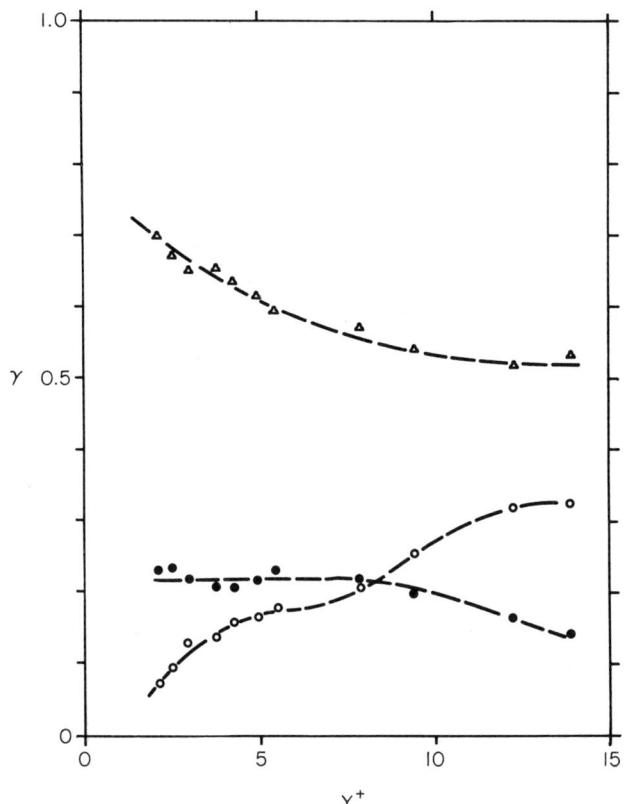

FIG. 7. Distributions of intermittency factors for the intermittent phases and the noncontaminated signal. ○, γ_B, ejection phase; ●, γ_I, inrush phase; △, $(1 - \gamma_B - \gamma_I)$, noncontaminated signal.

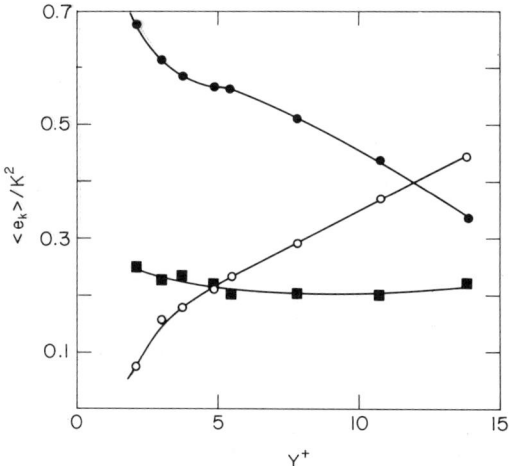

FIG. 8. Contribution of the different phases to the variance of the whole signal K^2. ○, $\langle e_k \rangle_B$, ejection phase; ●, $\langle e_k \rangle_I$, inrush phase; ■, $\langle e_k \rangle_0$, noncontaminated signal.

5. Conclusions

Probability density distribution analysis is found to be a very useful tool in statistical analysis of the wall turbulence. The overall shape of the probability distribution is found to represent a characteristic feature of the turbulence at a given nondimensional distance from the wall as well as to indicate the presence of intermittent processes going on inside the wall layers. However, conventional averaging analysis over a long period could only indicate a certain process but is unable to describe it in detail.

Conditional averaging analysis, proven to be an efficient technique in investigations of outer-layer intermittency, is found to be very useful in turbulence analysis of the inner layers also. While the criteria by which the presented results are obtained are not optimized, the results clearly indicate the relative importance of different intermittent phases as a function of nondimensional distance from the wall. Through the analysis presented, the shape of the probability distributions at different Y^+ could be explained. It is believed that further development of the analysis could eventually lead to a quantitative description of turbulent transport processes in wall layers.

References

Armistead, R. A., Jr., and Keyes, J. J., Jr. (1968). *J. Heat Transfer* **90**, 13.
Blackwelder, R. F., and Kaplan, R. E. (1972). *IUTAM Meet., 13th, Moscow*.
Blackwelder, R. F., and Kovasznay, L. S. G. (1972). *Phys. Fluids* **15**, 1545–1557.
Comte-Bellot, G. (1963). Thèse, Univ. de Grenoble, Grenoble.
Corino, E. R., and Brodkey, R. S. (1969). *J. Fluid Mech.* **37**, 1–30.
Corrsin, S., and Kistler, A. L. (1955). Rep. 1244 NACA (Nat. Adv. Comm. Aeronaut.)
Duhamel, P., and Py, B. (1972). *Colloq. Aerodin, Appl. AAAF, 9th. Paris*.
Eckelman, H. (1970). *Mitt. Max-Planck-Inst., Goettingen* No. 48.
Frenkiel, F., and Klebanoff, P. S. (1967). *Phys. Fluids* **10**, 507–520.
Grass, A. J. (1971). *J. Fluid Mech.* **50**, 233–255.
Gupta, A. K., Laufer, J., and Kaplan, R. E. (1971). *J. Fluid Mech.* **50**, 493–512.
Kim, H. T., Kline, S. J., and Reynolds, W. C. (1968). Rep. MD-20, Dept. Mech. Eng., Stanford Univ., Stanford, Calif.
Klebanoff, P. C. (1955). Rep. No. 1247 NACA (Nat. Adv. Comm. Aeronaut.)
Kline, S. J., Reynolds, W. C., Schraub, F. A., and Runstadler, P. W. (1967). *J. Fluid Mech.* **30**, 741.
Kovasznay, L. S. G., Kibens, V., and Blackwelder, R. F. (1970). *J. Fluid Mech.* **41**, 283–325.
Marcillat, J. (1964). Thèse, Univ. de Aix-Marseille, Marseille.
Meek, R. L., and Baer, A. D. (1970). *AIChE J.* **16**, 841–848.
Popovich, A. T. (1969). *Ing. Eng. Chem., Fundam.* **8**, 609–614.
Shlanchiauskas, A. A. (1972). *In* "Teplo- i Massoperenos" Vol. 1, Part 1, pp. 8–17. Inst. Teplo-massoobmena Minsk.
Townsend, A. A. (1949). *Proc. Roy. Soc., Ser. A* **197**, 124–140.
Wallace, J. W., Eckelman, H., and Brodkey, R. S. (1972). *J. Fluid Mech.* **54**, 39–48.
Willmarth, W. W., and Lu, S. S. (1972). *J. Fluid Mech.* **55**, 65–92.

Zarić, Z. (1969). *C. R. Acad. Sci., Ser. A* **269**, 986–989.
Zarić, Z. (1972a). *Advan. Heat Transfer* **8**, 285–350.
Zarić, Z. (1972b). In "Teplo- i Massoperenos" Vol. 9, Part 2, pp. 3–26. Inst. Teplomassoobmena Minsk.
Zarić, Z. (1972c). *C. R. Acad. Sci., Ser. A* **275**, 459–462.
Zarić, Z. (1972d). *C. R. Acad. Sci., Ser. A* **275**, 513–515.

CHARACTERISTICS OF TURBULENCE WITHIN AN INTERNAL BOUNDARY LAYER

R. A. ANTONIA AND R. E. LUXTON[1]

*Department of Mechanical Engineering
University of Sydney, Sydney, Australia*

1. INTRODUCTION

At the first conference in this series held in Oxford, England, in 1958, G. I. Taylor (1959) surveyed "The present position in the theory of turbulence diffusion." He began by describing his 1913 work as a meteorologist on the sailing whaler *Scotia* (Taylor, 1915) from which he flew kites holding temperature, humidity, and pressure recorders.[2] "It was not possible to get to any great height with this equipment but the very rapid changes in temperature with height which occur in the first few hundred feet in this part of the world made results of interest in spite of this limitation. The records showed that in most cases the temperature of the air was much higher at a few hundred feet than at sea level. This was clearly due to the fact that warm westerly winds flowing off the American continent during summer were rapidly cooled when they reached the iceberg-laden waters off Newfoundland." Schematically what was happening was that an "internal" cold boundary layer was forming under the "external" warm boundary layer. The interface between the two layers defined the distance to which information about the new surface condition had diffused and the rate of growth of the internal layer indicated the rate at which the new information was propagating.

Since Taylor led the way into this important area of information diffusion in a boundary layer—or more particularly, the response of a boundary layer flow to a change in boundary condition—many have followed because of the relevance of the problem in many fields. In this conference we are especially concerned with the atmosphere and this exists, broadly speaking, on three scales. Our present problem is of interest to the micrometeorologist in the lowest few meters of the atmosphere as the upwind terrain—"fetch" as he calls it—may influence heat and moisture transfer from the surface (Bradley,

---

[1] *Present address:* Department of Mechanical Engineering, University of Adelaide, Adelaide, South Australia.

[2] G. I. Taylor's writings have been paraphrased on so many occasions that we wish here to do him the courtesy of including his own evocative description.

1968). On the meso-scale, the turbulence structure and diffusion properties of an internal layer, if present, are likely to have a significant effect on the diffusion and advection of material discharged into the atmosphere from a flue or stack. On the synoptic scale, the active discussion of the paper by Lettau (1959) at the Oxford conference clearly evidenced the importance of the relation between surface drag and large scale atmospheric motions.

While measurements of mean field and time averaged turbulence quantities have been made in an internal layer, it would seem that the turbulence structure has not yet been studied. To this end we have reanalyzed data recorded in our 1968 laboratory experiments on the flow downstream of a step change in surface roughness (Antonia and Luxton, 1971a, 1972, which are from now on referred to as I and II, respectively). Surface roughness changes from smooth to rough (I), and, from rough to smooth (II) are considered. The skewness and flatness factor of the streamwise (u) and normal (w) velocity fluctuations and of their instantaneous product (uw) within the internal layer have been calculated. A significant difference in the behaviour of these quantities is found near the internal layer/external layer interface and it is suggested that this difference is related to an effective "new" turbulence/"old" turbulence ("signal/noise") ratio when the signals are considered as being a random switching between two different turbulent flows, analogous with the switching between turbulent and irrotational flow at the free stream boundary of a layer. Propagation velocities associated with the growth of the internal layer are found to be appreciably higher for the smooth to rough change than for the rough to smooth case, in agreement with the earlier finding that the internal layer grows faster in the first case. When seeking scaling relationships for spectra, it is found that the internal layer thickness is a significant length scale for most of the internal layer. We attempt to relate this to the "active" and "inactive" motion ideas of Townsend (1961) and Bradshaw (1967b).

2. Experimental Conditions

A detailed description of the experimental equipment and techniques used in the investigation may be found in I and II. Briefly, for the smooth to rough surface change, the tunnel floor boundary layer was tripped about 0.3 m from the start of the working section and then developed over a further 2.14 m at which point the surface roughness began. The roughness consisted of a 2.44-m long section with 3.2-mm square section bars placed transversely across the floor of the tunnel at a streamwise spacing of 12.7 mm (roughness height to spacing ratio 1 : 4) with the crests of the roughness elements aligned with the smooth surface. For the rough to smooth case, the last 1.22 m of the roughness section was removed and replaced by 1.22-m smooth surface. The free steam velocity used for all the measurements discussed here was approximately 5.58 m/s and the pressure gradient was set to zero. This gave a boundary layer thickness δ_0 at the beginning of the rough surface of approximately 48 mm and a momentum thickness Reynolds number R_θ of about 3000. At the end of the first 1.22 m of roughness the boundary layer was found to be nearly self-preserving with the 99.5 % thickness[3] equal to about 71 mm and a corresponding R_θ of 4300.

The measurements of u, w, and uw were made with an X-wire operated by nonlinearized constant temperature anemometers. The anemometer outputs were amplified, filtered by sharp cut-off low pass filters set at 1 kHz and digitized at a sampling rate of 3 kHz on the digital data acquisition facility mentioned in I. Final processing of the data was done on the English Electric KDF9 computer in the Basser Computing Laboratory at the University of Sydney. All digital records were analyzed for a duration of approximately 10 s. This was adequate to yield a standard deviation of about 10 % for the third and fourth moments presented in Section 3, and also for most of the spectral estimates (obtained via the FFT technique) of Section 5.

3. Skewness and Flatness Factors of u, w and uw

The skewness and flatness factors of an instantaneous quantity e are defined as

$$S = \overline{e^3}/(\overline{e^2})^{3/2} \quad \text{and} \quad F = \overline{e^4}/(\overline{e^2})^2$$

The skewness and flatness factors of the streamwise and normal velocity fluctuations u, w, and of $uw - \overline{uw}$, the fluctuations in Reynolds shear stress with respect to the mean $\overline{uw}$, have been measured across the internal layer for both the smooth to rough and the rough to smooth surface changes. For the smooth to rough change the distributions are shown in Figs. 1 and 2 for $x/\delta_0 = 2.10$ (closed symbols) and 3.14 (open symbols), where x is measured downstream from the step change and δ_0 is the smooth wall boundary layer thickness at $x = 0$ ($\delta_0 \simeq 4.82$ cm). The distributions of F_u, F_w, and F_{uw} in Fig. 1 are closely similar at the two stations when plotted against z/δ_i, where δ_i is the internal layer thickness determined in I from both mean velocity and Reynolds stress profiles ($\delta_i \simeq 0.88$ cm at $x/\delta_0 = 2.10$ and 1.14 cm at $x/\delta_0 = 3.14$ cm). This suggests that δ_i is a physically significant length scale in the internal layer turbulence. The distributions of F_u and F_w increase to a maximum value of about 3.50 near the edge of the internal layer. For $z/\delta_i > 1.0$, F_u and F_w decrease gradually to a level corresponding to that of the undisturbed smooth wall layer. Although not shown here, F_u, F_w, and F_{uw}

---

[3] Note that this is also the value of δ_0 used in the following sections in the context of the rough to smooth case.

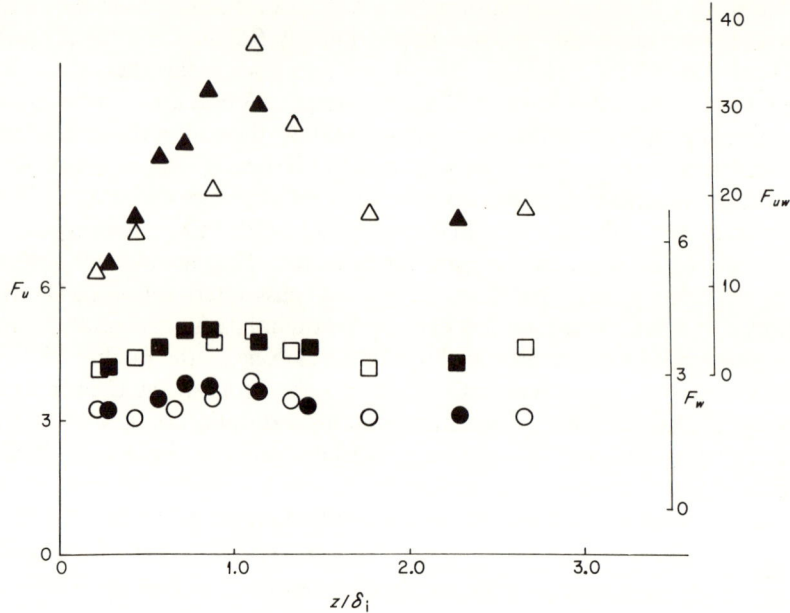

Fig. 1. Flatness factors in smooth to rough internal layer. $x/\delta_0 = 2.10$; ●, F_u; ■, F_w; ▲, F_{uw}. $x/\delta_0 = 3.14$; ○, F_u; □, F_w; △, F_{uw}.

increase again significantly as the outer edge of the boundary layer is neared (see Antonia and Atkinson, 1973; Gupta and Kaplan, 1972). At the smallest values of z measured here the values of F_u and F_w are close to the Gaussian 3.0 but Gupta and Kaplan (1972) found a significant increase of F_u, F_w, and especially of F_{uw} well within the viscous sublayer. The flatness F_{uw} in the present experiments increases from about 12 at $z/\delta_i = 0.25$ to near 35 at $z/\delta_i = 1.0$. The high value near the edge of the internal layer is related to the distinctly non-Gaussian character of u and w in this region of the flow. It is reasonable to attribute the departures from Gaussianity near $z/\delta_i = 1.0$ to an "intermittency" caused by the "switching" between two different types of turbulent flow, namely between the turbulence which is diffused outward from the rough wall and the turbulence associated with that part of the pre-existing smooth wall layer which is yet to be directly affected by the new surface. The breadth of the peaks of F_u, F_w, and F_{uw} indicate that some information about the new surface does penetrate the external layer to z/δ_i of about 1.6, which suggests that the instantaneous interface between the internal and external layers makes large excursions such that the "intermittent" region occupies a significant fraction of δ_i.

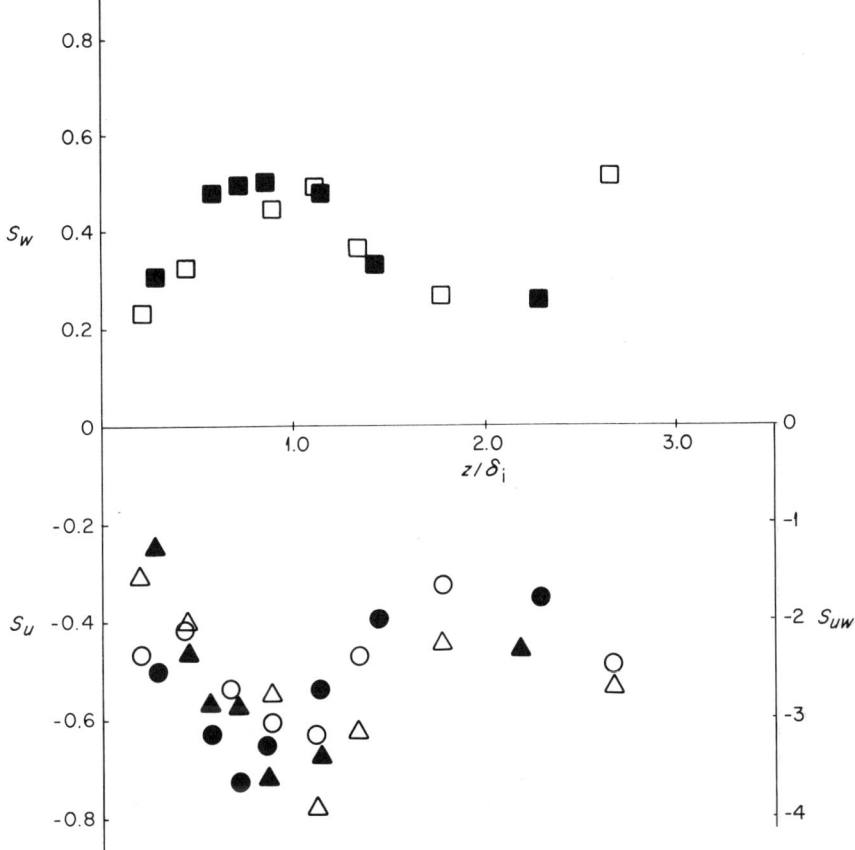

FIG. 2. Skewness in smooth to rough internal layer. $x/\delta_0 = 2.10$; ●, S_u; ■, S_w; ▲, S_{uw}. $x/\delta_0 = 3.14$; ○, S_u; □, S_w; △, S_{uw}.

The skewness of w (Fig. 2) remains positive across the internal layer and shows a broad maximum near $z/\delta_i = 1.0$ while S_u and S_{uw} are negative in this region. Near the outer edge of the external boundary layer ($z/\delta_i > 2.0$) the negative S_u and positive S_w would be consistent with bulges of low momentum fluid in the external part of the boundary layer making incursions into higher momentum free stream fluid. The comparable feature at around $z/\delta_i = 1.0$ similarly could be interpreted as incursions by the energetic "new" turbulence from the strongly retarded internal layer into the (tired?) "old" turbulence of the higher momentum external layer.

In contrast to the smooth to rough case, the skewness and flatness distributions in the rough to smooth internal layer (Figs. 3 and 4) at $x/\delta_0 = 0.71$

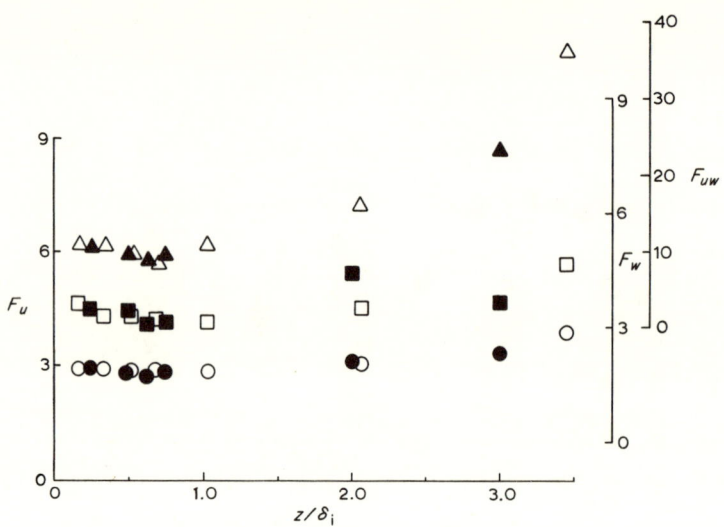

FIG. 3. Flatness factors in rough to smooth internal layer. $x/\delta_0 = 0.71$; ●, F_u; ■, F_w; ▲, F_{uw}. $x/\delta_0 = 1.42$; ○, F_u; □, F_w; △, F_{uw}.

and 1.42 show no noticeable features near $z/\delta_i = 1.0$. F_u and F_w remain close to the Gaussian value of 3.0 while F_{uw} is approximately 10 and only shows the expected rise for $z/\delta_i > 2.0$. A tentative explanation for the absence of a maximum in F or $|S|$ around $z/\delta_i = 1.0$ in Figs. 3 and 4 is that the rough to smooth change is accompanied by a *decrease* in turbulence intensity near the smooth surface and a gain of energy by this part of the flow through diffusion from the outer region of the internal layer (see II). It seems plausible that the statistics of the u and w signals within the internal layer are dominated by the relatively high intensity pre-existing rough wall turbulence. Effectively this results in a "signal" to "noise" (i.e., "new" turbulence to "old" turbulence) ratio of order unity and hence the F and S distributions do not reveal the interface between the internal and external layers. This does not negate the concept of an interface for the rough to smooth internal layer but suggests a relatively weak interaction between the internal layer smooth wall turbulence and the pre-existing rough wall turbulence which is in keeping with the slow growth of the internal layer found by other means in II. Further, as the "new" smooth wall turbulence and the "old" rough wall turbulence differ in their scales, it is likely that F and S of *bandpassed* signals could indicate interface characteristics clearly.

The probability densities of u, w, and $uw - \overline{uw}$ are shown in Fig. 5 for the smooth to rough internal layer at $x/\delta_0 = 3.14$ for $z/\delta_i = 0.22$ and 1.10. The

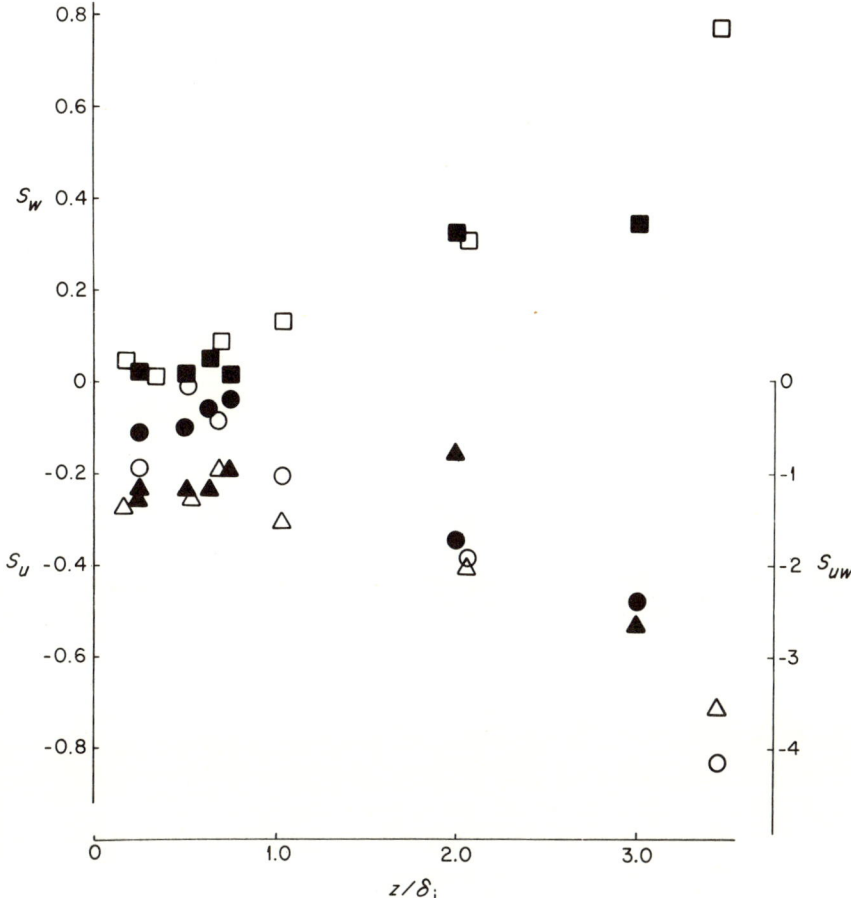

FIG. 4. Skewness in rough to smooth internal layer. $x/\delta_0 = 0.71$; ●, S_u; ■, S_w; ▲, S_{uw}. $x/\delta_0 = 1.42$; ○, S_u; □, S_w; △, S_{uw}.

probability densities p are plotted such that the area under all the experimental distributions in Fig. 5 is equal to unity, e.g.,

$$\int_{-\infty}^{\infty} u' p(u/u')\, d(u/u') = 1$$

where u' is the rms value of u. Although the plots of Fig. 5 only cover ± 3 standard deviations, the contributions to p were measured to ± 6 standard deviations, which is adequate to cover the dynamic range of the signals

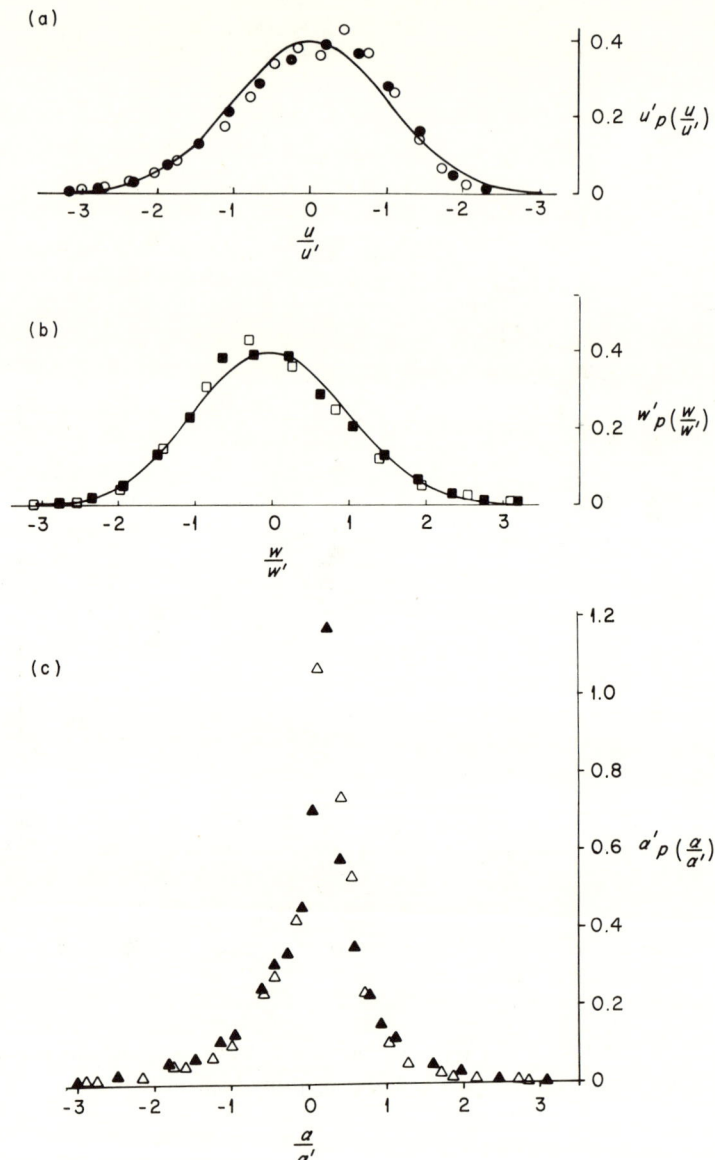

FIG. 5. Probability densities of u, w, and uw in smooth to rough internal layer. (a) u; (b) w; (c) $a = uw - \overline{uw}$. Open symbols refer to $z/\delta_i = 1.10$. Closed symbols are for $z/\delta_i = 0.22$.

considered here.[4] The distributions of $p(u)$ and $p(w)$ are not appreciably different for the two values of z/δ_i and do not deviate very much when compared with, say, $p(uw)$ from the Gaussian distributions shown in Figs. 5a and 5b. The probability density $p(u)$ at $z/\delta_i = 1.10$ is clearly negatively skewed and is slightly higher than $p(u)$ at $z/\delta_i = 0.22$ for near zero values of u/u', so leading to larger flatness factors. For the intermediate values of $|uw - \overline{uw}|$ shown in Fig. 5c, the probability $p(uw - \overline{uw})$ is appreciably smaller near the edge of the internal layer than at $z/\delta_i = 0.22$. At the high values of $|uw - \overline{uw}|$ the trend is reversed thus leading to the relatively high value of F_{uw} found at $z/\delta_i = 1.10$.

To predict adequately high order moments of u, w, or uw in the inner region of a self-preserving boundary layer it is necessary to include small departures from Gaussianity in the assumed probability densities of u or w (see Frenkiel and Klebanoff, 1973; Antonia and Atkinson, 1973). When predicting the high order moments of uw, the normal joint probability density of u and w

$$p(u, w) = \frac{1}{2\pi u'w'(1 - r^2)^{1/2}} \exp\left\{-\frac{1}{2(1 - r^2)}\left(\frac{u^2}{u'^2} - \frac{2ruw}{u'w'} + \frac{w^2}{w'^2}\right)\right\}$$

where r is the correlation coefficient $\overline{uw}/u'w'$, is not adequate for predicting F_{uw} even in the inner region of the layer. Antonia and Atkinson (1974) found that a fourth-order model for $p(u, w)$ was necessary to achieve satisfactory agreement with the measured F_{uw} values. They suggested that adequate predictions of F_{uw} in the intermittent region of the boundary layer can be obtained provided the intermittency characteristics of the flow are known and some of the statistics of u and w in both the turbulent and irrotational parts of the flow are available. This idea may be extended to the prediction of F_{uw} near the edge of the internal layer provided "intermittency" characteristics of this layer are known. With the experimental technique used here, an adequate definition of the "interface" between the two types of turbulent motion would seem almost impossible. However, it may be possible to tag this interface by adding heat to the internal layer by, for example, low intensity heating of the rough surface. Johnson (1959) observed intermittent temperature fluctuations near the edge of a thermal internal layer developing downstream of a step change in heat flux on a smooth surface. With the clearer definition of the interface the conditional sampling technique could then be used to determine the properties of the two turbulent fields.

---

[4] It should be noted that the values of S and F presented in this section were computed directly from the digital records and not via the probability density.

4. Diffusion Velocities of u and w

Bradshaw (1967a) defines a velocity of diffusion W_p of turbulent energy in the z direction as $W_p = (\frac{1}{2}\overline{q^2 w} + \overline{pw})/\frac{1}{2}\overline{q^2}$, where $\overline{q^2}$ is the turbulence intensity $(\overline{u^2} + \overline{v^2} + \overline{w^2})$ and p is the pressure fluctuation. At the edge of a self-preserving boundary layer W_p is equal to the entrainment rate provided that advection and diffusion are equal and that the shear stress, and therefore the production and dissipation of turbulent energy, is small. Turbulent energy balances in both the smooth and rough wall boundary layers show that, near the outer edge of the layers, the diffusion of turbulent energy by $\frac{1}{2}\overline{q^2 w}$ is very nearly equal to the advection term, so that, fortunately, it would appear that the pressure diffusion term is small and the above definition of W_p can be simplified to $W_p = \overline{q^2 w}/\overline{q^2}$, at least well away from the wall.[5] Near the edge of the internal layer, the diffusion by $\overline{q^2 w}$ is again approximately equal to the advection but the contributions from the production $\overline{uw}(\partial U/\partial z)$ and the dissipation are appreciably larger (see Fig. 19 of I and Fig. 12 of II). One cannot then simply equate W_p to the rate of increase of the internal layer thickness δ_i, but it is useful to compare the measured values of W_p for the two internal layers considered here. The distributions of W_p, normalized by the local mean velocity U, are shown in Fig. 6 as a function of z/δ_i, where δ_i is the internal layer thickness, at two streamwise stations for the smooth to rough change and four x stations for the rough to smooth change. In evaluating W_p the intensity $\overline{q^2}$ was approximated by $\frac{3}{2}(\overline{u^2} + \overline{w^2})$ as $\overline{v^2}$ was not measured. In the smooth to rough case, W_p/U reaches a maximum well inside the edge of the internal layer and, although not shown in Fig. 6, has another maximum near the outer edge of the external layer. This maximum in W_p well within the internal layer, when taken in conjunction with the broad peak in the flatness factor, suggests that most of the "entrainment" of "old" turbulence occurs near the "troughs" in the interface, i.e., at around $z/\delta_i \simeq 0.5$. For the rough to smooth change, on the other hand, W_p/U first decreases at small values of z/δ_i but then rises to a maximum near $z/\delta_i = 1.0$. The value of this maximum is only about half that of the maximum for the smooth to rough change which seems to be in qualitative agreement with the different rates of propagation of δ_i reported in II. The location of this peak is consistent with an "upside-down" intermittency with "upside-down" entrainment occurring at the crests of the lower energy internal layer, i.e., at the troughs of the higher energy external layer.

Bradshaw (1967a) has also suggested that the energy flux velocity defined

---

[5] On the rough wall the pressure diffusion term may be important in the inner 20 % of the layer (see Antonia and Luxton, 1971b).

FIG. 6. Diffusion velocity in z direction. Smooth to rough: $\bigcirc$, $x/\delta_0 = 2.10$; $\square$, 3.14. Rough to smooth: $\triangle$, $x/\delta_0 = 0.35$; $+$, 0.71; $\times$, 1.05; $\triangledown$, 1.42.

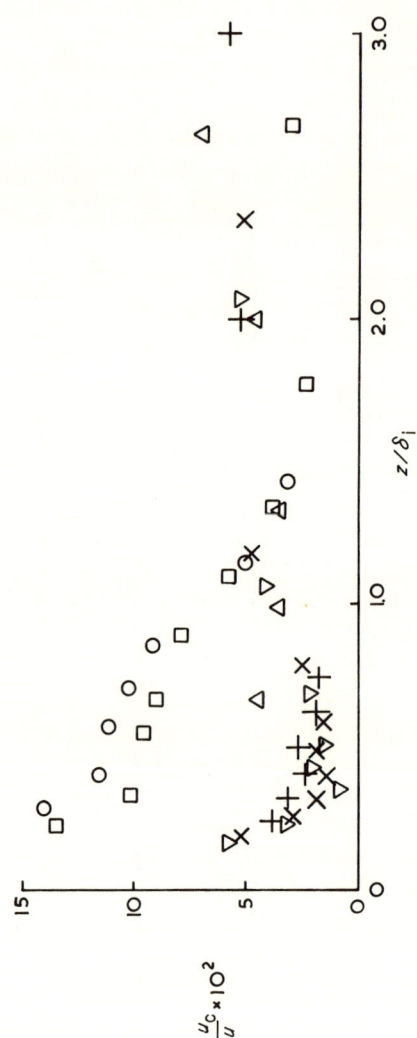

FIG. 7. x component convection velocity. Symbols same as for Fig. 6.

as $(\frac{1}{2}\overline{q^2}U + \frac{1}{2}\overline{q^2 u} + \overline{pu})/\frac{1}{2}\overline{q^2}$ may be a plausible measure of the convection velocity in the streamwise direction. The difference u_c between the energy flux velocity and the local mean velocity U is plotted in Fig. 7 neglecting the contribution from the $\overline{pu}$ correlation. In the smooth to rough case the energy flux convection velocity well within the internal layer can evidently be as much as 14% below the mean velocity. For the rough to smooth case, u_c/U values are close to zero in the central region of the internal layer which suggests a relatively passive convection of the turbulence field by the mean field. In contrast to this, the convection velocities deduced from space–time correlations were approximately the same as the local mean velocity throughout both the smooth to rough and the rough to smooth internal layers.

5. Spectra of u and w

For the inner part of a layer which is in energy equilibrium, Bradshaw (1967b) has suggested that the use of $\tau^{1/2}$ and z as, respectively, velocity and length scales when normalizing spectral densities of u and w and the uw cospectrum, should collapse the high frequency ends of the u and w spectra and the whole of the uw cospectrum. The low frequency ends of the u and w spectra receive contributions from the large-scale "inactive" motion (Townsend, 1961; Bradshaw, 1967b) which does not contribute to τ in the inner part of the layer and hence the $\tau^{1/2}$, z scaling will not collapse the low frequencies.

In the present experiments one would not necessarily expect the τ, z scaling to collapse spectral densities for the "active" (shear stress producing) motion. The shear stress τ varies appreciably across the internal layer and the energy budgets for the smooth to rough (I) and the rough to smooth (II) surface changes reveal that the internal layers are not in energy equilibrium as evidenced by significant advection and diffusion. The inadequacy of the τ, z scaling in these circumstances is clearly shown in Figs. 8–10 for two streamwise stations in the smooth to rough case and one only in the rough to smooth case. These figures are plotted in the form $\phi(k^*)$ versus k^* where $\phi_u(k^*)$, for example, is the dimensionless spectral density $\overline{u^2}(k^*)/\tau$ and k^* is the dimensionless wave number $k_1 z = \omega z/U$. The cospectrum $\phi_{uw}(k^*)$ has the obvious constraining property that

$$\int_0^\infty \phi_{uw}(k^*) \, dk^* = 1.$$

The general trend of the results is that for relatively high frequencies ($k^* > 3$, say) the spectral density increases with increasing z. The opposite trend is observed for $k^* < 3$.

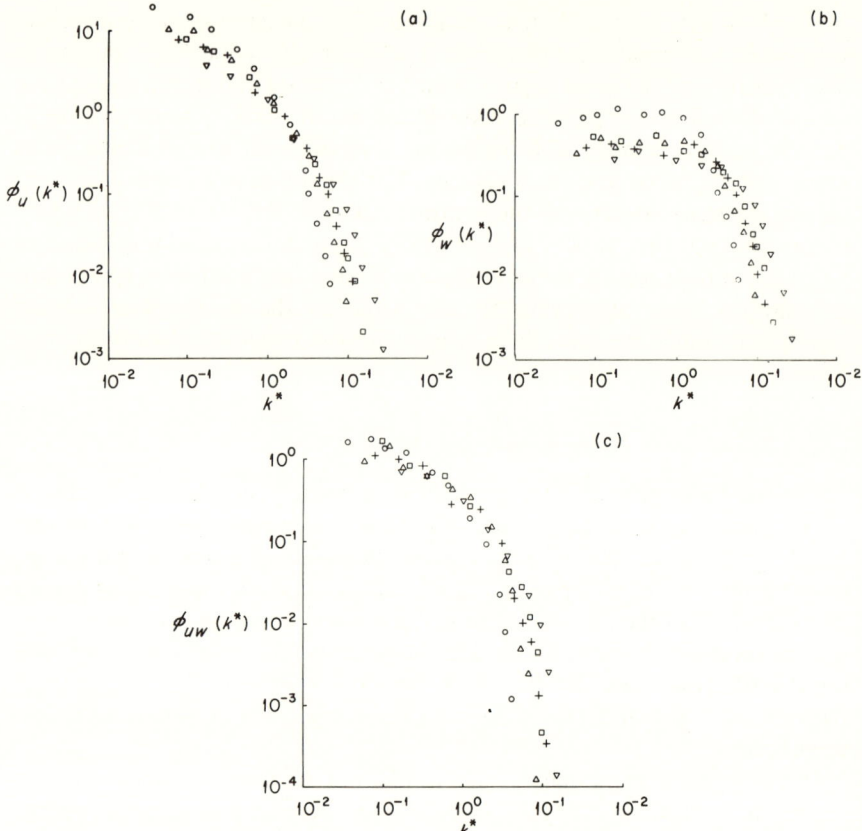

FIG. 8. Frequency spectra at $x/\delta_0 = 2.10$ (smooth to rough) normalized with τ and z. (a) u component; (b) w component; (c) uw. $\bigcirc$, $z/\delta_i = 0.29$; $\triangle$, 0.58; $+$, 0.87; $\square$, 1.16; $\triangledown$, 2.32.

In a fully developed nearly self-preserving boundary layer with zero pressure gradient on either a smooth or a rough wall, where advection and diffusion terms are small, the "active" and "inactive" motion ideas and the associated $\tau^{1/2}$, z scaling should be effective in collapsing spectra. A sample of such spectra for the u component obtained in the inner region ($y/\delta < 0.25$ approximately) of the nearly self-preserving smooth and rough wall layers is shown in Fig. 11. The changes in spectral shape with increasing distance from the wall are the same as those revealed in Figs. 8–10 for the nonequilibrium internal layers. Klebanoff's (1955) data and the densities ϕ_w and ϕ_{uw}, not shown here, also exhibit the same trends. The failure of these data to collapse on τ, z scaling as effectively as the data from the two equilibrium layers in adverse pressure gradients measured by Bradshaw (1967b)

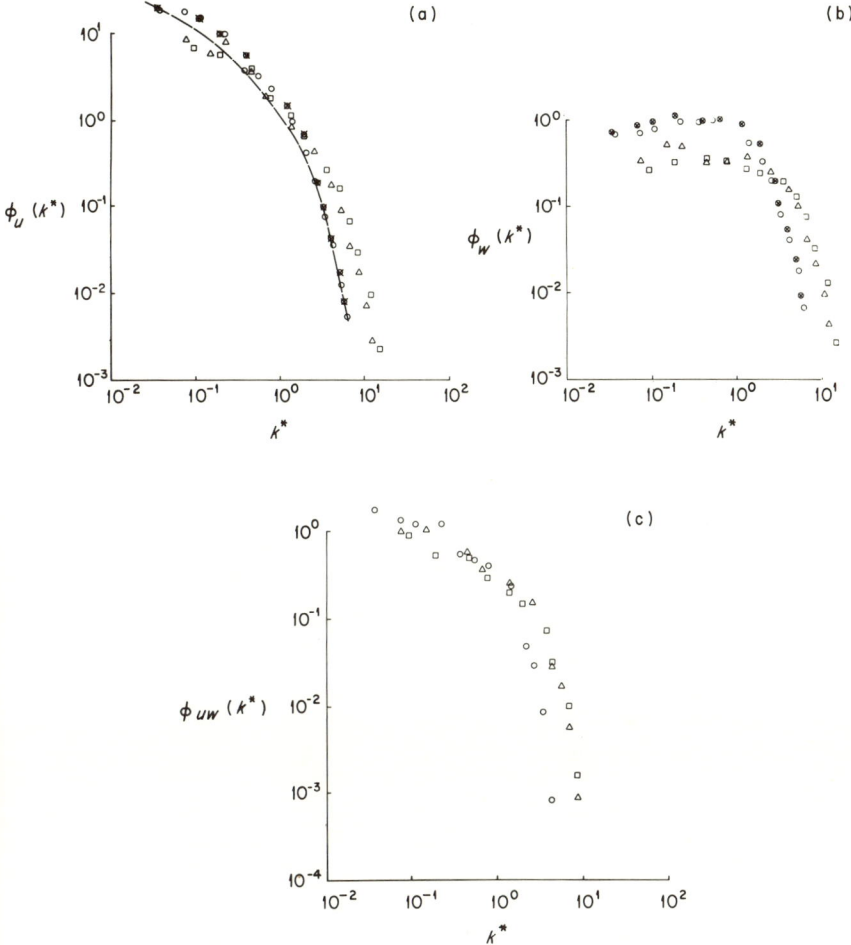

FIG. 9. Frequency spectra at $x/\delta_0 = 3.14$ (smooth to rough) normalized with τ and z. (a) u component; (b) w component; (c) uw. $\bigcirc$, $z = 0.25$ cm, $z/\delta_i = 0.22$; $\triangle$, 0.50, 0.44; $\square$, 1.0, 0.88. $\otimes$, $z = 0.25$ cm, $z/\delta_i = 0.29$, $x/\delta_0 = 2.10$. ---, smooth wall distribution, $z = 0.25$ cm, $z/\delta_0 = 0.079$.

probably arises from the absence of any significant Kolmogorov inertial subrange in the present spectra.[6] The inertial subrange law $\overline{u^2}(k_1) = \alpha_1 \varepsilon^{2/3} k_1^{-5/3}$, where α_1 is a universal constant and ε is the turbulent energy dissipation rate, can be written as $\phi_u(k^*) = (\alpha_1/\tau)\varepsilon^{2/3} z^{2/3}(k^*)^{-5/3}$. For

---

[6] The fairly high values of the spectral correlation coefficient R_{uw} (Figs. 14 and 15) are also consistent with the absence of local isotropy suggested by the negligible extent of "$-\frac{5}{3}$ law" in the spectra.

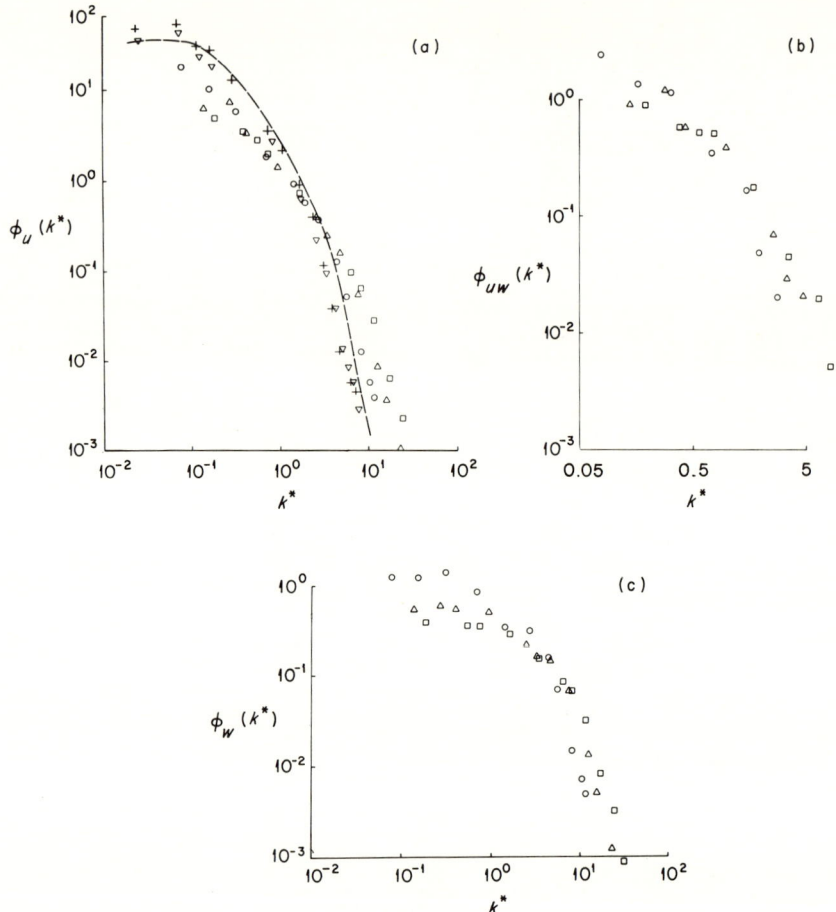

FIG. 10. Frequency spectra at $x/\delta_0 = 1.42$ (rough to smooth) normalized with τ and z. (a) u component; (b) uw; (c) w component. $+$, $z/\delta_i = 0.17$; $\bigcirc$, 0.34; $\triangle$, 0.68; $\square$, 1.02. $\triangledown$, $z = 0.25$ cm, $z/\delta_i = 0.67$, $x/\delta_0 = 0.35$. $---$, rough wall distribution, $z = 0.25$ cm, $z/\delta_0 = 0.036$.

energy equilibrium $\varepsilon = \tau^{2/3}/\kappa z$, where κ is the von Kármán constant, and thus $\phi_u(k^*) = (\alpha_1/\kappa^{2/3})(k^*)^{-5/3}$ which is independent of z as required to collapse the spectra. In the absence of an inertial subrange this argument does not apply and no collapse of spectra can be expected.

At the low wave number end of the u spectra as $k^* \to 0$, $\phi_u(k^*) = (\overline{u^2}/\tau) \times (L/z)$ where L is the integral length scale. In the inner equilibrium layer L is proportional to z to a good first approximation. As $\tau/\overline{u^2}$, or $\tau/\overline{q^2}$, also increases appreciably with z in that part of the layer (Bradshaw, 1967a) the low

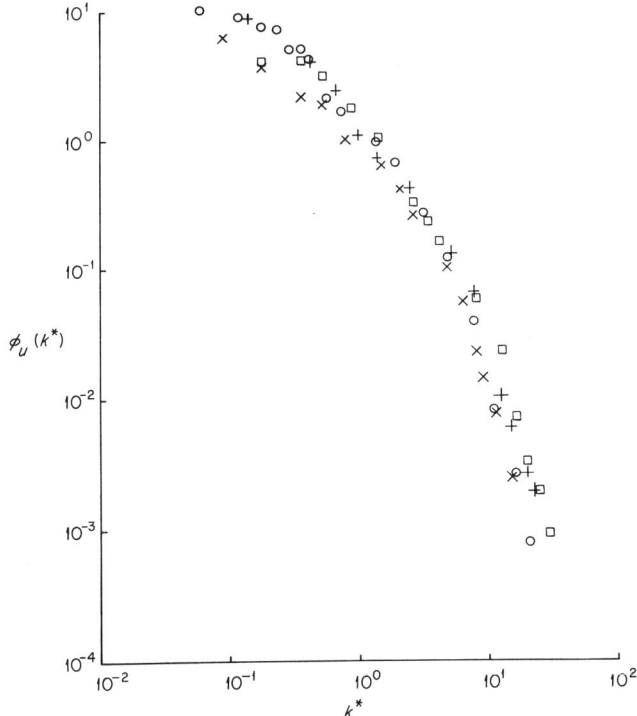

FIG. 11. u-Component spectra in fully developed smooth wall and rough wall boundary layers. Smooth wall: ×, $z/\delta_0 = 0.22$; rough wall: ○, $z/\delta_0 = 0.107$; +, 0.14; □, 0.21.

wave number end of $\phi_u(k^*)$ must decrease with increasing z. Figure 9a shows that for $z = 2.54$ mm, $\phi_u(k^*)$ as $k^* \to 0$ is reduced downstream of the smooth to rough change. Although this is in part due to an increase in τ/q^2, it is also consistent with the reduction in length scale reported in I. For the rough to smooth change (Fig. 10a), there is an increase in $\phi_u(k^*)$ as $k^* \to 0$ also consistent with the increase in length scale reported in II.

As an alternative to the τ, z scaling, the spectral densities ϕ_u and ϕ_{uw} in the internal layer have been replotted in Figs. 12 and 13 in the form $\phi(k^+)$ versus k^+ where k^+ is now $k_1 \delta_i$, i.e. the wave number normalized with respect to the internal layer thickness δ_i. This new scaling appears to give a reasonable collapse of the high frequency end of the spectra suggesting that the smaller scale motion in the outer part of the internal layer may be determined by the local shear stress and the internal layer thickness. For the region of the internal layer closest to the wall, one would still expect z to be the representative length scale, at least as far as the smaller scale motion is

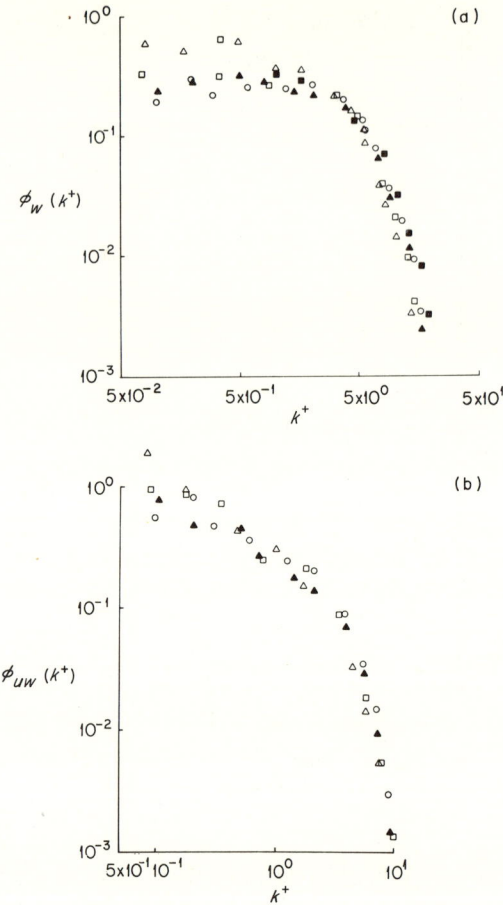

FIG. 12. Frequency spectra for smooth to rough internal layer, normalized with τ and δ_i. (a) w component; (b) uw. $x/\delta_0 = 2.10$: ○, $z/\delta_i = 0.57$; □, 0.86; △, 1.15. $x/\delta_0 = 3.14$: ■, $z/\delta_i = 0.67$; ▲, 0.89.

concerned. This is substantiated by the $\phi_u(k^*)$ results of Figs. 9a, 9b, and 10a for $z = 2.54$ mm and a range of X stations. The $\phi_w(k^+)$ and $\phi_{uw}(k^+)$ spectra of Fig. 13 show surprisingly good collapse even at low wave numbers as does the $\phi_u(k^+)$ which is not shown here. A major difference between the smooth to rough and the rough to smooth spectra is a substantial increase in the low wave number part of $\phi_w(k^+)$ across the smooth to rough internal layer but not across the rough to smooth internal layer. This is consistent with more energetic diffusion in the smooth to rough case leading to a more rapid growth of the internal layer.

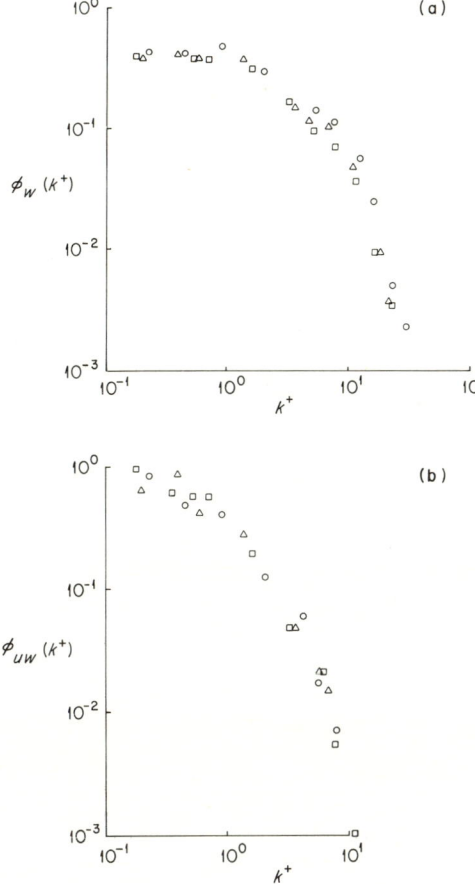

FIG. 13. Frequency spectra for rough to smooth internal layer, normalized with τ and δ_i. (a) w component; (b) uw. $x/\delta_0 = 1.42$: ◯, $z/\delta_i = 0.34$; △, 0.68; □, 1.02.

The correlation coefficient between a narrow band of u fluctuations and the same narrow band of w fluctuations is

$$R_{uw} = \overline{uw}(k_1)/[(\overline{u^2}(k_1))^{1/2}(\overline{w^2}(k_1))^{1/2}]$$

and is shown in Figs. 14 and 15 for the smooth to rough internal layer. Apart from the low values of R_{uw} at the station closest to the wall, there is little change in both the magnitude and shape of R_{uw} across the outer region of the internal layer. A broad maximum value of about 0.6 is obtained at low wave numbers. The uw spectrum can be written as $S_{uw} = C_{uw} - iQ_{uw}$ where C_{uw} is the cospectrum and Q_{uw} is the quadrature spectrum. Note that $\phi_{uw} = C_{uw}(k^*)/\tau$ and $R_{uw} = \phi_{uw}/[\phi_u \phi_w]^{1/2}$. The coherence between the u and

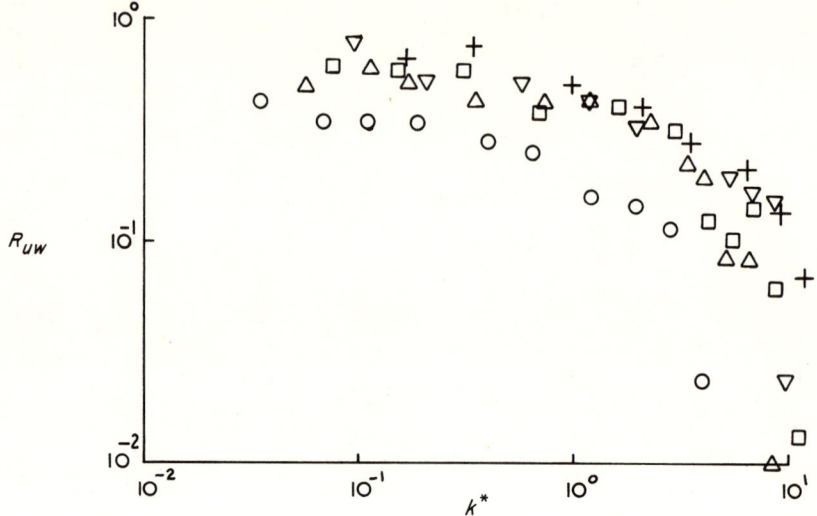

Fig. 14. Spectral correlation coefficient for smooth to rough internal layer at $x/\delta_0 = 2.10$. ○, $z/\delta_i = 0.29$; △, 0.58; □, 0.87; ▽, 1.16; +, 2.32.

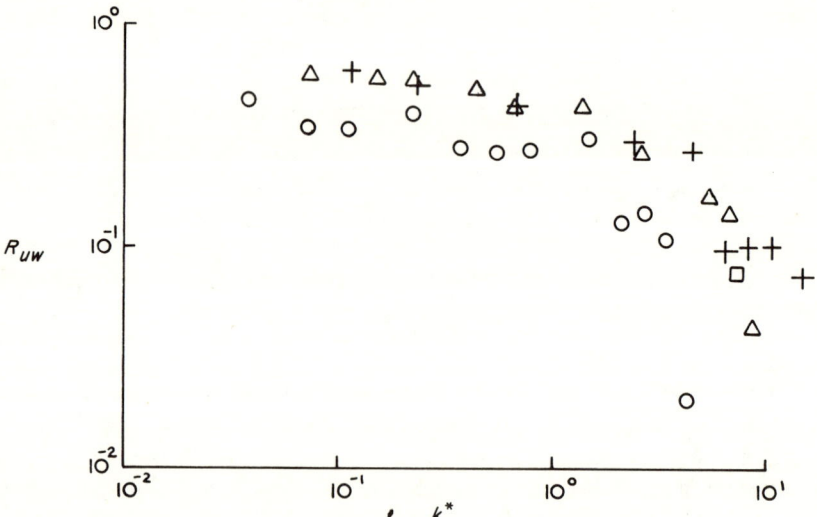

Fig. 15. Spectral correlation coefficient for smooth to rough internal layer at $x/\delta_0 = 3.14$. ○, $z/\delta_i = 0.22$; △, 0.44; +, 1.10.

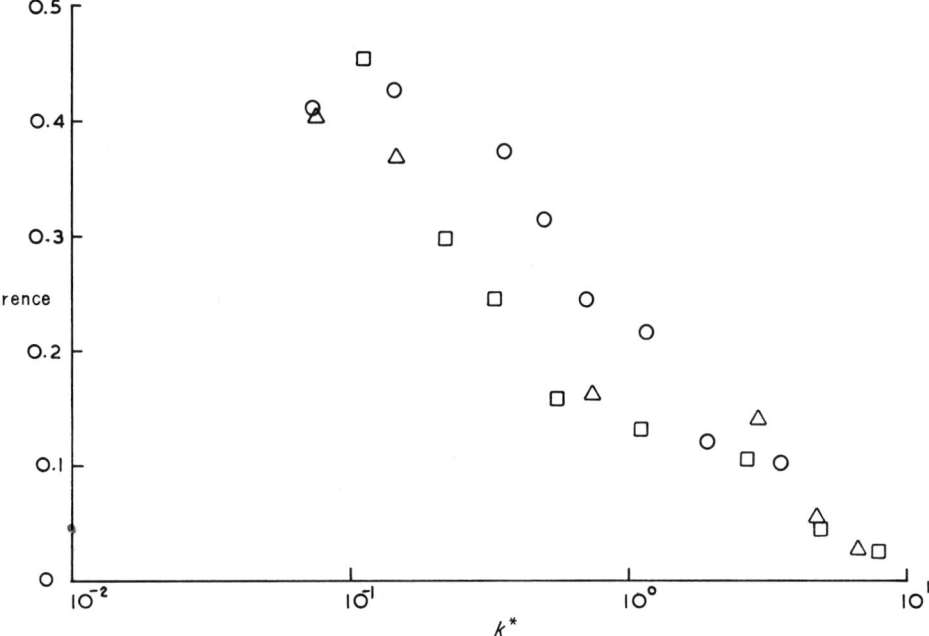

FIG. 16. Coherence of uw near the edge of the internal layer (smooth to rough). □, $x/\delta_0 = 3.14$, $z/\delta_i = 1.11$; △, 2.10, 0.86. ○, smooth wall, $z/\delta_0 = 0.16$.

w fluctuations can be defined as $R = (S_{uw} S_{uw}^*)^{1/2}/[\overline{u^2}(k_1)\overline{w^2}(k_1)]^{1/2}$, where S_{uw}^* is the complex conjugate of S_{uw}. At low frequencies, R is nearly identical to R_{uw} because the u and w fluctuations are nearly in phase and the contribution from Q_{uw} is therefore small. With increasing frequency (Fig. 16), R remains appreciably larger than R_{uw} due to the increased importance of Q_{uw}.[7]

6. Concluding Remarks

In this paper we have presented calculations of higher order moments, propagation velocities, and spectra of fluctuations in an internal layer which forms downstream from a change of surface roughness. The main feature of these calculations of the turbulence structure is that they lead to physically penetrating conclusions which are fully consistent with the conclusions drawn in I and II based on much less sophisticated data analysis. It is stressed that the raw data used for the present calculations is *identical* with that reported in I and II. Skewness and flatness factors near the edge of the

---

[7] It should be noted however that the accuracy of R_{uw} and R at the higher frequencies is poor since they result from taking small differences between relatively large signals.

smooth to rough internal layer suggest an "intermittency" of the interface associated with "switching" between two different turbulence fields. One has the impression of energetic smaller scale "new" turbulence eating into the tired larger scale "old" turbulence of the original smooth wall boundary layer. Such a picture suggests a relatively rapid propagation of the new turbulence as found earlier and as indicated by our present discussion of propagation velocity W_p. Although the two turbulence fields seem to have separate identities which could, with different techniques, be separated even where they interact at the interface, one may still expect some large-scale "stirring" of the internal layer by the external layer, as indicated by changes in the scaled spectra at low frequencies.

In the rough to smooth change situation, the picture is not nearly as clearly drawn but it is nevertheless similar in outline. The sharp, intermittently "switched" interface is no longer revealed by the overall distributions of flatness and skewness factors. This is probably because the relative intensities of the "new" and the "old" turbulence are comparable even though their scales are quite different. Flatness and skewness of band passed signals could reveal the intermittent switching at the interface but this has not yet been done. Temperature "tagging" of one of the layers by low intensity surface heating would also allow details of the switching to be determined as well as some statistical details of each of the turbulence fields (by conditional sampling). This information is needed if higher-order moments are to be predicted.

The present experimental situation, while providing much useful information, is probably too complicated to understand fully, given our present knowledge of turbulent shear flows. A similar investigation using a change of low intensity heat flux at the boundary, similar to that studied by Johnson (1959), and conditional sampling techniques to study in detail the internal/external layer interface would provide directly information about the nature of diffusive processes in a "simple" turbulent boundary layer. But the atmospheric boundary layer is not simple so more complex experiments are also required, e.g., the interaction of an internal layer with a density stratified external layer would be of direct interest. From the isothermal results of this paper it would appear that every effort should be made to discharge smoke into an external layer if one exists.

A major question which arises from this work concerns the scale on which the turbulent/turbulent interaction occurs at the interface. At the outer edge of a turbulent boundary layer we have a turbulent/irrotational interface which, presumably, involves molecular scales in a viscous superlayer. Can we conceive a comparable hierachy of scales in a turbulent/turbulent interaction? With the aid of conditional sampling the answer to this question does not appear to be unattainable.

Acknowledgment

The work described in this paper forms part of a programme of research on turbulent shear flows supported by the Australian Research Grants Committee, the Australian Institute of Nuclear Science and Engineering and the Commonwealth Scientific and Industrial Research Organisation.

References

Antonia, R. A., and Atkinson, J. D. (1973). *J. Fluid Mech.* **58**, 581.
Antonia, R. A., and Luxton, R. E. (1971a). *J. Fluid Mech.* **48**, 721.
Antonia, R. A., and Luxton, R. E. (1971b). *Phys. Fluids* **14**, 1027.
Antonia, R. A., and Luxton, R. E. (1972). *J. Fluid Mech.* **53**, 737.
Bradley, E. F. (1968). *Quart. J. Roy. Meteorol. Soc.* **94**, 361.
Bradshaw, P. (1967a). *J. Fluid Mech.* **29**, 625.
Bradshaw, P. (1967b). *J. Fluid Mech.* **30**, 241.
Frenkiel, F. N., and Klebanoff, P. S. (1973). *Phys. Fluids* **16**, 725.
Gupta, A. K., and Kaplan, R. E. (1972). *Phys. Fluids* **15**, 981.
Johnson, D. S. (1959). *J. Appl. Mech.* **26**, 325.
Klebanoff, P. S. (1955). *Nat. Adv. Comm. Aeronaut., Rep.* **1247**.
Lettau, H. H. (1959). *Advan. Geophys.* **6**, 241.
Taylor, G. I. (1915). *Phil. Trans. Roy. Soc. London. Ser. A* **215**, 1.
Taylor, G. I. (1959). *Advan. Geophys.* **6**, 101.
Townsend, A. A. (1961). *J. Fluid Mech.* **11**, 97.

STRUCTURE OF THE REYNOLDS STRESS AND THE OCCURRENCE OF BURSTS IN THE TURBULENT BOUNDARY LAYER

W. W. WILLMARTH AND S. S. LU[1]

*Department of Aerospace Engineering
University of Michigan, Ann Arbor, Michigan, U.S.A.*

1. INTRODUCTION

The past decade has been a period of rapid advance in our understanding of the nature of the structure of the fluctuating flow field in the turbulent boundary layer. The decade began at Stanford University with the perfection, by Runstadler *et al.* (1963), of a flow visualization method using small hydrogen bubbles produced by electrolysis of water along a fine current carrying wire. With this flow visualization method the Stanford group was able to identify a deterministic pattern and sequence of events called a burst during which large contributions to Reynolds stress occurred. The bursts that were observed (see Kim *et al.*, 1968) resulted in a violent and intense swirling and mixing of low speed fluid near the wall with fluid farther from the wall.

Corino and Brodkey (1969) have recently reported results of flow visualization studies in the boundary layer near the wall obtained by observing the motion of small neutrally buoyant particles suspended in water. Their observations confirmed the results of the Stanford group and contributed the important new information that when a burst occurred there was a violent small scale interaction in the region $7 < y^+ < 30$ between fluid moving unusually slowly near the wall and higher speed fluid slightly farther from the wall [here $y^+ = yu^*/\nu$ and $u^* = (\tau/\rho)^{1/2}$ where τ is the wall shear stress]. In addition, Corino and Brodkey also identified an event they called a "sweep" in which it appeared that after a burst of slower speed outward moving fluid occurred the fluid in the violent bursting interaction was replaced or swept away by high speed fluid moving toward the wall. Corino and Brodkey estimated that approximately 70 % of the contribution to the

---

[1] *Present address:* Department of Power Mechanical Engineering, National Tsing Hua University, Hsinchu, Taiwan.

mean Reynolds stress was produced during the bursting or ejection of low speed fluid from the wall region.

Grass (1971) also used the hydrogen bubble flow visualization technique to study instantaneous profiles of streamwise and normal velocities all across the boundary layer developed on smooth and rough walls. Grass used a digital computer to select pairs of instantaneous streamwise, u, and normal, v, velocity fluctuation profiles under the condition that the streamwise velocity was a maximum or a minimum at various specified distances from the wall. The profile pairs were then used to compute instantaneous profiles of the product uv (note that in this paper the notation is that $\bar{u} = \bar{v} = 0$). The results for smooth (or rough walls) showed clearly that when u was a minimum at any distance from the wall, uv was negative in that region and therefore contributions to the Reynolds stress occurred. Grass interpreted this to mean that bursts or eruptions of low momentum fluid exert a continued influence throughout the boundary layer. On the other hand, when Grass selected u and v profile pairs (measured on smooth walls) on the basis that the streamwise velocity was a maximum at a given height, he found that contributions to Reynolds stress, $uv < 0$, occurred only at points near the wall.

Using concepts obtained from the above flow visualization experiments as a guide we have devised a number of experiments using hot wire anemometers. These experiments are designed to provide additional quantitative information about the Reynolds stress during bursting which is difficult to obtain with flow visualization methods. In addition to the results we have obtained, a number of other investigations of Reynolds stress and bursting phenomena have been reported in which hot wire or hot film anemometers were used. These results appear in the papers of Blackwelder and Kaplan (1971), Wallace *et al.* (1972), and Rao *et al.* (1971). When appropriate the above papers will be discussed and compared with the present measurements as they are described in the body of the paper.

The present measurements have been analysed with the aid of a digital computer which was used to detect the occurrence near the wall of fluid eruptions or "bursts" and of inrush or "sweep" events. It has been possible to measure the convection speed and scale of these events near the wall. In addition, measurements of uv throughout the boundary layer have been made in which the outward flow of fluid with low streamwise momentum is shown to contribute more to the Reynolds stress than does the inward flow of fluid with high streamwise momentum. A unique method was devised to identify burst and sweep contributions to Reynolds stress throughout the boundary layer. This method is used to confirm that burst and sweep related events occur and scale with outer variables throughout the boundary layer.

A mechanism or model that we have previously proposed (Willmarth and Tu, 1967) for the deterministic structure responsible for bursts and sweeps is discussed in the light of the present results.

2. Experimental Apparatus and Methods

The experiments were conducted in a thick ($\delta \simeq 5$ in.) turbulent boundary layer in the 5 × 7 ft wind tunnel of the Department of Aerospace Engineering at The University of Michigan. Most of the measurements were made in a boundary layer with a thick sublayer that was produced at low free stream speeds, $U_\infty \simeq 20$ ft/sec. A few measurements were made at higher free stream speeds $U_\infty \simeq 200$ ft/sec. The boundary layer parameters are set forth in Table I.

A complete description of the experimental apparatus has been given in Willmarth and Lu (1971, 1972) and Lu and Willmarth (1972). This includes the mean flow in the boundary layer, the hot wire probes, the electronic equipment, the frequency modulated analog tape recorder and the analog to digital converter used to prepare the data (which were reproduced at one-eighth the recording speed) for digital processing. Almost all of the data reduction was performed with an IBM 360/67 computer using several simple FORTRAN programs and a few assembly language subroutines. We emphasize that the use of the digital computer for data reduction makes it possible to measure a great variety of different statistical parameters inexpensively and efficiently. Indeed, it would not have been possible to perform all the measurements that we shall describe using analog methods, owing to limitations of time, money, and manpower.

3. Experimental Measurements

The measurements are described in chronological order, which is naturally related to our increased understanding of the phenomena during the course of the investigation and to our increased skill in devising and writing new data reduction programs for the computer. The first exploratory measurements were made using analog methods to detect eruptions of low speed fluid near the wall and to select samples of the signal uv when eruptions occurred. It soon became apparent that analog data gathering and processing methods were extremely inefficient in comparison to digital methods. Using analog methods one must first construct the special circuitry and then operate it successfully over the long periods of time required to obtain stable average values. When sampling random data, the simplest analog averaging method, a capacitor fitted with a resistive leak, is very wasteful of time and

TABLE I. Properties of the actual and ideal turbulent boundary layer

U_∞ (ft/sec)	Re_θ	δ (ft)	δ^* (ft)	θ (ft)	δ^*/θ	U_τ/U_∞	Re_x	Remarks
19.7	4,230	0.405	0.0494	0.0363	1.365	0.0386	—	Transition location unknown
—	3,800	—	—	—	1.383	0.0387	2.1×10^6	Coles's ideal boundary layer
204	38,000	0.42	0.041	0.0315	1.30	0.0326	3.1×10^7	Willmarth and Tu (1967)
—	39,000	—	—	—	1.30	0.0318	3.2×10^7	Coles's ideal boundary layer

valid data because the data signal itself is usually used to develop the charge on the capacitor. In measuring average values rather large time constants were required to obtain stable averages during sampling measurements of infrequently occurring events. When digital methods were used, the amount of data required to obtain stable average values was approximately 5 % of the data required using analog methods.

We shall describe only results obtained using digital data reduction methods. All of the digital data reduction was done using prerecorded data reproduced at one-eighth the real time recording rate. After the initial exploratory measurements had been made and the experimental set up was designed, the actual data collection and recording process required only two weeks time. Approximately two years were spent writing programs and using them to reduce the digitized data.

3.1. Conditionally Sampled Measurements of Reynolds Stress

The method of conditional sampling was first used by Kibens (1968; for a summary, see Kovasznay et al., 1970) in the study of the motion and shape of the turbulent bulges in the outer intermittent region of a turbulent boundary layer. We have extended their sampling concepts to allow the extraction of individual contributions to Reynolds stress from a fully turbulent signal. To separate the Reynolds stress contributions from the background turbulence signals one needs a criterion to decide when the Reynolds stress contribution occurs. In our study of the flow field near the wall we were guided by the results of Kim et al. (1968) and Corino and Bodkey (1969) which show that before an eruption or bust occurs the fluid beneath the region of eruption attains an unusually low streamwise velocity.

We used the simple criterion that when the streamwise velocity u_w very near the wall becomes low and decreasing with negative slope (i.e., $\partial u_w/\partial t < 0$) an eruption or burst is likely to occur. To test the validity of this criterion we sampled the instantaneous record of the product uv obtained from an x configuration hot wire anemometer probe for a set of nine different values of u_w ($-2u'_w < u_w < 2u'_w$) for both positive and negative slope. Figure 1 is a sketch showing the configuration of the hot wire probes and the coordinates denoting the position of the hot wire probes.

The basic scheme we used is outlined in the sketch of Fig. 2 in which a single sample of the uv signal is acquired in a "window" centered about the point at which the signal u_w has attained, for example, a certain specified level with positive slope. The sampling was performed by an IBM 360/67 computer which was programed using FORTRAN to acquire samples from all nine levels of u_w with both positive and negative slopes in one pass through the data recorded on digital magnetic tape. The samples for each of

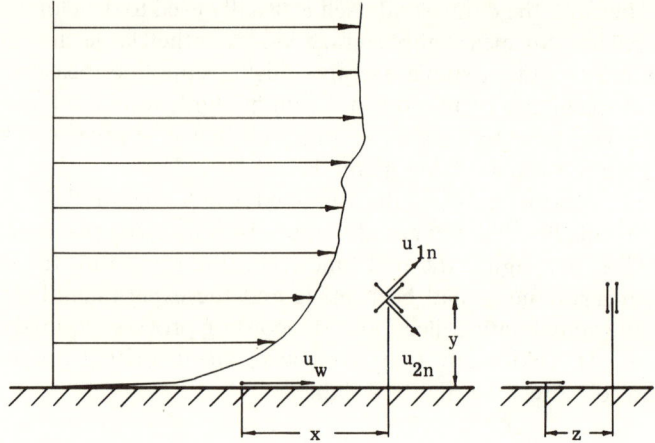

FIG. 1. Sketch of arrangement of hot wires for measurements of u_w, u_{1n}, and u_{2n} (with permission of *J. Fluid Mech.*).

the 18 different criteria were added together in 18 separate storage regions and an average value computed for each point in the sampling window for each of the 18 criteria. Our initial results showed slightly larger contributions to Reynolds stress in the middle of the sampling window for the criteria with $u_w < 0$ with negative slope.

With this encouraging result we began a search for improved criteria. An important improvement was obtained by passing the signal representing u_w through a third-order Butterworth low pass filter with half power point at 80 Hz. The filter removed high frequency components and apparently caused a reduction in the number of repeated or spurious samples. At any rate the contribution to Reynolds stress in the samples corresponding to

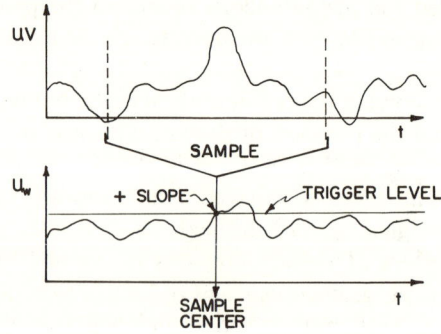

FIG. 2. Sketch of conditional sampling method (with permission of *J. Fluid Mech.*).

$u_w < 0$ with negative slope was greatly increased. A correction in the location of the center of the sampling window was made to account for the approximately constant time lag introduced in the signal u_w by the low pass filter. An example of the data we obtained in this way can be found in Willmarth and Lu (1972, Fig. 6). Here, we proceed to the next refinement before presenting the results of the sampling measurements.

3.2. Conditionally Sampled and Sorted Measurements of Reynolds Stress Contributions

The next improvement in the sampled data was made by sorting the samples we obtained at a given value of u_w and positive or negative slope into four categories depending upon the quadrant of the u-v plane in which the signal uv occurred. In order to describe the procedure, define $h_i(\tau)$ where

(1) $$h_i(\tau) = \begin{cases} 1 & \text{for any time } \tau \text{ that the point } (u, v) \text{ is in the } i\text{th quadrant in the } u\text{-}v \text{ plane,} \\ 0 & \text{otherwise,} \end{cases}$$

for $i = 1, 2, 3, 4$. The averaged values of the portions of the uv signal in each quadrant are

(2) $$\langle uv_i \rangle = N^{-1} \sum_{j=1}^{N} [h_i(\tau) uv(\tau)]_j ,$$

where N is the total number of samples and τ is the time relative to the beginning of the "window" in a given sample. In the measurements to be described below $\langle uv_2 \rangle$ comes from the second quadrant in the u-v plane and is associated with the outflow, burst, of low speed fluid, while $\langle uv_4 \rangle$ comes from fourth quadrant and is associated with the inflow, sweeps, of high speed fluid. $\langle uv_1 \rangle$ and $\langle uv_3 \rangle$ are the other interactions.

Figure 3 shows results of measurements of the sampled, sorted contributions to Reynolds stress as expressed in Eq. (2) in which the x wire is a distance $y^+ \simeq 30$ from the wall directly above the u_w wire at a distance $y^+ \simeq 15$ from the wall. In the measurement the criterion is that $u_w = -u'_w$ with negative slope, here the prime denotes the root-mean-square value. The use of this criterion in addition to sorting the uv signal results in detection of large contribution to the Reynolds stress caused by eruptions or bursts at the time when the sampling criterion is met at the center of the sampling window.

Figure 4 shows similar results of measurements as expressed in Eq. (2) with the hot wire probes in the same position as Fig. 3 but with the criterion that $u_w = u'_w$ with positive slope. In this case the detection criteria show the occurrence of sweeps $\langle uv_4 \rangle$ at the center of the sampling window when the criterion is satisfied.

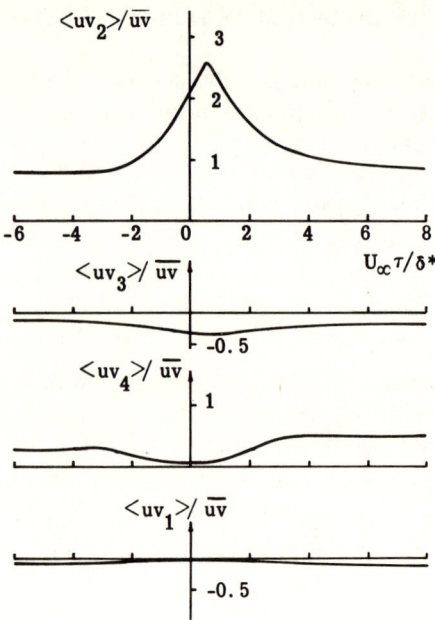

FIG. 3. Measurements of sampled sorted Reynolds stress. $u_w/u'_w = -1$, negative slope; $x/\delta^* = 0$, $y/\delta^* = 0.118$ ($y^+ \simeq 30$), $z/\delta^* = 0$ (with permission of *J. Fluid Mech.*).

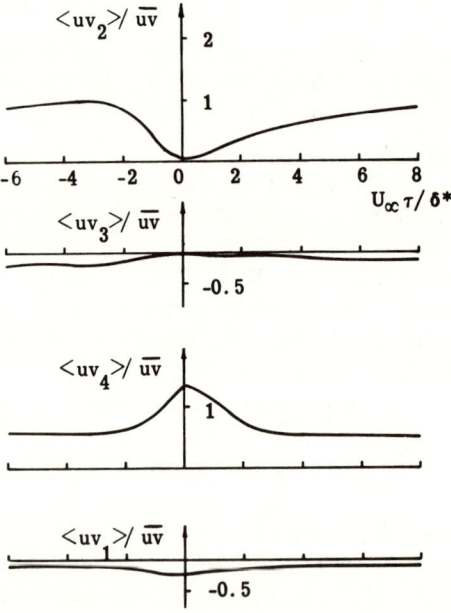

FIG. 4. Measurements of sampled sorted Reynolds stress. $u_w/u'_w = +1$, positive slope; $x/\delta^* = 0$, $y/\delta^* = 0.118$ ($y^+ \simeq 30$), $z/\delta^* = 0$ (with permission of *J. Fluid Mech.*).

3.3. Convection and Scale of Burst and Sweep Patterns near the Wall

Using these detection criteria we were able to study the convection of bursts and sweeps by measuring the sorted contributions to Reynolds stress at various distances downstream of the u_w wire used for detection. The sampled sorted Reynolds stresses $\langle uv_i \rangle / \overline{uv}$ were measured with the x wire probe at four positions directly downstream of the u_w wire at a distance of approximately $y = 0.15\delta^*$ or $y^+ \simeq 39$ from the wall. Figure 5 shows the results of the measurements plotted in a space–time format. As can be seen

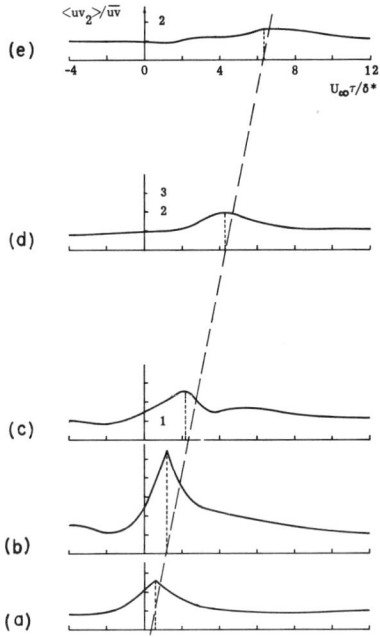

FIG. 5. Convection and decay of sampled sorted Reynolds stress. $\langle uv_2 \rangle / \overline{uv}$, with sampling conditions of $u_w / u'_w = -1$, negative slope; $y/\delta^* \approx 0.169$, $z/\delta^* = 0$, and $U_\infty \sim 20$ ft/sec; (a) $x/\delta^* = 0$, (b) $x/\delta^* = 0.34$, (c) $x/\delta^* = 0.84$, (d) $x/\delta^* = 1.69$, (e) $x/\delta^* = 2.53$ (with permission of J. Fluid Mech.).

in this figure there is a time lag required for the occurrence of the peak in $\langle uv_2 \rangle / \overline{uv}$ data. The origin of each plot in this figure is located vertically in proportion to the distance x of the x wire probe downstream from the u_w wire. Thus, the dashed line in the figure represents the space–time trajectory of the convected burst events. From the slope of this line the burst convection speed U_{CB} at this distance from the wall was found to be about 8.5 ft/sec, which is somewhat less than the local mean flow velocity ($U_{CB}/U \approx 0.8$ and $U_{CB}/U_\infty \approx 0.425$) at $y = 0.15\delta^*$.

Measurements of sampled but unsorted values of $\langle uv \rangle$ at greater distances from the wall (both downstream and to the side of the u_w wire) show that the magnitude of the peak in $\langle uv \rangle / \overline{uv}$ plot decreases as one travels outward from the wall or spanwise at a fixed downstream station from the u_w wire. The decrease in magnitude of the peak is also observed as one travels downstream at a fixed distance from the wall. From these more extensive measurements of $\langle uv \rangle / \overline{uv}$ (see Lu and Willmarth, 1972) it is apparent that the burst events are confined to a narrow region in the spanwise direction near the wall and downstream from the u_w wire.

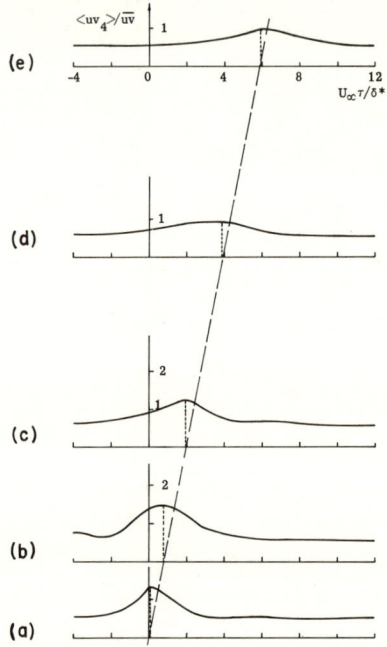

FIG. 6. Convection and decay of sampled sorted Reynolds stress. $\langle uv_4 \rangle / \overline{uv}$, with sampling conditions of $u_w/u'_w = +1$, positive slope; $y/\delta^* \approx 0.169$, $z/\delta^* = 0$, and $U_\infty \simeq 20$ ft/s; (a)–(e) same as Fig. 5 (with permission of J. Fluid Mech.).

The convection of sweep related events detected with the sampling criterion that $u_w = u'_w$ with positive slope was also studied. The sampled sorted Reynolds stress from sweeps $\langle uv_4 \rangle / \overline{uv}$ are displayed in Fig. 6 in the same format as used for Fig. 5 and at the same distances downstream and from the wall. The behavior of the magnitude of the peak in $\langle uv_4 \rangle / \overline{uv}$ is similar to that of the burst events; it decreases as one travels outward from the wall or downstream at a fixed distance from the wall. The speed of convection of the sweep events is represented by the dashed line in Fig. 6. The sweep convection speed U_{CS} was found to be nearly the same as the burst convection speed. Thus, $U_{CS}/U \simeq U_{CB}/U \simeq 0.425$ and $U_{CB}/U_\infty \simeq U_{CS}/U_\infty \simeq 0.8$.

Rough estimates of the speed of convection of the burst events were also made at various larger distances from the wall. The burst convection speed U_{CB}, was found to increase with the distance from the wall, however we have not been able to obtain accurate measurements of convection speed at greater distances from the wall. It appears from other measurement (Willmarth and Wooldridge, 1962) that wall pressure disturbances, for example, are convected at speeds ranging from $0.58 < U_C/U_\infty < 0.83$. The lower convection speeds are associated with small scale eddies near the wall. The present measurements therefore suggest that the small scale burst pattern emanates from the wall region, travels outward, and grows larger as it is carried downstream. As it enlarges it also is sheared and distorted because the convection velocity is higher farther from the wall. The evolving and enlarging burst pattern soon loses coherence with the detection criterion near the wall ($u_w = -u'_w$ with negative slope) so that the sampled values of $\langle uv_2 \rangle$ can no longer reveal the burst structure. The bursts still exist, however, as will become apparent below.

Before the detection criterion fails, the peak in the $\langle uv \rangle / \overline{uv}$ value which represents the region of disturbance caused by the burst that is coherent with the sampling criteria increases from a size of $y/\delta^* = 0.506$ at $x/\delta^* = 0$ to a size of $y/\delta^* = 0.912$ at $x/\delta^* = 1.686$ or more as one travels further downstream. There is still some contribution to $\overline{uv}$ even at a station of $x/\delta^* = 2.53$ downstream of the u_w detector wire. The spanwise extent of the region of disturbance is confined to a narrow swept back region with an included angle of approximately 20° centered upon the free stream direction.

3.4. Large Individual Contributions to Reynolds Stress near the Wall

In another attempt to improve the detection of bursts (and sweeps) two hot wires mounted side by side near the wall at $y^+ \simeq 15$ were used to measure u_{w_1} and u_{w_2}. The distance between the wire centers was $z^+ \simeq 20$ and the wire length was $l^+ \simeq 20$. Note that Kim et al. (1968) report that the typical spacing between streaks caused by the low and high velocity in the sublayer is of the order of $z^+ \simeq 100$. A wire spacing of $z^+ \simeq 20$ should be small enough to allow the detection of the center of a streaky region. To find the central region of a burst, two or more simultaneous criteria are required. After a number of trials of different detection schemes the following criteria were successful. A computer program was written that searched the u_{w_1} and u_{w_2} data for those few events in which both u_{w_1} and u_{w_2} reached the level $-u'_{w_1}$ and $-u'_{w_2}$ at approximately the same time with negative slope, remained below these levels for approximately the same time, and both contributed approximately the same value to the integrals of u_{w_1} and u_{w_2} with respect to time during the time u_{w_1} and u_{w_2} were below the level u'_{w_1} and u'_{w_2}. With this elaborate procedure we found ten events meeting these criteria in 52 sec of digitized data (4.1 × 10^5 digitized data points).

For each event, the uv signal was examined and it was found that in each case $uv/\overline{uv} > 20$ somewhere in the sampling window of 320 data points. An example of a single large contribution $uv/\overline{uv} \simeq 60$ is shown in Fig. 7.

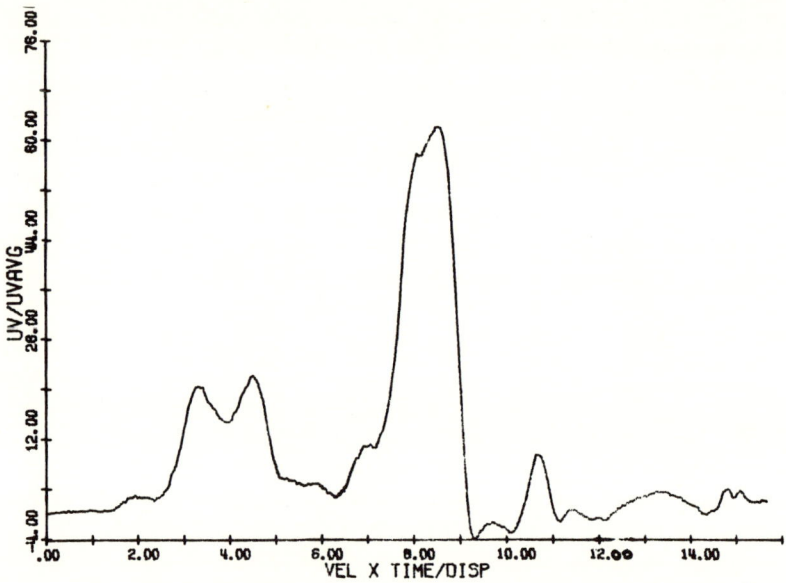

FIG. 7. Example of a very large single contribution to Reynolds stress (with permission of J. Fluid Mech.).

Another sample contained two equally large contributions. It is clear that extremely large contributions to Reynolds stress occasionally occur.

The variation of the uv signal near the wall as a function of time is shown in Fig. 8 and indicates that the contributions to Reynolds stress are very intermittent. In Section 3.5 a measure of the intermittency is discussed. The probability density distribution of the uv signal was also determined with the computer. Figure 9 displays β_{uv} as a function of $uv/\overline{uv}$, where $\beta_{uv} \, d(uv)$ is the probability that uv lies between uv and $uv + d(uv)$. Note that the distribution has a large "tail" for $uv/\overline{uv} > 1$ (i.e., $uv < 0$ since $\overline{uv} < 0$). This again indicates that large negative values of uv occur occasionally.

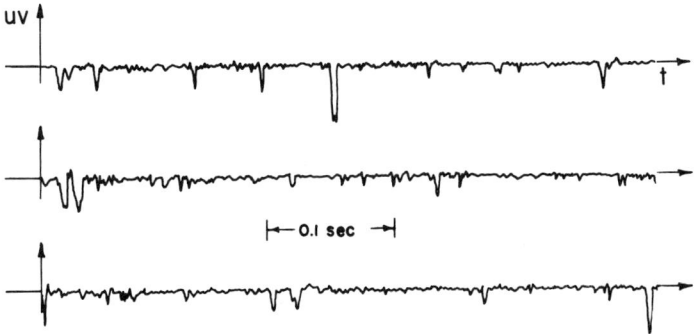

FIG. 8. The signal uv as a function of time at $y^+ \simeq 30$ in the low speed boundary layer (with permission of *J. Fluid Mech.*).

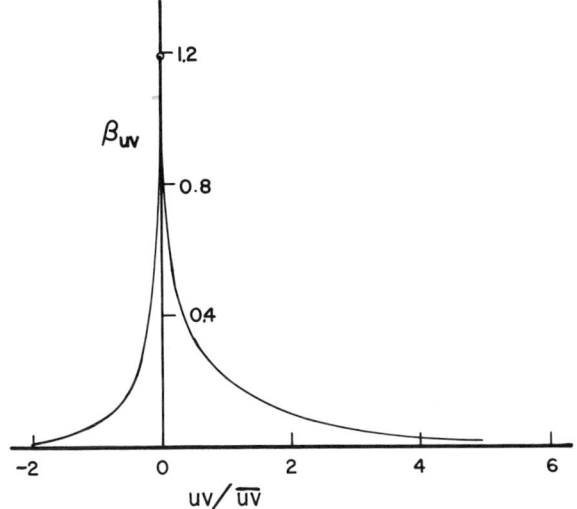

FIG. 9. Probability density of uv measured at $y^+ \simeq 30$ in the low speed boundary layer (with permission of *J. Fluid Mech.*).

3.5. Distribution of Reynolds Stress across the Boundary Layer and Contributions from Different Events

The distribution of Reynolds stress across the entire boundary layer was measured. Figure 10 displays measurements of $\overline{uv}/u'v'$ and shows that $\bar{R}uv/u'v' \simeq 0.43$ out to $y/\delta \simeq 0.7$. In order to study contributions to $\overline{uv}$ from different events, contributions to $\overline{uv}$ from different regions in the u-v plane were measured. The measurements were made with the x wire at various distances from the wall. The u-v plane was divided into five regions as shown in Fig. 11. In the figure, the cross-hatched region is called the "hole," which is bounded by the curves $|uv|$ = constant. The four quadrants excluding

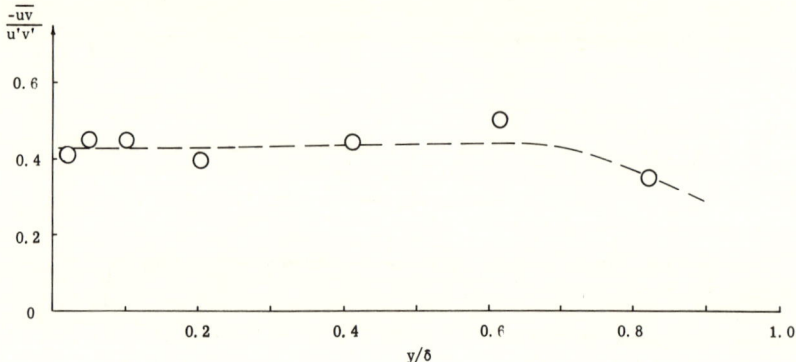

Fig. 10. Measurement of Reynolds stress distribution across the boundary layer $U_\infty \simeq 19.7$ ft/sec (with permission of *J. Fluid Mech.*).

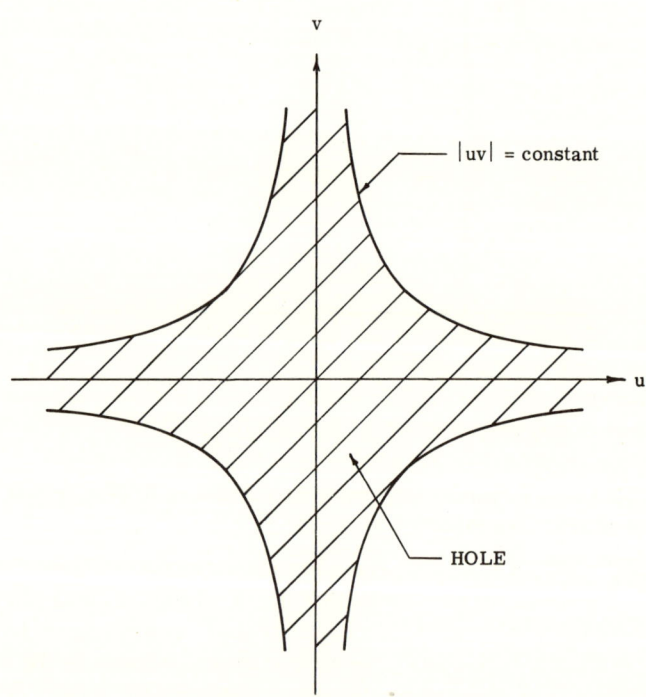

Fig. 11. Sketch of "hole" region in the u-v plane (with permission of *J. Fluid Mech.*).

the "hole" are the other four regions. The size of the hole is determined by the curves $|uv|$ = constant. Introduce the parameter H and let $|uv| = Hu'v'$, where u' and v' are the local root mean square values of u and v signals. The parameter H is called the hole size. With this scheme, large contributors to $\overline{uv}$ relative to $u'v'$ from each quadrant can be extracted leaving the smaller fluctuating uv signal in the hole. The contribution to $\overline{uv}$ from the hole would mean the contribution during the more quiescent periods, while the second quadrant represents the burstlike events and the fourth quadrant the sweeplike events.

The contributions to $\overline{uv}$ from the four quadrants were computed from the equations

(3) $$\frac{\widetilde{uv}_i(H)}{\overline{uv}} = \frac{1}{\overline{uv}} \lim_{T \to \infty} \frac{1}{T} \int_0^T uv(t) S_i(t, H)\, dt, \quad i = 1, 2, 3, 4,$$

where the subscript i refers to the ith quadrant and

(4) $$S_i(t, H) = \begin{cases} 1 & \text{if } |uv(t)| > Hu'v' \text{ and the point } (u, v) \text{ in the } u\text{-}v \text{ plane is in the } i\text{th quadrant,} \\ 0 & \text{otherwise.} \end{cases}$$

Contributions to $\overline{uv}$ from the hole region were obtained using Eq. (3) but with S_i replaced by S_h where

(5) $$S_h = \begin{cases} 1 & \text{if } |uv(t)| < Hu'v', \\ 0 & \text{otherwise.} \end{cases}$$

These five contributions, uv_i and uv_h, are all functions of the hole size H, and

(6) $$\sum_{i=1}^{4} \frac{\widetilde{uv}_i(H)}{\overline{uv}} + \frac{\widetilde{uv}_h(H)}{\overline{uv}} = 1.$$

Typical results of the measurements at low speed at various distances from the wall are shown in Figs. 12–14. There was only one measurement for the case of high speed flow which was made at a distance of $y^+ = 265$ from the wall. This result is shown in Fig. 15. The results were very similar for both high and low Reynolds number measurements and regardless of the x wire probe location in the turbulent boundary layer. In Figs. 12–15 curves representing the fraction of total time that uv signal spent in the hole region are also included.

For a large portion of the time, $|uv|$ is very small. Stated in another way, uv has an intermittency factor of 0.55 since in Figs. 12–15 99% of the contribution to $\overline{uv}$ occurs during 55% of the time. As a matter of fact, these two curves, i.e., the fraction of the total time in the hole and the contribution to $\overline{uv}$ from the hole region, can be derived from the assumption of joint

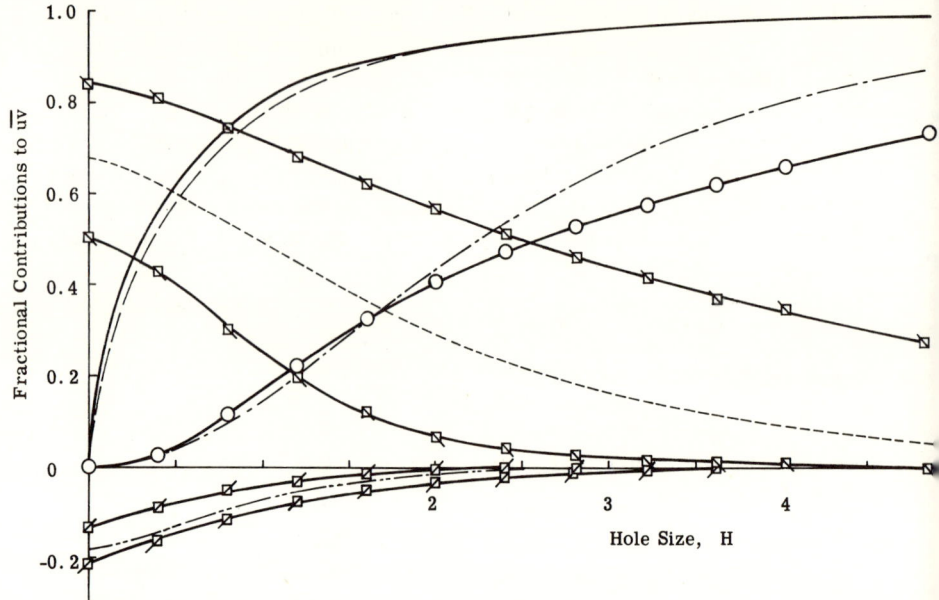

FIG. 12. Measurements of the contributions to $\overline{uv}$ from different events at various distances from the wall: ⊠ measured, ———— computed, $\tilde{uv}_1/\overline{uv}$; ⊠ measured, ———— computed $\tilde{uv}_2/\overline{uv}$; ⊠ measured, ———— computed, $\tilde{uv}_3/\overline{uv}$; ⊠ measured, ———— computed, $\tilde{uv}_4/\overline{uv}$; ○ measured, ———— computed, $\tilde{uv}_h/\overline{uv}$; ——— measured, ———— computed, fraction of time in hole. $U_\infty \simeq 20$ ft/sec, $Re_\theta \simeq 4230$, $y/\delta = 0.021$ (with permission of *J. Fluid Mech.*).

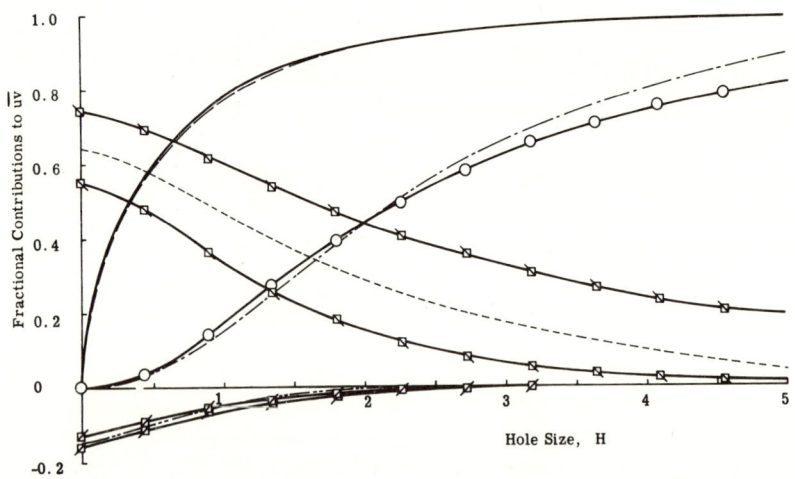

FIG. 13. Same as Fig. 12, except $y/\delta = 0.052$ (with permission of *J. Fluid Mech.*).

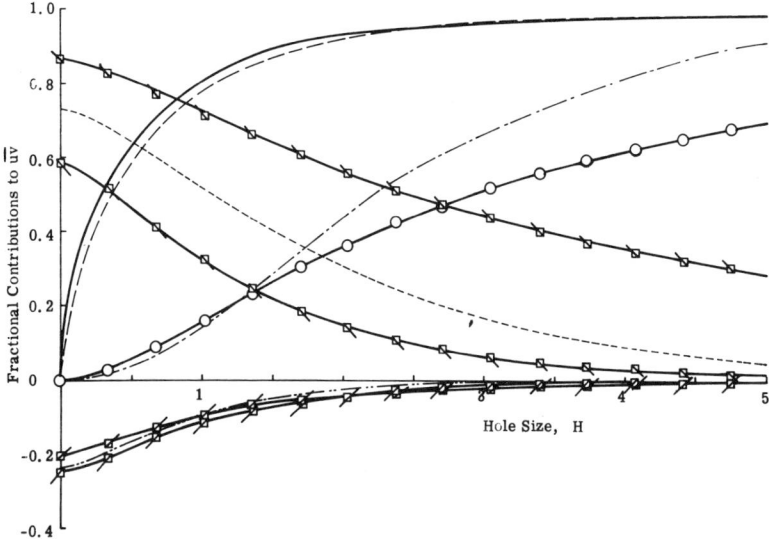

FIG. 14. Same as Fig. 12 except $y/\delta = 0.823$ (with permission of *J. Fluid Mech.*).

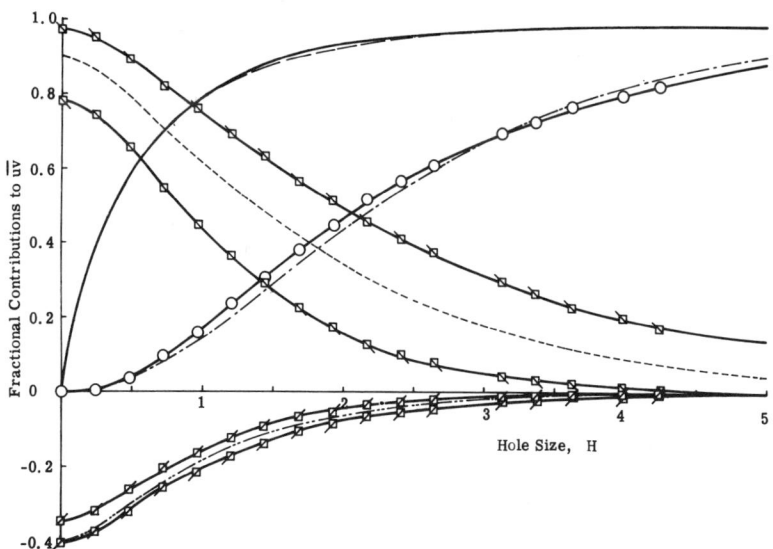

FIG. 15. Same as Fig. 12, except $U_\infty \simeq 200$ ft/sec, $Re_\theta \simeq 38,000$, $y/\delta \simeq 0.014$ ($y^+ \simeq 265$) (with permission of *J. Fluid Mech.*).

normality of u and v signals, for details see Willmarth and Lu (1971, 1972). The predicted curves are included in Figs. 12–15 for comparison.

The assumption of joint normality of u and v signals also implies that the contribution to $\overline{uv}$ from the second quadrant $\tilde{uv}_2$ should equal that from the fourth quadrant $\tilde{uv}_4$. Similarly, $\tilde{uv}_1 = \tilde{uv}_3$. The predicted curves for $\tilde{uv}_2$ and $\tilde{uv}_1$ are also shown on the figures. The deviation from joint normality is apparent since $\tilde{uv}_2 \neq \tilde{uv}_4$ and $\tilde{uv}_1 \neq \tilde{uv}_3$, regardless of the flow speed and the location in the turbulent boundary layer. As can be seen, the largest contribution comes from the second quadrant which is burstlike. The second largest contribution is $\tilde{uv}_4$ and is sweeplike. The contributions from $\tilde{uv}_1$ and $\tilde{uv}_3$ are negative and relatively small. When the hole size H becomes large, there are only two contributors. One is $\tilde{uv}_2$ and the other comes from the hole region. Thus the importance of the burstlike events in the turbulent boundary layer is obvious. At the hole size of $H = 4.5$, which amounts to $|uv| > 10|\overline{uv}|$, there is still a 15–30 % contribution to $\overline{uv}$ from the second quadrant, i.e., $\tilde{uv}_2/\overline{uv} \approx 0.15$ to 0.30. At this level there are almost no contributions from the other three quadrants.

3.6. Results for the Burstlike and Sweeplike Events

Results will be discussed here regarding the contributions to $\overline{uv}$ from burst- and sweeplike events with $H = 0$. Both $\tilde{uv}_2/\overline{uv}$ and $\tilde{uv}_4/\overline{uv}$ are nearly constant across the boundary layer except very close to the wall and near the edge of the boundary layer. It is found that $\tilde{uv}_2/u'v' \approx -0.34$ and $\tilde{uv}_4/u'v' \approx -0.24$, or $\tilde{uv}_2/\overline{uv} \approx 0.77$ and $\tilde{uv}_4/\overline{uv} \approx 0.55$. Thus, burstlike events account for 77 % of the local Reynolds stress, and the sweeplike events have 55 % to their account. This leaves 32 % of local Reynolds stress to the other two negative contributors.

The ratio of the contribution to $\overline{uv}$ from the burstlike events and from the sweeplike events is plotted in Fig. 16 as a function of y/δ. There is a sharp rise near the wall, while for most of the boundary layer the ratio is nearly constant with a value of 1.35. The single high speed measurement gave a value of 1.25, which was measured at $y/\delta = 0.014$ or $y^+ = 265$. The results are replotted in Fig. 17 as a function of y^+. In this figure, the results obtained by Wallace et al. (1972) at a much lower Reynolds number in a channel flow are included. It turns out that contributions to $\overline{uv}$ from the sweep period are approximately the same as in Wallace et al. (1972) but that the present results show larger contributions to $\overline{uv}$ during the burst period. The reason for this disagreement may be due to the difference in Reynolds number. The results for the present measurements (Fig. 17) seem to scale with the wall region variables even though the two flow conditions considered differ greatly in Reynolds number ($Re_\theta = 4230$ and $38,000$). Although there is only

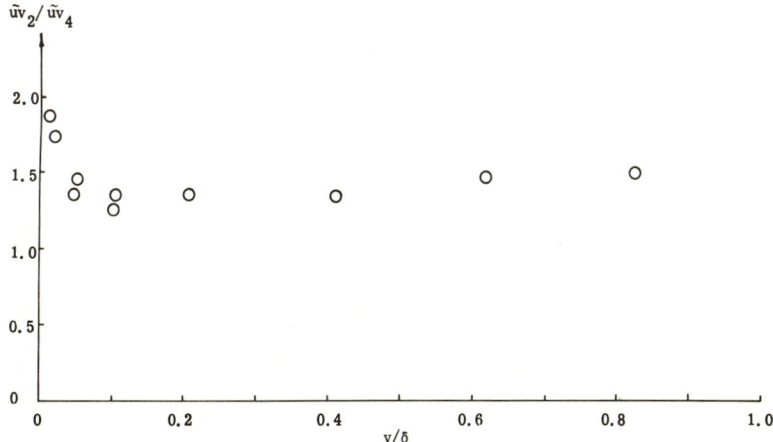

FIG. 16. Distribution of the ratio $\widetilde{uv}_2/\widetilde{uv}_4$ with $H = 0$, $Re_\theta \simeq 4230$ (with permission of J. Fluid Mech.).

one measurement for the high Reynolds number flow, it is conjectured that the Reynolds number similarity may hold. This would imply that the nature of the fluctuating turbulent structure in the wall region is similar above $Re_\theta \simeq 4000$ but that at very low Re_θ, as in the work of Wallace et al. (1972), significant changes in the structure of the turbulent fluctuations occur. It is interesting to note that at higher Reynolds numbers in the present work, the contribution to $\overline{uv}$ during the burst period is larger than that found by Wallace et al. (1972). However, a word of caution is in order; it is possible that significant measurement errors may be caused by operating hot wires or

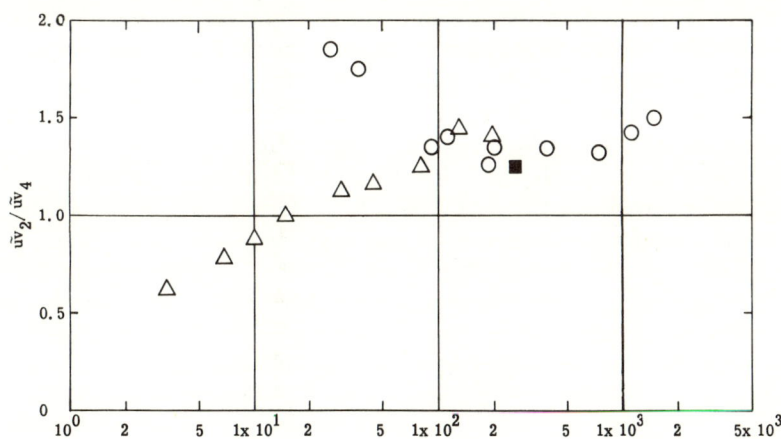

FIG. 17. The ratio $\widetilde{uv}_2/\widetilde{uv}_4$, with $H = 0$ as a function of y^+. ○, $Re_\theta \simeq 4230$; ■, $Re_\theta \simeq 38,000$; △ channel flow (Wallace et al., 1972), Re very low (with permission of J. Fluid Mech.).

hot film probes in the region very near the wall where the level of turbulent fluctuations is very high ($u'/U \simeq 0.4$) and the mean shear ($\partial U/\partial y$) is also very high. Accordingly, it remains a subject for further research to determine the reason for the difference between the present measurements and those of Wallace et al. (1972) near the wall.

3.7. Mean Periods of Bursts and Sweeps

In the visual studies of Runstadler et al. (1963), Schraub and Kline (1965), and Kim et al. (1968, 1971), the mean time intervals $\bar{T}_B$, between bursts were measured by visual counting of the violent events of ejection near the wall in a turbulent boundary layer. Kim et al. found that the mean time interval $\bar{T}_B$, was nearly the same as the time lag required to obtain the second mild maximum in the curve of the autocorrelation coefficient of the fluctuating streamwise velocity R_{uu}. By a complex processing of a hot wire signal in a turbulent boundary layer, Rao et al. (1969, 1971) were able to measure the mean time interval between bursts. They showed, in a summary of their own and other[2] data obtained over a wide range of Reynolds numbers, that the mean burst period $\bar{T}_B$, scaled with outer rather than inner boundary layer flow variables. Among their summarized data, there was only one measurement of $\bar{T}$ for high Reynolds number flow ($Re_\theta = 38{,}000$). This was obtained from the second mild maximum in R_{uu} measured by Tu and Willmarth (1966). We have recently determined (Lu and Willmarth, 1974) that the second mild maximum in R_{uu} in that data was produced by a low pass filter used during our (1966) measurements of R_{uu}. Therefore the value of $\bar{T}$ at $Re_\theta = 38{,}000$ that is quoted in Rao et al. (1971) is not valid.

In any process of counting the number of bursts, a definitive identification of bursts is required. In visual studies bursts of varying magnitude are observed embedded in a background of other turbulent fluctuations. When an event is not extremely violent and/or coherent, it is up to the observer to decide whether it is a burst or not. Very much greater difficulties are present in the measurement of the mean burst period from a trace of a single hot wire signal because only the velocity at one point in a complex bursting pattern can be observed.

In the hot wire measurements of Rao et al. (1969, 1971), the signal u was passed through a band pass filter to make "bursts" stand out more clearly (we are not certain from their paper whether or not the u signals were differentiated before filtering. First, as pointed out by Kim et al. (1971), it remains to be determined whether this process will show the phenomenon of

---

[2] Rao et al. (1971) summarized the results for the mean burst period T, obtained from papers by Kim et al. (1968), Schraub and Kline (1965), Runstadler et al. (1963), and Laufer and Badri Narayanan (1971).

"oscillatory motion" in the second stage of the bursting process observed by Kim et al. (1968, 1971). Second, even if this technique does make the burst stand out, counting the number of bursts using human eyes is somewhat arbitrary since the "bursts" are not too well organized or clearly identifiable in the traces of the processed u signal (see Rao et al., 1971, Fig. 1). However, Rao et al. using certain special procedures were able to arrive at a characteristic time, called $\bar{T}_m$, for the burst period.

In the present study we attempted to estimate, in a different manner, the characteristic times related to bursts and sweeps and their durations. Similar difficulties, mainly definitive identification of bursts and sweeps, were encountered. Extensive measurements were made, in a consistent manner, for the low speed flow across the turbulent boundary layer. A single high speed measurement was also made to study the Reynolds number effect on the burst and sweep rates.

As is evident from the measurements of sampled, sorted Reynolds stress, there is a large contribution to $\overline{uv}$ during the occurrence of bursts. Serious difficulties are encountered when it is desired to obtain definite identification of bursts from uv measured at a single point. Assume that if the uv signal reaches a certain specified level (i.e., hole size H) or larger in the second quadrant, a burst occurs. By counting the number of times the above conditions are detected in a given time interval, the mean time interval between burst contributions $\bar{T}_B$, at a given hole size can be found. The nondimensional mean time interval $U_\infty \bar{T}_B/\delta^*$ between bursts is shown in Fig. 18 as a function of the hole size H with the distance from the wall y/δ as a parameter. These data were obtained from the low speed ($U_\infty \approx 20$ ft/sec) measurements. The mean time interval between bursts exceeding a given H is nearly independent of the distance from the wall throughout the turbulent boundary layer. On the other hand, the mean time interval between bursts $\bar{T}_B$, exceeding a given value of H increases rapidly as H is increased; see Fig. 18. A satisfactory criterion for determining $\bar{T}_B$ should have the property that the value of $\bar{T}_B$ determined from the criterion is independent of small changes in the criterion. In Fig. 18, the absence of a plateau in the variation of $\bar{T}_B$ as a function of H indicates that the value of H alone is not an acceptable criterion for determining the actual value of the mean burst rate. However, upon close examination of the plots of the contributions to $\overline{uv}$ from different events at different distances from the wall (Figs. 12–15) a unique and consistent feature is observed. As the hole size becomes large, the contributions to $\overline{uv}$ from quadrants one, three, and four vanish more rapidly than contributions from the second quadrant. It is observed that when H reaches a value of between 4 and 4.5, only $\widetilde{uv}_2/\overline{uv}$ is not zero regardless of the distance from the wall. Contributions to $\overline{uv}$ above this value of H must have come from the large spikes in the uv signal related to the bursts. For a hole

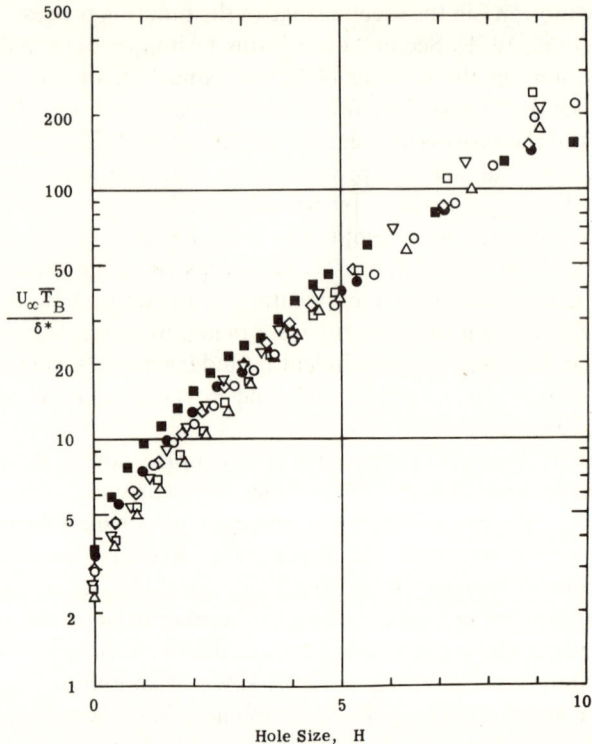

FIG. 18. Mean time interval between bursts as a function of hole size H and distance from wall, $\mathrm{Re}_\theta \simeq 4230$; ○, $y/\delta = 0.021$; △, $= 0.052$; □, $= 0.103$; ▽, $= 0.206$; ◇, $= 0.412$; ●, $= 0.618$; ■, $= 0.823$ (with permission of J. Fluid Mech.).

size of $H \simeq 4.5$, $|uv|$ is about ten times the absolute value of the local mean Reynolds stress. These bursts certainly are very violent relative to the value of $\overline{uv}$ at a given distance from the wall. Using this unique feature, applied consistently throughout the boundary layer, one can obtain a consistent measure of the characteristic time interval $\bar{T}_{\mathrm{CB}}$, between relatively large contributions to $\overline{uv}$ (which are larger than contributions to $\overline{uv}$ from any other quadrant at a given distance from the wall) by setting the specified level $H \simeq 4 \simeq 4.5$.

Using this scheme, a consistent estimate of the characteristic time interval between relatively large bursts is shown in Fig. 19, which was obtained from Fig.18 by setting a level of $H \simeq 4 \simeq 4.5$. A value of $U_\infty \bar{T}_{\mathrm{CB}}/\delta^* \approx 32$ is found for most of the boundary layer. Measurements from the single high Reynolds number run are also included in Fig. 19. The fact that the value of $\bar{T}_{\mathrm{CB}}$ determined as described above scales with the outer flow variables is in

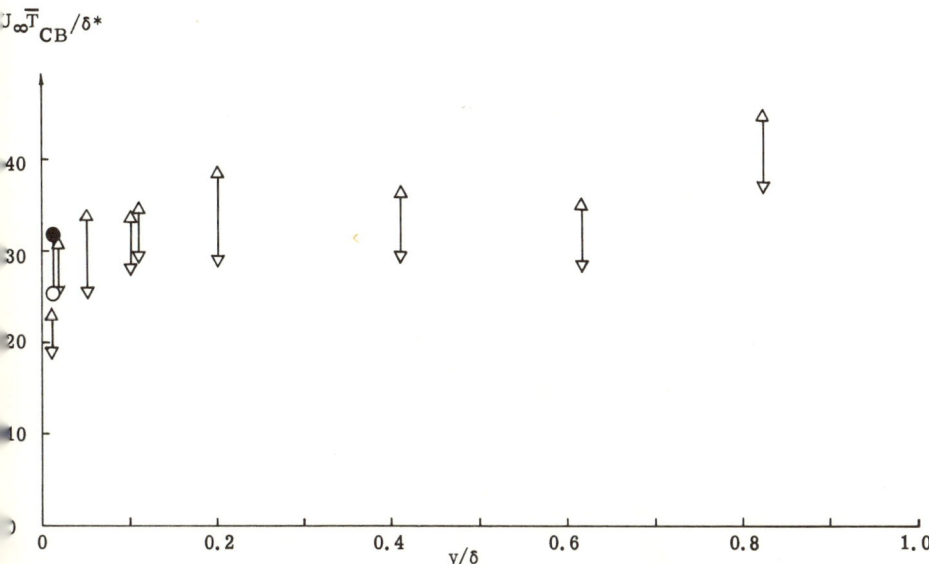

FIG. 19. Characteristic mean time intervals between large bursts. U_∞ (ft/sec): ●, 200, $H = 4.5$; ○ 200, $H = 4.0$; △ 20, $H = 4.5$; ▽ 20, $H = 4.0$ (with permission of J. Fluid Mech.).

accord with the scaling of the mean period between bursts reported by Rao et al. (1971). It must be regarded as a coincidence that the actual value of $U_\infty \bar{T}_{CB}/\delta^* \simeq 32$ (determined with $H \simeq 4 \simeq 4.5$) is almost the same as the value of the mean burst period determined by Rao et al. (1971). In this connection we again mention that the method used by Rao et al. (1971) to obtain the value $U_\infty \bar{T}/\delta^* \simeq 30$ from the high Reynolds number data of Tu and Willmarth (1966) is not correct; see Section 5.

A similar scheme was used to measure the mean time interval $\bar{T}_S$, between sweep contributions. A sweep is assumed to occur if the uv signal in the fourth quadrant reaches a specified value or larger. Thus, as in the case of bursts, the mean time interval $\bar{T}_S$, between sweeps is also a function of the hole size H, and the mean time interval between sweeps in nondimensional form $U_\infty \bar{T}_S/\delta^*$ is shown in Fig. 20 as a function of H, with y/δ as a parameter. The data are more scattered than those of Fig. 18, but the dependency of $U_\infty \bar{T}_S/\delta^*$ on the distance from the wall is not excessively large. As in the case of bursts, there is a no plateau in the mean time between sweep contributions $\bar{T}_S$, as H increases. Therefore, H alone cannot be used to determine the actual mean time between sweep contributions.

There is, however, another unique feature in the plots of the contributions to $\overline{uv}$ from different events (Figs. 12–15). At a hole size of $H \simeq 2.25 \simeq 2.75$, at any distance from the wall in the boundary layer, $\widetilde{uv}_1/\overline{uv}$ and $\widetilde{uv}_3/\overline{uv}$

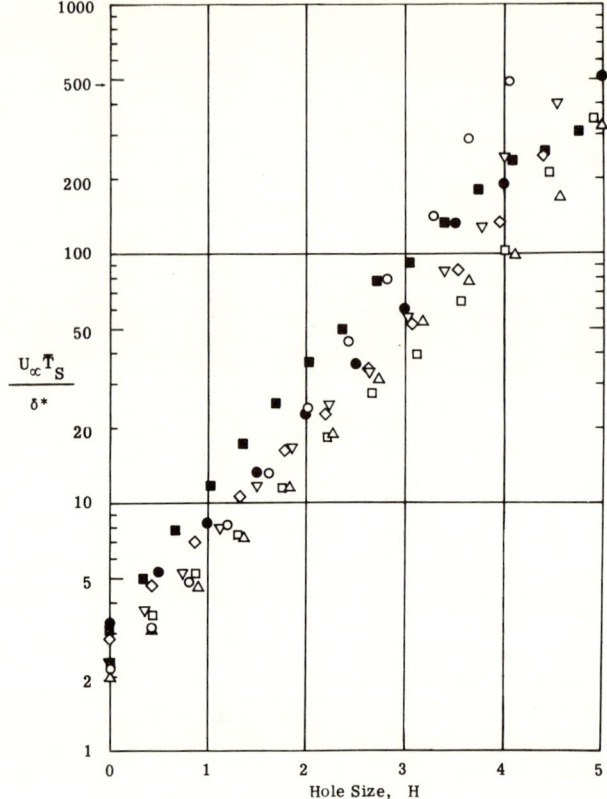

FIG. 20. Mean time interval between sweeps as a function of hole size H and distance from wall $Re_\theta \simeq 4230$. Symbols same as Fig. 18 (with permission of *J. Fluid Mech.*).

vanish. Thus, the characteristic time interval between relatively large sweeps $\bar{T}_{CS}$, can be obtained by setting the level at $H \simeq 2.25 \simeq 2.75$. When determined in this fashion, $\bar{T}_{CS}$ represents a consistent estimate of the mean time between sweeps which are larger than the largest positive contributions to $\overline{uv}$ at any given distance from the wall. Figure 21 shows the values of $U_\infty \bar{T}_{CS}/\delta^*$ obtained for H between 2.25 and 2.75 as a function of distance from the wall. A value of about 30 for $U_\infty \bar{T}_{CS}/\delta^*$ is found in most of the boundary layer. Thus, $U_\infty \bar{T}_{CB}/\delta^*$ and $U_\infty \bar{T}_{CS}/\delta^*$ are essentially equal. The same result was obtained for the high Reynolds number flow measurement. It appears that $\bar{T}_{CS}$ may also scale with the outer flow variables. Further studies of the sweep events at different Reynolds numbers and of their relationship to bursting are needed, preferably with measurements from more than one point in the flow.

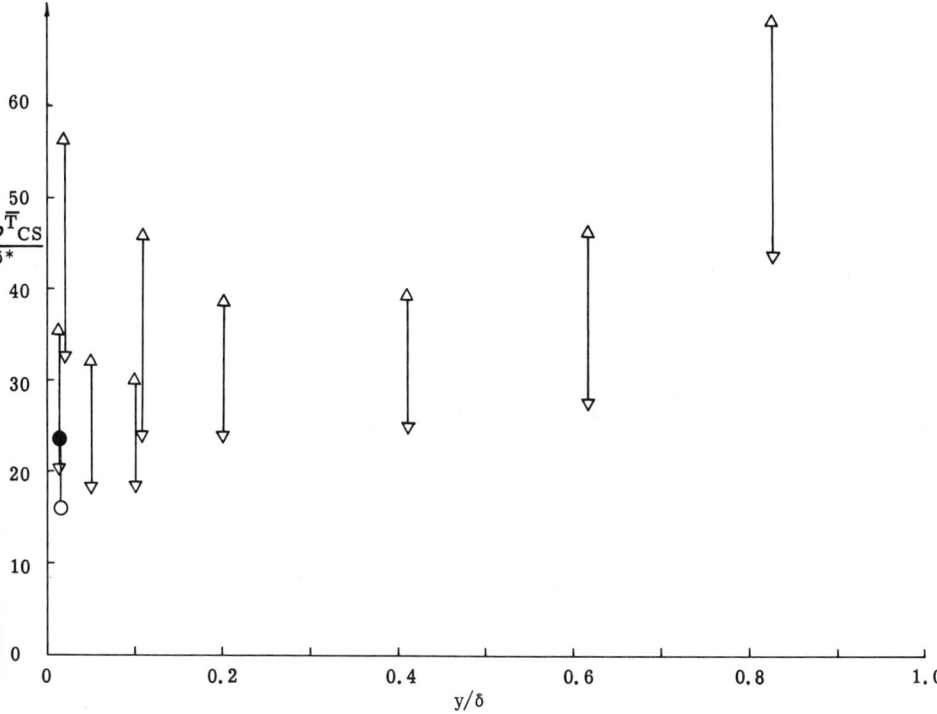

FIG. 21. Characteristic mean time intervals between large sweeps. U_∞ (ft/sec): ● 200, $H = 2.75$; ○ 200, $H = 2.25$; △ 20, $H = 2.75$; ▽ 20, $H = 2.25$ (with permission of *J. Fluid Mech.*).

4. Discussion of the Measurements

The ejection of low momentum fluid from the wall is a dominant feature of the structure of the turbulent boundary layer. The importance of the ejection or burst event is obvious from the results of Section 3.4 and Fig. 7 which show that large contributions to Reynolds stress occur occasionally near the wall. In Section 3.5 it is shown that the second quadrant of the u-v plane contains by far the largest contribution to Reynolds stress at any distance from the wall. In Section 3.2 the conditionally sampled measurements of $\langle uv_2 \rangle$ show that near the wall the various portions of the deterministic burst pattern are convected at somewhat less than the local mean speed (i.e., at $y \simeq 0.15\delta^* U_{CB}/U \simeq 0.8$). As the deterministic burst pattern is convected downstream, it enlarges both spanwise and in a direction normal to the wall. However, the rate of convection is much greater than the transverse rate of growth. For example, no contributions to $\langle uv_2 \rangle$ can be measured when the x wire probe is far downstream of the detector probe u_w, and a line between the x probe and detector probe forms an angle greater than approximately 10° with the stream direction. In addition, the contributions to $\langle uv_2 \rangle$ decay as one moves the x wire probe downstream at a constant small distance from the wall.

There is no doubt that these measurements indicate that an initially small deterministic burst pattern is growing in scale and is distorted by the shearing motion as it is convected downstream. The question of how large the evolving, deterministic bursting patterns ultimately become cannot definitely be answered until someone devises (if it is possible) a more reliable burst detection scheme. It seems likely that the effects of a violent ejection near the wall can, after evolution and convection, reach a station remote from the wall in the turbulent boundary layer. This conjecture is in agreement with the results of Grass (1971), the speculations of Kovasznay et al. (1970), and present results which show that $\bar{T}_{CB}$ is approximately constant throughout the boundary layer.

Although definitive identification of bursts and sweeps is difficult, some characteristic mean time intervals between bursts and sweeps have been found. The scaling of the mean time interval between large bursts $\bar{T}_{CB}$, with the outer flow variables at Reynolds numbers Re_θ, of 4230 and 38,000 is confirmed. As for the large sweep events, the mean sweep rates were also obtained for Reynolds numbers Re_θ, of 4230 and 38,000. The mean time interval between sweeps is roughly the same as that between bursts. There is not enough data to allow us to draw a firm conclusion about the scaling of the sweep rate, although at both Reynolds numbers we obtain roughly the same value of about 30 for $U_\infty \bar{T}_{CS}/\delta^*$. However, if large bursting events are indeed followed by large sweep events as suggested by Corino and Brodkey (1969), the mean sweep period must also scale with the outer flow parameters and $U_\infty \bar{T}_{CS}/\delta^* \approx U_\infty \bar{T}_{CB}/\delta^* \approx 32$.

It has been speculated that the bursts may have some bearing on the turbulent "bulges" in the outer intermittent flow region. See, e.g., Kovasznay et al. (1970) and Laufer and Badri Narayanan (1971). The present measurements of the mean time interval between bursts seem to confirm this idea. The mean burst period $\bar{T}_{CB}$, is approximately constant for most of the boundary layer. This suggests that after a burst originates near the wall it evolves into a larger convected disturbance with unchanged time interval between bursts $\bar{T}_{CB}$. In addition, our measurements of contributions to $\overline{uv}$ from different events as a function of H (Figs. 14–17) show that the contributions $\widetilde{uv}_i$ ($i = 1$–4, h) are very similar throughout the boundary layer. This again is consistent with the idea that the bursting events originating near the wall continue to produce, as they evolve, relatively the same (but larger scale and less intense) contributions to $\widetilde{uv}_i$ throughout the boundary layer.

The dominant feature of ejection near the wall in a turbulent boundary layer can be constrasted with many well-known statistical characteristics of the u and v signals throughout the boundary layer. This leads one to speculate that the turbulence in the inner part of the turbulent boundary layer may be considered as a "universal motion" plus an "irrelevant motion" as suggested by Townsend (1957, 1961). The universal motion may be considered as random occurrence (both temporally and spatially) of bursts, which is controlled by the outer flow, plus the ensuing more diffuse return flow, which may be related to the sweep events. The irrelevant motion may be considered as the accumulation of the remnants of what has happened upstream. The contribution to $\overline{uv}$ from the latter would be small.

From the measurements of sampled Reynolds stress $\langle uv \rangle$ near the wall using the sampling criteria that the velocity u_w, at the edge of the viscous sublayer is low and decreasing, it is found that there are burst events producing large contributions to the Reynolds stress that are convected at speeds lower than the local mean speed. In addition the line in the x-y plane on which the peak values of $\langle uv \rangle$ occur, at no time delay, travels outward from the wall at an angle of 16–20°. This may be explained by the convection past the measuring station of a certain deterministic pattern, for example, the hairpin vorticity model proposed by Willmarth and Tu (1967). This model was suggested and is consistent with the numerous space–time correlation measurements reported by Willmarth and Wooldridge (1962, 1963) and Tu and Willmarth (1966). As a matter of fact, this pattern of vorticity, if imagined to evolve to a larger scale, may also be used to describe the time sequence of the instantaneous velocity profiles near the wall as observed by Kim et al. (1968) (see Fig. 4.13 in their report) and could produce intermittent turbulent bulges at the outer edge of the boundary layer. This would provide the interaction between the inner and outer regions of the boundary layer that is implied by the scaling of the mean time between bursts with the outer flow variables. Since a large part of the Reynolds stress near the wall is produced during the times when our sampling procedure indicates that bursts occur, it is likely that a model like that of Willmarth and Tu (1967) may determine the flow structure near the wall and may well be a part of the universal motion mentioned above.

Acknowledgments

We gratefully acknowledge the financial support of the Fluid Dynamics Branch of the Office of Naval Research and the Engineering Division of the National Science Foundation.

References

Blackwelder, R. F., and Kaplan, R. E. (1971). Intermittent structure in turbulent boundary layer. *AGARD Conf. Proc.* **93**, 5.

Corino, E. R., and Brodkey, R. S. (1969). *J. Fluid Mech.* **37**, 1.

Grass, A. J. (1971). *J. Fluid Mech.* **50**, 233.

Kibens, V. (1968). The intermittent region of a turbulent boundary layer. Ph.D. Thesis, Johns Hopkins Univ., Baltimore, Maryland.

Kim, H. T., Kline, S. J., and Reynolds, W. C. (1968). "An Experimental Study of Turbulence Production near a Smooth Wall in a Turbulent Boundary Layer with Zero Pressure Gradient," Rep. MD-20. Thermosci. Div., Dep. Mech. Eng., Stanford Univ., Stanford, California.

Kim, H. T., Kline, S. J., and Reynolds, W. C. (1971). *J. Fluid Mech.* **50**, 133.

Kovasznay, L. S. G., Kibens, V., and Blackwelder, R. F. (1970). *J. Fluid Mech.* **41**, 283.

Laufer, J., and Badri Narayanan, M. A. (1971). *Phys. Fluids* **14**, 182.

Lu, S. S., and Willmarth, W. W. (1972). "The Structure of the Reynolds Stress in a Turbulent Boundary Layer," ORA Rep. 021490-2-T. Dep. Aerosp. Eng., Univ. of Michigan, Ann Arbor.

Lu, S. S., and Willmarth, W. W. (1974). Research note. *Phys. Fluids* **16**, 11, 2012.

Rao, K. N., Narasimha, R., and Badri Narayanan, M. A. (1969). "Hot Wire Measurements of Burst Parameter in a Turbulent Boundary Layer," Rep. 69 FM 8. Dep. Aeronaut. Eng., Indian Inst. of Sci., Bangalore.

Rao, K. N., Narasimha, R., and Badri Narayanan, M. A. (1971). *J. Fluid Mech.* **48**, 339.

Runstadler, P. W., Kline, S. J., and Reynolds, W. C. (1963). "An Investigation of the Flow Structure of the Turbulent Boundary Layer," Rep. MD-8. Thermosci. Div., Mech. Eng. Dep., Stanford Univ., Stanford, California.

Schraub, F. A., and Kline, S. J. (1965). "Study of the Structure of the Turbulent Boundary Layer with and without Longitudinal Pressure Gradients," Rep. MD-12. Thermosci. Div., Mech. Eng. Dep., Stanford Univ., Stanford, California.

Townsend, A. A. (1957). *In* "The Turbulent Boundary Layer," IUTAM Symp., Freiburg pp. 1–15. Springer-Verlag, Berlin and New York.

Townsend, A. A. (1961). *J. Fluid Mech.* **11**, 97.

Tu, B. J., and Willmarth, W. W. (1966). "An Experimental Study of the Structure of Turbulence Near the Wall Through Correlation Measurements in a Thick Turbulent Boundary Layer," Tech. Rep. ORA 02920-3-T. Univ. of Michigan, Ann Arbor.

Wallace, J. M., Eckelmann, H., and Brodkey, R. S. (1972). *J. Fluid Mech.* **54**, 39.

Willmarth, W. W., and Lu, S. S. (1971). Structure of the Reynolds stress near the wall. *AGARD Conf. Proc.* **92**, 3.

Willmarth, W. W., and Lu, S. S. (1972). *J. Fluid Mech.* **55**, 65.

Willmarth, W. W., and Tu, B. J. (1967). *Phys. Fluids* **10**(9), Part 2, S134.

Willmarth, W. W., and Wooldridge, C. E. (1962). *J. Fluid Mech.* **14**, 187.

Willmarth, W. W., and Wooldridge, C. E. (1963). *AGARD Rep.* **456**.

SOME FEATURES OF TURBULENT DIFFUSION FROM A CONTINUOUS SOURCE AT SEA

K. F. BOWDEN, D. P. KRAUEL,[1] AND R. E. LEWIS[2]

Oceanography Department, University of Liverpool, Liverpool, England

1. INTRODUCTION

The problem considered in this investigation is the dispersion which occurs in flow from a fixed source into a current of water, such as a tidal stream. In these conditions lateral and vertical eddy diffusion give rise to the formation of a plume. The properties of plumes arising from discharges into the atmosphere have been studied quite extensively and the results of these studies are applicable to some extent to plumes in fresh water or the sea. However, there are significant differences, one of which is the greater ratio of the lateral to vertical extent of the plume in water.

Dispersion in inland bodies of water was investigated in a comprehensive series of dye plume experiments in the Great Lakes (Csanady, 1970; Murthy, 1972). The conditions of flow in lakes differ from those in the sea mainly in the relative weakness of the currents, compared with the tidal streams in many coastal areas. Comparatively few observations of dye plumes from continuous sources have been made in the sea, although many experiments on instantaneous discharges have been carried out. Among the continuous release experiments which have been reported are those of Reinert (1965) in Long Island Sound and of Karabashev and Ozmidov (1965) in the Black Sea. Foxworthy (1968) and Foxworthy and Kneeling (1969) have given the results of several plume experiments, as well as a number of instantaneous releases, off the coast of southern California, where the conditions were characterised by high stability and weak currents.

2. EXPERIMENTAL METHOD

The experiments which provided the data for this paper were carried out in several areas of the Irish Sea, where the tidal currents had an amplitude of the order of 50–100 cm/sec, and in the Holy Loch, an inlet off the Firth of

---

[1] *Present address:* Bedford Institute of Oceanography, Dartmouth, Nova Scotia, Canada.
[2] *Present address:* I.C.I. Brixham Laboratory, Devon, England.

Clyde, where the currents were usually less than 10 cm/sec. Rhodamine B dye solution was used as the tracer material in each case, released continuously at the desired depth from a container mounted on a moored buoy. In a typical experiment the release lasted for a period of 3 to 4 hr, during a single flood or ebb tide, with the depth of release set at a value between 1 and 6 m. The resulting dye plume, which was normally visible at the surface, was traversed by a research vessel, using one or more sampling intakes at selected depths and pumping the water through a Turner model III fluorometer. In the initial experiments there was only one sampling channel but in the later series three intakes, at different depths and each connected to its own fluorometer, were used. The fluorometer outputs were recorded on strip charts, the readings of which were later digitized, and in some cases direct digital recording was utilised. A more detailed account of the experimental technique and method of analysis has been published elsewhere (Bowden and Lewis, 1973).

3. ANALYSIS

The position of a point in the plume may be represented by coordinates x, measured downstream from the source, y laterally across the plume, and z vertically downwards from the surface. The basic data from each traverse, at a given depth, consisted of a series of concentration values $C(y)$. In the later experiments three such series, at different values of z, were obtained simultaneously. The downstream distance x of the traverse was converted to a diffusion time t, knowing the velocity of the current.

For a particular traverse, the peak concentration C_P was found by examination and the integrated cross-plume concentration C_A, the position $\bar{y}$ of the centre of mass of the plume and the variance σ_y^2 about the centre of mass were computed. Most of the discussion in this paper is concerned with relative diffusion although some reference will be made later to the problem of absolute diffusion. The rate of relative lateral diffusion may be represented by a coefficient of eddy diffusion K_y, defined by

(1) $$K_y = \tfrac{1}{2}\, d\sigma_y^2/dt$$

Estimates of the vertical variance σ_z^2 and the corresponding eddy coefficient K_z were also made, but the values were less reliable as measurements at only three depths were available.

4. LATERAL DIFFUSION

The major problem in the interpretation of diffusion experiments, familiar to all workers in this field, is the great variability of the observed data. In the present case, part of the variability may be attributed to the experimental technique in that the source was not at a fixed point but attached to a moored buoy subject to movements due to waves and possibly yawing in the current. In addition the rate of discharge may have varied with time. It is thought, however, that these factors would not seriously affect the relative dispersion in a given cross section. There seems little doubt that much of the observed variability was due to inherent properties of the field of flow, including the presence of large-scale horizontal eddy motion which produced meandering and distortion of the plume on many occasions. It becomes essential to adopt a statistical approach, either by making a number of repeated crossings at each downstream distance x (corresponding to a given diffusion time t) or by considering the whole plume rather than separate traverses in deriving representative parameters.

Attempts to derive K_y directly from Eq. (1) in finite difference form, or from the slope of the tangent at a given point to a curve relating σ_y^2 to t, led to highly variable results. The procedure adopted, therefore, was to fit a power law curve of the form

(2) $$\sigma_y^2 = at^m$$

where a and m are constants, to all the data from a given release. This equation assumes a point source at zero time and cannot be expected to be valid close to the source. From (1), K_y is then given by

(3) $$K_y = \tfrac{1}{2}mat^{m-1}$$

so that K_y may be calculated for a given diffusion time t from the fitted values of a and m.

It follows from (2) and (3) also that

(4) $$K_y = c\sigma_y^r$$

where $r = 2(m-1)/m$ and c is a constant. If σ_y is regarded as a measure of the scale of the distribution, then Eq. (4) represents the dependence of K_y on scale for that particular release.

Before considering the observational values of m, the relation of this index to theoretical treatments of diffusion may be pointed out. Table I shows the values of $m - 1$ and r corresponding to the integral values $m = 1, 2,$ and 3. The case $m = 1$ is that of Fickian diffusion with K_y constant while $m = 2$

TABLE I. Values of power law indices

m	$m-1$	r	Comments
1	0	0	Constant K_y
2	1	1	Constant diffusion velocity
3	2	$\tfrac{4}{3}$	Inertial subrange

implies that K_y increases linearly with t or σ_y, corresponding to a constant diffusion velocity. The case $m = 3$ corresponds to K_y being proportional to $\sigma_y^{4/3}$, which is consistent with inertial subrange conditions in locally isotropic turbulence, if σ_y is identified with the scale of the process.

Figure 1 shows the areas of the Irish Sea in which experiments were made. In Fig. 2 a curve of σ_y^2 against t is shown for one of the releases in Red Wharf Bay in June 1969. The regression line fitted to the bilogarithmic plot corresponds to σ_y^2 proportional to $t^{2.1}$. The value of the index varied considerably from one release to another but the median value was approximately 2.

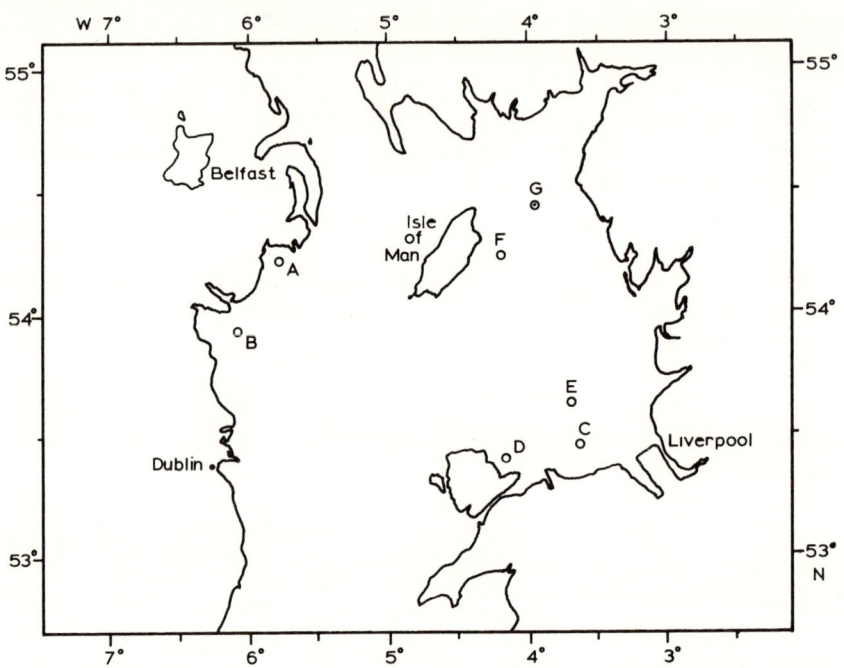

FIG. 1. Chart of the Irish Sea showing locations of experimental areas.

Figure 3 shows the relation of σ_y^2 to t for one of the releases in Liverpool Bay in September 1971. The scatter is somewhat accentuated in that points for the three sampling depths are plotted on the same diagram. The line shown is displaced but it has the slope $m = 2.3$, found by fitting a regression line to the points by a least squares method.

The 1970 and 1971 data were all obtained and analysed in the same way and most of the following statements refer to these data and the results derived from them. A total of 22 releases was made in the two years, with an average of nearly 20 crossings per release. In 10 releases the significance of the linear regression line was below the 5% level. Confining attention to the

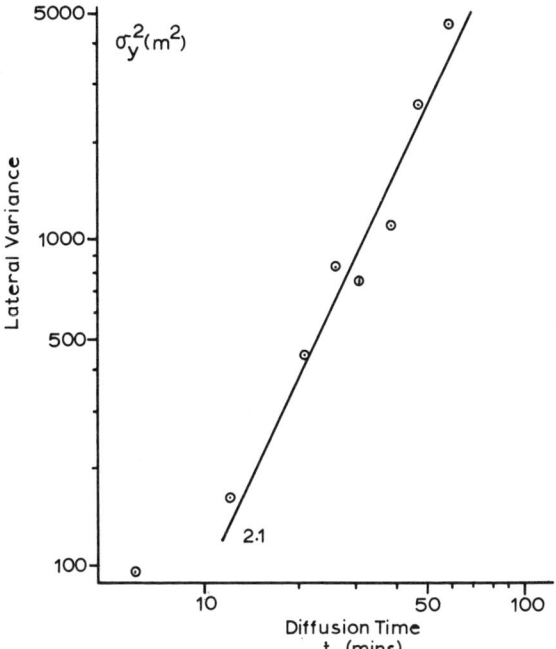

Fig. 2. Change of σ_y^2 with t for a dye plume in Red Wharf Bay, 21 June 1969.

other 12 releases, the value of m lay between 1.2 and 2.7. The constants in the power law equation were evaluated and the value of K_y computed for a standard diffusion time of 30 min, so that the values of K_y were comparable. They ranged from 240 cm²/sec to 4100 cm²/sec with a median value of 1200 cm²/sec.

In most of the experiments the data could be fitted to the power laws $m = 1.5$ or $m = 2$ without loss of significance, but a considerable loss of significance occurred if laws $m = 1$ or $m = 3$ were fitted. The implication of $m > 1$ is that diffusion took place at a faster rate than that corresponding to a constant K_y, i.e., to an eddy scale small compared with the width of the plume. On the other hand, since $m < 3$, diffusion was slower than would correspond to an inertial subrange in turbulence that was locally isotropic in the horizontal plane.

If $m = 2$, Eq. (2) may be written

$$\sigma_y^2 = B^2 t^2 \tag{5}$$

and (3) as

$$K_y = B^2 t \tag{6}$$

where B is a constant diffusion velocity.

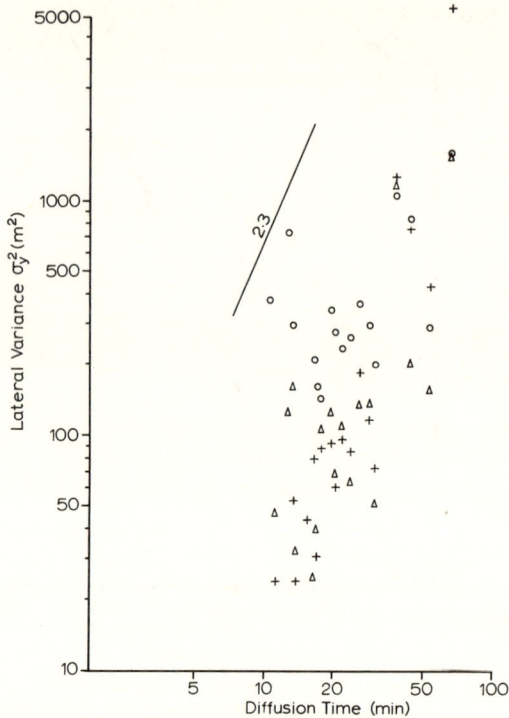

FIG. 3. Change of σ_y^2 with t for a dye plume in Liverpool Bay, 8 September 1971. Sampling depths: +1.3 m, △2.55 m, ○ 3.8 m.

TABLE II. Observed values of diffusion velocity

	Area	Year	Number of releases	Mean current (cm/sec)	Diffusion velocity B	
					Range (cm/sec)	Median (cm/sec)
	Holy Loch	1969	2	7	0.4–0.5	0.45
D	Red Wharf Bay	1969	4	52	0.7–1.8	1.1
D	Red Wharf Bay	1971	4	55	0.7–1.1	0.75
G	Off Cumberland	1969	6	83	0.3–2.6	1.2
C	Liverpool Bay	1970	2	58	1.3–1.6	1.45
E	Liverpool Bay	1971	4	72	0.8–1.3	1.2
F	Off Isle of Man	1971	2	22	0.4–0.8	0.6

Since this value of m could be fitted to the data without appreciable loss of significance, the value of B was determined in each case. For the twelve experiments in 1970–1971, referred to above, the values of B ranged from 0.4 to 1.6 cm/sec with a median of 1.0 cm/sec.

Table II summarizes the values of B for the various groups of experiments. It is seen that the Holy Loch values are appreciably lower than those for the Irish Sea except for the area off the Isle of Man where the currents were weak. Otherwise no clear distinction can be made between the different areas of the Irish Sea.

5. Vertical Diffusion

Determinations of the vertical variance σ_z^2 and the corresponding eddy diffusion coefficient K_z were dependent on measurements of concentration at only three depths and so were less reliable than those of σ_y^2 and K_y. No attempt will be made here to describe the methods of analysis used. It is sufficient to say that the standard deviation in the vertical was an order of magnitude less than in the horizontal. The value of σ_z was between 1 and 3 m for $t = 30$ min and values of K_z were in the range 1.2 to 33 cm^2/sec.

Attempts were made to fit a power law relation between σ_z^2 and t, i.e.

$$\sigma_z^2 = bt^n \tag{7}$$

analogous to Eq. (2) for σ_y^2. These led to very variable values of n and the method must be considered unreliable, not only in view of only three depths being available but also because the effects of a finite initial variance would be more severe.

6. Peak Concentrations

If the distribution of concentration were Gaussian in both the lateral and vertical directions, the peak concentration on the axis of the plume would be given by

$$C_P = Q/2\pi V \sigma_y \sigma_z \tag{8}$$

where Q is the mass rate of release of dye at the source and V is the current speed at the time of release. If σ_y^2 and σ_z^2 vary according to the power laws in Eqs. (2) and (7) respectively, then

$$C_P = At^{-q} \tag{9}$$

where $q = \tfrac{1}{2}(m + n)$ and $A = Q/[2\pi V(ab)^{1/2}]$.

From the Irish Sea data of 1970–1971, regression lines were fitted to bilogarithmic plots of C_P against t for each release, leading to values of q

which, for the significant cases, had a median value of 1.75. The median value of m in the power law for σ_y^2 was 1.85. The corresponding value of n in the power law for σ_z^2, computed from $n = 2q - m$ as in Eq. (9), would be 1.65, implying a K_z which increased with diffusion time t but at a somewhat slower rate than K_y.

7. Dependence of Diffusion on Environmental Conditions

It was not possible, from the Irish Sea data, to establish any clear dependence of either the lateral or vertical diffusion on stability. On the other hand

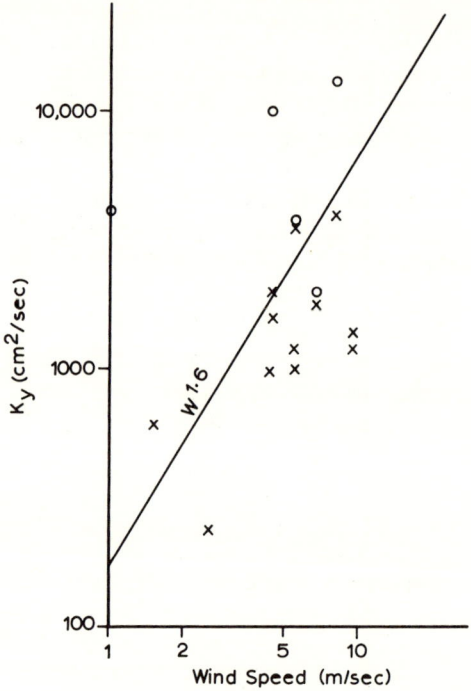

FIG. 4. Variation of lateral diffusion coefficient K_y with wind speed W: ○ 1970 data, × 1971 data.

the 1970–1971 data showed a strong dependence of both K_y and K_z on the current speed U and the wind speed W. Figures 4 and 5 show K_y and K_z respectively as functions of wind speed. The regression lines correspond to K_y increasing as $W^{1.6}$ at the 5% level of significance and K_z as $W^{3.0}$, significant at the 0.5% level. Both K_y and K_z also increased with the mean current speed $\bar{U}$ in a surface layer, 5 m deep, being proportional to $\bar{U}^{2.5}$ and

$\bar{U}^{3.4}$ respectively, significant at the 0.1 % level in each case. In fact, there was a direct correlation between K_y and K_z represented by

$$K_y \propto K_z^{0.55}$$

One possible explanation of the proportionality of K_y to K_z is that the apparent lateral diffusion arises from current shear interacting with vertical

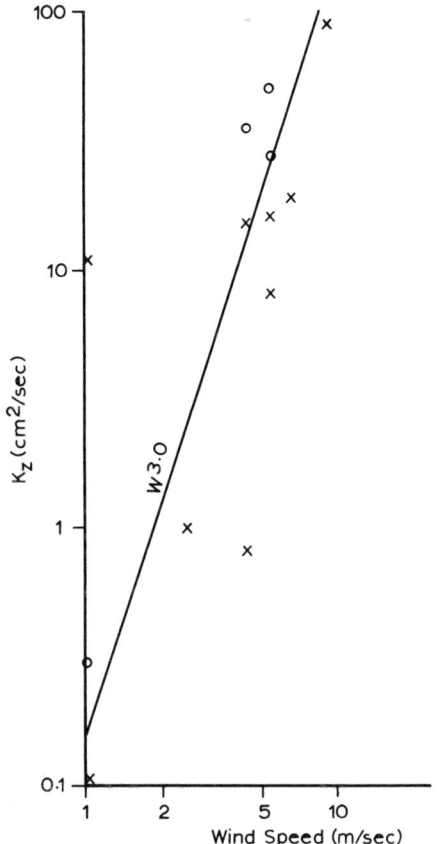

FIG. 5. Variation of vertical diffusion coefficient K_z with wind speed W: ○ 1970 data, × 1971 data.

diffusion. In an unbounded or semi-unbounded medium this would give a K_y increasing linearly with K_z (Saffman, 1962). The measurements of current shear were not accurate enough, however, to make a direct test of this hypothesis.

8. Absolute and Relative Diffusion

The foregoing discussion has been concerned with the relative diffusion of the dye plume about its centre of mass, neglecting any movement of the centre of mass itself. In most experiments this was unavoidable since the movements of the ship were not tracked with sufficient accuracy to enable absolute determinations of position to be made. An example will now be given of a release in which a large number of crossings were made at a constant distance downstream from the source and the necessary accuracy in position was attained to be able to locate the concentration measurements in absolute coordinates.

Figure 6 shows the concentration curves for 18 crossings at the same distance from the source (corresponding to a diffusion time of $1.03 \pm 0.15 \times 10^3$ sec), plotted against actual distance y perpendicular to the mean axis of the plume. The individual crossings show pronounced variability, in the position of the centre of mass, the relative variance, and the peak concentration. The broken line is the average concentration curve for the 18 crossings, having a lower peak value and a larger standard deviation than the average values of the individual curves. Figure 7 shows 10 of the 18 curves in the previous figure plotted with their centres of mass coincident. They are seen to vary considerably, both in the laterally integrated concentration and in the shape of the curve. The standard deviations of the 18 curves ranged from 7 to 16.5 m, i.e., by a factor of 2.4, while the peak concentration varied by a factor of 6.

Let σ_{ya} denote the absolute standard deviation of the meandering plume and σ_{yr} the mean standard deviation of the curves relative to their own centres of mass. Similarly, let C_{Pa} and C_{Pr} be the peak concentration of the absolute curve and the mean peak concentration of the individual curves respectively. If both the absolute and relative distributions were Gaussian in form and there were no meandering in the vertical one would expect

(10) $$C_{Pr}/C_{Pa} = \sigma_{ya}/\sigma_{yr}$$

as shown by Gifford (1959).

The observed values in the above experiment were

$$\sigma_{ya} = 34.2 \text{ m}, \qquad \sigma_{yr} = 10.6 \text{ m}$$

Thus $\sigma_{ya}/\sigma_{yr} = 3.2$. The ratio of the peak concentrations was

$$C_{Pr}/C_{Pa} = 2.4$$

The difference between the observed ratios, 3.2 and 2.4, is within experimental error. If meandering occurred in the vertical as well, the ratio

DIFFUSION FROM A CONTINUOUS SOURCE AT SEA 325

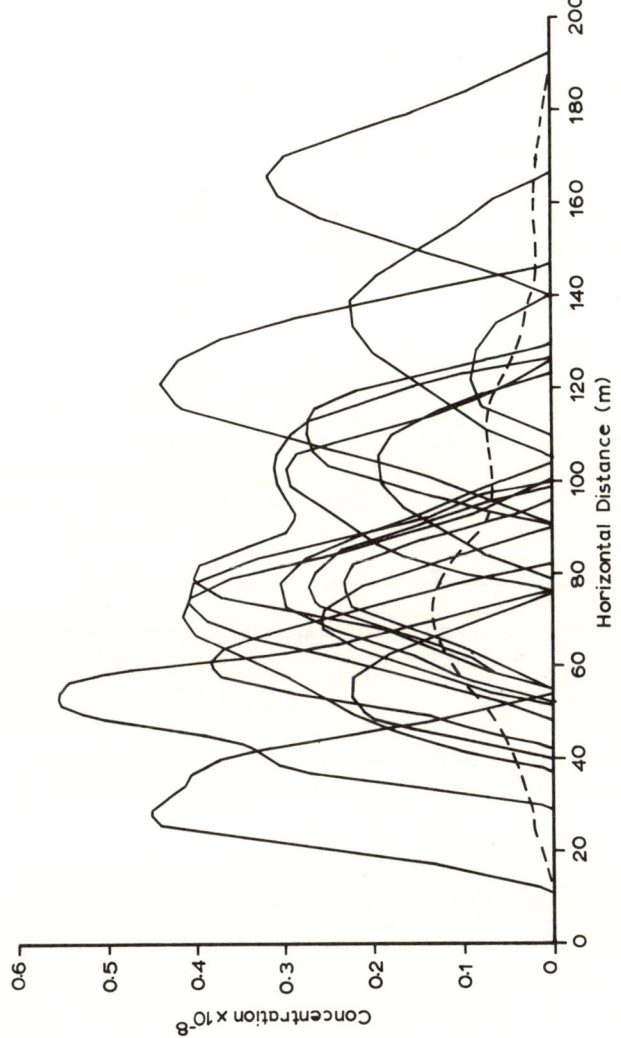

Fig. 6. A series of 18 cross-plume distributions, for the same diffusion time, plotted against absolute distance. Broken line indicates the average distribution.

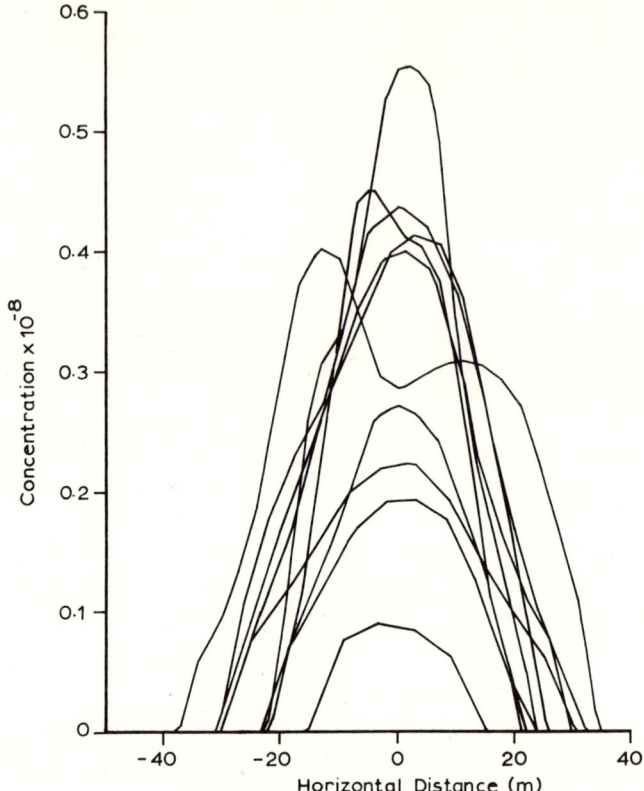

FIG. 7. A series of 10 cross-plume distributions, for the same diffusion time, plotted with centres of mass superimposed.

C_{Pr}/C_{Pa} would be greater than that of σ_{ya}/σ_{yr}, so that it is reasonable to conclude that, in this experiment, any vertical meandering was on a much smaller scale than that in the horizontal.

9. Discussion

The example given above is an indication of the variability of diffusion patterns even when the overall conditions of flow appear to be in an approximately steady state. Occasions during which the tidal current was changing rapidly in speed or direction, or when wind-induced shear was obviously producing a distortion of the plume, have been excluded from the foregoing analyses.

The present results may be compared with some other recent data obtained in similar conditions. Using the continuous discharge method near the Texel lightvessel in the southern North Sea, Meerburg (1972), for rather shorter diffusion times (4 to 12 min), derived values of K_y which increased with time. Analyzing his data in terms of a diffusion velocity B leads to a median value of $B = 2.7$ cm/sec, which is about twice the Irish Sea values. Recent experiments in the Liverpool Bay area, but using instantaneous discharges and on a larger scale, have been described by Barrett et al. (1972) and Talbot (1972). In Talbot's experiment, in which a patch of rhodamine B dye was traced for a 5-day period, the transverse coefficient K_y continued to increase with time for over 60 hr, eventually approaching an asymptotic value. For the first 12 hr the data, analyzed in terms of diffusion velocity, corresponded to a B of 1.05 cm/sec, in good agreement with the results given above. The experiments of Barrett et al., using a different technique, also gave values of K_y increasing with time but corresponding to a B of about half the value found above or that derived from Talbot's data.

The physical interpretation of the diffusion velocity B is uncertain: it may contain a component due to the interaction of current shear and vertical diffusion. An estimate of the possible magnitude of the effect on K_y of a shear in the transverse component of velocity V may be obtained from Okubo's (1967) analysis for the case of an unbounded sea, since the vertical spread in these experiments was relatively small. For a steady current and a uniform shear, this leads to an effective K_{ye} given by

(11) $$K_{ye} = K_z(dV/dz)^2 t^2$$

The effective diffusion coefficient would thus increase with t^2 instead of with t, as for a constant diffusion velocity B. However, matching the equations for a particular diffusion time t_0 leads to an effective B_e given by

(12) $$B_e^2 = K_z(dV/dz)^2 t_0$$

Taking $B_e = 1$ cm/sec, $K_z = 10$ cm^2/sec, and $t_0 = 30$ min as typical values, $|dV/dz| = 0.75$ sec^{-1}, corresponding to a shear in the transverse component of velocity of 3.75 cm/sec in 5 m. A shear of this magnitude may well have been present, although no current measurements of sufficient accuracy to detect it were made.

Another possible physical model is that of diffusion in a field of turbulence which differs from a state of horizontal local isotropy by having energy injected in the range of eddy scales concerned, i.e., from a few metres up to several tens of metres. The source might be in the action of surface waves, small-scale patterns of flow induced by the wind or eddies in the tidal currents. Okubo (1968, 1971) made an attempt on these lines to systematize the results of the much larger number of experiments with instantaneous releases, analyzed in terms of a model involving radially symmetrical diffusion. Although the present experiments were with continuous releases and on a time scale at the lower limit of the cases considered by Okubo, it is of some interest to relate our results to his. Okubo showed that the radial variances σ_r^2 of all the data, with diffusion times from 2 hr to 1 month, could be fitted approximately by the relation

(13) $$\sigma_r^2 = 0.0108 t^{2.34}$$

Comparing this equation with a model using a radial diffusion velocity P and matching the solutions at $t = 1$ hour gives $P = 0.42$ cm/sec, which is about half the values of B found in the Irish Sea.

Okubo pointed out that the relation predicted by the similarity law of turbulence, i.e.,

(14) $$\sigma_r^2 = C\varepsilon t^3$$

where ε is the rate of energy dissipation per unit mass through the turbulence and C is a constant of order unity, could be fitted quite well to the data, if ε varied from one range of time and length scales to another. At the small-scale end, Okubo (1968) showed that, putting $C = 1$, $\varepsilon = 9.7 \times 10^{-5}$ cm²/sec³ for t between 4×10^3 and 2×10^4 sec. Matching the corresponding equation for σ_r^2 to a diffusion model and putting $t = 1$ hour leads to $P = 0.59$ cm/sec, which is still only half the values of B found in the Irish Sea. Since B refers to transverse diffusion and the rate of longitudinal diffusion is usually several times greater, one might take $P = 2$ cm/sec as more typical of the Irish Sea and the corresponding value of ε would be an order of magnitude greater, say 10^{-3} cm²/sec³.

It would be interesting to have an independent estimate of the energy dissipation parameter ε in the conditions of the present experiments. Taylor (1918) made an estimate of the overall dissipation of tidal energy in the Irish Sea which, averaged over the whole volume, would correspond to approximately 0.22 cm²/sec³ per unit mass of water, i.e., about 200 times the value of ε required to account for the observed horizontal diffusivity. However, much of the dissipation of tidal energy, arising from tidal friction, probably takes place in a layer near the sea bed. The intensity of turbulence, due to this cause, is likely to be considerably lower in the surface layer and not all the energy there will be in the horizontal components of relevant scale. It would be unwise to pursue the analogy further than to say that there appears to be ample turbulent energy available, if it were in the right form, to provide for the observed rate of horizontal diffusion.

10. Conclusion

In general, where features of the apparent diffusion can be related to identifiable physical processes, such as lateral or vertical shear in the tidal current or wind-induced current, and can be treated quantitatively from this point of view it is desirable to do so. In many cases, however, there is likely to be a residual diffusive effect which one has to represent empirically by a suitable parameter. It is suggested that, in conditions similar to those of the experiments described in this paper, the lateral diffusion velocity B is a more useful parameter than the coefficient of eddy diffusion K_y, in that the range of values of B is much less than that of K_y. The physical interpretation of B remains uncertain, although some possible contributory effects have been discussed above.

References

Barrett, M. J., Munro, D., and White, K. E. (1972). *In* "Out of Sight, Out of Mind. Report of a Working Party on Sludge Disposal in Liverpool Bay," Vol. 2, pp. 145–169. Dep. Environ., HM Stationery Office, London.
Bowden, K. F., and Lewis, R. E. (1973). *Water Res.* 7, 1705–1722.
Csanady, G. T. (1970). *Water Res.* 4, 79–114.
Foxworthy, J. E. (1968). Rep. No. 68-1, 72 pp. Allan Hancock Found., Univ. of Southern California, Los Angeles, California.
Foxworthy, J. E., and Kneeling, H. R. (1969). Rep. No. 69-1, 176 pp. Allan Hancock Found., Univ. of Southern California, Los Angeles, California.
Gifford, F. (1959). *Advan. Geophys.* 6, 117–137.
Karabashev, G. S., and Ozmidov, R. V. (1965). *Izv. Akad. Nauk SSSR, Fiz. Atmos. Okeana* 1, 1178–1189.
Meerburg, A. J. (1972). *Neth. J. Sea Res.* 5, 492–509.
Murthy, C. R. (1972). *J. Phys. Oceanogr.*, 2, 80–90.
Okubo, A. (1967). *Int. J. Oceanol. Limnol.* 1, 194–204.
Okubo, A. (1968). *Chesapeake Bay Inst., Johns Hopkins Univ., Tech. Rep.* No. 38.
Okubo, A. (1971). *Deep-Sea Res. Oceanogr. Abstr.* 18, 789–802.
Reinert, R. (1965). *Symp. Diffus. Oceans Fresh Waters*, Lamont Geol. Observ. pp. 19–27.
Saffman, P. G. (1962). *Quart. J. Roy. Meteorol. Soc.* 88, 382–392.
Talbot, J. W. (1972). *In* "Out of Sight, Out of Mind. Report of a Working Party on Sludge Disposal in Liverpool Bay," Vol. 2, pp. 209–271. Dep. Environ., HM Stationery Office, London.
Taylor, G. I. (1918). *Phil. Trans. Roy. Soc. London, Ser. A* 220, 1–33.

DIFFUSION OF TURBIDITY BY SHEAR EFFECT AND TURBULENCE IN THE SOUTHERN BIGHT OF THE NORTH SEA

JACQUES C. J. NIHOUL

Institut de Mathématique, University of Liège, Belgium

1. INTRODUCTION

Concerned with the alarming state of pollution in the North Sea, the Belgian Government initiated in 1971 extensive experimental and theoretical investigations of the Southern Bight (51° N–53° N).

The Southern Bight is an irregular shallow area where very strong tidal currents (up to 1 m/s) are encountered. Observations indicated that the water column is fairly well mixed and that the *shear effect* (associated with the tidal currents) plays an essential role in the dispersion of pollutants (Nihoul, 1972).

The appellation *shear effect* is widely used to denote similar but not identical phenomena. In this context, it is meant in the following sense: space average concentrations (say over the depth or over the cross section) are governed by equations which are derived from the three-dimensional ones by space integration. In this process, the quadratic convection terms give two contributions, the first of which represents the advection by the mean motion while the second contains the mean product of deviations around the means and contributes to the dispersion.

This effect has been described by several authors in pipes, channels, and estuaries where, after integration over the cross-section, the flow—steady or oscillating—is essentially in one direction (Taylor, 1953, 1954; Elder, 1959; Bowden, 1965).

In the shallow waters of the Southern Bight, it is generally sufficient to consider the mean concentrations over the depth but, out at sea, no further averaging is possible and the dispersion mechanism is fundamentally two dimensional. Thus the models valid for estuarine diffusion must be generalized to account for the rotation of the tidal velocity vector (Nihoul, 1972).

In addition, to study and control the dumping of industrial wastes, the model must include the sedimentation of the solid suspensions. Some of the dumpings in the Belgian part of the Southern Bight release large size particles with a fairly important sedimentation velocity. The strong tides however produce high friction velocities at the bed and substantial recirculation.

In the following, the classical theory of the shear effect is extended to take into account the rotation of the tides and the sedimentation–recirculation process. A simple formulation is sought for the purpose of numerical prediction. The model appears to agree very well with experiments.

2. Diffusion of Turbidity

One considers the release of solid particles in a well-mixed sea of uniform density ρ_0. If $\mathbf{v}_0$ is the velocity of water, ρ_1 and $\mathbf{v}_1$ the density and velocity, respectively, of the solid contaminant and if α is the relative volume of the particles, one can define a density ρ and a velocity $\mathbf{v}$ for the mixture as follows:

(1) $$\rho = \rho_0(1 - \alpha) + \rho_1 \alpha$$

(2) $$\rho \mathbf{v} = \rho_0(1 - \alpha)\mathbf{v}_0 + \rho_1 \alpha \mathbf{v}_1.$$

It is convenient to introduce a sedimentation velocity $\boldsymbol{\sigma} = -\sigma \mathbf{e}_z$ by

(3) $$\mathbf{v}_1 = \mathbf{v} + \boldsymbol{\sigma}$$

the z axis is taken vertical upward and σ is assumed to be constant. The equation of mass balance for the sea water and the admixture take the form

(4) $$\partial \rho_0(1 - \alpha)/\partial t + \nabla \cdot \rho_0(1 - \alpha)\mathbf{v}_0 = 0$$

(5) $$\partial \rho_1 \alpha / \partial t + \nabla \cdot \rho_1 \alpha \mathbf{v}_1 = 0.$$

Dividing by ρ_0 and ρ_1 and adding, one gets

(6) $$\nabla \cdot [\alpha \mathbf{v}_1 + (1 - \alpha)\mathbf{v}_0] = 0.$$

Hence, using Eq. (2),

(7) $$\nabla \cdot \mathbf{v} = \frac{\rho_1 - \rho_0}{\rho_0} \sigma \frac{\partial \alpha}{\partial z}.$$

In practical situations, at least one of the three factors in the right-hand side of Eq. (7) is small and one may write, with a very good approximation,

(8) $$\nabla \cdot \mathbf{v} = 0.$$

Let the velocity $\mathbf{v}$ be separated into a mean part $\langle \mathbf{v} \rangle$ and a turbulent part and let

(9) $$\langle \mathbf{v} \rangle = \mathbf{u} + w \mathbf{e}_z$$

where $\mathbf{u}$ is the horizontal mean velocity vector.

Introducing a vertical eddy diffusivity λ, one obtains from Eq. (5),

(10) $$\frac{\partial \alpha}{\partial t} + \nabla \cdot \mathbf{u}\alpha + \frac{\partial}{\partial z} w\alpha = \frac{\partial}{\partial z}\left(\lambda \frac{\partial \alpha}{\partial z} + \sigma\alpha\right) + \text{h.t.d.}$$

where h.t.d. stands for the horizontal turbulent diffusion terms. There is no advantage in writing these explicity at this stage. If horizontal eddy diffusivities must be introduced, it is in fact more convenient to do so in the final depth-averaged equations and define eddy coefficients appropriate to these equations.

It is customary, in sedimentation analysis, to call α the *turbidity*. Equation (10) describes thus the diffusion of turbidity in the sea water.

3. Integration over Depth

Let $z = \zeta$ and $z = -h$ be the equations of the free surface and the bottom, respectively, and let

(11) $$H = h + \zeta.$$

Letting q stand for any variable $\mathbf{u}, \alpha, \ldots$, one defines

(12) $$\bar{q} = H^{-1} \int_{-h}^{\zeta} q \, dz$$

(13) $$q' = q - \bar{q}.$$

Integrating Eqs. (8) and (10) over depth, one gets (e.g., Nihoul, 1972):

(14) $$\frac{\partial H}{\partial t} + \nabla \cdot H\bar{\mathbf{u}} = 0$$

(15) $$\frac{\partial (H\bar{\alpha})}{\partial t} + \nabla \cdot (\bar{\alpha}\bar{\mathbf{u}}H) + \nabla \cdot \int_{-h}^{\zeta} \alpha'\mathbf{u}' \, dz = \left[\lambda \frac{\partial \alpha}{\partial z} + \sigma\alpha\right]_{-h}^{\zeta} + \int_{-h}^{\zeta} (\text{h.t.d.}) \, dz$$

Combining Eqs. (14) and (15), one obtains

(16) $$\partial \bar{\alpha}/\partial t + \bar{\mathbf{u}} \cdot \nabla \bar{\alpha} = S + Q + T$$

where

(17) $$S = -H^{-1}\nabla \cdot \left\{H \int_0^1 \overline{\alpha'\mathbf{u}'} \, d\eta\right\}$$

(18) $$Q = H^{-1}\{(\lambda/H)(\partial \alpha/\partial \eta) + \sigma\alpha\}_0^1$$

(19) $$T = \overline{\text{h.t.d.}}$$

(20) $$\eta = (z + h)/H$$

S represents the shear effect, Q is the mean flux of suspended particles, which depends essentially on the flux at the free surface—if any—and on the deposition on the bottom. It is observed that when the friction velocity at the bottom exceeds a critical value u_{*c}, the laminar boundary layer is disrupted and some of the sedimenting particles are ejected in the fluid above. When the friction velocity exceeds a further critical value u_{*e}, the flow is able to erode the bottom and more material is returned to the water column. Near the coast, waves are reported to have an important action on the processes of deposition and erosion.

It is assumed here that Q is known from sedimentation analysis. Expressions of Q appropriate to numerical modelling can be found in Nihoul (1973) with a discussion of the empirical formulas widely used in coastal engineering.

4. Shear Effect

Subtracting Eq. (16) from Eq. (10), one obtains

$$(21) \quad \frac{\partial \alpha'}{\partial t} + \bar{\mathbf{u}} \cdot \nabla \alpha' + \mathbf{u}' \cdot \nabla \alpha' + w \frac{\partial \alpha}{\partial z} + S - T' + \mathbf{u}' \cdot \nabla \bar{\alpha}$$

$$= \frac{\partial}{\partial z}\left(\lambda \frac{\partial \alpha'}{\partial z} + \sigma \alpha'\right) - Q.$$

If now the classical assumption is made that the deviation α' is small compared to the mean turbidity $\bar{\alpha}$, all terms in the left-hand side are small compared to the last one (note that almost everywhere the velocity deviation u' is of the same order as u) and Eq. (21) reduces to

$$(22) \quad \mathbf{u}' \cdot \nabla \bar{\alpha} = \frac{\partial}{\partial z}\left(\lambda \frac{\partial \alpha'}{\partial z} + \sigma \alpha'\right) - Q.$$

The physical meaning of this equation is clear: weak vertical inhomogeneities are constantly created by the inhomogeneous convective transfer of the admixture and they adapt to this transfer in the sense that the effects of convection, transverse diffusion, and unequal sedimentation are balanced for them.

Equation (22) can be used to calculate α' in terms of the gradient of the mean turbidity $\bar{\alpha}$. Multiplying the result by $\mathbf{u}'$ and integrating over depth, one obtains thus an estimate of the shear effect.

The result turns out to be fairly simple if one assumes (this may be legitimate in sufficiently shallow water) that

$$(23) \quad \mathbf{u}' = \bar{\mathbf{u}} \varphi(\eta)$$

$$(24) \quad \lambda = \kappa g(\eta).$$

Equation (22) becomes

(25) $$(\bar{\mathbf{u}} \cdot \nabla \bar{\alpha})\varphi = \frac{\sigma}{H} \frac{\partial \alpha'}{\partial \eta} + \frac{\kappa}{H^2} \frac{\partial}{\partial \eta}\left(g \frac{\partial \alpha'}{\partial \eta}\right) - Q.$$

Integrating twice, multiplying by $\mathbf{u}'$, and averaging over depth, one finds, assuming zero flux at the free surface and neglecting small-order terms:

(26) $$S = H^{-1} \nabla \cdot H\left\{\gamma_1 \frac{\sigma}{\bar{u}} \overline{\alpha \mathbf{u}} + \gamma_2 \frac{H}{\bar{u}} \overline{\mathbf{u}(\bar{\mathbf{u}} \cdot \nabla \bar{\alpha})}\right\}$$

where

(27) $$\bar{u} = \|\bar{\mathbf{u}}\|$$

(28) $$\gamma_1 = \int_0^1 \varphi \, d\eta \int_0^\eta g^{-1} \, d\xi$$

(29) $$\gamma_2 = -\int_0^1 \varphi \, d\eta \int_0^\eta g^{-1} \, d\xi \int_1^\xi \varphi \, d\beta$$

Among the small terms neglected in Eq. (26) is the contribution from the successive integration of the last term in the right-hand side of Eq. (25). This term is indeed of the order

$$H^{-1} \nabla \cdot \left(\frac{H^2}{\bar{u}} Q \bar{\mathbf{u}}\right)$$

and may be expected to be much smaller than Q because the horizontal length scale is much larger than the depth. Hence, after substitution in Eq. (16), this term can be neglected as compared to Q.

The coefficients γ_1 and γ_2 are of order 1. Their particular values depend upon the form of the functions φ and g. They can be estimated for instance by using Van Veen's power law ($u = 1.2\bar{u}\eta^{0.2}$); in which case γ_1 and γ_2 are found to be both very close to 1. Sometimes a better fit is obtained with a combined log-parabolic profile (Bowden and Fairbairn, 1952). Calculating γ_2, assuming a log-parabolic profile and a ratio of eddy diffusivity to eddy viscosity equal to 1.4 (Ellison, 1957) Nihoul found a value of 0.45, which compared very well with the observations (Nihoul, 1972).

The first term in the right-hand side of Eq. (26) represents a contribution to the shear effect due to the sedimentation. This contribution is negligible in most cases because the sedimentation velocity σ is usually rather small. It may become more important though when flocs are formed or when large solid lumps are released as observed for instance in some dumping grounds off the Belgian coast. In this case, its effect is more a reduction of the advection than an increase of the dispersion. This is easy to understand from a physical point of view. If the sedimentation velocity is important, the particles—although a high friction velocity may prevent their deposition and maintain them in suspension—will tend to concentrate in a region closer to the bottom where the velocity is smaller than the mean velocity. The second term in the left-hand side of Eq. (16) thus overestimates the real advection of those particles and this is corrected by the first part of the shear effect. (Of course the argument does not apply if the concentration becomes too large since the whole model is based on the hypothesis of weak inhomogeneities.)

The second term in the right-hand side of Eq. (26) may be interpreted as a dispersion term and appears as a natural generalization to variable depth and two-dimensional horizontal flows of Bowden's results (Bowden, 1965).

Combining Eqs. (16) and (26) and introducing a horizontal eddy diffusivity μ, one obtains

$$(30) \quad \frac{\partial \bar{\alpha}}{\partial t} + \bar{\mathbf{u}} \cdot \nabla \bar{\alpha} - H^{-1} \nabla \cdot \left(\gamma_1 \frac{H\sigma}{\bar{u}} \bar{\alpha} \bar{\mathbf{u}} \right)$$

$$= Q + H^{-1} \nabla \cdot \left[\gamma_2 \frac{H^2}{\bar{u}} \bar{\mathbf{u}} (\bar{\mathbf{u}} \cdot \nabla \bar{\alpha}) \right] + \nabla \cdot \mu \nabla \bar{\alpha}$$

This equation has been used extensively in modelling the dispersion of pollutants in the Southern Bight. The velocity $\bar{\mathbf{u}}$ and depth H are either measured or calculated by a separate numerical model combining the effects of winds and tides. The coefficients γ_1, γ_2, and μ can be estimated theoretically but they are as easily determined by numerical simulation if, as often, observations are available for reference. The calculated values agree well with the theoretical predictions.

Examining the second term in the right-hand side of Eq. (30), one can see that the shear effect produces a diffusion in the direction of the instantaneous velocity with an apparent diffusivity of the order of uH. The isotropic turbulent dispersion represented by the last term of Eq. (30) is usually much smaller as the eddy diffusivity μ can be two orders of magnitude smaller than uH. After one or two tidal periods, there results an enhanced dispersion in the direction of the maximum tidal velocity. Rapid variations of depth and strong winds can of course modify the situation but the tendency remains and the patches of pollutants have very often an elongated shape with a maximum dispersion roughly in the direction of the maximum current.

This prediction of the model is well verified by the observations (Nihoul, 1971).

One of the consequences of the cogent influence of the shear effect is that the second-order operator in Eq. (30) is nearly parabolic; the dominant terms being

$$(31) \quad \frac{H}{\bar{u}}\left\{\bar{u}_1^2 \frac{\partial^2 \bar{\alpha}}{\partial x_1^2} + \bar{u}_2^2 \frac{\partial^2 \bar{\alpha}}{\partial x_2^2} + 2\bar{u}_1 \bar{u}_2 \frac{\partial^2 \bar{\alpha}}{\partial x_1 \partial x_2}\right\}$$

As one might expect, this creates numerical difficulties and required a careful test of several techniques before a reliable one was perfected (Adam and Runfola, 1973).

In many cases, when one is only interested in a prediction of the pollutant's concentration every one or two tidal periods, it is advantageous first to average Eq. (30) over a tidal period. This generally reduces considerably the numerical difficulties and it is easy to see why it should be so. Indeed, if, in Eq. (31), $\bar{\alpha}$ was independent of time, the integration over a tidal period would make the operator elliptic by virtue of Swartz's inequality. Of course $\bar{\alpha}$ is not independent of time in general and in particular in the case of dispersion from an instantaneous source simulating a dumping. However it is usually possible to separate $\bar{u}$ and $\bar{\alpha}$ into slowly varying parts (over a tidal period) and rapidly oscillating parts (with the tidal period) and to apply the averaging technique of Krylov, Bogoliubov, and Mitropolsky with the same benefit (Nihoul, 1972).

The most important limitation of the model is due in most cases to the lack of precision in the observed or calculated values of $\bar{u}$. In situations where the velocity field was known with great accuracy, the spreading and distortion of the pollutant patches were always faithfully depicted by the model; confirming the importance of the shear effect dispersion in shallow open seas.

Acknowledgments

This study was supported by the Department of Scientific Politics in the scope of the *Programme National Belge sur l'Environment Physique et Biologique*.

References

Adam, Y., and Runfola, Y. (1973). Projet Mer, *N 14*. Programme Nat. Environ. Phys. Biol., Dep. Sci. Politics, Liège.
Bowden, K. F. (1965). *J. Fluid Mech.* **21**, 83–95.
Bowden, K. F., and Fairbairn, L. A. (1952). *Proc. Roy. Soc., Ser. A* **214**, 371–392.
Elder, J. W. (1959). *J. Fluid Mech.* **5**, 544–560.
Ellison, T. M. (1957). *J. Fluid Mech.* **2**, 456–466.
Nihoul, J. C. J. (1971). *Proc. North Sea Sci. Conf., Aviemore, Scotland* **1**, 89–108.
Nihoul, J. C. J. (1972). *Bull. Soc. Sci. Liege* **10**, 521–526.
Nihoul, J. C. J. (1973). *Proc. Liège Colloq. Ocean Hydrodyn.*, 5th pp. 1–10.
Taylor, G. I. (1953). *Proc. Roy. Soc., Ser. A* **219**, 186–203.
Taylor, G. I. (1954). *Proc. Roy. Soc., Ser. A* **223**, 446–468.
Van Veen, J. (1938). *J. Cons., Cons. Perma. Int. Explor. Mer* **13**, 7–36.

INVESTIGATION OF SMALL-SCALE VERTICAL MIXING IN RELATION TO THE TEMPERATURE STRUCTURE IN STABLY STRATIFIED WATERS

GUNNAR KULLENBERG

Institute of Physical Oceanography, University of Copenhagen
Copenhagen, Denmark

INTRODUCTION

The present experiments were designed to yield information on the rate of mixing in an internal layer of the sea with a view to investigating the dependence of the mixing on environmental conditions. A density-adjusted solution of rhodamine B was injected over an interval of 2-5 min in a subsurface layer at depths below the surface ranging from 20 to 50 m. The density of the solution was measured with hydrometers and adjusted within $\pm 0.2\sigma_t$-unit of the *in situ* density. The dye concentration distributions, both vertical and horizontal, were continuously recorded with an *in situ* fluorometer (Fig. 1) towed behind the ship and cycling in a certain depth range. The signal from a rapid response thermistor mounted on the fluorometer was simultaneously recorded, so that the dye and temperature distributions were observed with the same vertical resolution. The fluorometer and the tracing technique have been described by Kullenberg (1969).

The environmental observations were made with conventional equipment: water samples, *in situ* temperature-conductivity probes, moored and shipboard current meters, and in Lake Ontario electronic bathythermographs. The vertical resolution of these observations was 1-2 m. Profiles of salinity, temperature, and current were observed before and during the tracing period. In most cases one ship has been employed with the environmental observations and another with the tracing. The current measurements were made from a single anchored ship. Care has been taken to avoid the use of observations which apparently have been strongly influenced by the ship motion. The data have also been compared with data from the moored current meters.

From the profiles, average values of the mean vertical gradient of the

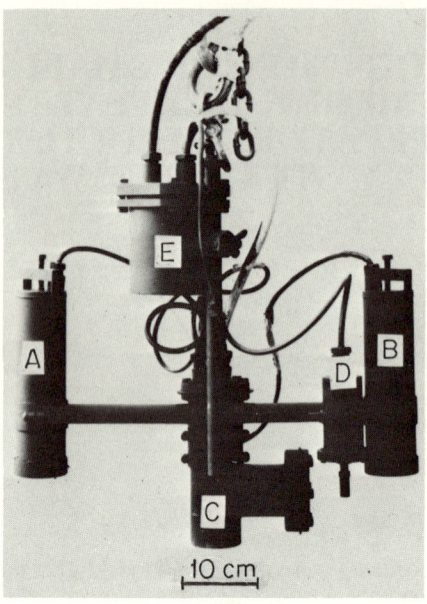

FIG. 1. The fluorometer: A, B photomultiplier units (B fluorescence detecting); C light source; D thermistor unit; E depth sensing unit.

density (σ_t) and the current vector (q) have been determined. The stratification is given by the parameter

$$N^2 = \left| \frac{g}{\rho} \frac{d\sigma_t}{dz} \right| \times 10^{-3}$$

Individual values of the gradients usually vary considerably; the mean variation around the average is 30 %. The variation is mainly due to natural variations of the observed parameters.

In all the present experiments, the mean wind has been weak, less than about 3 m s^{-1}, and the surface wave action very slight or absent. The experiments cover a wide range of environmental conditions (Table I). All the experiments were carried out in stable conditions. The duration of the tracing varies between 4 and 80 hr.

In many cases the temperature structure was observed using a specially designed (by Mr. H. Westerberg) temperature gradient meter. Mostly the temperature profiles showed several regions with typical small-scale structure interrupted by almost isothermal parts. The dye was injected in one of the regions with small-scale variations; the observed dye distributions are not artificially generated by the injection of the dye.

MIXING RATE AND TEMPERATURE IN STRATIFIED WATERS   341

TABLE I. Characteristics of the experiment

Date and type of area		Depth of dye (m)	Duration (hr)	Stratification parameter N^2 (sec^{-2})	Shear $|dq/dz| \times 10^{-3}$ (sec^{-1})	Overall Ri number
1970						
8 June	Sognefjord	16–19	7	1.2×10^{-3}	47	0.5
9–10 June	Sognefjord	21–26	14	1.8×10^{-3}	23	3.5
30–31 August	Baltic, Bornholm basin	49–53	25	2.0×10^{-3}	32	2.0
2–3 September	Baltic	26–28	10	5.3×10^{-4}	25	0.9
1971						
26 July	W. Mediterranean	20–30	9	5.3×10^{-4}	19	1.5
28 July	W. Mediterranean	23–26	9	2.0×10^{-3}	54	0.7
1972						
15–19 August	Lake Ontario 1	18–26	79	1.5×10^{-4}	12.4	1.4
29–30 August	Lake Ontario 3	35–45	30	2.3×10^{-6}	4.0	0.1
31 August–2 September	Lake Ontario 4	33–43	40	1.2×10^{-5}	4.0	0.6
6–8 September	Lake Ontario 5	25–35	60	3.6×10^{-5}	4.0	2.7
27–28 September	Lake Ontario 9	20–30	24	4.5×10^{-4}	—	—
17–20 October	Lake Ontario 12	25–50	72	3.7×10^{-6}	4.5	0.2

Profiles of the dye concentration and the temperature structure obtained with the fluorometer going down through the layer are used to interpret the data. When the instrument returns up through the layer the recorded profile is disturbed since the main body of the instrument passes through the layer before the sensors.

Although different kinds of distributions were observed in the experiments, they were very similar in nature. In every experiment pulse-formed layers with very sharp boundaries and almost homogeneous distribution were found. The thickness of this particular type of layer was virtually constant over considerable periods of time, and by its form, thickness, and position in the temperature structure a specific layer can often be identified and traced with increasing diffusion time. This yields the possibility of determining the vertical mixing intensity, as described by Kullenberg (1971). In general the persistency of the layers is considerable, proving the mixing to be very weak.

Primary data from the experiments are given in internal reports (Kullenberg *et al.*, 1972; Kullenberg, 1973).

The aim of the present investigation is: (1) to study the connection between the temperature and dye distributions and to compare the observations from the various areas; (2) to interpret the observed formations of the dye layers in the light of various possible vertical mixing mechanisms.

The Tracer Distribution in Relation to the Temperature Structure

In the Sognefjord the tracing was done in the interval 17–30 m, with a duration of 10–14 hr. The overall temperature distribution showed several regions with typical small-scale structure separated by rather thick layers with almost homogeneous temperature. The mean gradients of both temperature and salinity were very weak. After the initial period, the most common position of the dye layer was in a temperature gradient layer, or in the transition between an almost homogeneous layer and a sharp gradient. The dye was apparently trapped on a sheet of locally increased gradient or in the gradient layer. Only rarely was the dye found in a homogeneous layer, trapped between sharp gradients. These dye regions seemed to have a relatively small horizontal extent. Examples of profiles (Fig. 2) demonstrate the pulse-form and the sharp boundaries, as well as the possibility of identifying layers with increasing time. The thickness of the layers varied between 5 and 60 cm, the dominant thickness being 15–20 cm. The thickest layers were found in locally weak gradient layers or in homogeneous layers. No multilayered structures were found, only a few indications of a dividing layer.

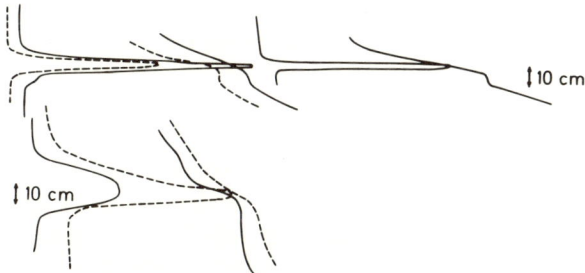

FIG. 2. Profiles observed in the Sognefjord 9–10 June around 25 m depth after, respectively, 11.7, 12.4 (- - - -), and 13.7 hr (*top*), and 12.5 and 14 (- - - -) hr of tracing. Dye trace (*left*); temperature trace (*right*), increasing to the right.

In the Baltic 30–31 August 1970 the tracer was injected 50 m below the surface in weak overall temperature and salinity gradients. The typical Baltic temperature minimum layer was situated at 50–52 m. The dye was found in this region and in the slightly more stable region immediately above. The temperature structure showed several sheets with strong gradients separated by isothermal layers. Occasionally several steps were observed. The dye was found in all types of temperatures, the dominant position being in a gradient layer. The dye sheets encountered in these regions were often trapped in a local weakening of the gradient. The most common form of the profile was pulselike with very sharp boundaries and a nearly homogeneous distribution. There were also several cases of profiles with ragged boundaries and with separated sheets or leaves of dye in the profile. These types of distributions were found in regions with no, or very weak temperature gradients, in the temperature minimum layer. The thickest layers, up to 200 cm thick, were invariably found in the homogeneous part of the temperature structure. In a multilayered temperature structure the dye was trapped in the homogeneous layers.

Even after 24 hr of tracing well defined, 50–80 cm thick layers with very sharp boundaries were detected. Profiles given in Figs. 3 and 4 demonstrate the above points.

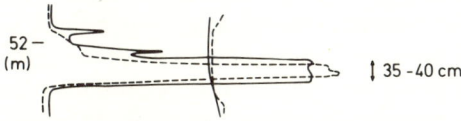

FIG. 3. Profiles observed in the Baltic 30–31 August after 8 and 21.9 (- - - -) hr of tracing. Note that depth increases upwards in all figures.

In the experiment 2–3 September 1970 in the same area, the dye was injected at 26–28 m depth in a marked overall temperature gradient (thermocline layer). The tracer was usually detected in rather thin layers situated in local features of the basic gradient, such as a step, a weakening, or a sharpening. The thinnest layers, of the order of 10 cm were found in regions with a locally increased gradient. Most commonly observed were 15–50 cm thick pulselike layers trapped in a local weakening of the gradient. Some divided layers were found, seemingly formed by a splitting of one layer. The boundaries were very sharp and the distribution often nearly homogeneous. No ragged boundaries were found as in the experiment at 50 m depth.

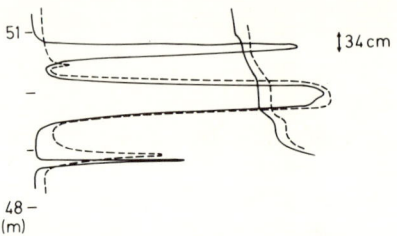

FIG. 4. Profiles observed in the Baltic 30–31 August after 12.7 and 13.5 (- - - -) hr of tracing.

In the western Mediterranean the observations were made at 25–30 m, in a weak overall thermal stratification. The temperature distribution displayed several disturbed sections interrupted by quiet regions. The tracer was mostly found in a temperature gradient layer in connection with a local weakening of the gradient, seemingly trapped there. The thickness was of the order 20–100 cm. Thicker layers, up to 200 cm thick, were found in weak gradients or homogeneous layers.

It is striking how similar the dye distributions are, although observed in widely different regions. The pulse-formed layers with an approximately homogeneous distribution are the most common ones. They are very persistent, and layers can, with a reasonable degree of certainty, be identified with increasing diffusion time. The sharp boundaries indicate the presence of shear.

The similarity of the observed distributions suggests that similar processes are responsible for mixing. The influence of the local wind during these particular experiments was either absent or very small. If the mixing is governed by local small- or micro-scale processes, the density as well as the current structure must be observed at these scales in order to arrive at a thorough understanding of the processes.

The experiments in Lake Ontario represent a step toward this goal in so far as the density structure can be determined from the temperature observations.

By and large the same type of layered structure is found in the lake. The dye layers are clearly related to the temperature distribution. The layers are sharp and often with an almost homogeneous concentration distribution (Fig.5). This holds true for all the experiments, and layers can be identified over considerable time intervals. The remarkable persistency of the layers is proof enough of the weak current and the low turbulence intensity. It should be noted that also the density stratification is considerably weaker in the lake than in the sea.

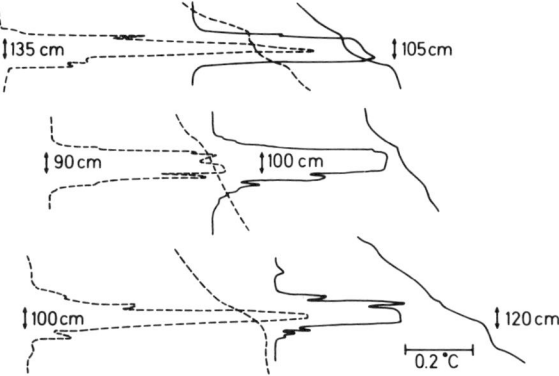

FIG. 5. Profiles observed in Lake Ontario experiment 4, around 40 m depth, after, respectively, 39 (- - - -) and 25 hr (*top*), 38 (- - - -) and 22 hr (*center*), 40.5 (- - - -) and 24.5 hr (*bottom*) of tracing.

There are, however, significant differences between the distributions found in the lake and the sea. The lake layers are generally thicker and the multi-layered structures are more common. Several distinct sheets above each other occur (Figs. 6 and 7). The microstructure is more pronounced, both as regards dye and temperature distribution. A step-formed distribution is also rather common with more or less pronounced steps. For instance, a pulse-formed profile can show small steps of the order one tenth of the overall thickness (Figs. 5 and 8), or a very well-defined symmetrical distribution can occur (Fig. 9). In several cases, concentration inversions are connected to such steps. The most remarkable step structures are those observed during experiment 12. There are several examples of both symmetrical (two-sided) and one-sided steps. The horizontal extension of such layers was several hundred meters. They are only found in weak temperature gradient layers and evidently a sharp gradient prevents the formation of steps (Fig. 9).

346   GUNNAR KULLENBERG

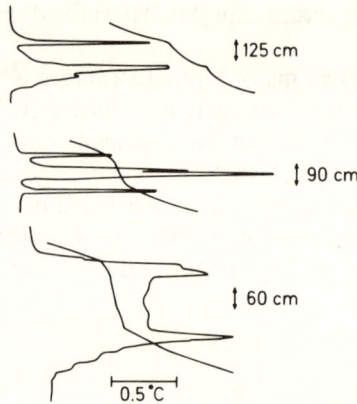

Fig. 6. Profiles observed in Lake Ontario experiment 1, around 24 m depth, after, respectively, 52.5 hr (*top*), 54.5 hr (*center*), and 73.5 hr (*bottom*) of tracing.

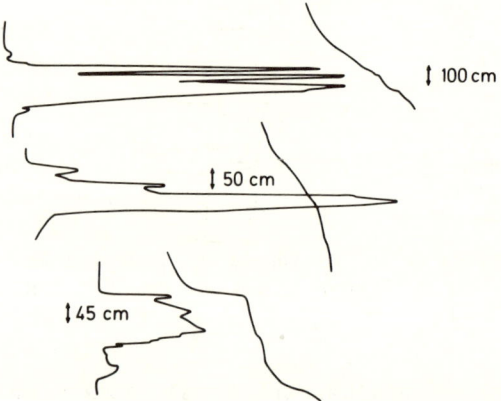

Fig. 7. Profiles observed in Lake Ontario experiment 9, around 23 m depth, after, respectively, 14.4 (*top*), 3.5 (*center*), and 16.2 (*bottom*) hr of tracing.

Interpretation of Various Layer Types

The various types of dye layers can be classified as follows:

(1) pulse-formed layers with sharp boundaries and nearly homogeneous concentration distribution,

(2) layers with ragged boundaries and/or concentration inversions,

(3) leaf structure with two or more successive thin sheets,

(4) layers with a clearly step-formed concentration distribution, either one- or two-sided.

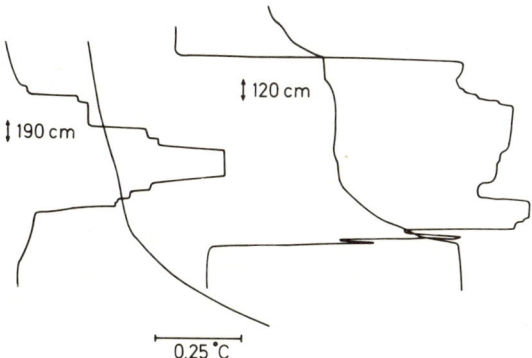

FIG. 8. Profiles observed in Lake Ontario experiment 12, after, respectively, 19.8 (*left*), and 8.8 (*right*) hr of tracing.

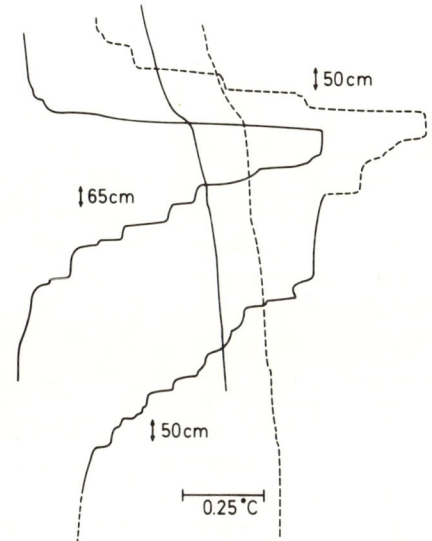

FIG. 9. Profiles observed in Lake Ontario experiment 12 after 31.5–32 hr of tracing.

It is tempting to speculate on the formation of the layers in the light of various possible generation mechanisms. Clearly the layers are related to the density and current structures. In most cases the overall stratification and current shear are considerable, the stratification being the dominating factor.

In cases of very weak stratification, it is conceivable that the density perturbation due to the tracer solution generates artificial effects. Double-diffusive phenomena could possibly be caused by a three or four part system consisting of acetic acid, methanol, dye, and heat. The differences in molecular diffusivities are present. The corresponding stability ratios (Turner, 1965) are calculated for some cases and found to be of the order 10^3. The maximum thickness of the convecting layers between successive interfaces estimated by the relation of Stern and Turner (1969) is at least two orders of magnitude less than the observed step thickness. Also Stern's (1960) criteria imply that salt-fingers cannot develop in the present conditions. These considerations show that double-diffusive phenomena can be ruled out. It is furthermore evident that the natural density variations in the water are much larger than the perturbation introduced by the tracer solution, at least after an initial period.

Possible layer-generating mechanisms are then advective processes, overturning instabilities and wave induced shear instabilities.

(1) Advective processes may generate a small-scale structure of interleafing water types reflected in the temperature–salinity distributions. There are always more or less pronounced natural horizontal inhomogeneities present in the water body which will cause advective motion possibly generating intermediate to small-scale vertical variations. A strong enough horizontal variation of temperature or salinity can give rise to horizontal layered convection as pointed out by Stern (1967) and demonstrated for instance by Wirtz et al. (1972). It is, however, not likely that advective processes are the direct cause of microstructure (Gregg and Cox, 1972).

(2) Overturning instabilities as described by Orlanski and Bryan (1969) are possible when the density distribution shows maxima or minima. An examination of the Lake Ontario records shows that the requirement is at hand in some cases, and the pictures suggest that overturning instabilities can occur.

Orlanski and Bryan neglect rotational effects, which, however, are included in Frankignoul's (1972) analysis of the stability of finite amplitude waves. He demonstrates the importance of the gravitational instability showing also that with rotation it will develop before the shear instability, except for waves of small amplitude.

Considering the records it is not likely that this is the dominating process in the present observations.

(3) The wave induced shear instability is believed to be of central importance. The model described by Woods and Wiley (1972) can explain the occurrence of interleafing structures (Figs. 6 and 7).

Recently Garrett and Munk (1972a) proposed a universal small amplitude internal wave spectrum which they applied to the mixing problem (Garrett and Munk, 1972b). In particular they consider the inertial frequency. They find that for this case, the wave-induced shear instability is much more likely to occur than overturning. This is especially the case in the presence of microstructure since the microstructure Richardson number always will be less than the overall Richardson number.

For the Lake Ontario observations the density structure can be determined with the same resolution as the dye distribution. The following discussion is therefore focused on these observations.

It seems likely that the shear instability mechanism can account for most of the layer formation. To investigate this, the Lake Ontario experiments have been examined and the shear required for the Richardson number $Ri = \frac{1}{4}$ has been calculated for a number of layers falling more or less clearly into the categories 1–4 defined above. It is found that (Table II):

(1) Pulse-formed layers with sharp boundaries are generally observed in connection with sharp density gradients where the required shear is large relative to the observed overall shear. If breaking occurs, the shear is so large that the dye outside the pulse is swept away rapidly and not detected. In this way the pulse form is maintained. The strong gradients prevent any step formation. The pulse-formed layers can be either trapped on a density sheet or between two density sheets. The overall shear in the layer between two sheets is generally large enough to create a homogeneous distribution in the layer, so that entrained clean water will be rather rapidly absorbed in the layer.

(2) The layers with ragged boundaries are situated in regions where breaking is more likely to occur. The required shear is not so large implying that its "cleaning" effect is not so pronounced, and indications of a step formation can be detected.

(3) Leaf structure, or separated sheets, are predominantly found in regions where shear instability is likely to occur. The persisting dye sheets are trapped in local disturbances in the density distribution. The dye between is more rapidly mixed due to the relatively more efficient shear in the locally weak density gradients.

(4) The well-defined, step-formed distributions are invariably connected to very weak density gradients. The shear required for breaking is correspondingly weak. The step-formed structure can be observed only because the pulsation and the shear of the horizontal current are not intense enough to destroy it. The one-sided step distributions are very significant. No steps are developed at the boundary connected to a sharp density gradient while well-marked steps are developed at the other boundary.

The above classification is somewhat simplified in the sense that layers of more complicated structure occur. Nothing definite can be said about the dominating processes in the sea. Considering the similarity of the distributions found in the sea and the lake it is likely that the same interpretation holds for the sea. It is noted that a step-formed structure like that in Fig. 9 has not been observed in the sea.

TABLE II. Characteristics of various layer types

Layer category	Figure number	Density distribution Type	Stratification $N^2 \times 10^6$ (sec^{-2})	Current shear $\times 10^3$ Required for breaking (sec^{-1})	Observed mean (sec^{-1})
1	5	Sheet	49.0	14.0	4.0
	(top right)	Layer	8.2	5.7	4.0
2	5	Sheet	18.4	8.6	4.0
	(top left)	Layer	4.9	4.4	4.0
2	5	Sheet	15.5	7.9	4.0
	(middle right)	Layer	8.2	5.7	4.0
3	6	Sheet	190	27.6	12.4
	(top)	Layer	72	17.0	12.4
3	6	Sheet	173	26.0	12.4
	(middle)	Sheet	160	25.0	12.4
		Layer	40	12.6	12.4
3	6	Sheet	180	27.0	12.4
	(bottom)	Sheet	250	32.0	12.4
		Layer	44	13.0	12.4
1 and 2	8	Sheet, shallow	33.4	11.5	4.9
	(right)	Sheet, deep	24.7	10.0	4.9
4	8	Smooth	2.1	2.9	4.1 (steps)
	(left)	Smooth	1.2	2.2	4.1 (steps)
1 and 4	9	Sheet	15.7	8.0	3.9 (no steps)
	(left)	Smooth	2.3	3.0	3.9 (steps)
4	9	Shallow	2.3	3.0	3.9 (steps)
	(right)	Deep	6.0	4.9	3.9 (steps)

Conclusions

It is concluded that (a) a contaminant injected in a subsurface layer of stratified waters will become distributed in layers related to the current and density conditions in the water; due to trapping of layers they can be very persistent; and (b) the vertical mixing rates are low; the observations favour the interpretation of the mixing as an intermittent process mainly governed by shear instabilities.

Acknowledgments

The work carried out in the sea has been supported by the Nordic University Group of physical oceanography and the Lake Ontario work by Canada Center for Inland Waters and the International Field Year for the Great Lakes. The most valuable assistance of Mr. H. Westerberg in these experiments is greatfully acknowledged, as is the excellent cooperation with the various ships' crews involved.

References

Frankignoul, C. J. (1972). Stability of finite amplitude internal waves in a shear flow. *Geophys. Fluid Dyn.* **4**, 91–99.

Garrett, C., and Munk, W. (1972a). Space-line scales of internal waves. *Geophys. Fluid Dyn.* **2**, 225–264.

Garrett, C., and Munk, W. (1972b). Oceanic mixing by breaking internal waves. *Deep-Sea Res. Oceanogr. Abstr.* **19**, 823–832.

Gregg, M. C., and Cox, C. S. (1972). The vertical microstructure of temperature and salinity. *Deep-Sea Res. Oceanogr. Abstr.* **19**, 355–376.

Kullenberg, G. (1969). Measurements of horizontal and vertical diffusion in coastal waters. *Acta Regiae Soc. Sci. Litt. Gothoburgensis, Geophys.* **2**, 52 pp.

Kullenberg, G. (1971). Vertical diffusion in shallow waters. *Tellus* **23**, 129–135.

Kullenberg, G. (1973). "Data Report on Diffusion Experiments." Inst. Phys. Oceanogr., Univ. of Copenhagen, Copenhagen.

Kullenberg, G., Murthy, C. R., Westerberg, H., and Miners, K. (1972). "An Experimental Study of Diffusion Characteristics in the Thermocline and Hypolimnion Regions of Lake Ontario," Data Rep. Can. Cent. Inland Waters,

Orlanski, I., and Bryan, K. (1969). Formations of the thermocline step structure by large-amplitude internal gravity waves. *J. Geophys. Res.* **74**, 6975–6983.

Stern, M. E. (1960). The "salt-fountain" and thermohaline convection. *Tellus* **12**(2), 172–175.

Stern, M. E. (1967). Lateral mixing of water masses. *Deep-Sea Res. Oceanogr. Abstr.* **14**, 747–753.

Stern, M. E., and Turner, J. S. (1969). Salt fingers and convecting layers. *Deep-Sea Res. Oceanogr. Abstr.* **16**, 497–511.

Turner, J. S. (1965). The coupled turbulent transport of salt and heat across a sharp density interface. *Int. J. Heat Mass Transfer* **8**, 759–767.

Wirtz, R. A., Briggs, D. G., and Chen, C. F. (1972). Physical and numerical experiments on layered convection in a density-stratified fluid. *Geophys. Fluid Dyn.* **3**, 265–288.

Woods, J. D., and Wiley, R. L. (1972). Billow turbulence and ocean microstructure. *Deep-Sea Res. Oceanogr. Abstr.* **19**, 87–121.

TURBULENT DIFFUSION OF HEAT AND MOMENTUM IN THE OCEAN

CARL H. GIBSON AND LUIS A. VEGA

*Department of Applied Mechanics and Engineering Sciences
and the Scripps Institution of Oceanography
University of California, San Diego, La Jolla, California 92037, U.S.A.*

AND

ROBERT BRUCE WILLIAMS

Nato Saclant ASW Research Center, La Spezia, Italy

1. INTRODUCTION

Radiation from the sun is absorbed in the upper surface of the oceans, changing the density of the nearby water and air in such a way that atmospheric turbulence is generated and oceanic turbulence is suppressed. Viscous dissipation rates differ by three orders of magnitude for distances a meter above and below the surface and four or five orders of magnitude for a hundred meters. Nevertheless, turbulent diffusion is crucial to many oceanic transport processes. This is suggested by the fact that the time constant L^2/D for the molecular diffusion of oxygen $(D \sim 10^{-5}$ cm^2/sec) through a meter of water is several decades, yet life abounds throughout oceans many kilometers deep.

Because oceanic turbulence is so weak, and because background noise, platform motion, and general inaccessibility is so strong, very few direct measurements of turbulence or turbulent mixing in the ocean exist. The problem is illustrated by Table I, which lists typical values of viscous dissipation ε and thermal dissipation χ encountered in laboratory, atmosphere, and ocean experiments. ε values encountered in water tunnel studies are seven orders of magnitude larger than those one would like to be able to measure in the ocean. The viscous scale in the ocean is nearly sixty times as large for this reason, but this only brings the Kolmogoroff length up to 3 mm, which is still very small for oceanographic current meters.

Temperature signal levels are equally difficult to detect in the ocean and occur at even smaller scales since the Prandtl number $\Pr \equiv \nu/D \doteq 10$ for temperature and the diffusive length is $(\nu^3/\varepsilon)^{1/4} \Pr^{-1/2}$ (Batchelor, 1959). Because millimeter spatial resolution is needed at towing speeds of order 100 cm/sec, frequency response of the temperature sensor to a kilohertz is needed to resolve the temperature fine structure fully.

Small scale, high frequency response insulated hot film anemometer

TABLE I. Viscous and thermal dissipation rates in laboratory, atmospheric, and oceanic turbulence experiments[a]

	Water tunnel[b]	Air jet[c]	Atmospheric boundary layer[d]	Ocean[e,f,g,h]
ε (cm²/sec³)	10–10^3	10^4	10^2	10^{-1}–10^{-4}
χ (°C²/sec)	10^{-3}–10^{-6}	10^{-1}	10^{-2}–10^{-4}	10^{-5}–10^{-9}

[a] $\varepsilon = 2\nu \langle e_{ij}^2 \rangle$; $e_{ij} = \frac{1}{2}(u_{i,j} + u_{j,i})$; $\chi = 2D \langle T_i^2 \rangle$; where $\mathbf{u}$ = velocity, T = temperature, ν = kinematic viscosity, and D = thermal diffusivity.
[b] Gibson et al. (1968).   [f] Nasmyth (1970).
[c] Friehe et al. (1972).   [g] Williams and Gibson (1973).
[d] Gibson et al. (1970a).   [h] Hacker (1973).
[e] Grant et al. (1962).

probes have been available for several years now and were used in the present study. They are not without difficulty, however. A recent laboratory and field evaluation (Frey and McNally, 1973) of such probes for oceanographic flow sensors concludes with the sentence, "Finally, caveat emptor." The situation is like comparing democracy to other forms of government; hot film probes are the worst possible sensors for oceanic turbulence, except for all others that have been tried.[1]

The earliest turbulence and temperature measurements in the ocean were carried out by a Canadian group working in a tidal channel (Grant et al., 1962) and open ocean (Stewart and Grant, 1962; Grant et al., 1968a,b; Nasmyth, 1970) with film and thermistor sensors towed by a ship and mounted on a submarine. Values of ε and χ as small as 10^{-4} cm²/sec³ and 10^{-7} °C²/sec were reported.

Kolesnikov (1960) describes a "turbulimeter" developed in 1955 which was apparently used for millimeter scale measurements of turbulence and temperature fluctuations in the ocean and under ice. Few details are given of the method of measurement or results. Ozmidov and his group (Ozmidov and Belyaev, 1972; see also Ozmidov, 1971) have towed hot film anemometer and conductivity probes to measure small scale turbulence and temperature fluctuations in the Atlantic and Indian Oceans on a number of cruises since 1969. Few details are given of the instrumentation or results, except that generally turbulence was found to exist only on scales less than a meter and with marginal shear Reynolds number.

In the present paper some results are presented from experiments with instruments towed in the Cromwell current at about 100 m depth, su-

---

[1] The hydroresistance anemometer probe recently developed by the Russians may prove to be the exception (Vorobjev et al., 1974).

spended at 20–120 m in the surface layer from the Scripps stable platform FLIP and towed in a mixed layer at about 20 m depth off San Diego. In part, the experiments were intended to develop reliable instruments and techniques for turbulence measurements at sea, as well as make the measurements themselves.

2. Measurement Techniques

2.1. Towing Methods

A major source of false velocity signal for a sensor towed by a ship is the surface motion transmitted through the suspending cable. Various methods might be used to reduce the amount of surface motion transmitted to the towed body such as a servo winch controlled by a pressure sensor on the body to maintain constant depth (Nasmyth, 1970), using springs or elastic material on the cable to act as a low pass filter of ship's motion, or by attaching lifting and/or suppressing bodies to the cable. Ozmidov describes an arrangement with the sensors mounted above a depressor and below a lifting body designed to keep the package on a vertical cable at constant depth (Ozmidov and Belyaev, 1972; see also Ozmidov, 1971).

The method employed for the present towed measurements was to take advantage of the fact that the ship induced motion is at relatively low frequencies compared to that of the smallest scale turbulence and temperature fluctuations. A typical wave frequency at sea is 0.1 Hz, which is small compared to either 1 Hz, corresponding to 1 m maximum length scale of the turbulence at a towing speed of 1 m/sec, or 1000 Hz, corresponding to a viscous scale of 1 mm. A 600-lb 5-in. diameter steel cylinder filled with lead was provided with a 2×5 ft fin and upstream instrument support. This "fish" was supported by a 600-m long $\frac{3}{16}$ in. diameter double lay armored cable on a winch. Four conductors inside the cable were used to transmit information to the laboratory in the ship. Further information concerning the arrangement during the Cromwell Current measurements is given elsewhere (Williams and Gibson, 1974; Gibson and Williams, 1973). Using a vibrotron pressure sensor it was found that the vertical fish motion was in phase with the surface waves and of smaller amplitude for depths of about 100 m at towing speeds of about 1 m/sec.

2.2. Universal Similarity Theory

Estimates of the turbulence and temperature dissipation rates ε and χ were obtained by fitting the measured high frequency velocity and temperature spectra to forms predicted by universal similarity hypotheses based on Kolmogoroff and Batchelor length and time scales (Gibson, 1968). The actual shapes of the universal spectra have been determined by laboratory tests (Gibson and Schwarz, 1963). Because of the various features and interdependence of the normalized temperature and velocity spectra, it is possible to make a variety of tests for internal consistency in attempting to extract ε and χ values.

Figure 1 illustrates the methods used to evaluate ε and χ from measured velocity spectra Φ_u and temperature spectra Φ_T. Frequency f is converted to wave number k by assuming frozen velocity and temperature patterns and setting $k = 2\pi f/U$, where U is the mean velocity measured on the fish with a ducted current meter. The smallest wave number portions of the spectra will be affected by either ship's motion or buoyancy, depending on how fast the

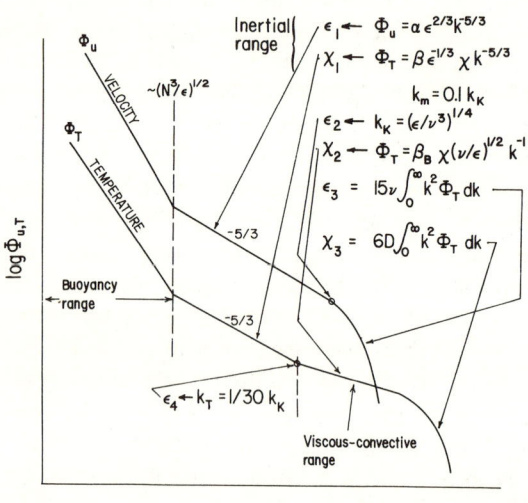

FIG. 1. Use of universal spectra to estimate ε_i and χ_j.

sensors are towed. The higher the towing speed, the better the low wave number portion of the spectrum is determined, possibly at the cost of having inadequate frequency response to resolve the high wave number part of the spectrum. Buoyancy effects are expected to occur at a wave number, proportional to $k_R \equiv 2\pi(N^3/\varepsilon)^{1/2}$, where buoyancy forces approach inertial forces of an inertial-subrange velocity spectrum in a stably stratified fluid with Vaisala frequency $N = [(g/\rho)(\partial\rho/\partial z)]^{1/2}$, where g is gravitational acceleration, ρ is density and z is the vertical direction. The measurements in Gibson and Williams (1973) indicated the proportionality constant was about one (see Table II below).

TABLE II. Turbulence and turbulent mixing parameters measured in the Cromwell current, *Aries IV* expedition, April 1971, 150°W[a]

	1°N	0°N	Units
ε	0.08	0.08	cm^2 sec^{-3}
χ	8 × 10^{-6}	7 × 10^{-5}	°C^2 sec^{-1}
N^2	6.7 × 10^{-4}	2.75 × 10^{-4}	sec^{-2}
dT/dz	2.75 × 10^{-3}	1.13 × 10^{-3}	°C cm^{-1}
v_T	12	25	cm^2 sec^{-1}
D_T	0.52	27	cm^2 sec^{-1}
Pr$_T$	22	0.91	—
L_R	0.7	1.4	meters
R_i	10	1	—
Q	−1.4 × 10^{-3}	−28 × 10^{-3}	cal cm^{-2} sec^{-1}
B	−3.6 × 10^{-4}	−7.5 × 10^{-3}	cm^2 sec^{-3}
τ	0.1	0.3	dynes cm^{-2}

[a] From Williams and Gibson (1974) and Gibson and Williams (1973).

If $\Phi_u \sim k^{-5/3}$, then ε can be calculated from $\Phi_u = \alpha \varepsilon^{2/3} k^{-5/3}$ where α is assumed to be a universal constant equal to $\frac{1}{2}$. This estimate of ε is denoted ε_1 as shown in Fig. 1. It was found in laboratory tests (Gibson and Schwarz, 1963) that the transition from the inertial to viscous subranges for Φ_u occurs at $k = 0.1(v^3/\varepsilon)^{1/4}$, from which a second estimate ε_2 can be calculated. A closely related estimate

$$\varepsilon_3 = 15v \int_0^\infty k^2 \Phi_u \, dk$$

can also be found by assuming the velocity field is locally isotropic.

If the various ε estimates converge to a reasonably consistent value, then $\Phi_T = \beta \chi \varepsilon^{1/3} k^{-5/3}$ in the inertial subrange can be used to calculate χ_1 using the universal constant β. Recent evidence (Gibson et al., 1970a) shows that β at very high Reynolds numbers may increase substantially, but for the present marginal Reynolds numbers a value of 0.4 should be satisfactory for β (Gibson and Schwarz, 1963).

It has been found (Gibson et al., 1970b) that the transition from inertial to viscous–convective subranges shown in Fig. 1 for Φ_T occurs at about $k = \frac{1}{30} k_K$, which gives an independent (although imprecise) method for estimating $\varepsilon = \varepsilon_4$. Φ_T in the viscous convective subrange is used to find χ_2 from

$\Phi_T = \beta_B \chi (\nu/\varepsilon)^{1/2} k^{-1}$ where β_B is another universal constant taken to be about 2. χ_3 is found from the assumption of local isotropy and the formula $6D \int_0^\infty k^2 \Phi_T \, dk$. Clearly ε_3 and χ_3 will be difficult to obtain since they demand very high frequency response and signal to noise ratios for velocity and temperature detection.

2.3. Electronic Circuits

A variety of circuits have been employed in our efforts to bring turbulence signals out of the noise. Casual observation of an oscilloscope trace of the anemometer output suggests there is no turbulence at all when the instrument is suspended in the mixed layer, since the velocity is generally dominated by the surface wave orbital motion. If $L_R = 1$ m and $\varepsilon = 10^{-3}$ cm^2/sec^3 we find the turbulence velocity $(\varepsilon L_R)^{1/3} = 0.2$ cm/sec compared to 10–100 cm/sec orbital velocity. Only by high pass filtering and increasing the gain would it be possible to observe such a low level signal. During the February 1972 FLIP measurements a Model 1010 Thermosystems constant temperature anemometer bridge was used with a 120-m probe cable. This arrangement was successful, although noise pickup in the long cable could not be avoided and reduced frequency response to achieve balance was required. For the November 1972 towed measurements off San Diego a constant temperature anemometer bridge was designed by Tom Deaton of our group using compact integrated circuit components which could be conveniently housed in a pressure case and submerged to reduce cable pickup. Figure 2 shows power spectra of hot film derivative signals from the microcircuit constant temperature anemometer bridge and the Model 1010 used alternately with the same probe at overheat ratio of 1.1 on the axis of the wake of a sphere in a water tunnel at Reynolds number of about 60,000. It was found that the measured power spectra and distribution function for the two circuits were identical, except that the frequency response and noise level of the microcircuit anemometer were slightly better than those of the commercial circuit.

Also included in the submerged electronics package were an ac bridge to measure small scale temperature fluctuations with hot film probes used as resistance thermometers, circuitry for the Marine Advisors ducted current meter, as well as cable driving circuitry to transmit the three signals up the four conductors with less than -80 dB cross talk; all designed by Tom Deaton. The circuit arrangement is shown in Fig. 3. The purpose of the ducted current meter was to provide continuous calibration of the hot film anemometer by matching power spectra for frequencies less than about 1.5 Hz which is the maximum response of the ducted current meter.

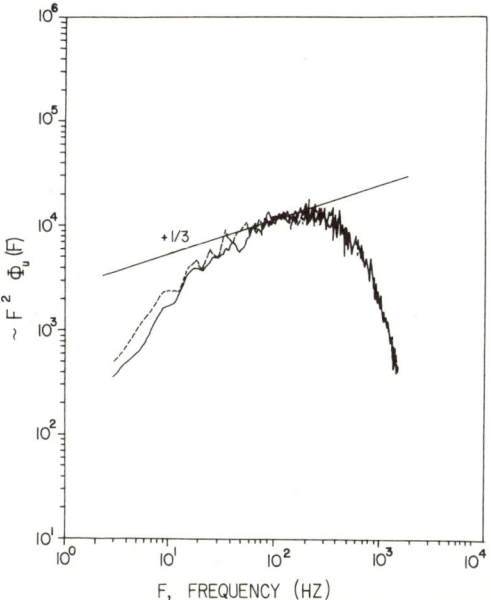

Fig. 2. Comparison of submersible microelectronic constant temperature anemometer circuit (---) and Thermosystems model 1010 velocity derivative spectra (—). Sphere wake axis nine diameters downstream, $\overline{U} = 150$ cm/sec, $D = 3.81$ cm, Re = 57,000, 0.5-mm conical wedge probe.

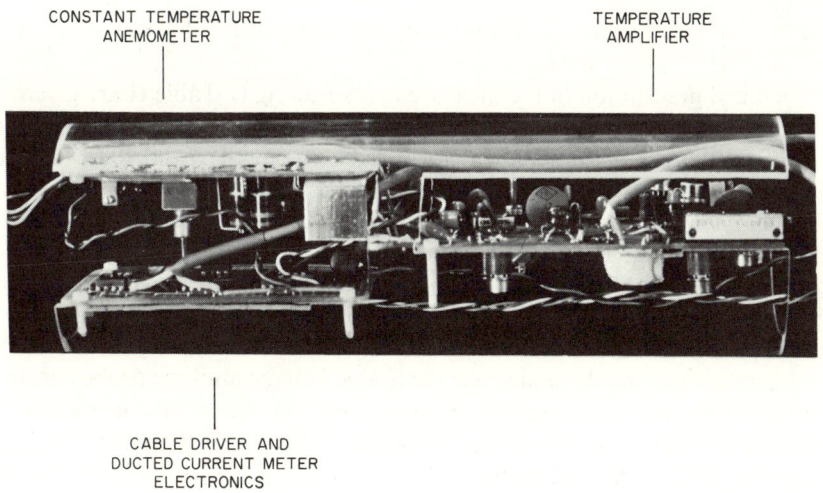

Fig. 3. Submersible microelectronic circuits for constant temperature anemometer, temperature bridge, ducted current meter, and cable telemetry. Length about 30 cm.

2.4. Sources of Ocean Turbulence

Turbulence may be generated in the ocean by a wide variety of mechanisms such as shear flow, breaking surface and internal waves, and buoyant instability due to surface cooling or internal "inversions" in the normally stably stratified water column. Less obvious mechanisms such as nonlinear wave–wave, wave–shear, and wave–stratification interactions may also occur, as well as phenomena resulting from the difference in molecular diffusivities of momentum, temperature, and salinity. One of the most challenging problems of physical oceanography will be to isolate the wide variety of such turbulence-related phenomena which no doubt exist in some parts of the diverse oceans and which will be revealed by increasingly sensitive and detailed measurements.

3. Results

3.1. Cromwell Current

The instrumented "fish" described previously was towed at the equator and at 1°N at depths just above the point of maximum velocity of the Pacific Equatorial undercurrent. Dissipation rates ε_1, ε_4, χ_1, and χ_3 were estimated from ducted current meter and cold film temperature spectra as described in Williams and Gibson (1974) and Gibson and Williams (1973). Mean velocity, temperature, and salinity profiles were determined by Taft et al. (1974) and were used to calculate the values of Vaisailla frequency N, Richardson number Ri, eddy viscosity v_T, and eddy diffusivity of sensible heat D_T shown in Table II.

Detailed description of the measurements leading to Table II are given in Williams and Gibson (1974). Substantial agreement was found between the various estimates of χ and ε, which ranged between 10^{-5} and 10^{-6} °C^2/sec for χ and were constant at 0.08 cm^2/sec^3 for ε. From these values and from the mean temperature profile it was possible to calculate v_T values of 12–25 cm^2/sec compared to 0.5–27 cm^2/sec for D_T, which are quite reasonable compared to other diffusivity estimates determined by previous oceanographic observations of mean quantities (see Table III for definition of v_T and D_T).

Attempts were made to increase the bridge voltage until the probe became velocity sensitive in a constant current mode, but these were unsuccessful because of high noise. By adjusting bridge voltage and frequency, and by experimenting with combinations of filters and differentiation of the signal before recording, taking on-line fast Fourier transform power spectra with the IBM 1800 computer on board the *Thomas Washington* it was possible to extract the rather noisy temperature spectra in Williams and Gibson (1974)

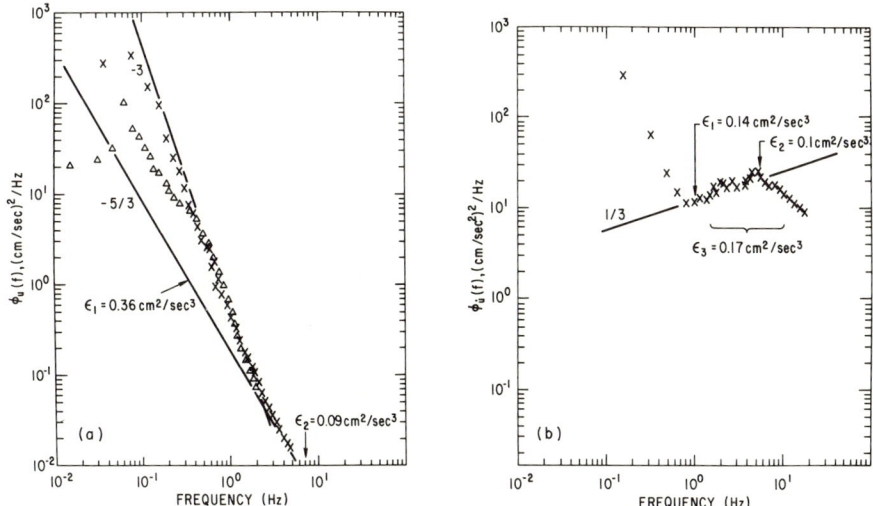

FIG. 4. Velocity spectra from FLIP. (a) Hot film (TSI 1010, 39 min) (×) calibrated by ducted current meter (52 min) (△) 2/27/72, 29 m, 7.3 cm/sec. (b) Derivative spectrum, hot film (10 min) (×) 2/28/72, 23 m, 18 cm/sec.

and Gibson and Williams (1973) from which the values ε_4, χ_1, and χ_2 could be derived. However, the temperature signal was too contaminated by harmonic noise at several frequencies to be able to say from a strip chart whether or not the signal was continuous or patchy.

3.2. FLIP Measurements

Figure 4a shows velocity spectra measured at a depth of 29 m from FLIP using the ducted current meter to calibrate the hot film probe in the TSI 1010 constant temperature anemometer bridge. Figure 4b shows a velocity derivative spectrum measured at 23 m the next day. The inertial subrange extends from about 0.9 to 5 Hz, corresponding to length scales of 20–3.6 cm from the mean velocity of 18 cm/sec observed. As indicated, ε_1 derived from the $k^{-5/3}$ region was 0.14 cm^2/sec^3, which was in good agreement with $\varepsilon_2 = 0.11$ cm^2/sec^3 computed from the viscous cut-off frequency. ε_3 for the same data was found to be 0.17 cm^2/sec^3, giving excellent, although probably fortuitous agreement between these techniques. The low frequency limit on the inertial subrange of 20 cm is probably determined by the wave frequency rather than the buoyancy length L_R due to the low mean velocity value, which brings the surface wave frequency of about 0.15 sec^{-1} close to the frequency of the viscous cut-off. The buoyancy length $L_R = (\varepsilon/N^3)^{1/2}$

was 370 cm at 23 m according to the temperature–depth profile obtained from an XBT taken at the time of the measurement.

Figure 5 shows a typical cold film temperature spectrum measured using the same bridge arrangement as in the Cromwell current experiment, except with a low impedance cable only 150 m long rather than the 600-m armor cable. A Princeton Applied Research HR 124 lock-in amplifier was used as a phase sensitive detector in the ac bridge. The spectrum in Fig. 5 shows a

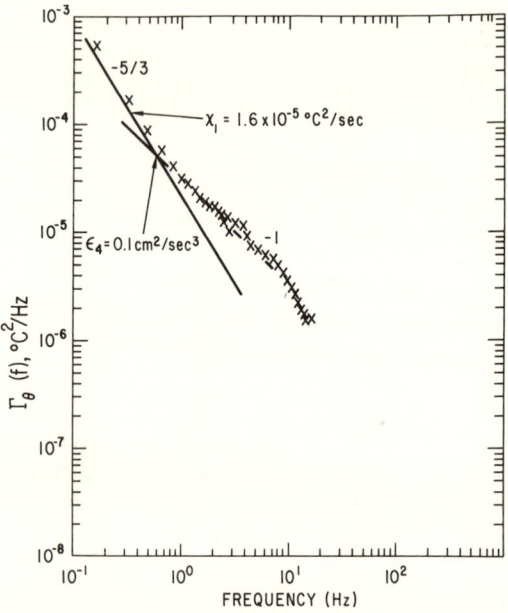

FIG. 5. Temperature spectrum from FLIP 2/27/72. Cold film (20 min) (×). Depth 120 m, $\bar{U} = 7.2$ cm/sec.

well-defined viscous convective subrange $\sim f^{-1}$ from an inertial subrange $\sim f^{-5/3}$. A χ value of 1.6×10^{-5} °C^2/sec is indicated as well as ε_4 of 0.1 cm^2/sec^3. These values may be somewhat high since the low frequency portion of the spectrum might be affected by surface wave motion at the low mean velocity of 7.2 cm/sec. The value of ε_1 for the ducted current meter was only 0.04 cm^2/sec versus $\varepsilon_4 = 0.1$. Reducing the $f^{-5/3}$ spectral level to a value corresponding to $\varepsilon_4 = 0.04$ only reduces χ to a value of 1.3×10^{-5} °C^2/sec, however.

Table III summarizes and defines various parameters determined in the FLIP experiments of 2/27/1972 and 2/28/1972. As indicated, diffusivity parameters v_T and D_T were both about 100 cm^2/sec for 23 m on 2/28 and

TABLE III. Estimates of turbulence parameters from FLIP

Date	Depth (m)	ε_1 (cm²/sec³)[a]	ε_2 (cm²/sec³)[b]	ε_3 (cm²/sec³)[c]	ε_4 (cm²/sec³)[d]	χ (°C²/sec)[e]	N (sec⁻¹)[f]	L_R (m)[g]	ν_T (cm²/sec)[h]	D_T (cm²/sec)[i]
2/27/72	29	0.36								
	120	0.04	0.09							
	120				0.1	1.62×10^{-5}	7.4×10^{-3}	2.7	55	66
2/28/72	23	0.14	0.11							
	23	0.14	0.1	0.17						
	23	0.14	0.03							
	23				0.4	3×10^{-5}	8.5×10^{-3}	3.7	100	94

[a] $\varepsilon_1 \equiv (2\pi/\overline{U})(1/\mathscr{A})^{3/2}\phi_u(f)^{3/2}f^{5/2}$; $\mathscr{A} = 0.5$.
[b] $\varepsilon_2 \equiv \nu^3(2\pi/\overline{U})^4(10f_u)^4$; $f_u \equiv$ frequency cutoff in velocity spectrum; $\nu = 1.2 \times 10^{-2}$ cm²/sec.
[c] $\varepsilon_3 \equiv 15\nu(\overline{u'^2}/\overline{U}^2)$; $\overline{u'^2} \equiv \int_0^\infty \phi_u(f)\,df$.
[d] $\varepsilon_4 \equiv \nu^3(2\pi/\overline{U})^4(30f_\theta)^4$; $f_\theta \equiv$ transition frequency from inertial to viscous convective subrange.
[e] $\chi \equiv \beta_K^{-1}\varepsilon^{1/3}(2\pi/\overline{U})^{2/3}T_\theta(f)f^{5/3}$; $\beta_K = 0.35$.
[f] $N \equiv [(g/\rho_0)|(\partial\rho/\partial z)|]^{1/2}$; $\rho \doteq \rho_0(1 - \gamma T)$.
[g] $L_R \equiv (\varepsilon/N^3)^{1/2}$.
[h] $\nu_T \equiv 0.1\varepsilon/N^2$.
[i] $D_T \equiv \frac{1}{2}[\chi/(\partial T/\partial z)^2]$.

55 cm²/sec at 120 m on 2/27. ε seemed to decrease with depth on 2/27 by about a factor of 10 from 29 to 120 m. χ was twice as large at 23 m on 2/28 than at 120 m on 2/27. Low values of ε_2 were found at 29 and 23 m on 2/27 and 2/28, but are probably low due to reduced frequency response of hot film probes fouled by plankton. Generally, the data seem reasonably consistent with what could be expected in the upper layer of the ocean. Again, however, noise levels were sufficiently high that strip chart records did not give a satisfactory definition of continuous or patchy turbulence.

3.3. Oconostota—Mixed Layer

The same "fish" used in the Cromwell current measurements was towed in November 1972 from the Scripps' ship *Oconostota* at several depths from 20 to 50 m in the mixed layer and thermocline a few miles off San Diego. As indicated above, the velocity and temperature electronics had been completely redesigned and packaged in a pressure case attached to the fish, and the various signals were sensed, amplified, prewhitened, isolated, and transmitted up the four-conductor armored cable to the tape recorder in the ship laboratory.

A strip chart of temperature, velocity, and their derivatives is shown in Fig.6, along with the corresponding XBT temperature versus depth profile. The instruments were towed at a depth of 18.3 m at 130 cm/sec with the ship moving with the waves. The XBT temperature profile shows $T = 16.90 \pm 0.05°C$ to about 25 m, so the data correspond to a horizontal sample three-quarters of the way down in a well-mixed surface layer. From the upper trace of Fig. 6 it can be seen that the temperature derivative signal $\dot{\theta}'$ was not continuous but occurred in bursts which are closely correlated with the periodic variations in the hot film anemometer output shown in the bottom trace. A similar correlation exists for the velocity derivative signal $\dot{U}'$ although the latter is more noisy and is contaminated by plankton "hits." Periodic variations in the velocity past the fish may be induced through the cable by the ship or may exist in the water, but in both cases should be in phase with the surface waves. The magnitude of the variations of 10–20 cm/sec is consistent with the expected amplitude of the wave orbital velocity at the depth of towing and the period of 5–10 sec is consistent with the observed period of the surface waves, rather than half the period which would be expected if the periodicity was due to heaving of the ship. Horizontal velocity variations of the ship due to the waves are generally rather poorly transmitted to towed bodies, but in any event should be in phase with the package motions since the wire angles were less than 5°. Therefore it seems reasonable to assume that most of the motion indicated by the lower trace in Fig. 6 is in phase with the surface waves and probably due to the orbital velocity rather than ship motion.

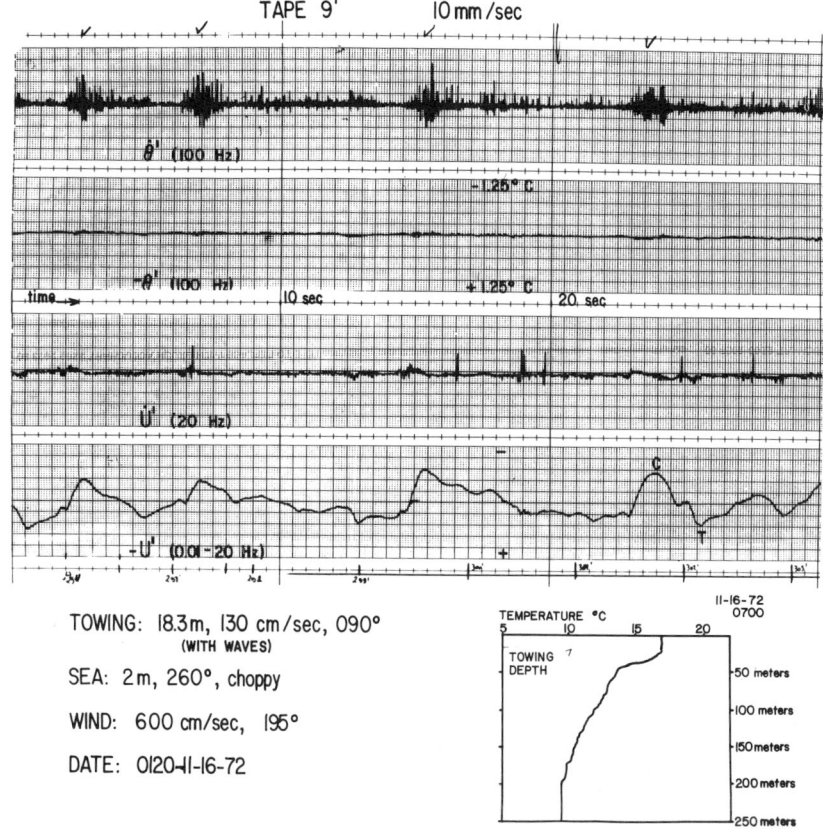

Fig. 6. Temperature and velocity in mixed layer towing from *Oconostota* 11/16/72. Depth 18.3 m, $\bar{U} = 130$ cm/sec, $\dot{T}$, T, $\dot{V}$, V (filter frequencies). C = crest, T = trough.

With this interpretation we can identify the location of maximum velocity and temperature derivative activity at a position under the wave crests, beginning at some point between the downwave trough and the crest.

Interpreting the derivative spectrum of the velocity signals shown in Fig. 7 as turbulence indicates a narrow range of turbulence scales, with transition to $k^{-5/3}$ from wave lengths of 20–30 cm and smaller (compared to 100–200 cm for the turbulence in the Cromwell current) and with ε_2 estimate of about 0.08 cm^2/sec^3. χ from the temperature spectrum was about 10^{-5} °C^2/sec. The indicated maximum length scales of the turbulence is small compared with the buoyancy length $L_R = (\varepsilon/N^3)^{1/2}$, which is about 200 cm based on the mean temperature gradient to 50 m and much larger based on 18.3-m depth. Therefore it would seem that if the velocity and temperature signals are indeed turbulence, then their maximum length scales are not limited by buoyancy. It is interesting to note that the turbulent energy scale indicated by Fig. 7 is close to the orbital diameter of the surface waves at this depth, which should be about 10–30 cm.

Having discussed the possibility that the signals observed are turbulence, it is well to point out that these results are quite preliminary and should be interpreted with caution. It would appear that the turbulence and temperature fluctuations are generated in some way by the surface waves at depths of nearly 20 m. No established mechanism is known to the authors by which surface waves can generate turbulence at such depths. Even though the maximum length scales are small, the corresponding time scales of the

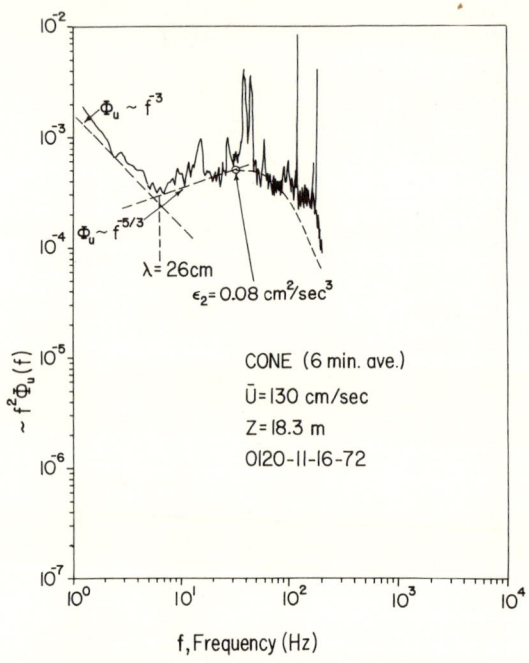

FIG. 7. Velocity derivative spectrum (conditions same as Fig. 6).

"turbulence" would seem to be too large. It would seem that whatever unknown mechanism is generating the "turbulence" may also be needed to damp it out. Other explanations of the observed signals would seem to be noise induced by the package motion or a periodic turbulence signal generated by the package moving up and down through a turbulence "layer." The latter explanation is not consistent with observation that the "turbulence" signal is generally periodic with the wave frequency rather than twice the wave frequency which would be expected if the package were moving in and out of a layer. Maximum turbulence under the wave crests would imply that the turbulence layer is grazed at the top of the package motion cycle, which seems improbable.

The former explanation, namely that the "turbulence" is simply noise induced by the package motion, seemed the most likely at first. However, there is another set of published temperature and velocity derivative data taken in a mixed layer where the "package" was a submarine and would not be expected to have much motion. Figure 8 shows a strip chart of data recorded at 15-m depth in a well-mixed surface layer 40 m thick by Grant et al. (1968a). The sinusoidal velocity signal is described in the paper as due to the surface waves. The velocity derivative signal is described as "continuous

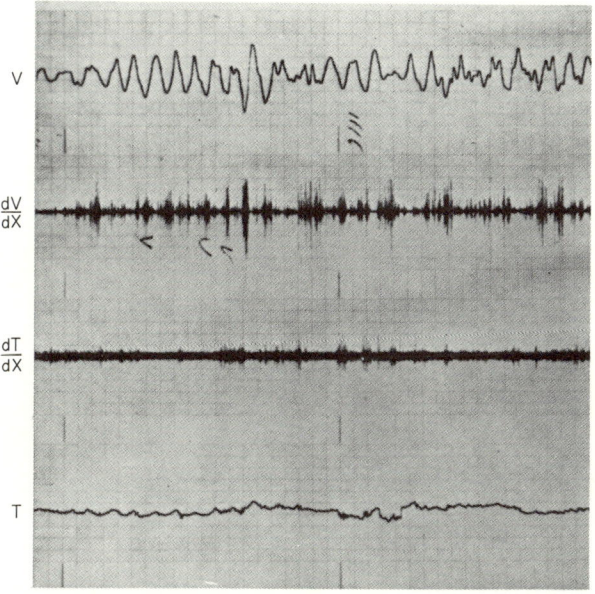

FIG. 8. Velocity and temperature in mixed layer from submarine (Grant et al., 1968a) V, $\dot{V}$, $\dot{T}$, T. Depth 15 ms, $\bar{U} = 100$ cm/sec.

turbulence"; however, close examination reveals very much the same correlation with the surface waves shown by the data of Fig. 6. Usually when the velocity is a minimum (presumably under a crest) the "turbulence" turns on, and somehow turns off before the next wave arrives. Generally there is only one burst of turbulence per wave period, which makes the hypothesis of periodic motion through a layer unlikely. Some of the bursts are marked by a "v," which H. L. Grant (personal communication) describes as associated with vibration noise detected by listening to the velocity signal with earphones. However, not all bursts were associated with such vibration, and it is conceivable that the vibration might be accompanied by turbulence. Unfortunately the data were obtained as the submarine cruised parallel to the wave crests, so it is not possible to tell whether the turbulence signals occurred under the crests or troughs of the surface waves, to compare with the indicated phase dependence observed from the present data. It should also be mentioned that similar bursts of temperature and/or velocity activity were observed with the *Oconostota* moving into the waves where the point of maximum activity occurred where the velocity was maximum; again indicating the activity was under the wave crests. Further experiments are clearly needed to better document this surprising phenomena.

4. Generation of Turbulence by Surface Waves

One mechanism that might generate turbulence in phase with surface waves is illustrated in Fig. 9, which shows streamlines in the wind and water

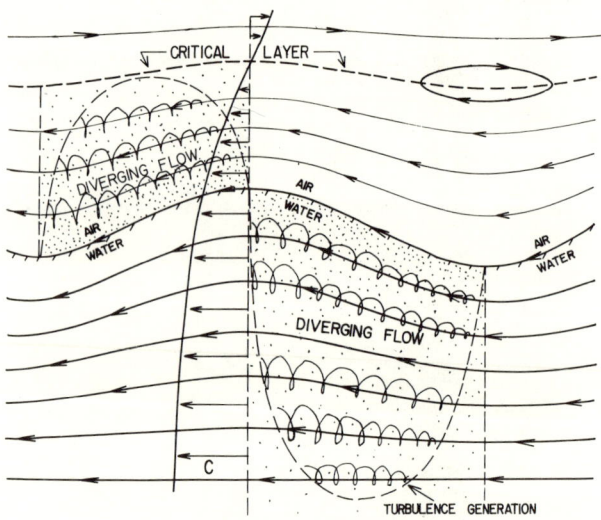

FIG. 9. Wave induced turbulence model. Coordinate system moving with wave phase velocity C.

in a coordinate system moving with the phase velocity C of the wave. As indicated by Fig. 9, regions exist in the wind and water where the streamlines diverge and converge. It is well known that diverging flows tend to be unstable and that converging flows tend to be stable; e.g., see the solution for diverging and converging flow between two planes by Hamel (1916) described by Landau and Lifshitz (1959). It is shown that the criterion for instability is a Reynolds number based on the volume flux per unit length divided by the kinematic viscosity. The Reynolds number must exceed $18.8/\alpha$ where α is the angle of divergence.

Estimating the Reynolds number of the present flow as VL/v, where V is the orbital velocity and L is the diameter of the orbit gives $R \sim 10^5$. If $\alpha = L/\lambda$, where λ is the wavelength of the wave, $R_{\text{critical}} \sim 10^4$ which shows that the mechanism is consistent with the apparent observation of marginally transitional turbulence. It also indicates the difficulty of ever observing such a phenomenon in laboratory waves.

Other mechanisms which could function include interaction of the surface waves with current shear, internal waves or variation in the local density. Since little is known of these phenomena for the present measurements, and since the parameters characterizing these mechanisms are not clear, little more can be said about them.

5. Summary and Conclusions

Attempts have been described to measure turbulence and turbulent temperature in the ocean and to relate the results to the vertical diffusivity of momentum and heat. Using millimeter scale sensors it was possible to estimate power spectra of velocity and temperature fields to high wave numbers in the viscous and thermal diffusive ranges, and to make use of universal spectrum functions to estimate the normalizing parameters of universal similarity theories; namely, the viscous and diffusive dissipation rates ε and χ. Consistency between different ε_i and χ_j values necessary to fit various i, j portions of the measured u and T spectra was found in some cases, and is taken as an indication that the fluid is indeed turbulent, and characterized by the $\varepsilon_i \to \varepsilon$ and $\chi_j \to \chi$ values that permit the fit. It was found that turbulent diffusivity parameters v_T and D_T estimated from ε and χ in the Cromwell current were consistent with those estimated by independent methods. The ratio v_T/D_T was dependent on the Richardson number, ranging from 1 to 25.

Measurements of turbulence in the mixed layer have not been fully analysed, but seem to show evidence of a small scale turbulence generation mechanism in phase with the surface waves. Similar evidence apparently exists in the data published by Grant et al. (1968a) although in both cases the possibility of an alternative explanation of the effect exists and further measurements under more controlled conditions are called for.

Acknowledgments

The authors are grateful for the assistance and advice of a number of their colleagues at UCSD in the course of this work, particularly John Clay and Carl Friehe who participated in both the FLIP and *Oconostota* measurements.

This research was supported by ONR Contract N000-4-69-A-0200-6006 and N000-14-69-A-0200-6002 and partially by the Advanced Research Projects Agency of the Department of Defense monitored by the U.S. Army Research Office-Durham under Contract DAHCO4-72-C-0037 and by the Advanced Research Project Agency of the Department of Defense monitored by ONR under Contract N000-14-69-A-0200-6039.

References

Batchelor, G. K. (1959). Small-scale variation of convected quantities like temperature in turbulent fluid. *J. Fluid Mech.* **5**, 113–139.

Frey, H. R., and McNally, G. J. (1973). Limitations of conical hot platinum film probes as oceanographic flow sensors. *J. Geophys. Res.* **78**, 1449–1461.

Friehe, C. A., Van Atta, C. W., and Gibson, C. H. (1972). Jet turbulence: Dissipation rate measurements and correlations. *AGARD Conf. Proc.* **93**, Pap. No. 18.1-7.

Gibson, C. H. (1968). Fine structure of scalar fields mixed by turbulence. II: Spectral theory. *Phys. Fluids* **11**, 2316–2327.

Gibson, C. H., and Williams, R. B. (1973). Measurements of turbulence and turbulent mixing in the Pacific equatorial undercurrent. *In* "Oceanography of the South Pacific 1972" (R. Frazer, comp.), pp. 19–23. New Zealand Nat. Comm. for UNESCO, Wellington, 1973.

Gibson, C. H., and Schwarz, W. H. (1963). The universal equilibrium spectra of turbulent velocity and scalar fields. *J. Fluid Mech.* **16**, 365–384.

Gibson, C. H., Chen, C. C., and Lin, S. C. (1968). Measurements of turbulent velocity and temperature fluctuations in the wake of a sphere. *AIAA J.* **6**, 642–649.

Gibson, C. H., Stegen, G. R., and Williams, R. B. (1970a). Statistics of the fine structure of turbulent velocity and temperature fields at high Reynolds number. *J. Fluid Mech.* **41**, 153–167.

Gibson, C. H., Lyon, R. R., and Hirschsohn, I. (1970b). Reaction product fluctuations in a sphere wake. *AIAA J.* **8**, 1859–1863.

Grant, H. L., Stewart, R. W., and Moilliet, A. (1962). Turbulence spectra from a tidal channel. *J. Fluid Mech.* **12**, 241–268.

Grant, H. L., Moilliet, A., and Vogel, W. M. (1968a). The spectrum of temperature fluctuations in turbulent flow. *J. Fluid Mech.* **34**, 423–442.

Grant, H. L., Hughes, B. S., Vogel, W. M., and Moilliet, A. (1968b). Some observations of the occurrence of turbulence in and above the thermocline. *J. Fluid Mech.* **34**, 443–448.

Hacker, P. (1973). The mixing of heat deduced from temperature fine structure mass in the Pacific ocean and Lake Tahoe. Ph.D. Thesis, Univ. of California, San Diego.

Kolesnikov, A. G. (1960). Vertical turbulent exchange in a stably stratified sea. *Izv. Akad. Nauk SSSR, Ser. Geofiz.* **II**, 1614–1623.

Landau, L. D., and Lifshitz, E. M. (1959). "Fluid Mechanics," pp. 81–86. Pergamon, Oxford.

Nasmyth, P. W. (1970). Oceanic turbulence. Ph.D. Thesis, Phys. Dep., Univ. of British Columbia,

Ozmidov, R. V. (1971). *Akad. Nauk CCCP* **9**, 60–66.

Ozmidov, R. V., and Belyaev, V. S. (1972). Some peculiarities of turbulence in the stratified ocean. *Int. Symp. Stratified Flows, Novosibirsk, USSR*, pp. 3–9.

Stewart, R. W., and Grant, H. L. (1962). Determination of the rate of dissipation of turbulent energy near the sea surface in the presence of waves. *J. Geophys. Res.* **67**, 3177–3180.

Taft, B. A., Hickey, B. M., Wunsch, C., and Baker, D. J., Jr. (1974). The equatorial undercurrent and deeper flows at 150°W in the central Pacific.

Vorobjev, V. P., Kuznetsov, E. T., Pelevich, L. C., Paka, V. T., (1974). *Proc. IAPSO–IAMAP Congr., Melbourne, January 1974*.

Williams, R. B., and Gibson, C. H. (1974). Direct measurements of turbulence in the Pacific equatorial undercurrent. *J. Phys. Oceanogr.* **4**, 104–108.

TURBULENT DIFFUSION AND BEACH DEPOSITION OF FLOATING POLLUTANTS[1]

G. T. CSANADY[2]

Department of Mechanical Engineering, University of Waterloo, Waterloo, Canada

I. INTRODUCTION

Some pollutants float on the free surface of the sea; the practically most important example is perhaps oil, which may be released, e.g., by a shipwrecked tanker. In the Great Lakes, large numbers of alewives (*Alosa pseudoharengus*) are regularly killed by thermal shock early in the season (Stanley, 1969). Masses of these small dead fish (weighing about 25 gm each) may often be observed floating on the lake surface. They are subsequently deposited on beaches and create a rather obnoxious nuisance.

The horizontal dispersion of buoyant pollutants exhibits certain peculiarities, on account of the fact that these are confined to a two-dimensional boundary of a three-dimensionally turbulent medium. At such a boundary, convergent and divergent zones occur randomly (Okubo, 1970). Nonbuoyant fluid wells up in divergences and dives down in convergences; floating particles are prohibited by their buoyancy from participating in vertical motions, with the result that they congregate in convergences, thin out in divergences (Csanady, 1963, 1970). The transport, turbulent diffusion, and beach deposition of such floating particles has received very little systematic attention in the literature so far.

When a moderately strong wind blows over a water surface, it generates a characteristic big-eddy field in the wind-driven current consisting of "Langmuir circulations" which are marked by long parallel lines of foam ("windrows"). Such flow structures have been the subject of a number of studies since the pioneering investigation of Langmuir (1938; for a recent review, see Scott *et al.*, 1969). The windrows are known to form at lines of confluence parallel with the wind, where the water sinks. They have characteristic spacings of several metres to several tens of metres. They collect not only foam,

---

[1] Much of this work was carried out while the author was on leave at Woods Hole Oceanographic Institution.

[2] *Present address:* Woods Hole Oceanographic Institution, Woods Hole, Massachusetts.

but any other floating material such as seagull feathers, oily substances, and alewives. Because windrows are present more often than not, it is of some interest to consider the diffusion of a cloud of buoyant particles in a horizontally infinite or semi-infinite field of windrows.

II. Experimental Evidence on Windrows

So far no satisfactory theory has been proposed to explain how the wind stress acting at the sea surface sets up a field of Langmuir circulations (Scott et al., 1969). The principal physical facts relating to windrows are, however, well known and provide sufficient background for our projected discussion of turbulent diffusion of floating particles. Figure 1 illustrates schematically the flow structure of Langmuir circulations; these may be regarded as secondary flows or big eddies in the sense of being an order of magnitude weaker

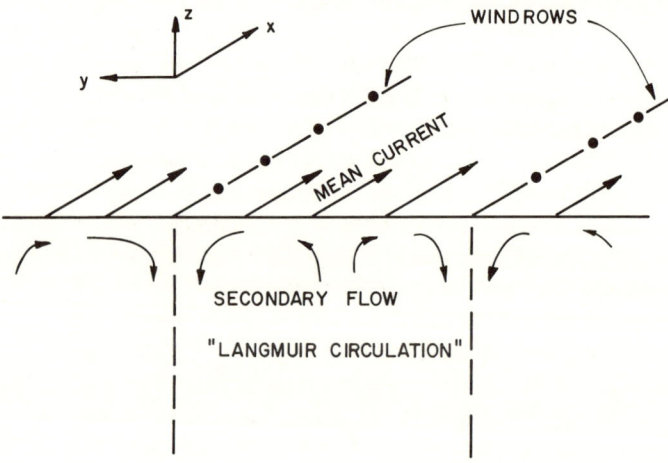

Fig. 1. Flow field of Langmuir circulations, schematic illustration.

than the main, wind-driven current. Although we do not know how precisely they are generated, Lighthill (1963, p. 99) has shown that similar big eddies are *necessary* to maintain the large mean vorticity observed near a solid surface in a turbulent boundary layer. The same argument presumably also applies to the wind-driven surface current, where large velocity gradients have also been observed in the top centimetre or so of water. McLeish (1968) has pointed out the similarity of the eddy structures in a boundary layer and at the free surface.

Dynamically, the difficult point to explain is how the longitudinal vorticity resident in Langmuir circulations is generated. Perhaps the simplest way

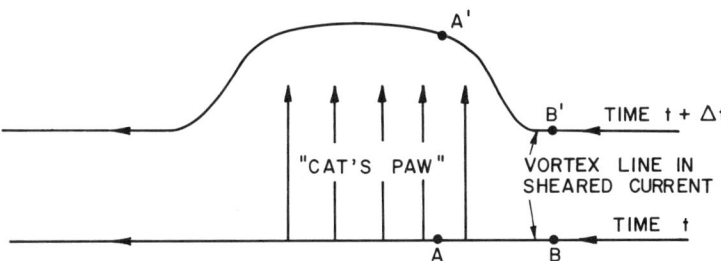

Fig. 2. Displacement of vortex lines in wind-driven surface current by local gust ("cat's paw").

in which this may happen is illustrated in Fig. 2; a low level gust accelerates a limited area of the water surface. That such localized high stresses in fact occur is illustrated by the presence of "cat's paws" on a water surface, and has been systematically demonstrated by recent work of Dorman and Mollö-Christensen (1973). The strong vortex lines of the mean wind driven current, which lie just below the free surface, are distorted by the locally applied gust. As points A and B in Fig. 2 move to A' and B', considerable longitudinal vorticity is formed. As is well known, a vortex line close to a flow boundary moves laterally [it may be imagined convected by the "image vortex" (Fig. 3) (see also the discussion in Lamb, 1932, p. 224)]. Thus a longitudinal vortex line moves away from the gust area, presumably until it meets another, opposing vortex line, generated by a neighbouring gust somewhere. As two such opposing vortices approach each other, a strong convergence line is presumably generated.

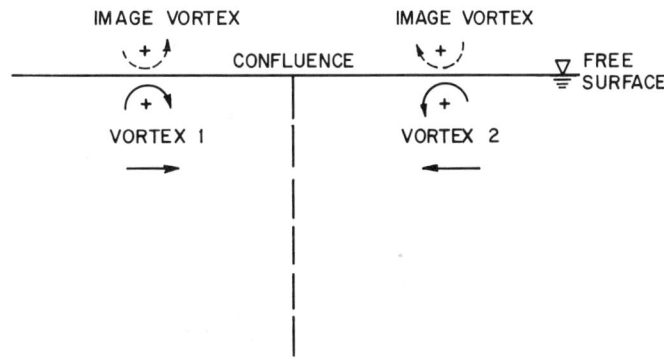

Fig. 3. Movement of vortex lines near free surface. "Image vortices" satisfy the boundary condition of zero normal flux across the surface.

Observed spacings of windrows are irregular. When a batch of floating objects are thrown overboard, they first form closely spaced, short rows, but after 10 min or so they are found in more widely spaced and well-defined, longer straight lines. These better defined lines presumably mark confluences between dominant big eddies, their spacing L ranging according to location from a few metres to a few tens of metres, say $L = O(10^3$ cm). Characteristic drift velocities u (of particles floating laterally into convergences, or of nonbuoyant particles sinking below convergences) have been observed to be a few centimetres per second. We conclude that the time scale of these big eddies is of order $T = L/u = O(10^3$ sec), and indeed individual windrows can be observed to persist for intervals of precisely that order.

During the summer of 1965 we carried out systematic windrow observations near the Baie du Dore Research Station on Lake Huron, about 5 km off shore. Five observations were taken daily (weather permitting), a total of 246 between May and July. Aluminium powder was thrown overboard to mark windrows. Because windrows are generally irregular in appearance, the observer often had doubts as to whether what he saw were windrows or not. Further details of the work are given in a report of limited circulation (Csanady, 1965). Here we give in Tables I and II the number of occasions when windrows were certainly or possibly observed, showing that the occurrence of windrows was apparently not tied to upward convective heat flux, but certainly tied to moderately strong winds. Overall, the frequency with which windrows were observed is seen to be quite high, and it would have been higher still if the fair weather bias of these data could have been avoided.

TABLE I. Number of cases of windrow observations in various categories[a]

		Windrows observed, number of occasions		
		Yes	Doubtful	No
Temperature	$t_s > t_w$	28	14	44
Difference	$t_s < t_w$	32	62	60

[a] t_w = wet bulb air, t_s = surface water temperature.

In a later season, 1968, we made observations on the diffusion of floating objects in a field of windrows (Csanady and Page, 1969). A net, some 200 m long, was spun across the current and caught drift bottles (equipped with hooks) released 200 m upstream of the net, as a "line source," parallel to the net. The time of travel from source to net was just barely enough for the

TABLE II. Number of cases of windrow observations in various categories against wind speed

		Windrows observed, number of occasions		
		Yes	Doubtful	No
Wind speed (m/sec)	0 –2.0	2	3	57
	2.1–4.0	5	25	30
	4.1–6.0	21	33	11
	6.1–8.0	22	15	5
	8.1–	15	1	0

drift-bottles to form into windrows (when these existed). Given a typical current speed of 20 cm sec^{-1}, this also verifies our previous estimate of time scale. When windrows were present, the bottles arrived at the net in well-defined batches (Fig. 4) separated by some 10 or 20 m. When they were absent, a more even distribution was observed (Fig. 5). Near the edges of the cloud the concentration of floating objects per windrow decreased, but only over 1 or 2 rows, see Fig. 6 for an example.

From the observed facts we may form the following physical picture of lateral diffusion of floating particles: as a cloud is released, it rapidly forms itself into a number of windrows. Given a typical lifetime of 10^3 secs, each windrow acts as a new source with about this frequency, redistributing its

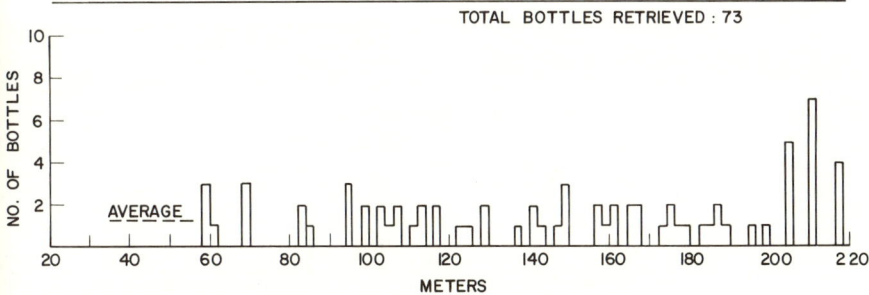

FIG. 4. Density of drift bottles in 2-m intervals, caught 200 m downstream of a "line source." Case with well-defined windrows.

EXP. NO. II DATE and HOUR: 15/7/1968 ···· 15·00
WIND SPEED: 6·5 m/sec DIRECTION: SOUTH
AIR TEMPERATURE AT 2m DRY BULB: 71·8°F WET BULB: 69·2°F
WATER SURFACE TEMPERATURE: 67·0°F
CLOUD COVER: CLEAR

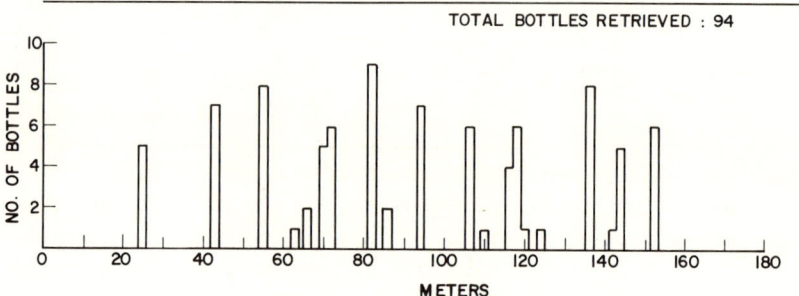

FIG. 5. As Fig. 4, case with no visible windrows.

load over a few successor windrows. The outcome over a period long compared to the lifetime of windrows is very much like a classical random walk process with step length of order L (10^3 cm) and time of step T (10^3 sec), i.e., a diffusivity of order $K = L^2/T = 10^3$ cm^2 sec^{-1}. This quantity is of the same order as the observed horizontal diffusivity of nonbuoyant dye (Csanady, 1970) and it is indeed not unreasonable to suppose that the

EXP. NO. 10 DATE and HOUR: 15/7/1968 ···· 10·45
WIND SPEED: 6 m/sec DIRECTION: SOUTH
AIR TEMPERATURE AT 2m DRY BULB: 70·4°F WET BULB: 68·5°F
WATER SURFACE TEMPERATURE: 66·4°F
CLOUD COVER: CLEAR

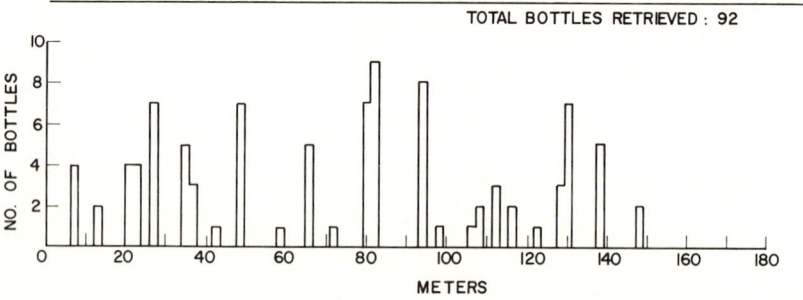

FIG. 6. As Fig. 4, case with windrows, note edge of cloud.

same big eddies which move floating particles horizontally into confluences, also govern the dispersal of marked fluid. From our point of view the main conclusion is that for floating particles we may use the classical diffusion equation in two dimensions, for phenomena of a time scale much larger than 10^3 sec, and using an estimated diffusivity of the order of 10^3 cm^2 sec^{-1}.

III. BEACH DEPOSITION OF *Alosa pseudoharengus*

While the horizontal diffusion in a field of windrows of a batch of any kind of floating objects is likely to proceed in an identical manner according to the above arguments, differences are certain to arise at the location of greatest practical interest, namely, the beaches where floating objects are washed up. Globules of oil are likely to be retained on beaches according to different laws than larger flotsam. To render the discussion more concrete, we shall here focus our attention to the beach deposition of *Alosa pseudoharengus* in the Great Lakes.

I have made a few observations over the years 1967–1973 on the deposition of alewives on a sandy beach fronting our cottage on Lake Huron. One finds larger quantities washed up at random times in May and June, but never later in the season. Thus the large scale die-off of these fish is definitely tied to the "spring" thermal regime of the Great Lakes, when a warm band of water (10°C or so) at the shores rings a cold (~ 4°C) central mass. The width of the warm band is of the order of 5 km (see, e.g., Csanady, 1972). On the Victoria day long weekend (Monday nearest 24th May) one may often observe cottagers collecting and incinerating quantities of alewives on their beaches.

On this year's Victoria day (May 21, 1973) the following could be observed. The water temperature near shore was between 10° and 12°C, but it must have been much less one or two kilometres off shore, where on the preceding day the water surface was mirror-smooth, in contrast to the rougher surface (covered by capillary waves) closer to shore. Such differences in surface appearance are caused by the stability of the air over cold water, may frequently be seen in the Great Lakes, and have been reported in other lakes by Ragotskie (1962). On the 20th May, one could see the full mirror image of boats travelling beyond the sharp boundary of the near-shore rough band of water.

About noon on the 21st a school of alewives, many dead, but some just dying, floated slowly toward the shore. There was a light onshore breeze, by which the lake surface was very slightly roughened (wave amplitude estimated at 5 cm). When within 100 m or so of the shore, the density of the alewives floating belly up was 1 per 10 square meters or so, in some places 1 per m^2. As they arrived at the shore, the alewives concentrated within a band about 1 m wide, their linear density being there about 1 per 3 m length of beach, in some places 3 per 1 m. Although the waves often placed a fish on the beach, they soon removed it again, not one of them being retained on the beach for more than a fraction of a minute. The beach was sand to fine gravel, having a slope of about 1 : 15. At the same time there was a band of dead alewives, a few days old, lying about 15 m from the shore, on land about a meter above lake level. These had evidently been deposited by larger waves, in the high linear density of 10 to 30 alewives per meter.

Those alewives floating in from the lake which were not quite dead yet could be seen to bleed and exude a jelly-like substance either at the gills or at one of their fins. They also swam in counterclockwise circles of about 30 cm diameter as if completely disoriented. Presumably, these were the effects of thermal shock which they experienced on swimming into warm coastal waters.

Later during the day the waves became somewhat higher (10 and even 15 cm amplitude), but there was still no permanent retention of fresh alewives on the beach. The entire school had by late afternoon concentrated into a nearshore band some 10 m wide (and several kilometres long, the exact extent being unknown) with a linear density of 5 alewives per meter or more. A longshore current was moving these northward at a speed of the order of 10 cm sec^{-1}.

In the evening (6 P.M. or so) the wave amplitude decreased again to a few cm. At this time all at once some alewives were permanently deposited in a density of about 1 per m, in about one hour. One may legitimately conclude that, as the waves were diminishing in height, some alewives could be left high and dry by a larger wave which was no longer repeated.

There can in fact be little doubt that the beach deposition of alewives is accomplished by wave-orbital water movements. The *retention* of each such floating object on the beach is a chance event, the probability of which must be proportional to the concentration of alewives (number per square meter χ) near the shore. The intensity I of permanent deposition (number per meter shoreline per second) also depends on the wave amplitude, and presumably on the rate at which this decreases. In analogy with other mass-transfer problems we may write

$$(1) \qquad I = v_s \chi$$

where v_s is a mass transfer coefficient or deposition velocity (cm sec^{-1}) about which we can say little. The observations just described suggest a deposition velocity of about 0.05 cm sec^{-1}, caused by such very small waves as they died out. Presumably, larger waves give rise to a much higher deposition velocity at their time of decay; the band of alewives 15 m from shore suggests $v_s = 1$ cm sec^{-1}, if the surface concentration was the same when these washed up as on the observed occasion (a hypothesis entirely without justification). Further evidence is clearly needed on this point, but we may conclude for the time being that the deposition velocity of alewives on a beach is often zero, but it reaches values of 0.1 to 1.0 cm sec^{-1} on those occasions when waves are diminishing in amplitude.

IV. A Simple Diffusion Model

We are now in a position to elucidate the role played by Langmuir circulations in the dispersal of dead alewives. Suppose that a batch of alewives experience thermal shock at a distance h from shore, of sufficient intensity to kill or at least disorient them entirely, so that they will float as passive objects thereafter. We assume that there is a shore-parallel current of constant velocity (in space and time) U. A straight shore will be taken to be the x axis, the y axis pointing into the lake.

According to earlier remarks, we may take the concentration $\chi(x, y, t)$ of alewives (per unit surface area) to be subject to the classical diffusion equation with constant diffusivity K:

$$(2) \qquad \frac{\partial \chi}{\partial t} + U \frac{\partial \chi}{\partial x} = K \left(\frac{\partial^2 \chi}{\partial x^2} + \frac{\partial^2 \chi}{\partial y^2} \right)$$

The boundary condition at $y = 0$ is given by Eq. (1), with v_s = constant and

$$(3) \qquad -I = -K \, \partial \chi / \partial y \Big|_{y=0}$$

The alewives are supposed released instantaneously at $t = 0$, $x = 0$, $y = h$. Our main interest lies in the total deposit D during the passage of a cloud, defined by

$$(4) \qquad D(x) = \int_0^\infty I(x, t) \, dt$$

The solution of this mathematical problem is well known: it is identical with the deposition of particles released from a chimney, and the solution also represents a heat conduction problem with a radiation boundary condition. Carslaw and Jaeger (1959, p. 358) give the basic solution, Wipperman (1959) its application to atmospheric diffusion. For our purposes, it is sufficient to consider the "slender plume" approximation ($y \ll x$ at all points of interest, where the concentration differs appreciably from zero). Provided that the deposition velocity is appreciable, or more precisely that

$$(5) \qquad v_s h/K > 2$$

the intensity of deposition at any instant is given by

(6) $$I = v_s \chi_0 = \frac{Qv_s}{4\pi Kt} \exp\left\{\frac{(x - Ut)^2}{4Kt}\right\} \frac{2}{1 + 2v_s t/h} \exp\left(-\frac{h^2}{4Kt}\right)$$

where Q is the number of alewives released. On substituting into Eq. (4) we find the total deposit D:

(7) $$\frac{DhU}{Qv_s} = \pi^{-1/2}\left(\frac{Uh^2}{Kx}\right)^{1/2}\left(1 + 2\frac{v_s x}{Uh}\right)^{-1} \exp\left(-\frac{Uh^2}{4Kx}\right)$$

This result shows the nondimensional deposit DhU/Qv_s to be mainly a function of the nondimensional distance variable

(8) $$\xi = Kx/Uh^2$$

The exponential term in Eq. (7) is only significant for values of ξ between about 0.05 and 1.0. At such values of ξ, the other nondimensional variable on the right of Eq. (7), $v_s x/Uh$ may be large:

(9) $$v_s x/Uh = (v_s h/K)\xi$$

The combination $v_s h/K$ already appeared in Eq. (5), and was supposed sufficiently large. Given $v_s = 0.1$ to 1.0 cm sec^{-1}, $K = 10^3$ cm^2 sec^{-1}, and a release spot of $h = 1$ to 5 km from shore, the range of this combination is $v_s h/K = 10$ to 500. The bracketed expression in Eq. (7), $(1 + 2v_s x/Uh)^{-1}$, then becomes $(1 + 1)^{-1}$ to $(1 + 1000)^{-1}$. For some of this range one may neglect 1 in comparison with $2v_s x/Uh$ and write Eq. (7) in a simplified form

(10) $$DUh^2/QK = \tfrac{1}{2}\pi^{-1/2}\xi^{-2}\exp(-1/4\xi) \qquad (v_s h/k \text{ large})$$

This relationship is illustrated in Fig. 7.

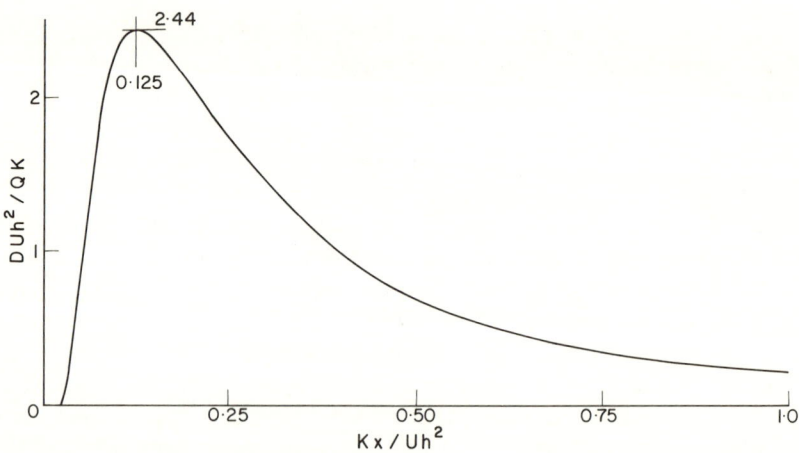

FIG. 7. Beach deposition of floating objects, released as a point source at a distance h from shore into a current of speed U.

By way of an example consider $v_s = 1$ cm sec^{-1}, $h = 5$ km; $Q = 1$ ton = 10^6 gm, $U = 20$ cm sec^{-1}, $K = 10^3$ cm^2 sec^{-1}. This yields a maximum deposit of $D_{max} = 50$ gm/km, or about 2 alewives per kilometre shoreline, at a distance of about $x = 2500$ km from the point of release, clearly a ridiculous result. It shows that diffusion alone will not cause significant beach deposition for release distances of a few kilometres: an onshore drift must bring the alewives in closer, or they must die much closer to shore if they are to be washed up anywhere. The physical reason is of course that a cloud of floating objects grows only slowly; the diffusive power of Langmuir circulations is quite modest.

References

Carslaw, H. S., and Jaeger, J. C. (1959). "Conduction of Heat in Solids," 510 pp. Oxford Univ. Press, London and New York.
Csanady, G. T. (1963). Turbulent diffusion in Lake Huron. *J. Fluid Mech.* **17**, 360–384.
Csanady, G. T. (1965). Windrow studies. "Baie du Doré Report, 1965," PR 26, pp. 60–82. Univ. of Toronto Great Lakes Inst., Toronto.
Csanady, G. T. (1970). Dispersal of effluents in the Great Lakes. *Water Res.* **4**, 79–114.
Csanady, G. T. (1972). The coastal boundary layer in Lake Ontario, Part I, the Spring regime. *J. Phys. Oceanogr.* **2**, 41–53.
Csanady, G. T., and Page, B. (1969). Windrow observations. "Dynamics and Diffusion in the Great Lakes," pp. 3–38. Dep. Mech. Eng., Univ. of Waterloo, Waterloo, Canada.
Dorman, C. E., and Möllo-Christensen, E. (1973). Observations of the structure of moving gust patterns over a water surface ("cat's paws"). *J. Phys. Oceanogr.* **3**, 120–132.
Lamb, H. (1932). "Hydrodynamics," 738 pp. Cambridge Univ. Press, London and New York.
Langmuir, I. (1938). Surface motion of water induced by wind. *Science* **87**, 119–123.
Lighthill, M. J. (1963). *In* "Laminar Boundary Layers" (L. Rosenhead, ed.). Oxford Univ. Press, London and New York.
McLeish, W. (1968). On the mechanism of wind-slick generation. *Deep-Sea Res. Oceanogr. Abstr.* **15**, 461–469.
Okubo, A. (1970). Horizontal dispersion of floatable particles in the vicinity of velocity singularities such as convergences. *Deep-Sea Res. Oceanogr. Abstr.* **17**, 445–454.
Ragotzkie, R. A. (1962). Effect of air stability on the development of wind waves on lakes. *Limnol. Oceanogr.* **7**, 248–251.
Scott, J. T., Myer, G. E., Stewart, R., and Walther, E. G. (1969). On the mechanism of Langmuir circulations and their role in epilimnion mixing. *Limnol. Oceanogr.* **14**, 493–503.
Stanley, J. G. (1969). Seasonal changes in the electrolyte metabolism in the alewife. *Alosa pseudoharengus* in Lake Michigan. *Proc. Conf. Great Lakes Res., 12th (Int. Ass. Great Lakes Res., Ann Arbor, Mich.)* pp. 91–97.
Wipperman, F. (1959). Zur Erdboden-Randbedingung bei atmosphärishen Diffusionsprozessen. *Beitr. Phys. Atmos.* **32**, 43–52.

ESTIMATION OF THE DIFFUSION COEFFICIENT OF THERMAL POLLUTION

K. Takeuchi[1]

Osaka District Meteorological Observatory, Osaka, Japan

AND

S. Ito

Japan Meteorological Association, Tokyo, Japan

1. Introduction

The thermal pollution as well as the air pollution caused by large power plants has recently become a serious problem in Japan. However, since it has been rather difficult to measure the temperature distribution of a water surface over a wide area, studies on thermal pollution have not made rapid progress.

We have recently observed the surface temperature distribution of thermally polluted water with an infrared radiation thermometer installed on an airplane. After some processing of the data thus obtained, the distribution of temperature was displayed as a picture of different coloration according to temperature. An analysis was then made of the turbulent diffusion of thermal pollution.

The case studied here may be considered typical of the diffusion from a continuous point source and the simplest kind of oceanic pollution. The aim of the present study is to find clues to the complicated problems of oceanic diffusion.

2. Instrument and Observation

The use of an infrared radiometer for measuring the surface temperature of an object is well known. Ochiai (1973) has recently used the radiometer for observing the temperature and pollution of the sea surface. In the present study, too, an airborne infrared radiometer was used to observe the temperature of a wide area of sea surface, thus saving manpower and time. The

---

[1] *Present address:* Meteorological Research Institute, Tokyo, Japan.

infrared line scanning system (Line Scanner, Model DEI-1200, made by Daeduls Co. Ltd., U.S.A.) was installed on the airplane, a Cessna 207. The system was free from undesirable movement of the plane within ±5° by the aid of a gyroscope. The main characteristics of the system are shown in Table I.

Close attention was paid in order to keep the plane on its right course with a prescheduled constant velocity. A series of flights at an altitude of 1000 m above sea level was started at about 10:00 JST, August 19, 1972. After some processing, the data were displayed like an aerial photograph of different coloration according to the energy intensity of the radiation, namely, the temperature of the sea surface.

TABLE I. Main characteristics of line scanning system of infrared radiation

Sensor	InSb
wave length	1–6 μm
filter	4.5–5.5 μm
measurable range	−10 to +50°C
sensitivity	0.1°C
Scan system	Rotary mirror type
scan rate	80 Hz
total field of view	77°20'
instantaneous angle of view	2.5 mrad
Reference source	2 controllable thermal black bodies
temperature range	−10 to +40°C
Data recorder	7-Track tape recorder
band width	DC to 100 kHz (FM)
tape speed	30 in. sec^{-1}
Gyrostabilizer	
rolling correction	±5°

After comparison with an existing precise map, some correction was made of the picture which was a little distorted due to unavoidable swing and speed change of the plane. The surface temperature distribution is depicted in Fig. 1.

For reference, meteorological data near the site at the time of the flight were as follows: weather: clear; wind: 7.2 m s^{-1} SW, temperature: 30.1°C.

As shown in Fig. 1, the x axis is taken on the sea surface along the center line of the plume of thermal pollution, with the y axis perpendicular to the x axis. The lateral spread of the plume is then obtained every 100 m from the source along the center line. Readings are tabulated in Table II, to the right and left separately.

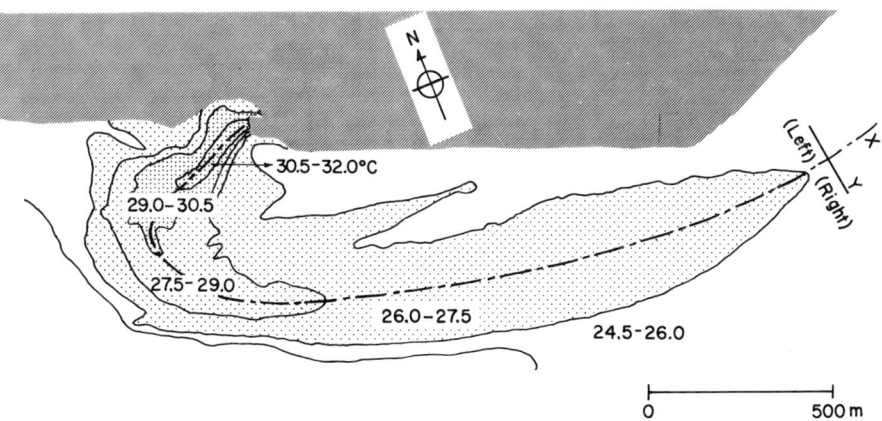

FIG. 1. Surface temperature of the thermally polluted sea. This diagram is produced from the original picture of different coloration.

3. SURFACE TEMPERATURE DISTRIBUTION OF THE SEA

It is said that the lateral spread of the plume in the sea may be expressed by a Gaussian distribution (e.g., Ozmidov, 1968; Okubo, 1970). When the difference of temperature at a point (x, y) from that far enough away is denoted by $T(x, y)$, then

(1) $$T(x, y)/T_0(x) = \exp(-y^2/2\sigma_y^2)$$

TABLE II. Spread of the plume of thermal pollution[a]

x (m)	y (left) (m)				y (right) (m)			
	Temperature (°C)				Temperature (°C)			
	26.0	27.5	29.0	30.5	30.5	29.0	27.5	26.0
100	96	60	41	27	28	60	99	—
200	173	105	79	30	21	87	122	150
300	282	158	117	—	—	100	154	190
400	345	190	60	—	—	40	107	176
600	328	107	—	—	—	—	88	165
800	275	47	—	—	—	—	45	104
1000	220	—	—	—	—	—	—	127

[a] The temperature very far from the source is 24.5°C.

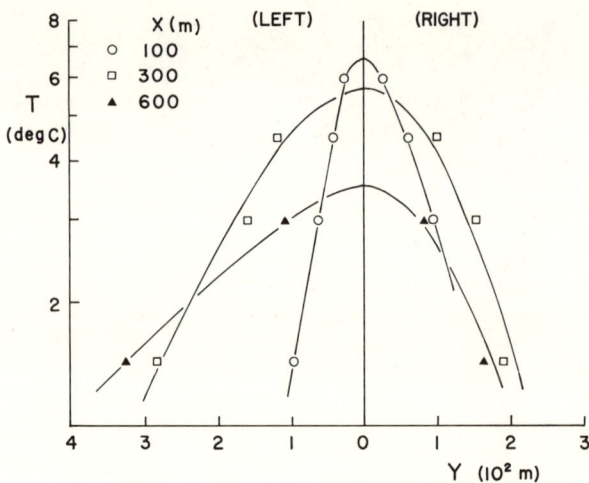

FIG. 2. Temperature difference T as a function of lateral distance from the center line y.

where σ_y^2 is the variance of the lateral spread and $T_0(x) \equiv T(x, 0)$, i.e., the temperature difference from the center line of the plume. Figures 2 and 3 show, respectively, the relationships between T and y and between T and y^2 for some x's. Then T_0 and σ_y^2 can be estimated, taking Eq. (1) into account. These are tabulated in Table III.

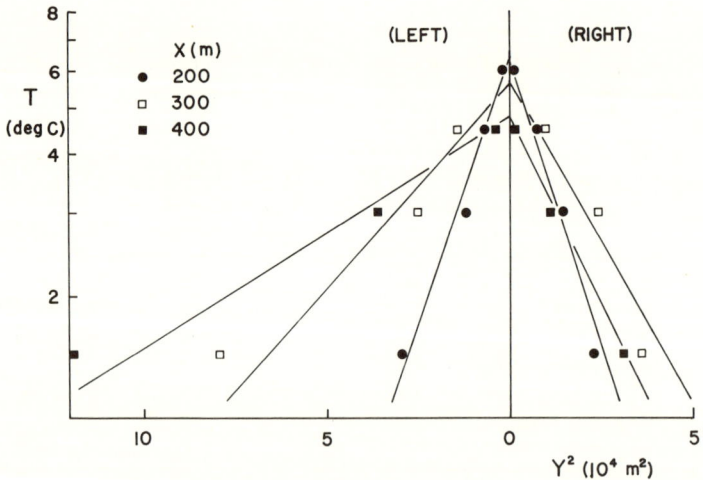

FIG. 3. Temperature difference T as a function of y^2.

TABLE III. T_0, σ_y^2, and σ_y

x (m)	T_0 (°C)	σ_y^2 (10^4 m^2)		σ_y (10^2 m)	
		Left	Right	Left	Right
100	6.6	0.26	0.54	0.51	0.74
200	6.3	0.91	0.91	0.95	0.95
300	5.7	2.47	1.60	1.57	1.26
400	4.8	4.53	1.36	2.13	1.17
600	3.6	5.44	1.60	2.33	1.26

Figure 4 shows the estimated T_0 as a function of the distance from the source x. The temperature seems to decrease almost linearly with distance at first (i.e., to about 300 m from the source). However, the decrease of temperature later on is likely to tend to be inversely proportional to the distance.

The lateral spread of the thermal pollution is plotted in nondimensional form (see Fig. 5). The figure shows that the dots are well represented by the Gaussian distribution.

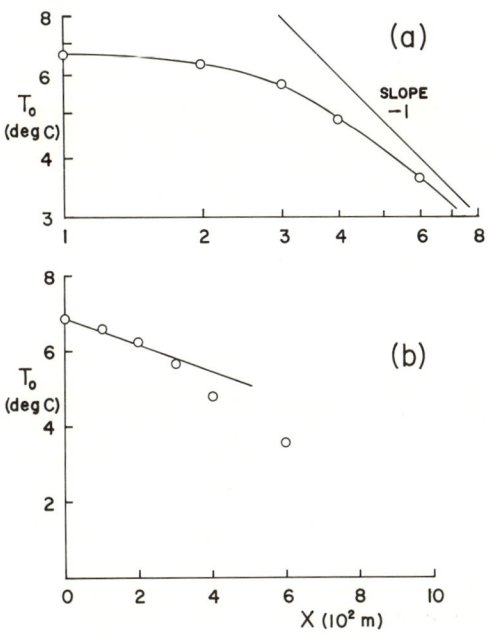

FIG. 4. Temperature difference from the center line T_0 as a function of distance from the source along the center line x: (a) logarithmic representation, (b) linear representation.

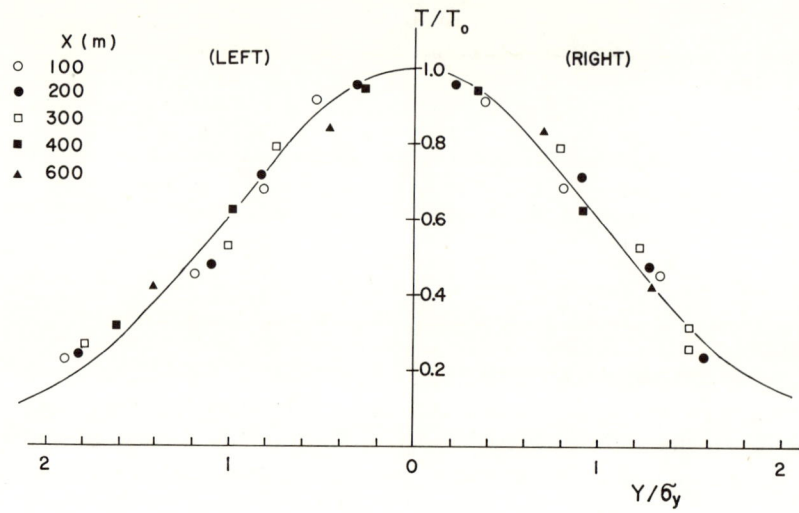

FIG. 5. T/T_0 as a function of y/σ_y. Solid line indicates $T/T_0 = \exp[-(y/\sigma_y)^2]$.

4. ESTIMATION OF DIFFUSION COEFFICIENT

The relationship between the diffusion coefficient of turbulent flow and the spread of plume has been well known since Taylor's work (1921). Namely, the lateral diffusion coefficient K_y is expressed as

(2) $$K_y = \tfrac{1}{2} d\sigma_y^2/dt$$

where t is time. When the average speed of the current is denoted by u, Eq. (2) can be written approximately:

(3) $$K_y = \tfrac{1}{2} u \, d\sigma_y^2/dx$$

Therefore, the gradient of σ_y^2 along the center line can give the diffusion coefficient. The relationship of σ_y^2 with x is depicted in Fig. 6. The coefficient is then estimated to be 10 m² s⁻¹ and 20 m² s⁻¹ respectively for the right and left with the average current speed being, respectively, taken as 0.5 m s⁻¹ and 0.3 m s⁻¹, which were obtained during the experiment.

Incidentally, the dissipation rate of turbulent energy ε is then estimated. Many studies on the turbulent diffusion have been carried out since Richardson (1926) and the following formula is now familiar:

(4) $$K_y \propto \varepsilon^{1/3} \sigma_y^{4/3}$$

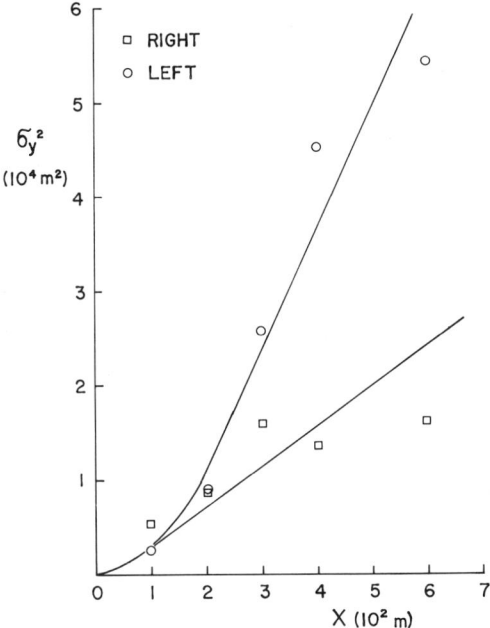

Fig. 6. σ_y^2 as a function of x.

When representative values are here adopted for K_y and σ_y, then ε can be estimated. It is found that ε is of the order of 10^{-1}–10^{-2} cm^2 s^{-3}. This is quite similar to the results obtained by Stewart and Grant (1962).

5. Results and Discussion

By the use of the airborne line scanning system of infrared radiation, the surface temperature of the thermally polluted sea was observed, saving manpower and time. Results obtained from an analysis of the data can be summarized as follows:

(1) The lateral distribution of temperature can be expressed by the Gaussian function though it is not symmetrical with respect to the center line; its nondimensional form agrees with the Gaussian distribution.

(2) The temperature along the center line near the source seems to decrease linearly with distance, but after some distance (beyond 500 m, say, from the source) it is likely to become inversely proportional to the distance.

(3) The diffusion coefficient is estimated to be of the order of 10^5 cm^2 s^{-1}, based on the relationship with the downstream gradient of lateral variance of the spread. The dissipation rate of turbulent energy is also estimated to be of the order of 10^{-1}–10^{-2} cm^2 s^{-3}.

In the present study, an airborne infrared radiometer was used for measuring the surface temperature of the sea. We have here some problems to consider. Emissivity of the sea surface and attenuation of infrared radiation during passage through the air will affect the readings of the radiometer. For instance, the infrared imagery of the sea surface polluted with pulp sewage from a factory is quite different from one not polluted, even though the temperature is not so different between them (see Ochiai, 1973). In the present experiment, therefore, the sea-surface temperature was directly measured at several points. In addition, attention was paid so that reflection of the sunshine would not spoil the readings.

The diffusion of thermal pollution is more or less affected by air temperature, wind velocity, depth of the sea, current velocity, exhaust velocity of the pollution, and other factors. However, since these were measured at only a few points in the present study, the data are not enough for analysis. This is a problem to be solved in the future.

Lastly, it should be pointed out that the plume of thermal pollution here studied is "instantaneous," and the diffusion coefficient obtained from it may be different from that averaged for some time (e.g., Ito, 1972).

References

Ito, S. (1972). The atmospheric diffusion in the horizontal direction in relation to finite sampling period. *J. Meteorol. Soc. Jap.* **50**, 59–64.

Ochiai, M. (1973). On aerial measurement of infrared imagery of sea surface with infrared line scanner. *J. Oceanogr. Soc. Jap.* **29**, 8–15. (In Jap. with Engl. abstr.).

Okubo, A. (1970). Oceanic turbulence and diffusion. *In* "Oceanic Physics" (J. Masuzawa, ed.) pp. 265–381. Tokai Univ. Press, Tokyo. (In Jap.).

Ozmidov, R. V. (1968). "Horizontal Turbulence and Turbulent Exchange in the Ocean," 199 pp. Nauka, Moscow. (In Russ.).

Richardson, L. F. (1926). Atmospheric diffusion shown on a distance-neighbour graph. *Proc. Roy. Soc. Ser. A* **110**, 709–737.

Stewart, R. W., and Grant, H. L. (1962). Determination of the rate of dissipation of turbulent energy near the sea surface in the presence of waves. *J. Geophys. Res.* **67**, 3177–3180.

Taylor, G. I. (1921). Diffusion by continuous movements. *Proc. London Math. Soc.* **20**, 196–212.

ACTIVITIES IN, AND PRELIMINARY RESULTS OF, AIR–SEA INTERACTIONS RESEARCH AT I.M.S.T.

MICHEL COANTIC AND ALEXANDRE FAVRE

Institut de Mécanique Statistique de la Turbulence
Marseille, France

1. INTRODUCTION

Continuing interest in air–sea interactions appears to be largely justified by their dominating influence upon our atmospheric and oceanic environment, as well as by the varied and basic scientific problems they raise. These interactions are, at all scales, mainly ruled by turbulent transfer processes, while they, themselves, are one of the essential agents in the generation of the mean and turbulent motions in the atmosphere and the oceans. This is especially true for the air and water layers near the interface, where small-scale momentum, heat, and mass transfers are intimately linked and establish the conditions for turbulent mixing and diffusion of natural and artificial pollutants.

Small-scale, air–sea interactions have been mainly studied, these last years, through field experiments and semi-empirical theories, laboratory work being generally restricted to the wave generation problem (see e.g., Pond, 1971). The basic idea underlying our research is to complement full-scale field work by laboratory experiments executed under controlled conditions, so that extensive, repeatable investigations can be performed wherein individual parameters can be studied separately.

The success of this approach obviously depends on the degree to which the natural processes can be usefully modelled in the laboratory facility. Preliminary studies have led to the conclusion that a partial simulation of the air–sea exchange processes could be obtained provided that a sufficiently large facility could be built. A 60 m overall long micrometeorological wind–water tunnel has been designed for that purpose, tested on a one-fifth scale model, then constructed and fitted with the various equipment necessary for controlling experimental conditions, for performing appropriate measurements, and for acquiring and processing of the scientific data. Most of that work has been already reported (Coantic *et al.*, 1969; Coantic and Favre, 1971) and will be only reviewed here.

After the completion of working tests and the addition of some minor improvements, the I.M.S.T.'s air–sea interactions simulation facility has been put into scientific use. Several measurement programs are presently underway, and the present paper will be mainly devoted to a description of some preliminary experimental results.

2. Review of Preliminary Work

2.1. Consideration of the Physical Processes

Small-scale energy transfer between the atmosphere and the oceans results from four main processes: radiative transfer (both short-wave and long-wave), evaporation and turbulent transfer of water vapour and latent heat, sensible heat transfer, and kinetic energy transfer. These processes are intimately linked together through important reciprocal interactions, the most obvious of which lies in the control of heat and mass transfer rates by turbulence obtained primarily from the mean kinetic energy. When compared to the classical problem of simultaneous turbulent momentum, heat, and mass transfers, air–sea interactions display two new characteristics: firstly, scalar contaminants no longer remain passive, instead they now introduce important buoyancy effects; secondly, wave generation by wind alters the flow structure, and correspondingly the transfer mechanisms, themselves.

2.2. Relevant Dimensionless Parameters

If, as can be done in a first approximation, the effects of radiation are taken as additive to those of the turbulent transfers, then the exchange rates can be expressed through friction coefficients, Nusselt numbers, and Sherwood numbers, which depend, themselves, upon the relevant dimensionless parameters of the problem. Significant modelling will be achieved only if the values attained in the laboratory by these parameters are of the same order as those encountered in nature. Thus, very high Reynolds numbers have to be obtained for a proper representation of the dynamical structure of the atmospheric surface layer. Richardson numbers of order unity have to be reached to observe meaningful stability or instability effects. Sufficiently high Froude numbers and wave ages are necessary to investigate the effects of wind-generated gravity waves.

2.3. General Conception of the Simulating Facility

Keeping the above considerations in mind, the facility can be described as a combination of a micrometeorological wind tunnel with a wind-wave tank, the water surface in the latter forming the floor of the test section in the former. Large dimensions have been chosen to meet with the requirements of sufficiently high dimensionless numbers: the test section is 40 m long, 3.2 m wide, 1.5 m high for the airflow, and 0.75 to 1.0 m deep for the water flow. The global effects of solar radiation are simulated by heating water, and those of infrared radiative transfer by temperature control of the test section ceiling. Evaporation, and convective heat transfer, are generated through turbulent humidity and temperature boundary layers resulting from control of the wet- and dry-bulb conditions for the incoming airflow. Lastly, dynamical processes are modelled through generation of interfacial turbulent boundary layers under controlled velocity and preturbulence conditions, while wave motions are produced naturally as well as artificially with the aid of a wavemaker.

3. Description of Experimental Equipment

3.1. Laboratory Facilities

Owing to the importance and the exploratory nature of this long range project, a one-fifth scale model of the air–sea interaction simulation facility has been built and used to check and improve upon the original design characteristics for the large facility. Results from these preliminary tests and a detailed description of the large wind–water tunnel can be found in the above-mentioned references.

The general scheme of the facility and a diagram of its temperature and humidity control equipment are given in Figs. 1 and 2. The realization of this facility encompassed a period of approximately five years from preliminary drafts to the first scientific experiments in early 1972. Due to its size, a new laboratory has been built which houses, not only the large facility and its auxiliary equipment, but also the one-fifth scale tunnel which is still being used for research work. In addition, the laboratory contains probe-calibrating systems together with the necessary workshops, offices, and other supplementary work areas that are required by a research effort of this scale.

The various components of the facility are equipped with transducers, the electrical outputs of which are connected to analogue regulators and recorders controlled from a central console. A recently installed digital process control system will, in the future, allow automatic, continuous supervision and operation of the entire facility.

3.2. Instrumentation and Data Processing

Experimental studies of small-scale, air–sea interactions require the recording of various physical variables, some of which raise new instrumental problems. For example, the measurement of turbulent velocity and temperature fluctuations in water necessitated the development of specific techniques

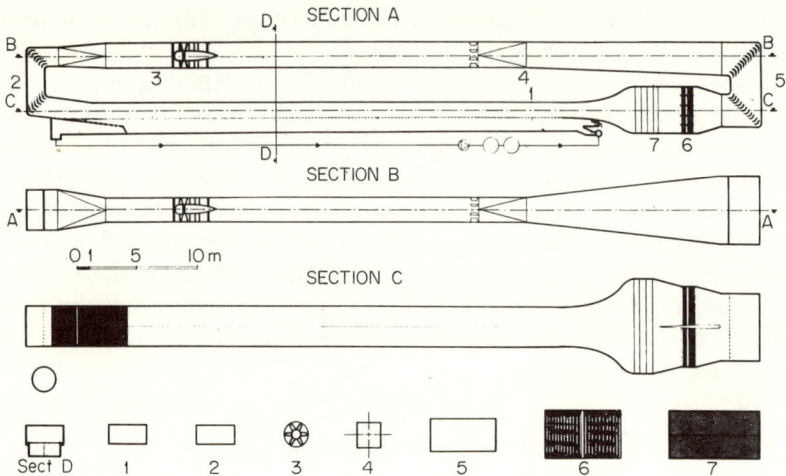

FIG. 1. General scheme of the facility.

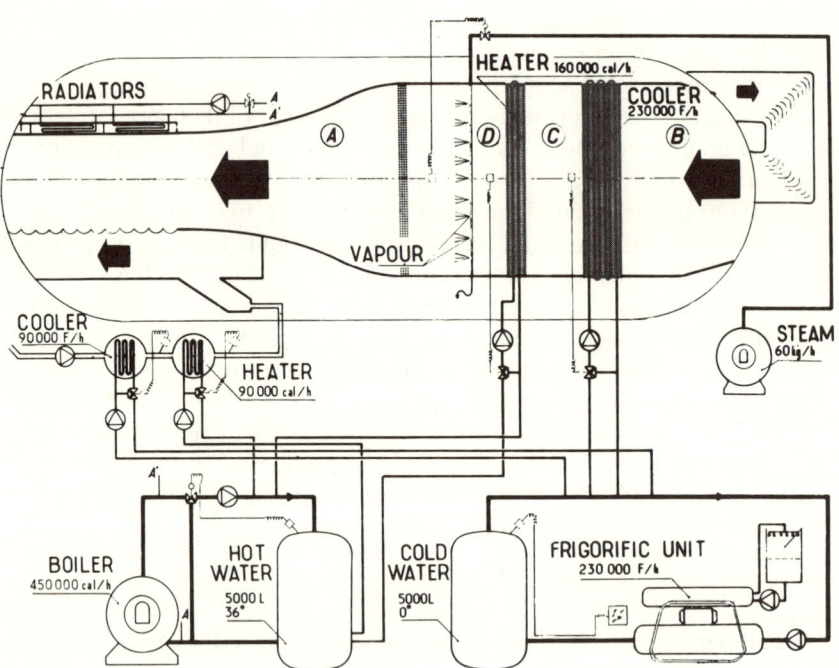

FIG. 2. Diagram of temperature and humidity control equipment.

(Resch, 1968), the recording of surface motions required the use of sensitive capacitance wave gauges (Ramamonjiarisoa, 1971), and the determination of mean values and turbulent fluctuations of humidity raises difficult problems (Coantic and Leducq, 1969). While instrument research is still going on concerning this last point, reasonably accurate determinations have been obtained using an automatic dew-point instrument for the mean values, and a small microwave refractometer (supplied by J. Lane, of R.S.R.S., and operated by R. Bolgiano and Z. Warhaft) for the turbulent fluctuations.

After conditioning and visual control, data are recorded and analyzed using either analogue statistical processing equipment or a high performance digital data acquisition system allowing direct preprocessing and control of information, and the editing of digital magnetic tapes for detailed processing by a computing center.

4. Results Concerning Wind-Generated Waves

An experimental study of the statistical properties of the wavy interface has been undertaken as a first step in a more general investigation of the wave-generating mechanism. Some typical results will now be presented and discussed.

The frequency spectrum of the wave profile has been determined in the small (Ramamonjiarisoa, 1971) as well as in the large facility (Ramamonjiarisoa, 1973). Results have been analysed in terms of various theoretical findings. For frequencies lower than the dominant wave's frequency n_0, they are found to obey Kitaigorodskii's (1961) similarity law:

$$\log(F(\omega) \cdot \omega^5/g^2) = a + b(u^+ \omega/g).$$

From these, and other experimental data sources, the constants seem to obey Mitsuyasu's (1968) relations:

$$a \simeq -8.78; \quad b \simeq 1.71(gX/u^{+2})^{0.283}.$$

Phillips' (1958) equilibrium range has been effectively observed for frequencies between approximately $2n_0$ and 10 Hz. In the intermediate zone, i.e., between n_0 and $2n_0$, the usual overshoot and undershoot effects noted by other investigators are displayed.

An interesting property appears when wave spectra from different sources are plotted (see Fig. 3) as $F(n)/F(n_0)$ as a function of n/n_0. It is clear that the relative width of the spectral peak increases when the fetch increases. The reason for this kind of "spectral ray widening effect" can be attributed to various linear or nonlinear modulation processes inside the liquid phase. Another explanation could lie in the turbulent wind's variability. As recently

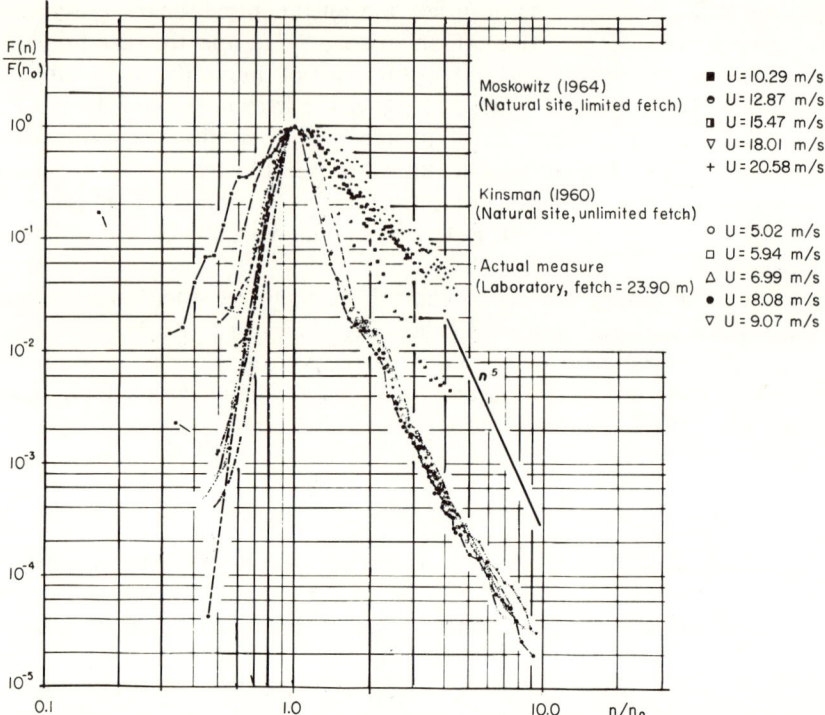

FIG. 3. Comparison of wave spectra at different fetches, in the laboratory and in the field.

emphasized by Dorman and Mollo-Christensen (1972), wave generation is a highly nonlinear, and therefore intermittent, process, so that the cumulative effects of the wind's temporal and spatial randomness could, at least partially, explain the spectral widening with increasing fetch.

Waves generated by wind over short fetches constitute a typical example of a random variable with a relatively narrow spectrum, and possess therefore the property that randomness appears only for extremely long observation times. Thus an autocorrelation function computed from a 20-s sample displays a quasi-deterministic, periodic behaviour, which is still apparent for the largest time lags in an autocorrelation curve computed from a 800-s sample (see Fig. 4). The wave field can therefore be approximately represented in such a situation as a succession of slightly perturbed dominant waves, whose frequency and amplitude are modulated over time intervals much larger than their characteristic period. Such a locally quasi-deterministic behavior should provide justification for use of theoretical models of wind action on purely periodic waves.

The probability distribution function of the instantaneous level of the moving water surface has also been determined. It shows a distinctly skewed

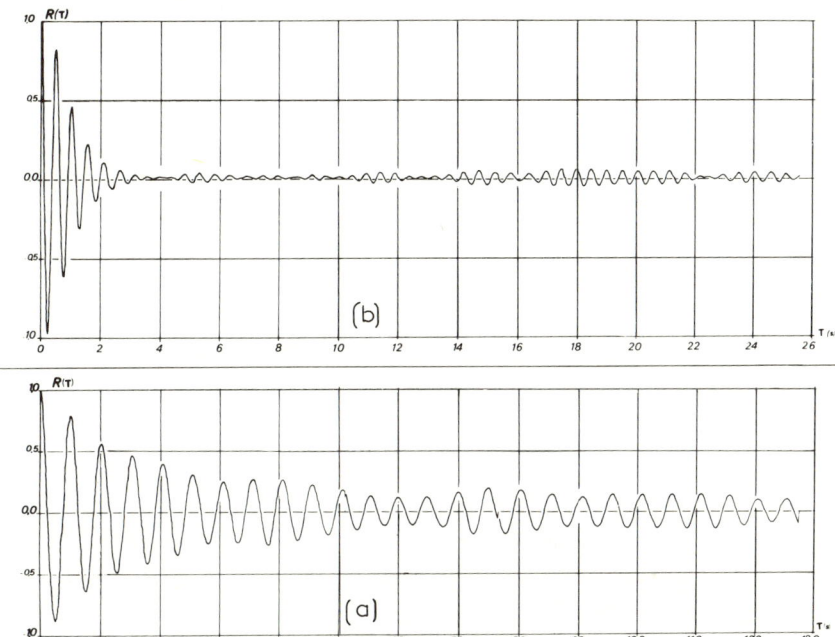

FIG. 4. Typical autocorrelation curves of surface displacement, from short and long samples.

non-Gaussian behavior, which can be related to the well-known fact that wind-wave crests are sharper than their troughs. Moreover, wave spectra display characteristic secondary peaks (or, at least, flattenings) corresponding to the second, and even higher-order, harmonics of the dominant wave frequency.

All these data agree with the picture of a wave field dominated by a train of slightly modulated, finite amplitude, nonlinear waves. Such a model already emerged from the observations of other authors (e.g., see Chang *et al.*, 1971). If confirmed, it could lead to reconsidering some of the hypotheses of wave–wave interaction theories (e.g., that each frequency independently obeys the classical dispersion relation) and could be used as a basis for the physical description of the adjacent wind field.

5. Mean Profiles and Fluxes

The development of interfacial velocity, temperature, and humidity turbulent boundary layers has been investigated in both the small and the large facilities, for a hitherto limited number of experimental situations. Figures 5, 6, and 7 display the evolution, along the 40-m length of the test section, of

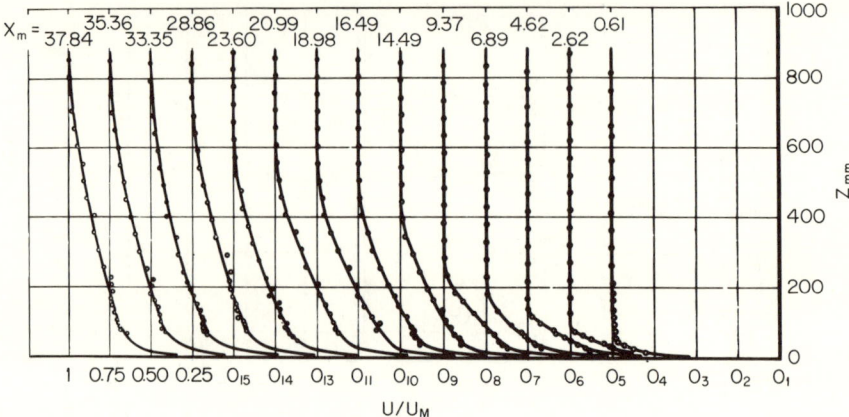

FIG. 5. Evolution of dimensionless velocity profiles.

the corresponding dimensionless profiles. Here, the mean wind velocity is 10 m s^{-1}, and the temperature and specific humidity differences are, respectively, about 10°C and 7×10^{-3}.

Turbulent momentum, heat, and evaporation fluxes can be obtained from such profiles, either using the results of the semiempirical theory for near-neutral layers (the "profile" method), or by means of integral relations of the von Kármán type (the "integral" method). Average evaporation rates are also determined by continuous weighing of the amount of water that has to be condensed to maintain a constant humidity level inside the facility.

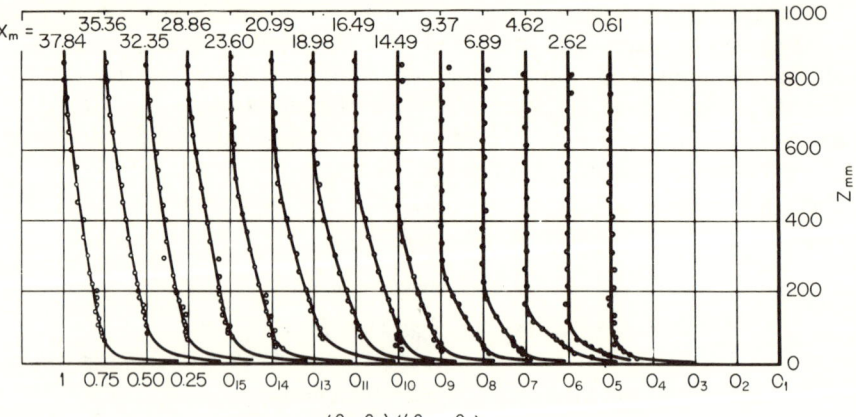

FIG. 6. Evolution of dimensionless temperature profiles.

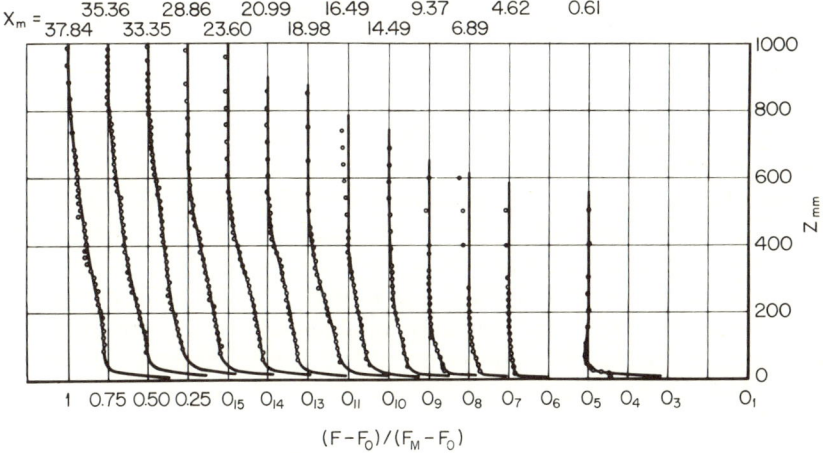

FIG. 7. Evolution of dimensionless humidity profiles.

Average evaporation Nusselt numbers determined in that way (for the large facility) are plotted in Fig. 8 as a function of length Reynolds numbers, together with results obtained in the small facility and in other laboratories (Lai and Plate, 1969; Delvaux, 1967). Although these data are very preliminary, they gather themselves reasonably well, in log-log coordinates, around a line with equation

$$\mathrm{Nu} = 0.08 \, \mathrm{Re}^{0.75}.$$

Use has also been made of isotopic techniques to investigate the evaporative process in the small facility. The measured fractionation rates of various isotopes tend to confirm the previously proposed dependence of the eddy mass diffusivity coefficient, as varying with the one-half power of the molecular diffusivity coefficient (Merlivat and Coantic, 1972).

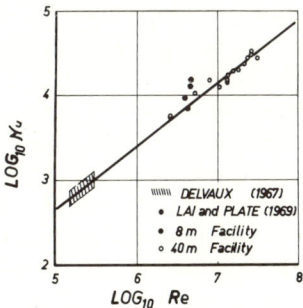

FIG. 8. Average evaporation Nusselt numbers, $\mathrm{Nu} = \tilde{J}_0 L / \rho \mathscr{D} \, \Delta F$, as a function of length Reynolds numbers, $\mathrm{Re} = UL/\nu$.

6. Influence of Stratification upon Turbulent Processes

The mean field and the turbulent structure of velocity, temperature, and humidity have been investigated by R. Bolgiano, Z. Warhaft, and R. Ezraty, under such conditions that appreciable stratification effects were likely to be observed. Most of the data have been recorded at a station 25 m downstream of the entrance section, for velocities ranging from 0.5 to 5 m s^{-1}, and for such temperature and humidity differences as to cover a Richardson's number's range from $+0.8$ to -0.7.

Characteristic variation in level and structure of all turbulent variables have been effectively observed as a function of stability or instability. Detailed statistical analysis of recorded data is now being performed, and Fig.9 shows, for a given situation, an example of the respective spectral

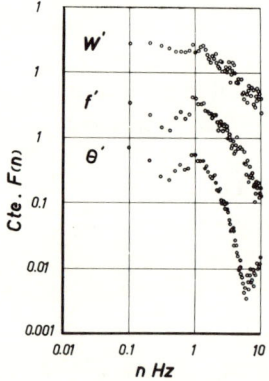

Fig. 9. Comparison of vertical velocity, temperature, and humidity spectra, for $\bar{U} = 1$ m s^{-1}, $\theta_{\text{air}} = 35.5°C$, $\theta_{\text{water}} = 19.5°C$, $\theta_{\text{sat.}} = 9.5°C$ (unpublished data kindly supplied by R. Bolgiano and Z. Warhaft).

distributions of vertical velocity, specific humidity, and temperature. One of the most interesting observations that has been made is the intermittent occurrence of intense turbulent bursts separated by rather quiescent intervals, for Richardson numbers of order $+0.3$, characterizing what has been qualified a "critical" situation. A typical example of simultaneous records of w', θ', and f' in such a case is displayed in Fig. 10. The existence of turbulent bursts in a stably stratified surface layer had already been noticed by Okamoto and Webb (1970). It could imply that turbulence suppression by buoyancy forces is not occurring through a gradual decrease of the level of all turbulent motions, but rather through a reduction in the number of the highly energetic turbulent structure that is able to "escape" from the lowest layers.

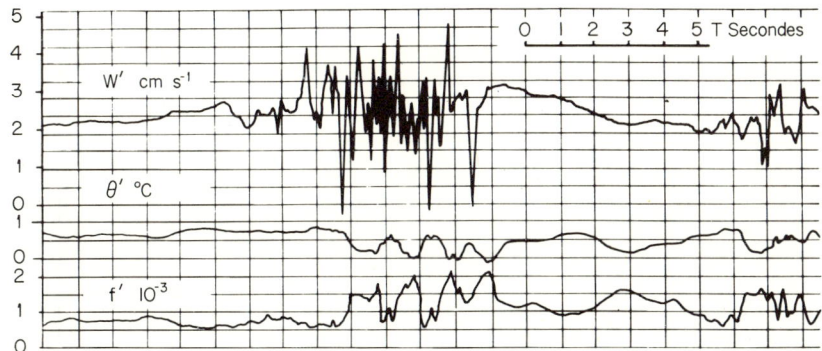

FIG. 10. Examples of simultaneous velocity, temperature, and humidity traces in a "critical" situation (Ri ≃ 0.3).

7. Statistical Properties of High Reynolds Number Turbulence

Considering the relatively high Reynolds numbers attainable in these experiments, it can be reasonably hoped that some degree of local isotropy will be effectively reached, so that comparisons will be feasible between the obtained data and certain theoretical forecasts. A preliminary investigation has been undertaken, by P. Mestayer, of the spectral properties of turbulence in the boundary layer. The data presented here concern the longitudinal component of velocity, recorded 30 cm above mean water level, in a section 33.4 m from the entrance, and for a reference velocity of 10 m s^{-1}. Thus the length Reynolds number of this experiment amounts to $R_x = 2.2 \times 10^7$; while the local mean and root-mean-square velocities are, respectively, $\overline{U} = 9.04$ m s^{-1} and $(\overline{u'^2})^{1/2} = 0.985$ m s^{-1}.

The spectrum of energy and the spectrum of dissipation are plotted in Fig. 11, using an area-conserving representation. It can be seen that the bulk of the turbulent energy corresponds to wavelengths ranging from 5 cm to 5 m, while dissipation is mainly restricted to wavelengths between 0.5 and 5 mm. The dissipation rate is then evaluated in the usual way to give $\varepsilon = 1.29$ m^2 s^{-3}, and the spectrum is again plotted, in log-log coordinates, in Fig. 12 using Kolmogorov's universal representation. The spectrum naturally flattens toward low wave numbers, but is in good agreement with field observations for all dimensionless wave numbers above 10^{-3}. The straight line describing the $-\frac{5}{3}$ power law can be fitted to the experimental points over nearly two decades, and the corresponding value for Kolmogorov's one-dimensional universal constant is $K' = 0.57$, which is in excellent agreement with some recently reported data (see Boston and Burling, 1972).

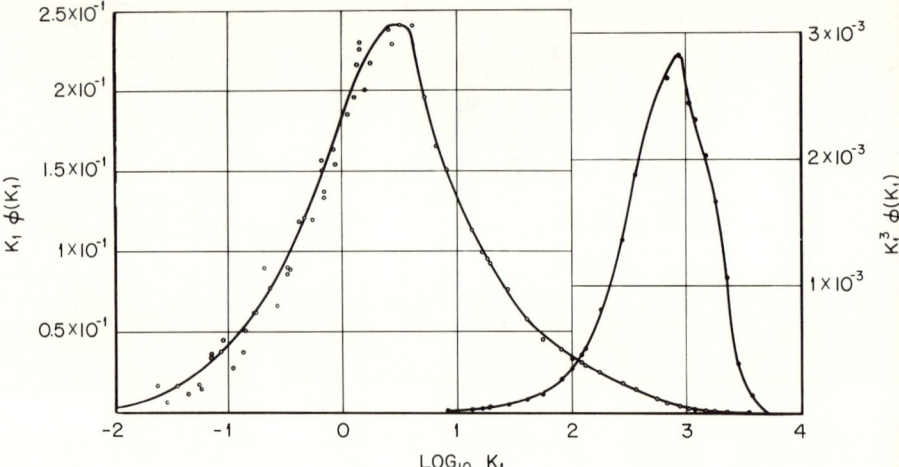

FIG. 11. Energy spectrum and dissipation spectrum for high Reynolds number boundary layer turbulence.

A preliminary analysis has also been undertaken, using newly developed numerical methods, of the probability distribution of turbulent velocity fluctuations. The result of that analysis of the same experimental record as above is presented in nondimensional coordinates in Fig. 13, and displays rather sensible departures from the Gaussian distribution plotted for comparison. Skewness and flatness factors amount, respectively, to $S = -0.096$ and $F = 2.51$.

8. Conclusions

As has been previously emphasized: the success of our approach depends largely on the degree to which natural processes can be usefully modelled in the laboratory. In that respect, definite conclusions cannot as yet be put forth, but the preliminary results reported herein seem rather encouraging. These are that:

(a) well-defined velocity, temperature, and humidity boundary layers have been generated under a variety of controlled experimental situations;

(b) sufficiently high Reynolds numbers have been reached to observe some indications of local isotropy;

(c) three-dimensional random gravity waves have been generated from the wind;

(d) appreciable stratification effects have been established for unstable as well as for stable conditions.

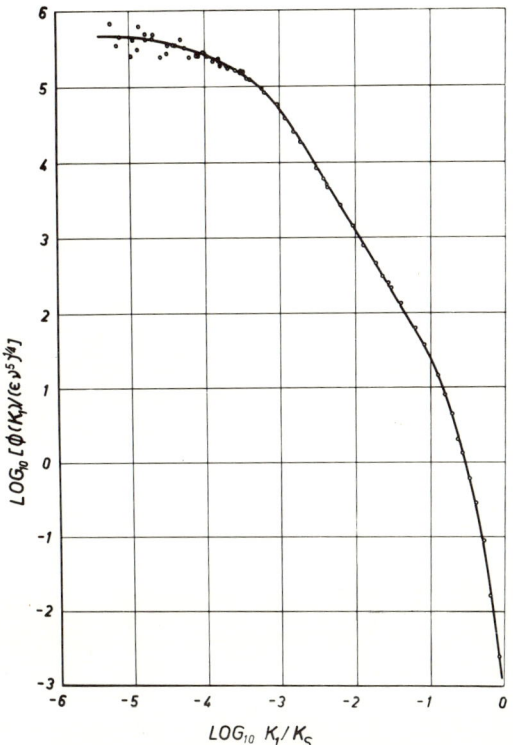

FIG. 12. Universal representation of high Reynolds number turbulence energy spectrum (Kolmogorov's velocity scale $u_k = 0.0665$ m s^{-1}, length scale $\eta = 2.27 \times 10^{-4}$ m.).

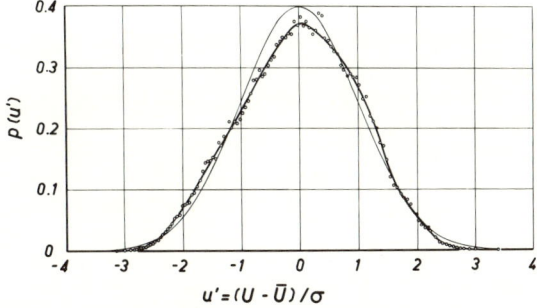

FIG. 13. Comparison with a Gaussian law of the probability distribution for high Reynolds number turbulent velocity fluctuations.

It is clear that much work still has to be done. More data than those presented here are about to be processed and interpreted. New experiments have been prepared, and will be performed during the next few months. The future experimental program is related, not only to the basic physical aspects of air–sea interaction which have been described above, but also to other problems, such as wave propagation in the lower maritime atmosphere and the important gaseous exchanges that occur across the air–sea interface, that can be usefully investigated by means of the facilities which are now available.

Since the very beginning of our activities, it has been recognized that the capabilities of the I.M.S.T. air–sea interaction simulation facility should not be used solely by any one scientific group. Consequently, cooperation with other groups has been encouraged. Cooperative efforts, some results of which have been reported above, have already proven to be most fruitful and will be actively pursued in the future.

Acknowledgments

This research program has benefited from the help and cooperation of many individuals and organizations, to which we want to express here our deep acknowledgments. It has been financially supported by a number of French governmental agencies: Délégation Générale à la Recherche Scientifique et Technique, Direction des Enseignements Supérieurs du Ministère de l'Education Nationale, and Université d'Aix-Marseille, Centre National pour l'Exploitation des Océans, Centre National de la Recherche Scientifique, Direction des Recherches et Moyens d'Essais. The cooperative work described in Section 6 has been partially supported by a NATO Research Grant. The scientific and technical team in charge of the program included a number of research workers, doctoral students and technicians all of whose names have not been cited in this paper. Here, we shall mention specially: P. Bonmarin, D. Leducq, B. Pouchain, B. Lhomme, B. Marmottant, G. Péri, J. Quaccia, F. Laugier, P. Chambaud, M. Bourguel, A. Laurence, B. Zucchini, R. Vaudo. Their active and efficient cooperation has been vital to the progress of the project and to the scientific results achieved to date.

References

Boston, N. E., and Burling, R. W. (1972). *J. Fluid Mech.* **55**, 473–492.
Chang, P. C., Plate, E. J., and Hidy, G. M. (1971). *J. Fluid Mech.* **47**, 183–208.
Coantic, M., and Favre, A. (1971). *Proc. Symp. Nav. Hydrodyn., 8th, Pasadena, Calif.* Off. Nav. Res. ARC-179, pp. 37–69.
Coantic, M., and Leducq, D. (1969). *Radio Sci.* **4**, No. 12, 1169–1174.
Coantic, M., Bonmarin, P., Pouchain, B., and Favre, A. (1969). *AGARD Conf. Proc.* No. 48, Paper No. 17.
Delvaux, L. (1967). Contribution à l'étude de l'évaporation par convection forcée turbulente à partir d'une surface plane horizontale. Thèse, Fac. Sci., Univ. de Paris, Paris.
Dorman, C. E., and Mollo-Christensen, E. (1972). "Observation of the Structure of Moving Gust Patterns over a Water Surface," Rep. Dep. Meteorol., Mass. Inst. Technol., Cambridge, Massachusetts.

Kinsman, B. (1960). "Surface Waves at Short Fetches and Wind Speeds." Tech. Rep. 19, Chesapeake Bay Inst., The Johns Hopkins University.
Kitaigorodskii, S. A. (1961). *Izv. Akad. Nauk SSSR, Ser. Geofiz.* pp. 105–117.
Lai, J. R., and Plate, E. J. (1969). "Evaporation from Small Wind Waves," Tech. Rep. CER 68-69 JRL 35. Fluid Dyn. Diffus. Lab., Colorado State Univ., Fort Collins, Colorado.
Merlivat, L., and Coantic, M. (1972). *Trans. Amer. Geophys. Union* **53**, 411. (Abstr.)
Mitsuyasu, H. (1968). *Rep. Res. Inst. Appl. Mech., Kyushu Univ.* **16**, No. 55; **17**, No. 59.
Moskowitz, L. (1964). *J. Geophys. Res.* **69**, No. 24.
Okamoto, M., and Webb, E. K. (1970). *Quart. J. Roy. Meteorol. Soc.* **96**, No. 410, 591–600.
Phillips, O. M. (1958). *J. Fluid Mech.* **4**, 426–434.
Pond, S. (1971). Air-sea interaction. *Gen. Assem. I.U.G.G., 15th* U.S. Nat. Rep., pp. 389–394. Amer. Geophys. Union, Washington, D.C.
Ramamonjiarisoa, A. (1971). *Cah. Oceanogr.* **23**, No. 12, 957–979.
Ramamonjiarisoa, A. (1973). *Colloq. Ocean Hydrodyn., 5th Liège Univ.* (to be published).
Resch, F. (1968). Études sur le fil chaud et le film chaud dans l'eau. Thèse Dr. Ing., Univ. d'Aix-Marseille, Marseille.

EDDY DIFFUSION COEFFICIENTS IN THE PLANETARY BOUNDARY LAYER

F. K. WIPPERMANN

Institut für Meteorologie der Technische Hochschule Darmstadt
Darmstadt, Germany

1. Introduction

In the past few years a model of the atmospheric boundary layer has been developed, closing the system of equations by a mixing-length hypothesis. Numerical integrations with that model show *inter alia* the variation with height of the eddy diffusion coefficients as dependent on thermal stratification, baroclinicity, and the diurnal time variation.

2. A Short Description of the Boundary Layer Model

The planetary boundary layer (PBL) of the atmosphere is defined as a steady state and horizontally homogeneous boundary layer; it is described by the set of equations (1)–(4):

(1) $$\frac{d^2 Y}{dZ^2} - \frac{X}{K_m} + \lambda_x = 0$$

(2) $$\frac{d^2 X}{dZ^2} + \frac{Y}{K_m} - \lambda_y = 0$$

(3) $$K_m = (L/\kappa)\{X^2 + Y^2\}^{1/2}$$

(4) $$L = L(Z, L_\infty) \quad \text{or} \quad L = L(X, Y; L_\infty)$$

Equations (1) and (2) are the two averaged equations of motion, the velocities of which have been eliminated using the flux–gradient relation. Equation (3) is a combination of the flux–gradient relation with Prandtl's mixing-length relation; Eq. (4) stands for a mixing-length hypothesis which is necessary in order to close the system of equations. Four different hypotheses for the nondimensional mixing-length L have been used:

(5a) $$L(Z) = \kappa Z/[1 + \kappa(Z - Z_0)^{5/4}/L_\infty]$$

(5b) $$L(Z) = \kappa Z/[1 + \kappa(Z - Z_0)/L_\infty]$$

(5c) $$L(Z) = L_\infty - (L_\infty - \kappa Z_0)\exp[-\kappa(Z - Z_0)/L_\infty]$$

(5d) $$L = L_\infty - (L_\infty - \kappa Z_0)T^p$$

$$p = -\kappa \frac{1 - Z_0\{2(dT/dZ)_0^2 + (dT/dZ)_0/2\}}{(dT/dZ)_0\{L_\infty - \kappa Z_0\}}, \quad T = \{X^2 + Y^2\}^{1/2}$$

In all of these four hypotheses appears the parameter L_∞, which stands for the asymptotic mixing length to be approached for $Z \to \infty$, except for (5a) where L_∞ is only a parameter since $L \to 0$ for $Z \to \infty$. The parameter L_∞ controls the thermal stratification of the PBL which is described by the internal stratification parameter μ,

(6) $$\mu = H/L_*$$

where $H = \kappa u_*/f$ is the internal scale height of the PBL and L_* the Monin–Obukhov stability length. An empirical relationship between L_* and μ has been given by Wippermann (1972):

(7) $$\mu = -53.4 - 28.8 \log_{10}(L_\infty)$$

The hypothesis (5a) has been given for the neutral case ($\mu = 0$, i.e., $L_\infty = 0.014$) by Lettau (1962) where κ/L_∞ was chosen to be 33.62 (or $L_\infty = 0.012$). The hypothesis (5b) is that one by Blackadar (1962) in a somewhat modified form; in the original one Blackadar has the factor $6.75 \times 10^{-4}/C_g$ instead of L_∞ showing that his asymptotic mixing-length depends on external parameters and does not satisfy the requirements of Rossby-number similarity. The hypothesis (5c) is that one by Ruzin et al. (1963) and later on used by Appleby and Ohmstede (1964); it is nondimensionalized and slightly modified. Equation (5d) is a hypothesis by Wippermann (1971a) which relates L with the (not yet known) dependent variable T instead of with the independent variable Z as do hypotheses (5a), (5b), and (5c); therefore a model using the hypothesis (5d) is less a priori. However this hypothesis has a defect too: $L(Z \to \infty) \neq L_\infty$ in a baroclinic PBL.

The set of equations (1)–(4) is a closed system for the four variables $X(Z)$, $Y(Z)$, $K_m(Z)$, and $L(Z)$. The boundary conditions are

$Z = Z_0$:

(8a) $\quad X = 1$

(8b) $\quad Y = 0$

(8c) $\quad K_m = Z_0$

(8d) $\quad L = \kappa Z_0$

$Z \to \infty$:

(9a) $X \to 0$ $X \to \dfrac{L_\infty^2}{\kappa^2} \lambda_x (\lambda_x^2 + \lambda_y^2)^{1/2}$ $X \to X(L_\infty, \lambda_x, \lambda_y)$

(9b) $Y \to 0$ $Y \to \dfrac{L_\infty^2}{\kappa^2} \lambda_y (\lambda_x^2 + \lambda_y^2)^{1/2}$ $Y \to Y(L_\infty, \lambda_x, \lambda_y)$

(9c) $K_m \to 0$ $K_m \to \dfrac{L_\infty^2}{\kappa^2} (\lambda_x^2 + \lambda_y^2)^{1/2}$ $K_m \to K_m(L_\infty, \lambda_x, \lambda_y)$

(9d) $L \to 0$ $L \to L_\infty$ $L \to L(L_\infty, \lambda_x, \lambda_y)$

   (5a)   (5b), (5c)   (5d)

The boundary conditions (8a), (8b) show that the coordinate system is orientated with the x axis parallel to the surface stress.

An iterative method for solving the system of the equations (1)–(4) with the boundary conditions (8a)–(8d), (9a)–(9d) has been given by Wippermann (1971b).

Three free parameters appear in the equations: one for the thermal stratification (L_∞ or μ) and two for the baroclinicity (λ_x and λ_y). These three parameters are internal ones; they do not contain any external parameter. Only in the boundary conditions (8c) and (8d) does an external parameter appear, namely Z_0, the nondimensional roughness length, which is according to the definitions made

(10) $$Z_0 = (\kappa C_g \, \mathrm{Ro}_0)^{-1}$$

It reflects the influence of the external parameters $|\hat{v}_{g0}|, f, z_0$. However, for $Z \gg Z_0$, the solutions become independent of Z_0, i.e., independent of any external parameter (Rossby number similarity); they depend only on the three internal parameters $\mu, \lambda_x, \lambda_y$:

$X(Z, \mu, \lambda_x, \lambda_y)$
$Y(Z, \mu, \lambda_x, \lambda_y)$ $Z \gg Z_0$
$K_m(Z, \mu, \lambda_x, \lambda_y)$
$L(Z, \mu)$ or $L(Z, \mu, \lambda_x, \lambda_y)$

and are called universal. From these, the universal profiles of other variables (for instance the velocity defect or the energy dissipation rate) can be derived.

We note, that in a PBL the profile $K_m(Z)$ is universal for $Z \gg Z_0$; such profiles will be studied in the following sections.

2. K_m Profiles in the Barotropic PBL

The results of numerical integrations are shown in Fig. 1 for four different thermal stratifications ($\mu = +20, 0, -20, -40$). The integrations have not been carried out for larger stabilities since an upper limit exist in the PBL for stable stratification, i.e., for the downward turbulent heat flux. This matter is discussed by Wippermann (1972). The curves in Fig. 1 represent the universal profiles $K_m(Z)$ obtained with different mixing-length hypotheses. The

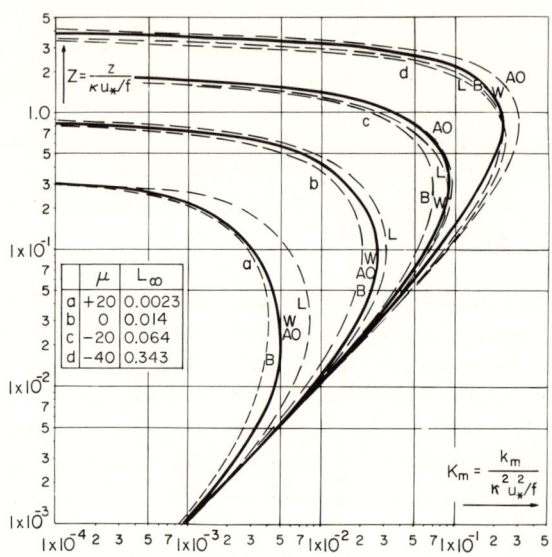

FIG. 1. Universal vertical profiles of the eddy diffusion coefficient varying with the thermal stratification μ. Different mixing-length hypotheses are used: AO = Appleby–Ohmstede, B = Blackadar, L = Lettau, W = Wippermann.

heavy curves (W) are the profiles obtained with hypothesis (5d); for a comparison, the other profiles are given also: "L" with hypothesis (5a), "B" with (5b) and "AO" with (5c). The differences in the results vary with the thermal stratification.

Two empirical formulae have been derived from these results, they are useful possibly for practical purposes:

(11) $\quad K_{m,\,\max} = 0.27\, Z_{K\,\max}, \qquad \lambda_x = \lambda_y = 0$

and

(12) $\quad K_m(Z) = Z \exp\{-c_\mu Z^{0.764}\}, \qquad \lambda_x = \lambda_y = 0$

TABLE I

μ	+20	+10	0	−10	−20	−30	−40
L_∞	0.0023	0.0060	0.014	0.031	0.064	0.154	0.343
c_μ	27.8	14.8	7.8	4.8	3.0	2.1	1.6

Some values of the coefficient c_μ are given in Table I. A comparison with observations is given in Fig. 2 for the neutral case ($\mu = 0$). The Leipzig wind profile data (dotted curve), which is commonly considered to be neutral and barotropic, fit the universal K_m profile very well. This profile (broken line) has been obtained by the PBL model using the mixing-length hypothesis (5d).

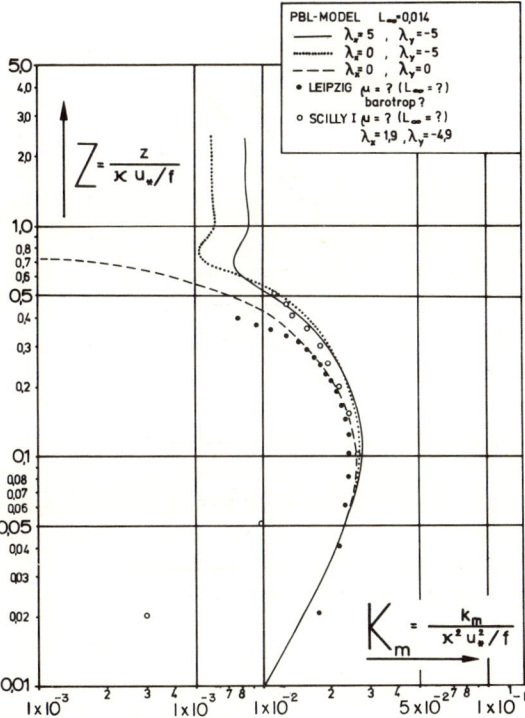

FIG. 2. Universal vertical profiles of the eddy diffusion coefficient in the neutral case (results of the PBL model and observations).

3. K_m Profiles in Baroclinic Cases

The baroclinicity has been determined for only a very few of the measured wind profiles. Figures 2, 3, and 4 show three examples, which are described in a paper by Wippermann and Yordanov (1972). Also the K_m profile for Scilly I (circles in Fig. 2) fits the universal profiles obtained by the PBL model for two neighbouring baroclinicities; only the lowest two points differ very much.

As discussed in the paper mentioned before, a sharp minimum can occur in the K_m profile when a certain baroclinicity is given (the thermal wind has to have a component in the direction of the geostrophic wind at the ground). Figures 3 and 4 give two examples of boundary layers in which a rather thin

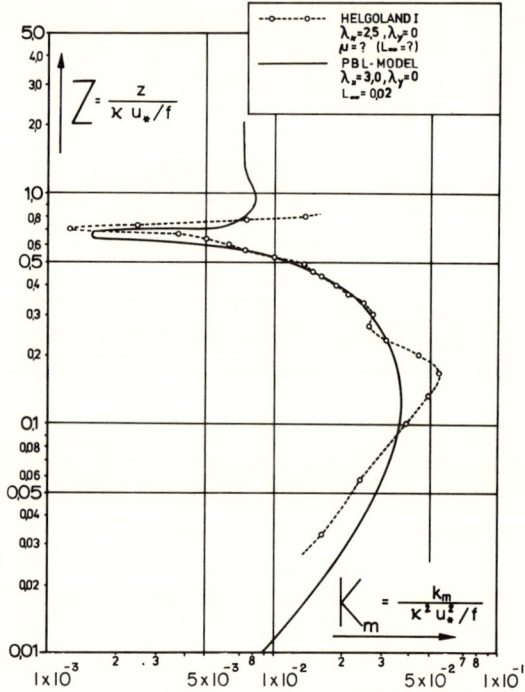

FIG. 3. The minimum (due to baroclinicity) in the vertical profile of the eddy diffusion coefficient as obtained by the PBL model and by observations.

layer is formed with a reduced eddy diffusion coefficient K_m due to the baroclinicity. For the K_m profile Helgoland II (Fig. 4) this minimum is situated at 325 m, it may act as a barrier against turbulent mixing in the vertical direction; such minima could be important for the diffusion from very tall stacks. They are due to the baroclinicity of the PBL only, they are not related to a temperature inversion.

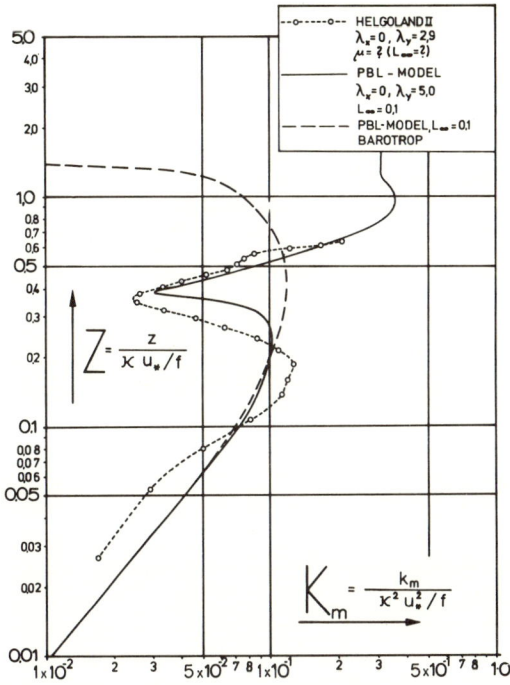

Fig. 4. The same as Fig. 3.

4. The Effect of Nonstationarity on the K_m Profiles

In order to get a time-dependent boundary-layer model, Eqs. (1) and (2) have been completed on the left-hand side by $-\partial/\partial F(Y/K_m)$ and $-\partial/\partial F(X/K_m)$, respectively; also the vertical derivatives in these equations are partial now. An uncentered (backward) difference scheme with time has been used; the integrations are carried out with $\Delta F = 0.2$, i.e., with 48 time steps during 24 hr. A detailed description is given in the paper by Wippermann et al. (1973). The input into the model is the diurnal variation of the stratification parameter μ (or L_∞) or the equivalent turbulent heat flux at the ground. This heat flux q_0 has been chosen symmetrical around noon as shown in Fig. 5, it corresponds to the conditions of a clear day. The computations have been carried out for three successive days with the same input in order to become independent of the initial state.

The results for $K_m(Z, F)$ are given by the isopleths in Fig. 6. The heavy line gives the position in height of the maximum of K_m; this is lifted from $Z \sim 0.03$ during the night to $Z \sim 1.1$ at noon. The maximum at noon amounts to about 350 times the value of the maximum during the night.

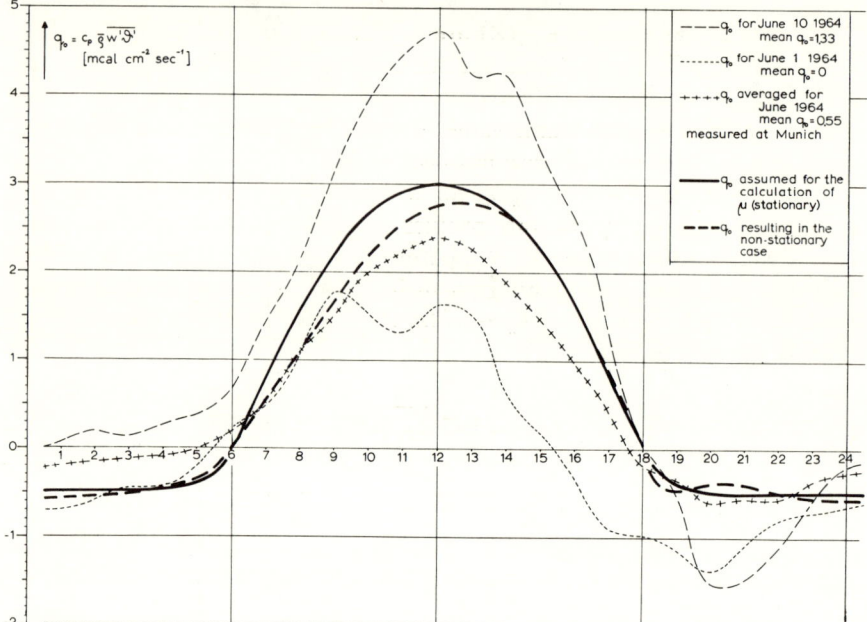

FIG. 5. The turbulent flux of sensible heat at the ground as chosen for the input into the time-dependent boundary layer model. Comparisons with observations. (After Wippermann et al., 1973.)

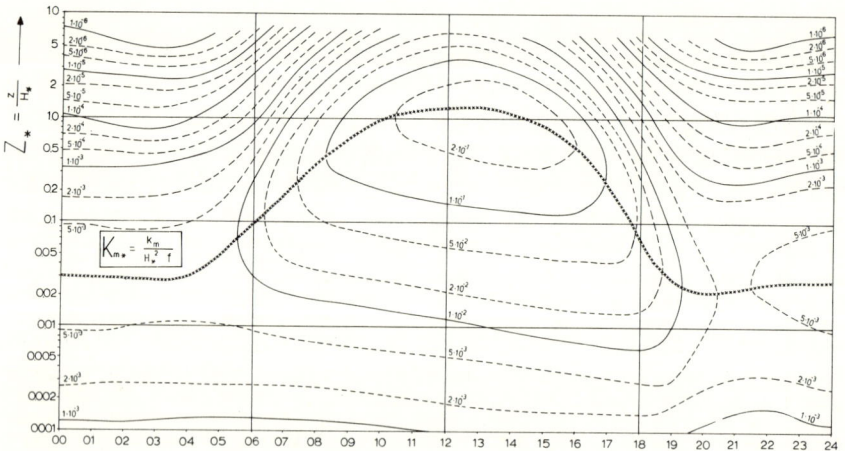

FIG. 6. The variation of the vertical K_{m*} profile during a clear day, obtained with a time-dependent boundary layer model. (After Wippermann et al., 1973.)

Of course, the profiles $K_{m*}(Z)$ are no longer universal now. Since the friction velocity varies with time, the nondimensionalisation would also vary with time. The values are therefore comparable only if they are all nondimensionalized with the same value; for that purpose the scale height H_* for $\mu = 0$ in the stationary case has been used.

Figure 7 shows the ratio k_h/k_m for four different times of day. For the computation of the temperature profiles $\Theta(Z)$ and of the k_h profiles, reference is made to the paper by Wippermann et al. (1973). Since in a PBL (without radiative effects) the turbulent heat flux q_0 is constant with height (but $\partial\Theta/\partial Z \to 0$ for $Z \to \infty$), $k_h \to \infty$ for $Z \to \infty$. The large increase of k_h/k_m with height is seen in Fig. 7.

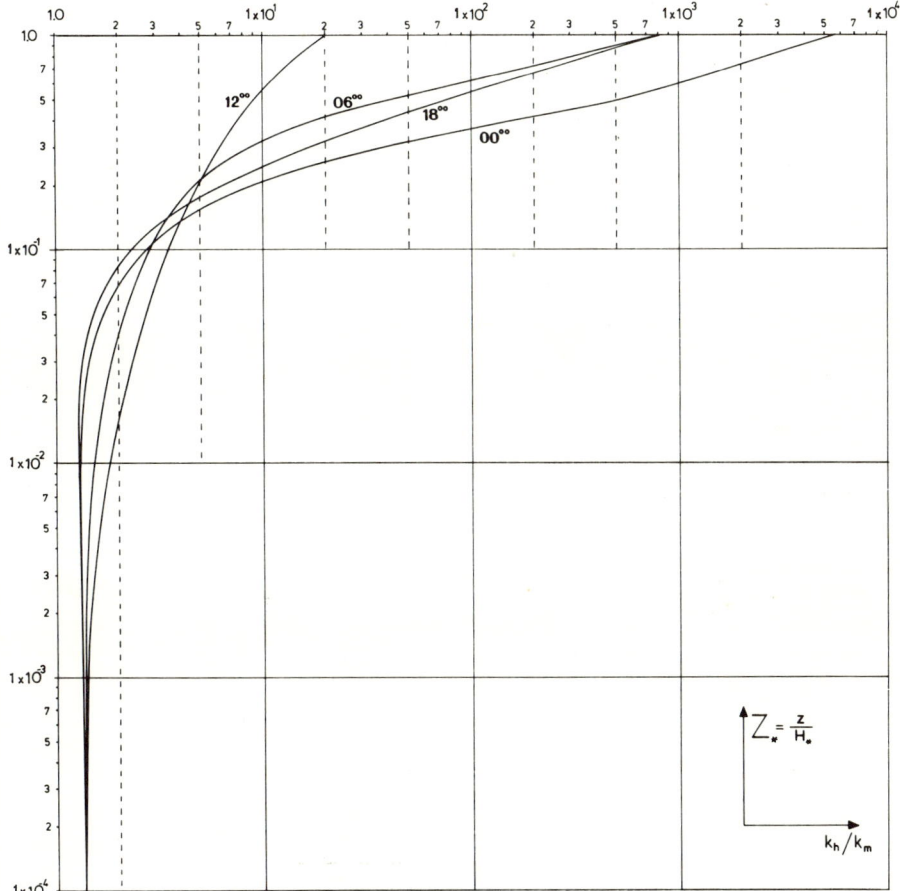

FIG. 7. The vertical profiles of k_h/k_m for various times of the day, obtained with a time-dependent boundary layer model. (After Wippermann et al., 1973.)

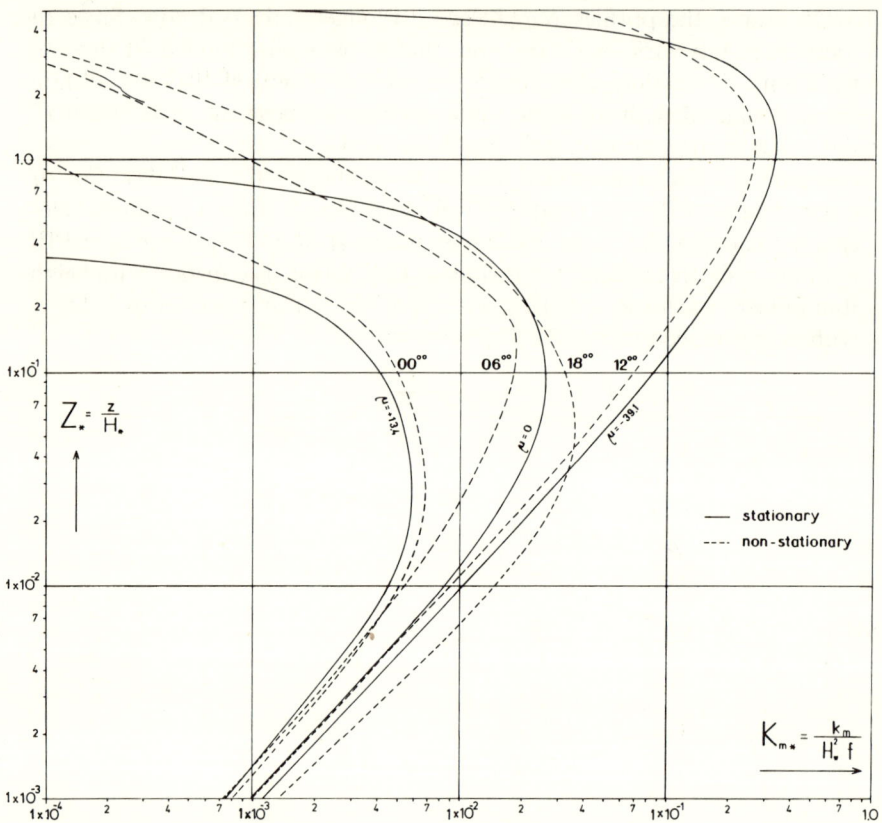

FIG. 8. The differences of the vertical K_{m*} profiles in the nonstationary case compared with the corresponding stationary cases (i.e. cases having the same thermal stratification).

It is interesting to compare the turbulent diffusion coefficients with those which are obtained in the stationary case having the same thermal stratification μ. Figure 8 shows the result of such comparisons. The nonstationary diffusion coefficients are smaller than the corresponding stationary ones at noon, but they are larger at night.

5. The Error in the Maximum Ground Concentration due to Neglecting Nonstationarity

The concentration $\bar{s}(x, y, 0)$ at the ground has a maximum in the stationary case when material is dispersed from an elevated (continuous) point source at $z = h$. It is given by the usual formula (K-theory)

(13) $$\bar{s}_{\max} = (2Q/e\pi h^2 \bar{u})[k_z/k_y]^{1/2}$$

In order to study the effect of nonstationarity, k_z and $\bar{u}$ have been computed for different times of the day from the results of the time-dependent boundary-layer model (clear day); vertical averages have been taken between the ground and 170 m. Then the concentration $\bar{s}_{max}$ obtained by Eq. (13) has been compared with that of the corresponding (same μ) stationary case; Fig. 9 shows the result. The numbers on the curve give the time of the day. For instance the same thermal stratification occurs in the chosen example (clear day) for 7 A.M. and 5 P.M., however the ratio $\bar{s}_{max, 07}/\bar{s}_{max, 17}$ is about 1.5; it shows that considerable error can be made by using the stationary diffusion formulae.

A detailed interpretation of the results is given by Wippermann (1973).

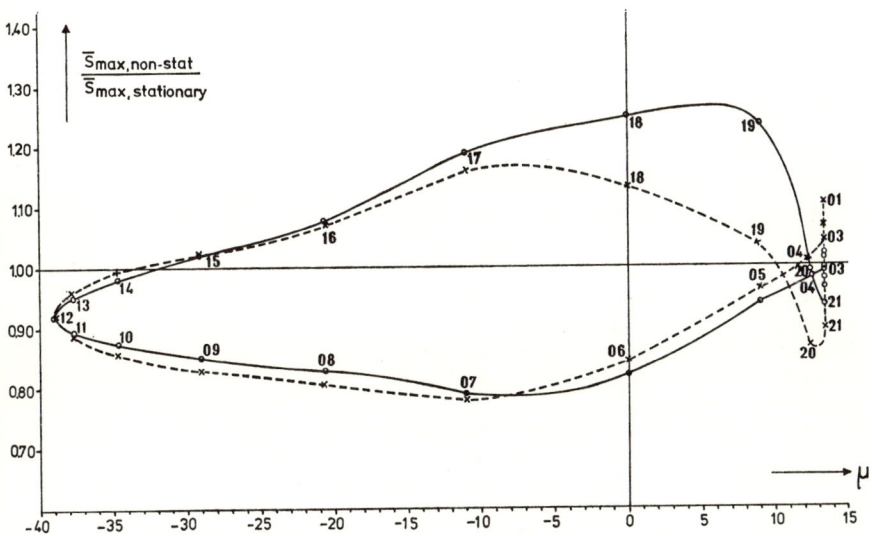

FIG. 9. The ratio of the maximum ground concentration in the nonstationary case to that in the corresponding stationary case (same thermal stratification) by applying the common diffusion formula.

List of Symbols

- C_g $u_*/|\hat{v}_{g0}|$ Geostrophic drag coefficient
- F ft, nondimensional time
- f Coriolis parameter
- H $\kappa u_*/f$, internal scale height of the PBL
- h Elevation of a continuous point source
- K_m $k_m/(H^2 f)$, nondimensional eddy viscosity
- k_m Eddy viscosity
- k_h Turbulent diffusion coefficient for heat
- L l/H, nondimensional mixing length

l Mixing length
L_* Monin–Obukhov stability length
PBL   Planetary boundary layer
Q Strength of a continuous point source (g sec^{-1})
q Turbulent vertical heat flux
Ro_0 $|\hat{v}_{g0}|/(fz_0)$, surface Rossby number
$\bar{s}$ Concentration of an emitted material
T τ/τ_0, nondimensional Reynolds stress
t Time
u_* Friction velocity
$|\hat{v}_{g0}|$ Geostrophic wind speed at the ground
X τ_x/τ_0, nondimensional stress component
Y τ_y/τ_0, nondimensional stress component
Z z/H, nondimensional vertical coordinate
Z_0 z_0/H, nondimensional roughness length
Θ Nondimensional potential temperature
κ von Kármán's constant
λ_x, λ_y Internal parameters for the baroclinicity
μ H/L_* Internal parameter for the stratification
ρ Density
τ_x, τ_y Components of the Reynolds stress

References

Appleby, J. F., and Ohmstede, W. D. (1964). Meteorol. Res. Note No. 8. Meteorol. Dep. USAERDA, Fort Huachuca, Ariz.
Blackadar, A. K. (1962). *J. Geophys. Res.* **67**, 3095–3102.
Lettau, H. H. (1962). *Beitr. Phys. Atmos.* **35**, 195–212.
Ruzin, M. I., Boldizeva, N. A., and Sabeleva, T. A. (1963). *Tr. L.G.M.I.*, *B* **15**, 66–68.
Wippermann, F. K. (1971a). *Beitr. Phys. Atmos.* **44**, 215–226.
Wippermann, F. K. (1971b). *Beitr. Phys. Atmos.* **44**, 293–296.
Wippermann, F. K. (1972). *Beitr. Phys. Atmos.* **45**, 305–311.
Wippermann, F. K. (1973). *Meteorol. Rundsch.* **26**(1), 11–18.
Wippermann, F. K., and Yordanov, D. (1972). *Beitr. Phys. Atmos.* **45**, 267–275.
Wippermann, F. K., Etling, D., and Leykauf, H. (1973). *Beitr. Phys. Atmos.* **46**(1), 34–56.

LAGRANGIAN–EULERIAN TIME-SCALE RELATIONSHIP ESTIMATED FROM CONSTANT VOLUME BALLOON FLIGHTS PAST A TALL TOWER

JAMES K. ANGELL

Air Resources Laboratories, National Oceanographic and Atmospheric Administration Silver Spring, Maryland 20910, U.S.A.

1. INTRODUCTION

Between September 25 and October 12, 1971, 56 free-floating constant volume balloons or tetroons (Fig. 1) were flown past the instrumented 460-m WKY television tower located on the northern outskirts of Oklahoma City. While the primary purpose of the experiment was to compare Reynolds stress obtained from tetroons and tower, the experiment also provided a unique opportunity to compare Lagrangian and Eulerian time scales of motion in the vertical. Summaries of earlier attempts to define the Lagrangian–Eulerian time-scale ratio β have been presented in "Meteorology and Atomic Energy" (Slade, 1968).

In this paper the Eulerian and Lagrangian time scales of motion are represented by the period or frequency of the vertical-velocity spectral peak where, as is customary, the absolute spectral density $S(n)$ has been multiplied by the frequency n (in cycles sec^{-1}), and the spectral peak determined from the maximum in $nS(n)$, as measured in m^2 sec^{-2}. Inasmuch as the pertinent theory deals with the integral time scale, there is some danger in this procedure, but it is the most feasible one owing to the difficulty in determining representative Lagrangian and Eulerian integral scales from sampling times of only 1–2 hr. In this connection, recall that the frequency of the $nS(n)$ spectral peak is proportional to the reciprocal of the integral time scale, with the proportionality factor equal to $1/2\pi$ when the spectrum corresponds to a simple exponential correlogram (Pasquill, 1963).

Figure 2 shows the location, with respect to Oklahoma City, of the WKY tower, the tetroon-tracking radar, and the usual tetroon launch site. Because of the prevailing south-southwesterly winds in that area, in most of the Eulerian–Lagrangian comparisons the air moved across Oklahoma City

Fig. 1. An example of the constant volume balloon (tetroon) used in the Oklahoma City experiment.

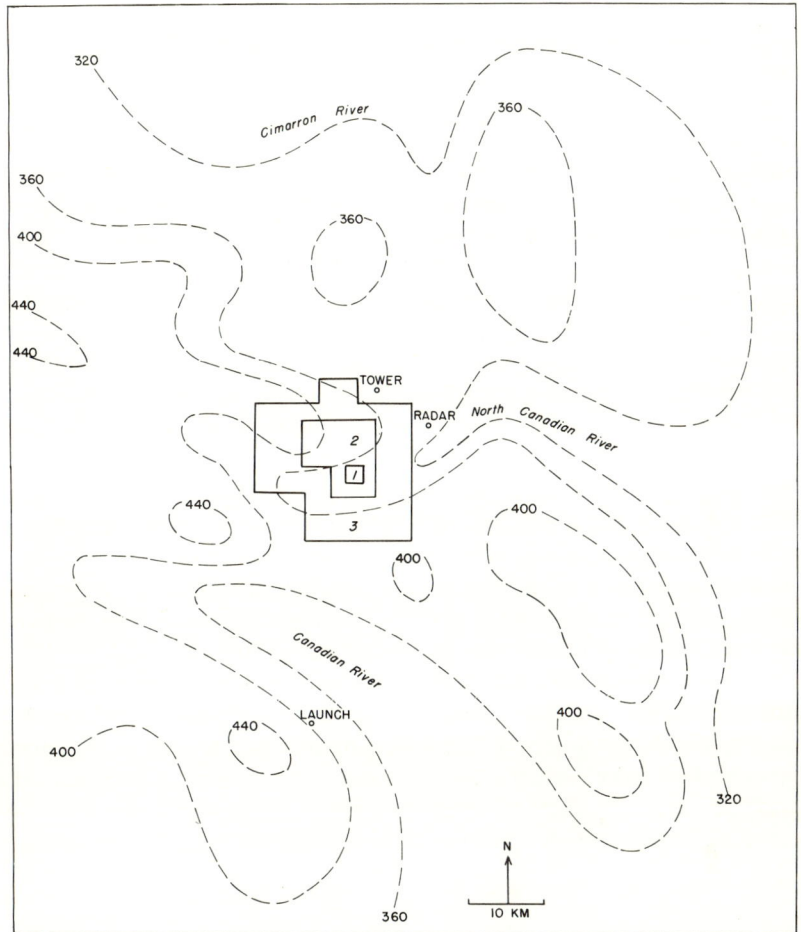

FIG. 2. Map of the Oklahoma City area showing the location of the instrumented WKY television tower, the tetroon-tracking radar, and the usual tetroon launch site. The solid lines divide Oklahoma City into downtown (1), industrial–commercial (2), and residential (3) areas. Terrain height in meters.

before passing by the tower, and as we shall see, this resulted in a shift of the Eulerian spectral peak toward higher frequency. Gill anemometers mounted on the tower at heights of 355 m and either 177 or 44 m provided vertical velocities in the Eulerian time system at intervals of $\frac{1}{10}$ sec. Eulerian vertical-velocity spectra for 67-min sampling intervals (to avoid zero filling) were then determined through the use of a Fast Fourier Transform (FFT) program, the variances being evaluated at equally spaced logarithmic frequency intervals by means of a subprogram developed by G. Herbert of the Air Resources Laboratories. The Eulerian spectra so obtained encompass periods of oscillation ranging from 17 min to 2 sec.

Tetroon positions were obtained (and stored on magnetic tape) at 1-sec intervals by means of a M-33 tracking radar from the Air Resources Field Research Office, National Reactor Testing Station, Idaho Falls. Tetroon launch mobility was provided by a large truck fitted out as an inflation van. As usual, transponders were attached to the tetroons to avoid the problem of ground clutter and hence to permit accurate tetroon positioning at very low elevation angles. The tetroons were inflated to float at 300 m in order that realistic comparisons could be made with the tower data, but because of the updrafts induced by the city, the mean tetroon height for the flights used herein was 360 m. On the average, each tetroon was positioned for 125 min, corresponding to an average tracking distance of about 80 km.

The tetroon positions at 1-sec intervals were averaged over 30 sec, yielding vertical velocities at 30-sec intervals in the Lagrangian (actually quasi-Lagrangian) system. The radar positioning was not thought sufficiently accurate to provide meaningful vertical velocities for shorter time intervals. Lagrangian vertical-velocity spectra, based on data for the entire length of each tetroon flight, were determined by means of a conventional spectrum analysis with a maximum lag of 60 min. The Lagrangian spectra so obtained encompass periods of oscillation ranging from 60 to 3 min.

Since all the tower data have not yet been reduced, and since each tetroon flight was tracked for about twice the tower sampling time (67 min), the simultaneous data sample for the following analysis consist of 21 tetroon spectra and 36 tower spectra. Accordingly, for purposes of comparison, a tetroon spectrum is often matched with two successive tower spectra. All spectra are based on vertical-velocity data obtained essentially in daytime.

2. Composite Eulerian and Lagrangian Spectra

Figure 3 presents composite Eulerian and Lagrangian $nS(n)$ vertical-velocity spectra based on the 10 hourly intervals when the tower was upwind of the city and the associated seven tetroon flights did not pass over the city. At a height of 355 m the tower (Eulerian) spectral peak is found at a period of 5.9 min (the exact period is determined by comparison of the variances each side of the peak variance), while the tetroon (quasi-Lagrangian) spectral peak is found at a period of 14.0 min, yielding a mean value for the Lagrangian–Eulerian time scale ratio β of 2.4. Since the mean tetroon height was 310 m, and the dominant Lagrangian period of oscillation increases somewhat with increasing height (Fig. 4), the value of 2.4 is a slight underestimate.

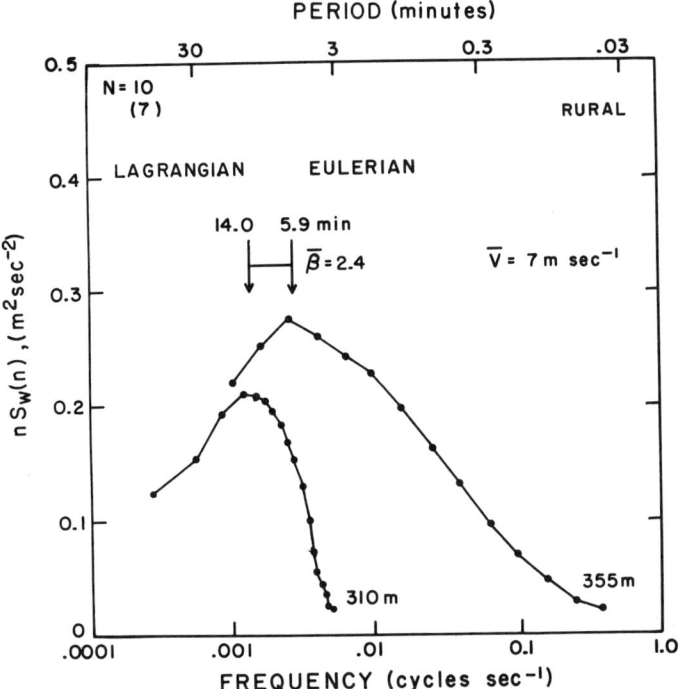

FIG. 3. Composite Eulerian and (quasi)-Lagrangian $nS(n)$ vertical velocity spectra based on tower data upwind of the city and tetroon flights which did not pass over the city (hereafter labeled "rural"). Indicated are the respective spectral-peak periods and the derived Lagrangian–Eulerian period-ratio β. The number of hourly tower samples and simultaneous tetroon flights (parenthesis) entering into the composite are given at upper left, with the respective mean heights at bottom.

The greater area under the Eulerian curve shows that the tetroons are underestimating the vertical velocity variance by about one-third, a result to be expected due to the restoring force acting to return the tetroon to its equilibrium float surface. A computer program has been developed (Hoecker and Hanna, 1971) which, based on the equation of motion, estimates the "air parcel" vertical velocity at the position of the tetroon given the tetroon vertical velocity, the displacement of the tetroon from its equilibrium float surface, and the drag coefficient (about 0.75) of the tetroon. This program was run for all the Oklahoma City tetroon flights, and while the low-frequency variance was increased thereby, there was no appreciable shift in the (composite) Lagrangian spectral peak, and accordingly in the following we use the tetroon vertical velocities directly.

With the assumption that the turbulent eddies are frozen and move with

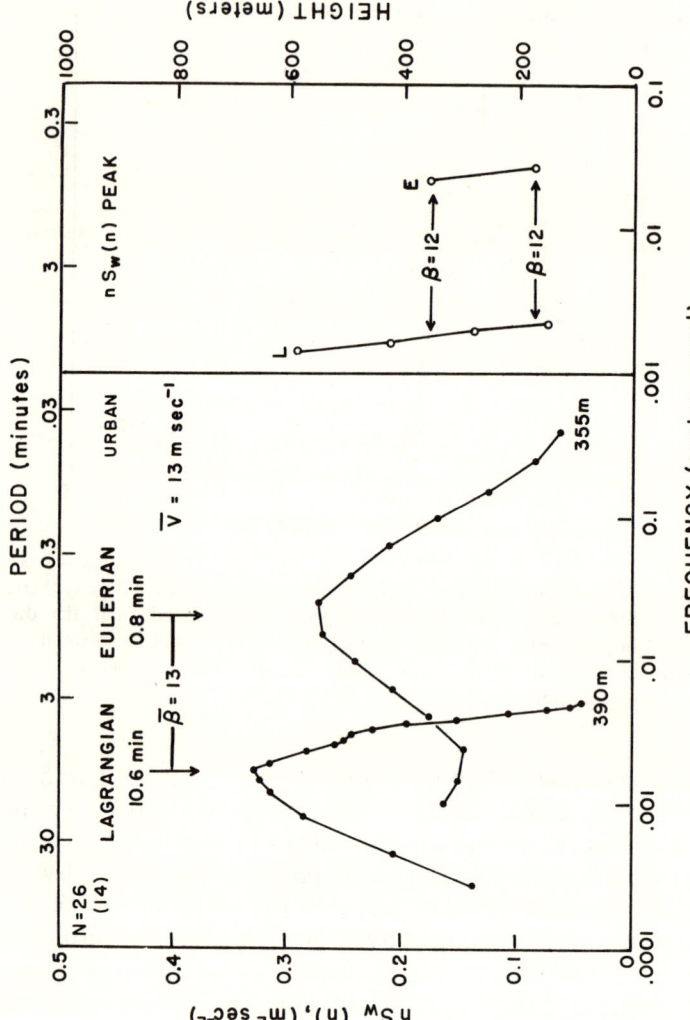

FIG. 4. Composite Eulerian and (quasi)-Lagrangian $nS(n)$ vertical velocity spectra based on tower data downwind of the city and tetroon flights which passed over the city (hereafter labeled "urban"). Indicated at right is the mean variation with height of the respective spectral peaks. Otherwise, see Fig. 3 legend.

the mean wind speed, the above combination of wind speed and dominant Eulerian period of oscillation implies a mean eddy diameter of 1200 m, or slightly more than three times the height above ground. Since other investigators have found a similar ratio (Pasquill, 1962), the "rural" Eulerian spectral-peak frequency appears fairly representative, and this is confirmed by the location of the spectral peak at a nondimensional or reduced frequency (nZ/V, where V is wind speed at height Z) of 0.15, rather close to the mean value of 0.30 obtained by Busch and Panofsky (1968).

Figure 4 shows the composite Eulerian and Lagrangian spectra based on the 26 hourly intervals when the tower was downwind of the city and the associated 14 tetroon flights passed directly over the city from the launch site indicated in Fig. 2. The mean tetroon height of 390 m (90 m above equilibrium level) is the result of the updrafts induced by the city, but these updrafts, reflecting a low-frequency contribution to the total variance, should not significantly influence either the Lagrangian or Eulerian spectral-peak frequencies. The Eulerian spectral peak is now found at a period of 0.8 min (in comparison with 5.4 min in the rural case) and the Lagrangian spectral peak at a period of 10.6 min (in comparison with 14.0 min in the rural case). A shift toward higher frequency of the Eulerian spectral peak would be anticipated because of the twofold greater wind speed in the urban case (13 versus 7 m sec^{-1}), but most of the observed sevenfold shift in dominant period of oscillation must be due to the presence of relatively small-scale eddies formed by the urban roughness elements located 5–10 km upwind of the tower (Pasquill, 1972). Inasmuch as the tetroon is within such an urban boundary layer for only a short time, the derived mean β of 13 is believed to be unrepresentative.

In this connection we note that, with the assumption of frozen turbulence, when the tower is downwind of the city the mean eddy diameter at a height of 355 m is about 300 m, or approximately the height above ground. However, as pointed out earlier, other investigators have found the eddy size to be about three times the height above ground, and this strongly suggests that the "urban" Eulerian spectral-peak frequency is atypical, as does the value of 0.57 for the reduced frequency of this spectral peak. It is interesting, though, that the magnitude of the spectral peak appears fairly representative, the mean stress of 4 dynes cm^{-2} resulting in a value of 0.7 for the ratio of (spectral peak) vertical velocity variance and square of the friction velocity, similar to the mean value of 0.5 obtained by Busch and Panofsky.

One should also consider the representativeness of the Lagrangian data obtained at Oklahoma City. The dominant period of 14 min in rural areas at a mean height of 310 m is in good agreement with the value of 18 min obtained at a mean height of 460 m near Las Vegas, Nevada, (Angell *et al.*, 1971) and the value of 16 min obtained at a mean height of 800 m at

Cardington, England (Angell, 1964). However, the 10-min period obtained when the tetroons pass across Oklahoma City appears rather small, possibly pointing up subtle urban influences on the Lagrangian period of oscillation as well.

The right-hand diagram of Fig. 4 illustrates the mean variation with height of Lagrangian and Eulerian spectral-peak frequencies, as estimated from tetroon flights across the city and from tower data downwind of the city. In both cases the dominant period of oscillation increases with increasing height, but equally so, implying little change in β with height. Because the mean tetroon height was 35 m greater than the anemometer height of 355 m, the mean value of β for the urban sample is really 12 rather than 13. In the following, the individual Lagrangian spectral-peak frequency values have been "normalized" to a height of 355 m by assuming a variation with height given by the L line in Fig. 4. This, of course, also results in the normalization of β to the 355-m level.

3. Variation of β with Meteorological Parameters

Figure 5 illustrates the variation of β with wind speed (measured at a height of 355 m on the tower), as determined from individual Lagrangian–Eulerian spectral comparisons. When the tower is downwind of the city and the tetroons pass over the city, the trace is labeled "urban", whereas when the tower is upwind of the city and the tetroons do not pass over the city the trace is labeled "rural." In both cases, β increases with increase in wind speed, as would be anticipated from the well-known shift of the Eulerian spectral peak toward higher frequency with increase in wind speed (the turbulence elements are transported more rapidly past the fixed point in strong winds), and the observation that the frequency of the Lagrangian spectral peak is almost independent of wind speed (being related more to atmospheric stability).

Note that while there is a discontinuous jump in the absolute value of β between urban and rural areas, the percentage change in β per unit (1 m sec^{-1}) increase in wind speed remains nearly the same, with average values of 11 and 13% in urban and rural areas, respectively. Thus, the increased surface roughness of the urban areas does not appreciably affect the *relative* variation of β with wind speed. The small circles in Fig. 5 represent our estimate of the actual mean values of β for wind speeds exceeding 10 m sec^{-1}, obtained on the assumption that the derived rural values of β are representative and that the percentage change in β (per unit increment of wind speed) is 12%. With this assumption, β is indicated to vary from a mean value of 2 at a wind speed of 6 m sec^{-1} to a mean value of 6 at a wind speed of 15 m sec^{-1}. The letters in Fig. 5 show the mean values of β at given

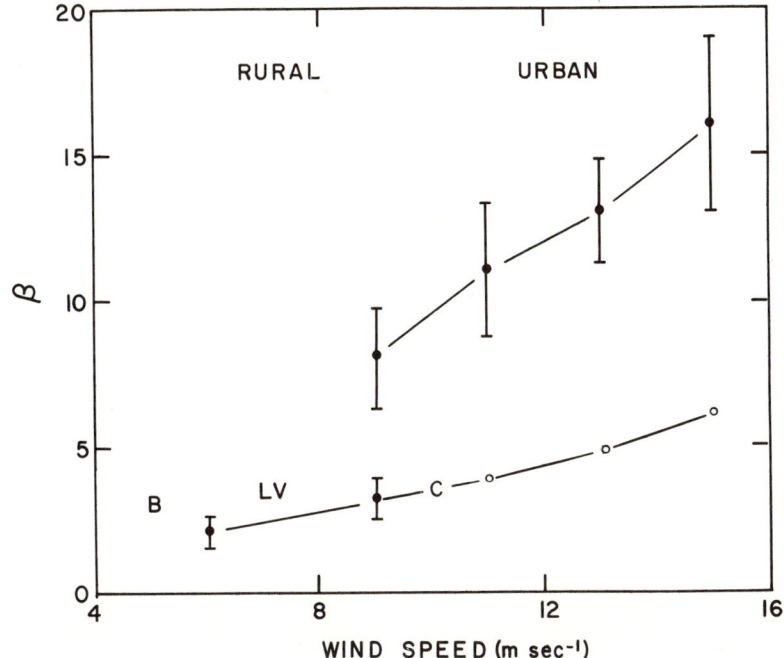

FIG. 5. Variation of β with wind speed under urban and rural conditions, with the vertical bars extending one standard deviation of the mean either side of the mean value. The small circles represent the attempt to normalize for urban effects (see text). Indicated are the mean β values obtained from earlier free-floating balloon experiments at Brookhaven (B), Las Vegas (LV), and Cardington (C).

mean wind speeds determined from earlier free-floating balloon experiments at Brookhaven (B), New York (Gifford, 1955), Las Vegas (LV), and Cardington (C). With the above assumption, the mean Cardington value falls exactly on the trend of β with wind speed obtained at Oklahoma City.

Figure 6 illustrates the variation of β with vertical turbulence intensity (measured at a height of 355 m on the tower), as determined from individual Lagrangian-Eulerian spectral comparisons. Since β is found to increase with decrease in root mean square vertical velocity (σ_w) at a given wind speed (for example, β increases by 40 % as σ_w decreases from 1.6 to 0.8 m sec^{-1} at a wind speed of 13 m sec^{-1}), β would be expected to be quite sensitive to vertical turbulence intensity (σ_w/V), and Fig. 6 shows that this is indeed the case. As with wind speed, there is a discontinuous jump in the absolute value of β between urban and rural areas, but again the percentage change in β per unit (.01) increase or decrease in vertical turbulence intensity remains nearly the same, with average values of 13 and 11 % in urban and rural areas,

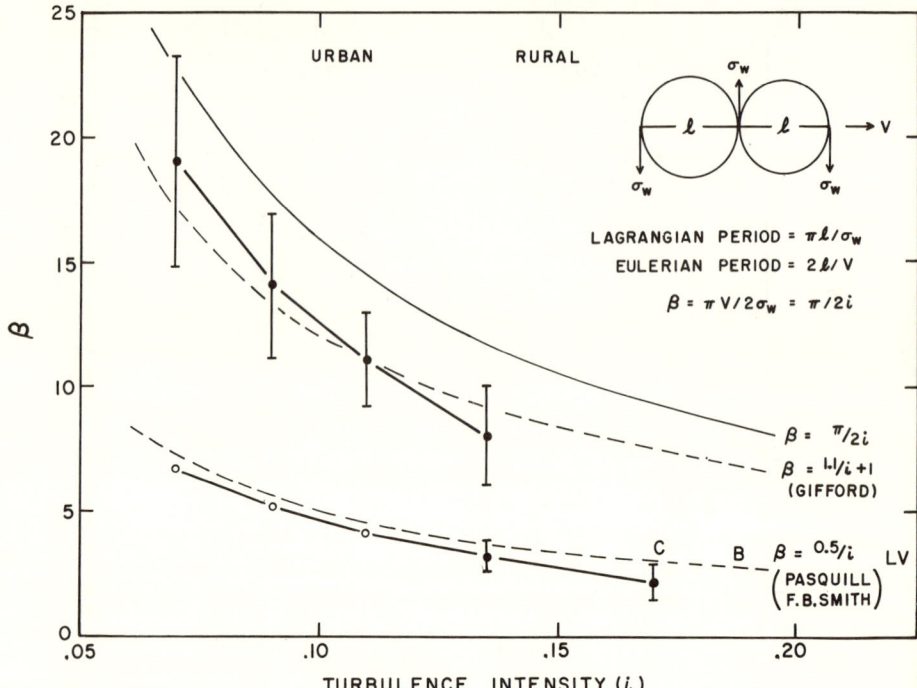

FIG. 6. Variation of β with vertical turbulence intensity under urban and rural conditions, with the vertical bars extending one standard deviation of the mean either side of the mean value. The small circles represent the attempt to normalize for urban effects (see text). Indicated for comparison is the variation derived from the steady-state "cartwheel" concept of eddies (light solid line) and from consideration of eddy decay and rebuilding and similarity theory (light dashed lines), as well as the mean β values obtained from earlier free-floating balloon experiments at Cardington (C), Las Vegas (LV), and Brookhaven (B).

respectively. The small circles in Fig. 6 represent our estimate of the actual mean values of β for turbulence intensities less than .12, obtained on the assumption that the derived rural values of β are representative and that the percentage change of β (per unit increment of turbulence intensity) is 12%. With this assumption, β is indicated to vary from a mean value of 7 at a turbulence intensity of .07 to a mean value of 2 at a turbulence intensity of .17.

It is of interest to compare the above relation between β and vertical turbulence intensity with that derived from the simple steady-state "cartwheel" concept of eddy structure. As shown at upper right in Fig. 6, if atmospheric eddies are visualized as circular vortices moving with the mean wind speed, a particle moving along the perimeter of the vortex would have a (Lagrangian) period of oscillation $\pi l/\sigma_w$ (where l is vortex diameter and σ_w is the typical perimeter velocity), whereas a fixed instrument on the track of the vortex would observe a (Eulerian) period of oscillation $2l/V$, where V is wind speed. The ratio of these two periods, β, is $\pi/2i$, where i is the vertical intensity of turbulence σ_w/V. The light solid line in Fig. 6 is the expression of this relationship, and it is apparent that even the large (and presumably unrepresentative) β values determined when the tower was downwind of the city and the tetroons passed over the city are not this big, implying that the "cartwheel" concept of eddies is an unsatisfactory one.

This brings up the question of the extent of eddy decay and rebuilding during the time the eddy moves past the fixed point instrument and during the time it takes the particle (tetroon) to move around the eddy perimeter. This problem has been considered by Gifford (1955) and Smith (1967), as well as others, and the light dashed lines in Fig. 6 represent the variation of β with turbulence intensity (i) derived from their theoretical work assuming a nonsteady eddy structure. Gifford obtained the expression $\beta = (1.1/i) + 1$, and while this expression agrees well with the "urban" β values found at Oklahoma City, again these particular values are almost assuredly not representative. Smith obtained several expressions for β depending on the assumed form of the correlogram, but in general they can be expressed by $\beta \approx 0.5/i$. This expression agrees well with the Oklahoma City data as normalized for city effects. Furthermore, note that the normalized Oklahoma City data are in almost exact agreement with Pasquill's (1968) expectation, based on similarity theory, that $\beta = 0.44/i$.

Finally, since β exhibits such a strong dependence on vertical turbulence intensity, the only satisfactory way to estimate the dependence of β on lapse rate (which, of course, is highly correlated with vertical turbulence intensity) is by means of a two-dimensional plot with coordinates of lapse rate and turbulence intensity. Such a plot shows that, at a given turbulence intensity, β tends to increase with decrease in lapse rate, with the increase becoming quite rapid in near-isothermal conditions. For example, β is found to increase by 60 % as the lapse rate changes from dry adiabatic to isothermal at a vertical turbulence intensity of .10. This result is in overall agreement with that obtained by Gifford (1967) from a survey of the then available data.

4. Comparison with Previous Results

Previous estimates of β have usually been based on Taylor's diffusion equation, and thus involved determination of the Lagrangian autocorrelation coefficient from a double differentiation of the mass-tracer concentration in the lateral (or vertical) direction, and comparison with the Eulerian correlation coefficient obtained from a fixed-point wind instrument. It has been shown by Haugen (1966) that the double-differentiation requirement introduces considerable uncertainty in the results obtained.

When β is plotted as a function of turbulence intensity, it is found that the Oklahoma City "rural" trace of Fig. 6 (including the normalization for urban effects) is pretty well bracketed by the values obtained from mass tracer experiments by Hay and Pasquill (1959), Panofsky (1962), and Haugen (1966), with Haugen obtaining β values about two-thirds as large, and Hay and Pasquill and Panofsky values about five-fourths as large (at a given turbulence intensity) as the values obtained at Oklahoma City. None of the above data agree with the results of Islitzer and Dumbauld (1963) since they found essentially no variation of β with turbulence intensity. In comparisons of this kind it should be kept in mind that the mass-tracer experiments generally yield β values for the lateral dimension whereas the tetroons, as used herein, yield them for the vertical dimension, and there is no guarantee that β is the same in the two directions.

5. Conclusion

The Oklahoma City tetroon and tower data are of high quality, but nevertheless, unrepresentative values of the Lagrangian–Eulerian scale ratio β have been obtained because, when the tower is downwind of the city, the tower data always reflect the relatively high frequency oscillations set up by the urban roughness elements whereas the tetroons spend only a brief time within this urban boundary layer. Because of the uncertainty surrounding the procedure used herein to correct for the urban influence on β, attempts will be made to isolate these portions of the tetroon trajectories which might reasonably be expected to be influenced by the city, and to obtain representative Lagrangian urban spectra for comparison with the Eulerian urban spectra measured on the tower. In general, it is believed desirable to analyze and understand these observations more fully before embarking on any new tetroon experiments aimed at evaluation of the Lagrangian–Eulerian scale ratio.

Acknowledgments

This experiment could not have been carried out without the able assistance of C. R. Dickson, G. White, L. Wendell, G. Start, J. Young, and H. Boen of the Air Resources Field Research Office, National Reactor Testing Station, Idaho Falls, and of W. Hoecker and A. Giarrusso of the Air Resources Laboratories' office in Silver Spring, Md. G. Herbert (then) of the latter office oversaw the acquisition of the tower data and its reduction. D. H. Pack has provided encouragement and pertinent comments throughout the analysis, and F. Pasquill uncovered a serious error in the interpretation of the tower spectra and provided useful guidance concerning the material worthy of presentation.

References

Angell, J. K. (1964). *Quart. J. Roy. Meteorol. Soc.* **90**, 57–71.
Angell, J. K., Pack, D. H., Hoecker, W. H., and Delver, N. (1971). *Quart. J. Roy. Meteorol. Soc.* **97**, 87–92.
Busch, N. E., and Panofsky, H. A. (1968). *Quart. J. Roy. Meteorol. Soc.* **94**, 132–148.
Gifford, F. (1955). *Mon. Weather Rev.* **83**, 293–301.
Gifford, F. (1967). *In* "USAEC Meteorological Information Meeting" (C. A. Mawson, ed.), pp. 485–502. Chalk River Nucl. Lab., Ontario.
Haugen, D. A. (1966). *J. Appl. Meteorol.* **5**, 646–652.
Hay, J. S., and Pasquill, F. (1959). *Advan. Geophys.* **6**, 345–365.
Hoecker, W. H., and Hanna, S. R. (1971). *NOAA Tech. Memo* ERL ARL-31, 31 pp. U.S. Dep. Commerce, Boulder, Colorado.
Islitzer, N. F., and Dumbauld, R. K. (1963). *Int. J. Air Water Pollut.* **7**, 999–1022.
Panofsky, H. A. (1962). *Quart. J. Roy. Meteorol. Soc.* **88**, 57–69.
Pasquill, F. (1962). "Atmospheric Diffusion," 297 pp. Van Nostrand-Reinhold, Princeton, New Jersey.
Pasquill, F. (1963). *Weather* **18**, 233–246.
Pasquill, F. (1968). *In* "Symposium on the Theory and Measurement of Atmospheric Turbulence and Diffusion in the Planetary Boundary Layer" (F. V. Hansen and J. D. Shreve, eds.), pp. 17–30. Sandia Lab., Albuquerque, New Mexico.
Pasquill, F. (1972). *Quart. J. Roy. Meteorol. Soc.* **98**, 469–494.
Slade, D. H., ed. (1968). "Meteorology and Atomic Energy," 445 pp. U.S. At. Energy Comm., Oak Ridge, Tennessee.
Smith, F. B. (1967). *In* "USAEC Meteorological Information Meeting" (C. A. Mawson, ed.), pp. 476–484. Chalk River Nucl. Lab., Ontario.

EFFECT OF MOLECULAR DIFFUSION ON THE STRUCTURE OF A TURBULENT DENSITY INTERFACE

P. F. LINDEN AND P. F. CRAPPER[1]

*Department of Applied Mathematics and Theoretical Physics
University of Cambridge, Cambridge, England*

1. INTRODUCTION

The dispersal of pollutants when released into the environment is often affected by the action of buoyancy forces either because the pollutant has a different density from its immediate environment, or because there is stratification already present in the environment, or both. Consider, for example, the dispersion of smoke from a chimney stack; in general the smoke is hotter (and therefore less dense) than its surroundings and the buoyancy force so produced causes the smoke to rise in a turbulent plume, mixing with the environment as it does so. If there is an inversion present in the atmosphere above the stack, this rise is inhibited by the stable density stratification. In the ocean a similar phenomenon occurs with the oceanic thermocline tending to prevent the downward turbulent mixing of pollutant released above it.

On a smaller scale there are many observations of density stratification, particularly in the ocean (see, e.g., Gregg and Cox, 1972), which show that density microstructure does exist down to the scale of a few centimeters. A feature of a large proportion of this microstructure and of the larger scale structure mentioned above is that there exist relatively deep uniform layers separated by thin interfaces across which there is a significantly larger change in density. For structure with a vertical scale of the order of a few centimeters, it is clear the molecular diffusion may play an important role in its development. The determination of the role of molecular diffusion at such an interface is the aim of this investigation.

This paper considers the results of laboratory experiments in which turbulent motions produced by the vertical oscillation of horizontal grids are

---

[1] *Present address:* Department of Mechanical Engineering, University of Toronto, Toronto, Canada.

imposed on a density interface. This method of production of turbulence is convenient because it allows the properties of the turbulence and the interface to be set independent of one another. There are, on the other hand, two drawbacks: first the properties of the turbulence may be related to the special geometry of the tank and grids and second, the experiments are not completely steady. However, we will see that these drawbacks do not appear to be major and are certainly inconsequential when compared with the convenience of the experimental arrangement.

The rate at which fluid is entrained across the interface has been measured by Turner (1968). (We have used the same experimental arrangement as Turner so that our results may be compared with his.) He found that the rate of entrainment is reduced as the stability of the interface increased. A measure of this stability can be given in terms of a Richardson number $Ri = g \Delta\rho \, l/\rho u^2$, where u and l are the velocity and length scale of the turbulent motions near the interface, and $\Delta\rho/\rho$ is the fractional density step between the two layers. Turner considered two cases in detail; first when the density step was produced by a temperature step and second when it was produced by salinity difference between the layers. He expressed his results in the form

(1) $$u_e/u \propto Ri^{-1} \quad \text{(heat)},$$

(2) $$u_e/u \propto Ri^{-3/2} \quad \text{(salt)},$$

where u_e is the rate of entrainment (defined as the amount of fluid entrained/unit area/unit time). Turner further noted that (1) is equivalent to assuming that the rate of change of potential energy due to the mixing is proportional to the rate of kinetic energy production by the grid, and then discussed the possible reasons for the reduced salt transport at large Ri.

Clearly the main difference between the two situations described by (1) and (2) is that the coefficient of molecular diffusion of heat is much larger than the corresponding coefficient for salt. The appropriate nondimensional parameter which describes the effect of molecular diffusion is the Peclet number $Pe = ul/\kappa$. In terms of this parameter (1) is the entrainment rate for $Pe = O(10^2)$ while (2) is for the case $Pe = O(10^4)$.

In this paper we look at the detailed structure of a density interface for a range of Ri and Pe covering essentially the same values as those appropriate to Turner's (1968) experiments. We find that for the lower values of Pe molecular diffusion plays an essential role in determining the structure of the interface and that the energy argument given by Turner (1968), which neglects all diffusive effects, is not correct. The consequences of the results for the geophysical situations mentioned above are then discussed.

2. The Experimental Setup

Only a very brief description of the experimental setup is given here; for more details the reader is referred to Turner (1968) and Crapper and Linden (1974). Basically the apparatus consists of a tank containing two horizontal grids which may be oscillated vertically at varying frequencies and amplitudes. The details of the turbulence so produced have been determined by Thompson (1969) who made measurements of the rms velocity u and integral length scale l of the turbulence in a homogeneous fluid. Consequently all the results are presented in terms of u and l directly, the values used being the ones appropriate to the turbulence near the interface in each experiment.

Vertical profiles of temperature and salinity were taken by traversing either a thermocouple or a conductivity probe through the interface. Each probe had a vertical resolution of about 0.1 cm and was sampled approximately every 0.05 cm of its traverse. The output was digitized and stored on magnetic tape and then analysed at leisure. The accuracy of the temperature and salinity measurements were found to be 0.3 % and 0.1 % of full-scale, respectively.

As was mentioned in the previous section this experimental configuration is not completely steady since the density difference between the layers decreases with time. However, the rate of change of $\Delta\rho$ with time was very slow and even when the entrainment was most vigorous a typical change of 0.001 gm/ml in $\Delta\rho$ took about 60 min. Consequently, even measurements taken over a period of 2 min which were required for mean profiles can be considered as representative of the quasi-steady structure of the interface.

3. The Experimental Results

3.1. Temperature Interface

Some examples of temperature–depth profiles for varying Ri and Pe are shown in Fig. 1. It is seen that all the profiles show two well-mixed layers separated by a region of essentially constant temperature gradient. On occasions the layers contain inversions (see Fig. 1b) presumably due to a mixing event taking place. It is seen that the interface becomes thinner as Pe increases.

In order to quantify this last statement we define an "interface thickness" by fitting a straight line to the central 50 % of the data points in the interface and extrapolating this line until it intersects the mean temperatures of the upper and lower layers. The vertical distance between these two intersections is defined as the interface thickness h.

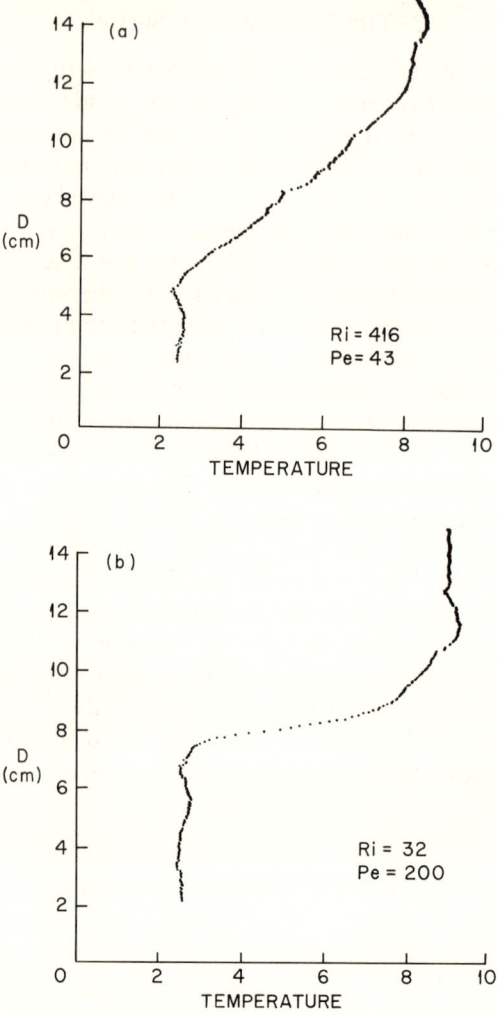

FIG. 1. Instantaneous profiles of temperature against depth for different values of Ri and Pe. Note the increased sharpness of the interface as the Peclet number increases.

Figure 2 shows the variation of h with ΔT (in the range $3 \leq \Delta T \leq 24°C$) when u and l are held fixed. There appears to be some evidence for a decrease in h as ΔT decreases, but in general there is little variation over the range. Consequently the decrease in h as Pe increases appears to be due to an increase in the turbulent intensity u (l remains fixed in these experiments).

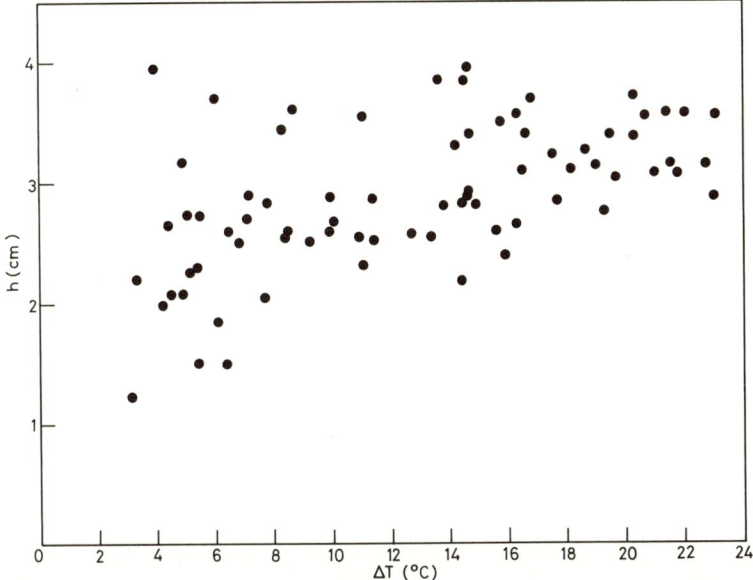

FIG. 2. The interface thickness plotted against the temperature difference across the interface for fixed u and l ($u = 0.22$ cm/sec, $l \approx 1$ cm).

This feature is shown in Fig. 3 which shows plots of h against u, for fixed $\Delta\rho$. There is evidence that at higher values of u the interface thickness appears to be approaching a constant value of approximately 2 cm.

3.2. Salinity Interfaces

The instantaneous profiles of salinity with depth shown in Fig. 4 exhibit the same basic features (two well-mixed layers separated by a region of constant salinity gradient) as the temperature profiles. However, one immediate difference is noticeable, viz there is little variation of h over the range of Ri and Pe shown. (The interface thickness h is defined in the same way as for the temperature interfaces.) This is shown in a quantitative way in Fig. 5 where h is plotted against u for $0.002 \leq \Delta\rho \leq 0.02$ gm/ml.

In general the experiment was set up in such a way that the two layers were allowed to diffuse for several days until a thick interface (≈ 10 cm) had formed; then the stirring was begun. It was found that the edges of this interface were eroded by the turbulent motions until eventually a "steady" value was reached. Figure 6 shows the interface thickness plotted against time during this erosion process for three different Richardson numbers; the

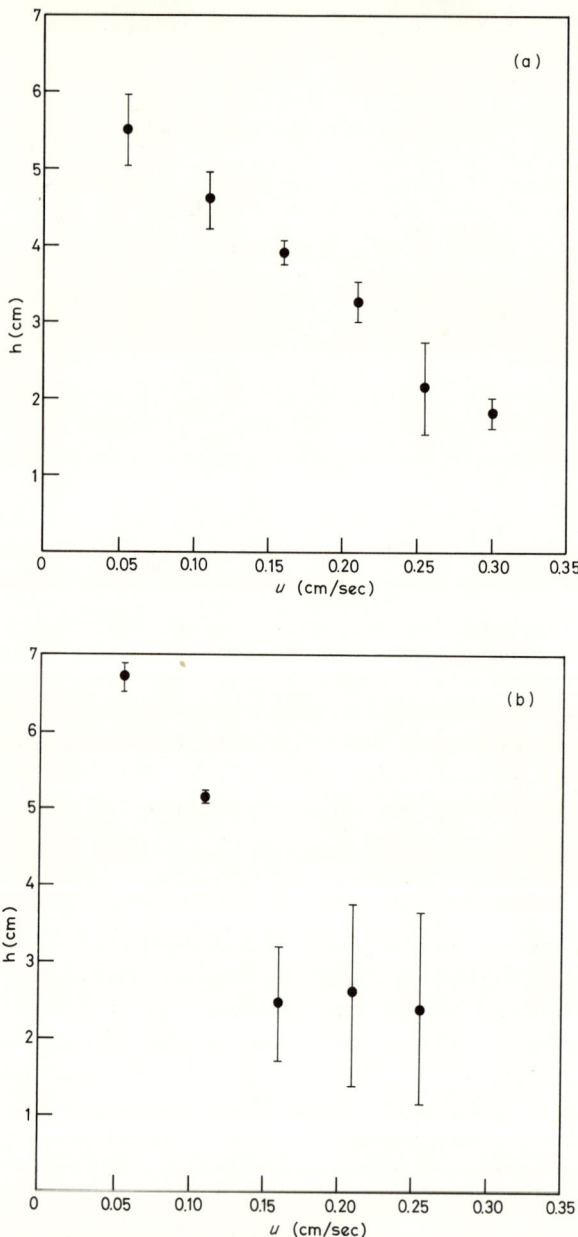

FIG. 3. The interface thickness plotted against the turbulent velocity for fixed values of $\Delta\rho$: (a) $\Delta\rho = 0.005$ gm/ml; (b) $\Delta\rho = 0.0017$ gm/ml.

STRUCTURE OF DENSITY INTERFACES

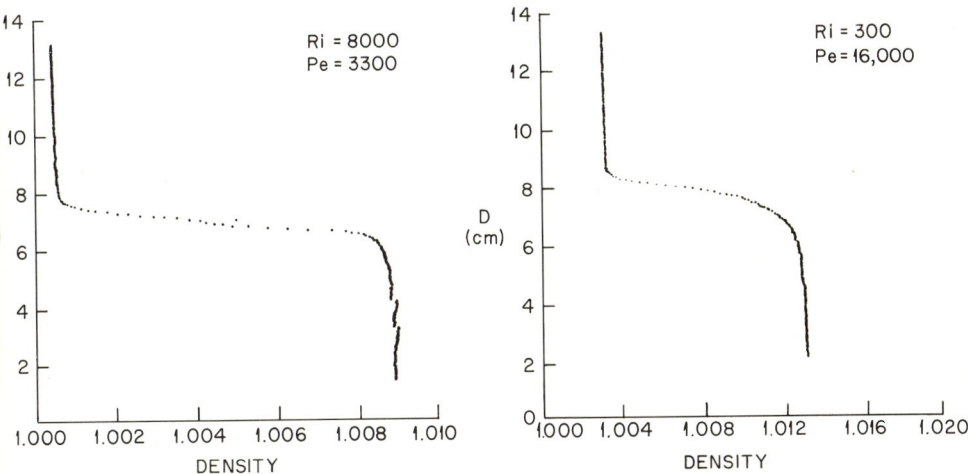

FIG. 4. Instantaneous profiles of density, produced by salinity, versus depth for various values of Ri and Pe.

erosion is most rapid when the Richardson number is least. The fact that the interface thickness then remains constant with further increase in time is also shown in Fig. 6. Visual observation with a shadowgraph showed that while the erosion was taking place the centre of the interface remained undisturbed while the edges were eroded by the turbulence. This was confirmed by profiles such as those shown in Fig. 7 where the original gradient can be seen to exist in the centre of the interface even though the edges have been eroded considerably.

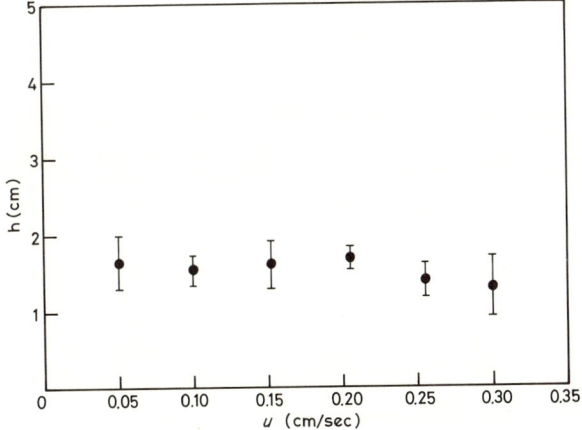

FIG. 5. The thickness of a salinity interface plotted against the turbulent velocity u. The values represent averages of all profiles at each value of u with Δp varying from 0.002 to 0.02 gm/ml.

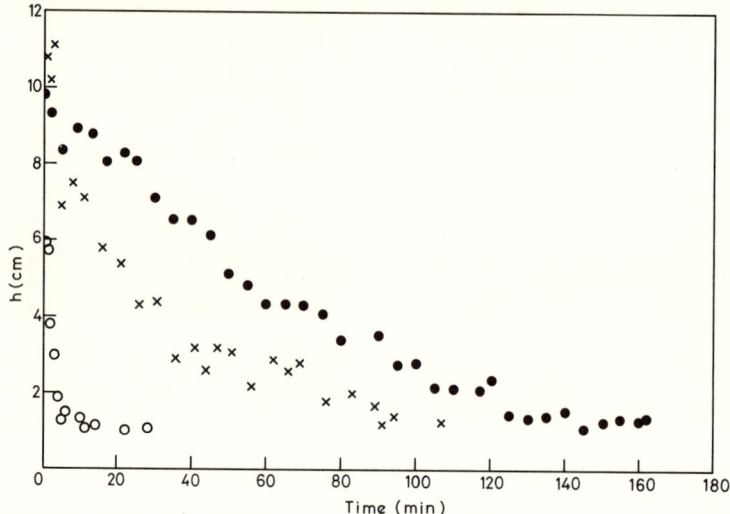

FIG. 6. The interface thickness plotted as a function of time for various values of the Richardson number. ● Ri = 280; × Ri = 97; ○ Ri = 31.

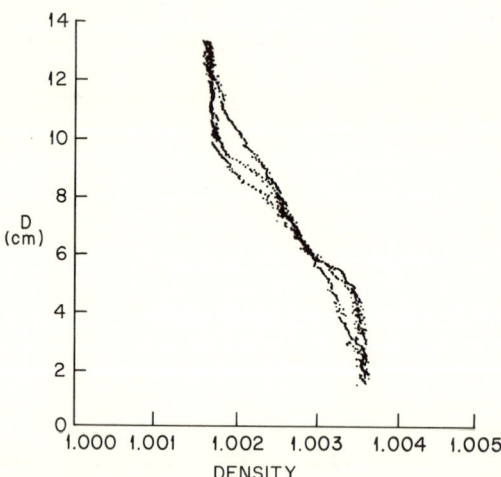

FIG. 7. Profiles of salinity interfaces taken at time intervals of 30 min. Note the erosion of the edges of the interface leaving the original gradient unchanged in the centre.

4. Discussion

The work described above was essentially motivated by the measurements of the entrainment rate across a density interface made by Turner (1968) which showed that fluid was entrained at different rates depending on whether the density step was produced by heat or salt. The two nondimensional parameters describing the turbulent flow (ignoring molecular viscosity) viz the Peclet number $Pe = ul/\kappa$ and the Richardson number $Ri = g\, \Delta\rho\, l/\rho u^2$ both depend on the turbulent velocity u in such a way that higher is Ri associated with lower Pe. Thus the divergence in the entrainment rates found by Turner (1968) at high Ri occurs when the Peclet number for the salt interface is much larger than the Peclet number for the heat interface.

The structure of the interface shows why the observed entrainment rates should differ at high Ri. At high values of Pe it was found that the thickness of the interface remained constant (and independent of Ri) at about 1.5 cm; we shall describe this interface as dynamically "thin" in that eddies in one layer were observed to penetrate through the interface and interact directly with fluid in the other layer. At low values of Pe ($\lesssim 200$) the interface was "thick" with the eddies not penetrating to the centre of the interface. In this latter case we conclude that the centre of the interface consists of a diffusive core which provides a flux, by molecular diffusion alone, which balances the entrainment flux from the edges of the interface into the turbulent layers. On the other hand, when the interface is thin it is not possible to provide a large enough flux by diffusion and mechanical breakdown with direct mixing of fluid from each layer must occur.

The conclusion that a thick interface contains a diffusive core is supported by two observations. First, estimates of the flux by diffusion along the gradient in the core showed that this was comparable to the observed fluxes into the well-mixed layers. Second, the gradient in the centre of the interface always adjusted very quickly when the interface changed in thickness implying a balance between the diffusive and entrainment fluxes; this is in contrast to the salt interface where the original gradient could be maintained in the interior of the interface for a considerable period of time.

At high Pe, when the interface is thin, and mechanical breakdown of the interface is required to produce the required amount of entrainment, it would seem reasonable to attempt to explain the entrainment process in terms of inertial processes ignoring the effects of diffusion. A mechanistic explanation along these lines has been provided by Linden (1973) which accounts for the observed $Ri^{-3/2}$ mixing rate. At lower values of Pe ($\lesssim 200$) the eddies can no longer penetrate through the interface and it is essential to consider the effects of molecular diffusion. Consequently the energy argument used by Turner (1968) cannot be valid; an alternative satisfactory explanation of the Ri^{-1} law has not yet been provided.

The above results indicate that when the Peclet number is large and molecular diffusion is unimportant the thickness of the interface is independent of the Richardson number at about 1.5 cm. This equilibrium thickness must result, in a way not yet understood, from a balance between the scouring of the interface due to the turbulent eddies and the production of intermediate density fluid by mixing across the interface. A concept often used to describe this type of balance when the turbulence is produced by shear at the interface is one of a critical gradient Richardson number. The appropriate gradient Richardson number for the geometry of these experiments is $Ri' = g\,\Delta\rho\,h/\rho u^2$, where h represents the interface thickness. It is clear from an examination of the data (see Fig. 5) that Ri' does not remain constant for a high Peclet number interface. This fact points to one of the major differences between these experiments described here and those involving a mean shear between the layers on either side of the interface (e.g., Kato and Phillips, 1969). It is conceivable that the results obtained in the absence of mean flows may be applied to the case where mean shears produce turbulence due to instability of the interface provided that the energies of the turbulence in both situations is the same. However, in the latter case the interface thickness is also affected by the likelihood of shear instability and this essential feature is missing from the experiments described in this paper.

We return now to the geophysical examples cited in the introduction. In the case of mixing across an atmospheric inversion or the oceanic thermocline, typically $Pe \gtrsim 10^8$. At these high values of the Peclet number molecular diffusion can be neglected. In the case of oceanic microstructure reported by Gregg and Cox (1972) they reported temperature interfaces of thickness varying between 0.8 and 4.1 cm (see their Table 2). In the latter case it seems that molecular diffusion would be an essential determining feature of the structure and fluxes across that interface.

One aspect of pollutant disposal discussed recently by Fischer (1971) is the effect of double-diffusion convection on sewage deposited in the sea. Fischer pointed out that in some circumstances vigorous salt-finger convection could occur, enhancing the vertical spread of the sewage cloud. Linden (1971) discussed the effect of turbulent motions on salt-fingers and it is appropriate to mention some of the results in relation to the work discussed in this paper. In particular, Linden showed that the length of the fingers formed at an interface decreased as the intensity of the turbulence increased (or equivalently as the Peclet number of the motions increased). His measurements indicated values of the interface thickness, as determined from vertical temperature profiles, very similar to those obtained for a simple temperature interface. We can define a "diffusivity" for the finger interface as the ratio of the heat flux to the temperature gradient. Linden's (1971) data then are consistent with the notion that is appropriate to discuss the thickness of the interface in terms of a Peclet number based on this value of the diffusivity. This implies that in circumstances where the finger convection is vigorous (as in some examples of sewage disposal, see Fischer, 1971) we may find that even in the ocean the interface acts as though it is a low Peclet number structure.

Acknowledgments

This work was supported by a grant from the Natural Environment Research Council.

References

Crapper, P. F., and Linden, P. F. (1974). The structure of turbulent density interfaces. *J. Fluid Mech.* (in press)

Fischer, H. B. (1971). The dilution of an undersea sewage cloud by salt fingers. *Water Res.* **5**, 909.

Gregg, M. C., and Cox, C. S. (1972). The vertical microstructure of temperature and salinity. *Deep-Sea Res.* **19**, 355.

Kato, H., and Phillips, O. M. (1969). On the penetration of a turbulent layer into a stratified fluid. *J. Fluid Mech.* **37**, 643.

Linden, P. F. (1971). Salt fingers in the presence of grid-generated turbulence. *J. Fluid Mech.* **49**, 611.

Linden, P. F. (1973). The interaction of a vortex ring with a sharp density interface: a model for turbulent entrainment. *J. Fluid Mech.* **60**, 467.

Thompson, S. M. (1969). Turbulent interfaces generated by an oscillating grid in a stably stratified fluid. Ph.D. Thesis, Univ. of Cambridge, Cambridge, England.

Turner, J. S. (1968). The influence of molecular diffusivity on turbulent entrainment across a density interface. *J. Fluid Mech.* **33**, 639.

RENORMALIZATION FOR THE WIENER–HERMITE REPRESENTATION OF STATISTICAL TURBULENCE

W. C. MEECHAM

School of Engineering and Applied Science
University of California, Los Angeles, California 90024, U.S.A.

1. INTRODUCTION

One of the methods of treatment of statistical turbulence involves the use of the Wiener–Hermite representation. The representation has been discussed in previous publications (Imamura *et al.*, 1965; Meecham and Jeng, 1968; Meecham, 1969). It has been used with some success in the treatment of diffusion problems by Saffman (1969). Early mathematical work on such representations was done by Cameron and Martin (1957). The treatment is based upon the white noise process—a process with a constant spectrum. Linear combinations of such white noise processes are sufficient to represent a Gaussian process. Non-Gaussian processes may be represented through the use of polynomial combinations of the white noise process. The favored forms are patterned after Hermite polynomials (hence the name of the representation). Some of the earliest work involving the use of white noise processes was in connection with Brownian motion problems. For a review of this work, see Wiener (1958).

The Wiener–Hermite (W–H) representation has been applied in a number of different problem areas; we shall be interested here in fluid mechanics applications of course. The methods have been used in connection with the full Navier–Stokes equations (see Meecham and Jeng, 1968). A number of investigators have used the method in treating a model equation described by Burgers (1950); for this work see, for instance, Meecham and Siegel (1964). The results of this previous work may be summarized as follows: the method gives fairly accurate and useful results for initial fluctuation Reynolds numbers up to about 20 for computation times of order 2 or 3 (in units of scale size divided by rms velocity fluctuation). For larger Reynolds numbers or longer times, the method becomes difficult to manage in its simplest form. This will be discussed more fully below. The reason is somewhat as follows: For non-Gaussian processes the Wiener–Hermite expansion involves in general a large number of terms in the series. For initial-value problems involving processes near Gaussian, one need *initially* employ only the first two terms of the series (continuing to higher order terms is in general computationally impractical). However, as time goes on (in particular for larger Reynolds numbers) the processes require progressively more terms for adequate representation: the representation becomes less rapidly convergent. The difficulty was discussed by Orszag and Bissonette (1967). If one employs a white noise process in the representation, which process is fixed for all time, then as time progresses the physical process becomes less and less well related to the initial white noise process. The representation has been shown by Wiener to be complete for processes of physical interest. In order to represent the later process in terms of the early white noise it turns out that more and more terms in the series are required.

To remedy this difficulty various investigators have proposed altering the white noise process with time so as to reduce the effect of this out-of-date characteristic: see Bodner (1969), Canavan and Leith (1968), and Clever and Meecham (1972). The result of this work shows a considerable improvement over earlier results, and examples from Burgers' model are presented below.

One may deduce from this discussion that the W–H representation is not unique. Thus we may represent a process using the white noise process appropriate for it at the initial time. Or alternatively we may use a white noise process appropriate at a later time, in order to improve the convergence of the representation. Of course, one often says that usually a process at later time becomes substantially statistically independent of its earlier values. For most purposes, that statement may be taken to be accurate. However, the late process cannot literally be statistically independent of the early process; the late process must in some way grow out of the early process. In fact, one can plausibly argue (see below) that certain high order moments of the early process will be well correlated with the later process (no practical use for this interesting result is offered here).

It is seen that there are many different ways to represent a given random process using the W–H expansion. These different modes of representation are related to one another by what are known as measure-preserving transformations. In the present paper we propose using some of the manifest arbitrariness of the representation to improve convergence. It is proposed that this be done by calculating for the process using one of the presently available (time-dependent) representations for the white noise process. One calculates for some time, then stops the calculation, readjusts the terms in the series, preserving certain necessary physical characteristics of the process. The readjustment is done in such a way as to improve convergence. Then the calculation is started with these renormalized values, continued again for some time, stopped, and the process repeated. Preliminary evidence will be presented to suggest that this should considerably improve the convergence of the W–H representation, as it is applied to nonlinear random processes.

2. Burgers' Model Turbulence

Burgers (1950) proposed a model equation similar to the Navier–Stokes equation

$$\partial u/\partial t + u\, \partial u/\partial x = \frac{1}{\mathrm{Re}} \partial^2 u/\partial x^2 \tag{1}$$

where the Reynolds number Re is defined in terms of u_0, the rms value of the velocity at the initial time, and of an initial characteristic length scale x_0. This equation can be treated through the use of the W–H expansion. Define a white noise process which is statistically independent point to point, Gaussian at every point, with co-variance given by

$$\langle H^{(1)}(x_1) H^{(1)}(x_2) \rangle = \delta(x_1 - x_2) \tag{2}$$

Proceeding as in Meecham and Siegel (1964), we have for the first two terms

$$u(x) = \int_{-\infty}^{\infty} K^{(1)}(x - x_1) H^{(1)}(x_1)\, dx_1 \tag{3}$$

$$+ \iint_{-\infty}^{\infty} K^{(2)}(x - x_1, x - x_2) H^{(1)}(x_1, x_2)\, dx_1\, dx_2$$

with

$$H^{(2)}(x_1, x_2) = H^{(1)}(x_1) H^{(1)}(x_2) - \delta(x_1 - x_2)$$

Here time is implicit. It can be shown that the difference arguments guarantee that the process will be statistically homogeneous (which we suppose to be appropriate to our problem). Take the Fourier transforms, substitute Eqs. (3) in (1), and apply suitable averaging to find

$$\left(\frac{\partial}{\partial t} + \frac{k^2}{\mathrm{Re}}\right) K^{(1)}(k) = \frac{ik}{\pi} \int_{-\infty}^{\infty} K^{(1)}(k_1) K^{(2)}(-k_1, k)\, dk_1 \tag{4}$$

$$\left[\frac{\partial}{\partial t} + \frac{(k_1 + k_2)^2}{\mathrm{Re}}\right] K^{(2)}(k_1, k_2) = \frac{i}{2}(k_1 + k_2) K^{(1)}(k_1) K^{(1)}(k_2)$$

where we neglect a term proportional to $(K^{(2)})^2$ assuming that this is negligible when the expansion converges properly. Once the kernels have been obtained by integration of Eq. (4) one can find the energy spectrum and the transfer term. The spectrum is related to the transfer by

$$(5) \qquad \left(\frac{\partial}{\partial t} + \frac{2}{\text{Re}} k^2\right) E(k) = T(k).$$

These functions can be written in terms of the kernels,

$$(6) \qquad E(k) \equiv E_1(k) + E_2(k) = (2\pi)^{-1} K^{(1)2} + (2\pi^2)^{-1} \int_{-\infty}^{\infty} K^{(2)2}(k_1, k - k_1) \, dk_1$$

$$(7) \quad T(k) = -\pi^{-2} \int_{-\infty}^{\infty} k K^{(1)}(k) K^{(1)}(k_1) K^{(2)}(k_1, k) \, dk_1$$

$$- (2\pi^2)^{-1} \int_{-\infty}^{\infty} k K^{(1)}(k_1) K^{(1)}(k - k_1) K^{(2)}(-k_1, k_1 - k) \, dk_1$$

This expansion is based upon a fixed white noise process (which was used in the initial description of the velocity field). As a result we can expect that as time goes on the expansion will converge less and less rapidly. The initial phases of this shifting in the terms of the expansion may be seen in Eq. (4). Suppose that $K^{(2)}$ is initially small compared with $K^{(1)}$ as is the case when the process is initially nearly Gaussian. We see that the changes in $K^{(2)}$ are relatively large in the initial phases of the development of the process. This shifting of energy continues as time goes on; later one would need still higher order terms in order to describe the development of the process adequately. This can be so even if the process remains at all times *strictly* Gaussian (as is not the case with Burgers' model under discussion here).

It has been recognized that some alteration in the representation is necessary in order to retain the convergence of that representation. Allowing the white noise process to change as time progresses is one way in which one can secure an improvement in convergence. Such are called time dependent representations: see Bodner (1969), Canavan and Leith (1968), and Clever and Meecham (1972). The results of the use of the fixed time base and of the use of a time dependent base proposed by Bodner are shown in Figs. 1 and 2, those results being taken from Clever and Meecham (1972). We shall not enter into a discussion here of details of the time dependent base other than to cite these results. In Fig. 1 results for the total energy [the integral of the energy spectrum given in Eq. (6)] are presented, normalized in terms of the initial value of that energy. The energy is shown plotted against dimensionless time as a function of the Reynolds number as parameter. The initial value of the energy spectrum is a modified Gaussian function. The process

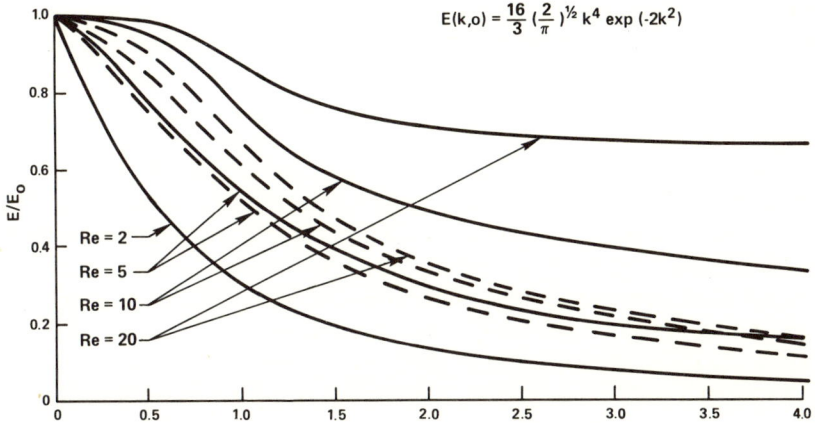

FIG. 1. Energy decay of Burgers' turbulence with a fixed base expansion. Solid curves, two-term fixed base Wiener–Hermite expansion; dashed curves, numerical experiment.

was chosen to be initially Gaussian. In Fig. 1 we see that the energy is given correctly for moderate times for Re ≤ 5. For larger Reynolds number, the representation is poor. Even this achievement is noteworthy; a Reynolds number 5 gives markedly nonlinear behavior in the initial stages of the process. It is evident that the W–H representation is capable of representing such nonlinear activity adequately by this energy criterion. The use of a time dependent base noticeably improves the situation as is seen

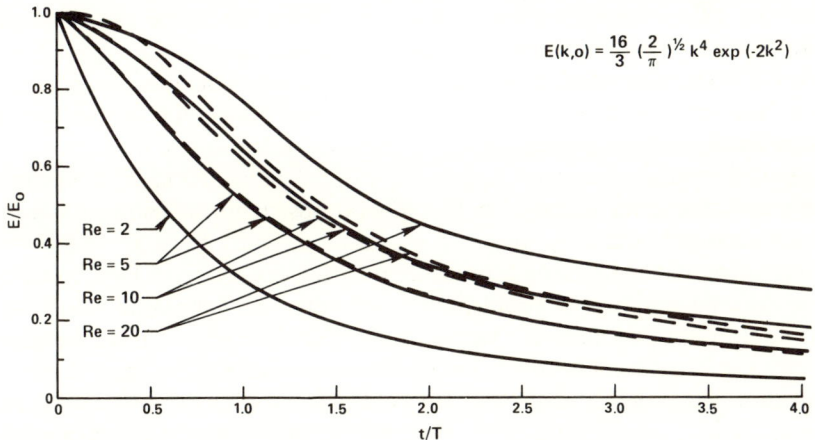

FIG. 2. Burgers' turbulence decay with a time-dependent random base. Solid curves, two term Wiener–Hermite time varying base; dashed curves, numerical experiment.

in Fig. 2. There the representation gives virtually perfect results for moderate times through Re = 10. Marked deviations occur for this modification at Re = 20. The W–H results are compared with numerical experiments performed on the model [Eq. (1)].

This work has shown, as discussed above, that the second term in the expansion becomes comparable to the first, and ultimately exceeds it in value; the expansion no longer converges sufficiently rapidly so that the first two terms are adequate to represent the process. We are faced then with the problem of altering the expansion in such a way that the higher order terms remain small compared with the first term. Of course, if the process develops large non-Gaussian parts, this is not possible even in principle. Burgers' model (see Clever and Meecham, 1972) may not be ideally suited in this way as a testing ground since it shows substantial non-Gaussian portions as the process develops. Nevertheless we propose to test the renormalization suggestion offered here against this model equation. (It is worth noting that driven random Burgers' processes are probably more easily treated using the W–H expansion than are decaying processes. This may be so because one can use the white noise process associated with the current value of the forcing term in the representation. We are thus not dependent upon a white noise process determined at an initial time. One might expect for this reason better convergence properties for such a forced random process. Many diffusion processes may be treated in this way.)

3. A Renormalization for Burgers' Model

Starting from given initial values for the kernels [using Eq. (4)] we calculate the values of the kernels up to a later time. Then the calculation is stopped; the transpired time is chosen in such a way that the higher order term $K^{(2)}$ remains small compared with $K^{(1)}$. Thus one may reasonably hope that the expansion is still rapidly convergent. Once the calculation is stopped, one will in general find that the higher order term in the representation of the energy spectrum E_2 [see Eq. (6)] has increased in size compared with its value at the initial time. We now propose renormalizing—changing the kernels $K^{(1)}$ and $K^{(2)}$—in such a way as to minimize E_2, at the same time minimizing the error in the transfer term $T(k)$ given by Eq. (7) and preserving $E(k)$. To simplify the problem we shall adjust $K^{(2)}$ choosing new values $\tilde{K}^{(2)}$ and using the calculated values for $K^{(1)}$. The problem then is a standard quadratic variational problem for the determination of $\tilde{K}^{(2)}$. Once this minimization has been accomplished, we use Eq. (6) to determine $\tilde{K}^{(1)}$ using the value of $E(k)$ found from the unmodified kernels. One could iterate this process using now $\tilde{K}^{(1)}$ in place of $K^{(1)}$; we will not pursue this further modification here. Let $\tilde{T}$ represent the value of T when $K^{(2)}$ is replaced by

$\tilde{K}^{(2)}$; $\tilde{E}_2(k)$ represent $E_2(k)$ with $K^{(2)}$ replaced by $\tilde{K}^{(2)}$. We see that we have great arbitrariness in the description of the functions E and T because the kernel function $K^{(2)}$ is a function of *two* independent arguments.

We choose a system of functions to represent the renormalized kernel $\tilde{K}^{(2)}$, where the representation will include a set of constants to be determined through the variational process. This kernel must be symmetric under the interchange of its two wave number arguments and should be an odd function of the two arguments taken together. We have

(8) $$K^{(2)}(k_1, k_2) = (k_1 + k_2) a_{m'n'} L_{m'}(k_1) L_{n'}(k_2)$$

with $a_{m'n'} = a_{n'm'}$ and $m', n' = -2, 0, 2, \ldots$. The L are even functions of their arguments and we assume summation on repeated indices over the whole set of functions to be used. It is convenient to define subsidiary functions

(9) $$M_{m'}(k) = K^{(1)}(k) L_{m'}(k),$$

$K^{(1)}$ is the function calculated by integration of Eqs. (4), prior to the stop for renormalization. We represent the modified kernels by Hermite polynomials with their (Gaussian) weighting function

(10) $$M_m(k) = He_m(2^{1/2}k) \exp(-k^2/2); \quad m, n = 0, 2, \ldots$$

The first function, M_{-2}, is described below. The functions M_m are mutually orthogonal. We shall adhere to the convention on range indicated for primed and unprimed indices, Eqs. (8) and (10).

In using the representation given by Eq. (8) it will be desirable to preserve the asymptotic form of the transfer for large wave numbers in order to preserve the inertial subrange. Consider the first term in the expansion of Eq. (8) and suppose that the coefficient $a_{-2-2} = 1$; this single could represent the transfer T exactly for all wave numbers as we shall see [from Eq. (7)]. Substitute that single term in Eq. (7), use the definition given in Eq. (9), take the Fourier transform of Eq. (7) after substitution to find for the function M_{-2} the following relation

(11) $$M^2_{-2}(x) - 2M_{-2}(x)M_{-2}(0) - 2\pi^2 \mathcal{T}(x) = 0$$

with

$$\mathcal{T}(x) = \int e^{ikx} k^{-2} T(k) \, dk \quad \text{and} \quad M_{-2}(x) = \int e^{ikx} M_{-2}(k) \, dk$$

(It is easily shown that T begins like k^2, for small k, so that the Fourier transform described here is convergent.) Equation (11) can be solved,

(12) $$M_{-2}(x) = M_{-2}(0) - [M^2_{-2}(0) + 2\pi^2 \mathcal{T}(x)]^{1/2}$$

$M_{-2}(0)$ is found from Eq. (11). From Eq. (9) we may reasonably expect that the transfer function falls off more slowly than M_m for large k. In such a case the first function, $M_{-2}(k)$, will be the dominant term in the representation of the transfer for large wave numbers. Since M_{-2} exactly represents the transfer process, we can guarantee that the behavior of the transfer will be correctly given for large asymptotic values of its argument.

Continue now with the variational problem. As with the three-mode process define a weighting parameter Λ and minimize by varying the coefficients a_{mn} [where $m, n = 0, 2, 4 \ldots$; we adhere to the notation that the primed subscripts include the lowest order term given by Eq. (12) whereas the unprimed include only the functions given by Eq. (10)]. One finds then for the variational problem

$$(13) \qquad (1 - \Lambda)\, \partial \tilde{E}_2/\partial a_{mn} + \Lambda\, \partial \mathscr{E}^2/\partial a_{mn} = 0$$

where the higher order contribution to the energy spectrum $\tilde{E}_2(k)$ is defined by replacing $K^{(2)}$ by $\tilde{K}^{(2)}$ in Eq. (6). We define

$$(14) \qquad \tilde{E}_2 = \int \tilde{E}_2(k)\, dk, \qquad \mathscr{E}^2 \equiv \int_{-\infty}^{\infty} \mathscr{E}^2(k)\, dk$$

with $\mathscr{E}^2(k) = [T(k) - \tilde{T}(k)]^2$. Using these definitions one finds for the terms in Eq. (13)

$$(15) \qquad \frac{\partial \tilde{E}_2}{\partial a_{mn}} = \frac{1}{2\pi^3} a_{m'n'} \int dx \left[\frac{\partial}{\partial x} L_{m'}(x) L_{n'}(x) \right] \left[\frac{\partial}{\partial x} L_m(x) L_n(x) \right]$$

$$(16) \qquad 4\pi^5 \frac{\partial \mathscr{E}^2}{\partial a_{pq}} = a_{mn} A_{mn,\,pq} - 2 a_{mn}[A_{m,\,pq} M_m(0) + A_{p,\,mn}]$$

$$+ 4 a_{m,\,n} M_n(0) M_q(0) A_{m,\,p}$$

where $L_{m'}(x)$ and $M_{m'}(x)$ are Fourier transforms of the functions $L_{m'}(k)$ and $M_{m'}(k)$. We define

$$(17) \qquad A_{mn,\,pq} = \int \left[\frac{\partial^2}{\partial x^2} M_m(x) M_n(x) \right] \left[\frac{\partial^2}{\partial x^2} M_p(x) M_q(x) \right] dx$$

and similarly for the coefficients $A_{m,\,pq}$ and $A_{m,\,n}$. Equation (13) in conjunction with Eqs. (15) and (16) yields a linear system to be solved for the unknown coefficients $a_{m,\,n}$. Once this minimization is accomplished, the renormalized value for the higher order contribution to the energy spectrum $\tilde{E}_2$ can be calculated; so also can one calculate the error in the transfer term $\mathscr{E}^2$. Again, using Eq. (8), substituting in Eq. (6), one can obtain the renormalized value for the first kernel $\tilde{K}^{(1)}$. With the renormalized values for the kernels chosen as new initial values we now return to the calculation using Eq. (4). One continues calculating, starting from those values and proceeding to another, later time (chosen short enough so that the higher order contribution to the energy spectrum E_2 still remains small). Then the calculation is stopped, and the kernels renormalized once again following the same procedure. This process is continued.

4. Preliminary Numerical Example of the Renormalization for Burgers' Model

The functions M_m have been chosen so as to take the fullest advantage possible of integral relations for the Hermite polynomials. We consider a simple preliminary example to test the ideas presented. We shall restrict to two functions to represent the renormalized kernel $\tilde{K}^{(2)}$. Thus let

(18) $M_0(x) = (2\pi)^{1/2} \exp(-x^2/2); \; M_2(x) = -(2\pi)^{1/2} \exp(-x^2/2) He_2(2^{1/2}x)$

and

(19) $$M_{-2}(k) = \tfrac{1}{2}\pi^{-1/2} \exp(-k^2/4)$$

so from Eq. (12)

(20) $T(k) = (k^2/4\pi^3)[(\pi/2)^{1/2} \exp(-k^2/8) - 2\pi^{1/2} \exp(-k^2/4)]$

From Eqs. (18) and (19) we see that the transfer term will be correctly given for large values of k, no matter what choice we make for the coefficients of the function in Eq. (18). The calculated value for the first kernel function we suppose to be

(21) $$K^{(1)}(k) = He_2^{-1}(k)$$

We have but two adjustable constants a_{00} and a_{22}.

Using these proposed values it is found that in order to properly weight the two elements of the variational problem seen in Eq. (13) we must choose $1 - \Lambda$ quite small. The choice used here is

(22) $$\Lambda/(1-\Lambda) = 100$$

We proceed to calculate the integrals needed for Eq. (13) using Eqs. (15) and (16) in the definitions. We find for the coefficients of the functions given in Eq. (18)

(23) $\quad a_{00} = 3.15 \times 10^{-3}; \quad a_{22} = -1.82 \times 10^{-3}$

Substituting these various quantities in the expression for the renormalized kernel, Eq. (8), and in turn substituting that in the second term right-hand side of Eq. (12) we find for the integral of the renormalized higher order energy [see Eq. (14)]

(24) $$\tilde{E}_2 = E_2 - 0.0477; \qquad E_2 = 0.1295$$

With just two adjustable constants the higher order term in the expansion has been reduced by 40 %, by this criterion. Similarly from these determined values we calculated the error in the transfer term. These values are given in Table I. It is seen that the fractional error in the transfer has been held to 10–20 % over the range of variables examined. We note as expected from the analytical work above that the transfer term for large values of its argument shows negligible error.

TABLE I. Error in the transfer term

k	$\mathscr{E}(k) = T(k) - \tilde{T}(k)$	$T(k)$
0	0	0
0.5	1.19×10^{-4}	-42.6×10^{-4}
1	7.68×10^{-4}	-133.0×10^{-4}
2	27.3×10^{-4}	-175×10^{-4}
4	27.8×10^{-4}	135×10^{-4}
5	11.3×10^{-4}	97.2×10^{-4}
7	0.22×10^{-4}	10.8×10^{-4}

5. Conclusions

It had been found that the W–H representation was adequate, for moderate Reynolds numbers and for moderate times, for the representation of (decaying) initial value problems of interest in turbulence theory. When these conditions were not met the representation became progressively more poorly convergent. This resulted in the buildup of higher order terms in the expansion at the expense of the lower order terms and would in principle require the retention of such higher order terms in the calculation. This was not computationally feasible. The introduction of time-varying white noise processes in the representation noticeably improved the convergence situation. Still it is evident that further development is necessary if the decaying processes are to be examined at later times and/or at higher Reynolds numbers. The renormalization proposal here offers hope of considerable improvement in the convergence of the representation under these more demanding circumstances. It was seen in the numerical example of the previous section that only two terms in the renormalization process were sufficient to reduce the higher order contributions to the energy spectrum noticeably without seriously harming the transfer function. It is expected that a computer treatment of this problem could easily embrace a large number of such adjustable constants. Estimates suggest that the improvement in the higher order contribution to the energy spectrum will be proportional to the number of terms in the expansion. It seems clear from this simple example that we may hope that the transfer function need not be seriously affected by the renormalization process.

If the renormalization procedure does in fact yield significantly improved results in a full scale computation, it will next be applied to the real fluid mechanics problem and to diffusion problems. There is more flexibility in the choice of the functions for such problems and it might be anticipated that the renormalization procedure will be even more effective in their treatment.

REFERENCES

Bodner, S. E. (1969). *Phys. Fluids* **12**, 33.
Burgers, J. M. (1950). *Proc., Kon. Ned. Akad. Wetensch.* **53**, 247–260.
Cameron, R. H., and Martin, W. T. (1957). *Ann. Math.* **48**, 385–392.
Canavan, G. H., and Leith, C. E. (1968). *Phys. Fluids* **11**, 2759–2761.
Clever, W. C., and Meecham, W. C. (1972). *Phys. Fluids* **15**, 244–255.
Inamura, T., Meecham, W. C., and Seigel, A. (1965). *J. Math. Phys. (N.Y.)* **6**, 695–706.
Meecham, W. C. (1969). *J. Statist. Phys.* **1**, 25–40.
Meecham, W. C., and Jeng, D. T. (1968). *J. Fluid Mech.* **32**, 225–249.
Meecham, W. C., and Siegel, A. (1964). *Phys. Fluids* **7**, 1178–1190.
Orszag, S. A., and Bissonette, L. R. (1967). *Phys. Fluids* **10**, 2603–2613.
Saffman, P. G. (1969). *Phys. Fluids* **12**, 1786.
Wiener, M. (1958). "Nonlinear Problems in Random Theory." MIT Press, Cambridge, Massachusetts.

SUBJECT INDEX

A

Air-sea interactions
 mean profiles and fluxes in, 397–399
 research in, 391–404
Alosa pseudoharengus, beach deposition of, 377–379
Angular velocity, in turbulent shear flow, 218–220
Anistropy, weak, in turbulent transport, 177
Arbritrary P, Q, particle statistics for, 66
Atmosphere
 interactions with ocean, 391–404
 surface layer of in unstable conditions, 131–138
Atmospheric boundary layer
 closure in, 195–198
 computational techniques in, 198–199
 Deardoff's model of, 198–208
 eddy-diffusivity models in, 203
 geostrophic drag law in, 199
 modeling of, 193–211
 steady-state neutral structure of, 199–204
 steady-state unstable structure in, 204–210
 turbulence distributions in, 199–203
Atmospheric diffusion
 see also Diffusion; Turbulent diffusion
 defined, 1–2
 turbulent, 1–21

B

Balloon flights, Lagrangian-Eulerian time-scale relationships for, 419–430
Baroclinic planetary boundary layer, 412
Barotropic planetary boundary layer, 410–412
Beach pollution, turbulent diffusion and, 371–381
Binary random walks, 63–65
Binary velocity fields, Markovian, 61–71
Boundary layer
 in air-sea interactions, 397–399
 internal, 263–284
 planetary, 407–417

Boundary layer flow, boundary change and, 263
Boundary layer height, as function of stability and eddy viscosity, 88–89
Bulk convection transfer, 93
"Burgerlence," 244
Burgers model turbulence, 447–450
 numerical example for, 453–454
 renormalization for, 450–453
Bursts and sweeps, mean time interval or period of, 306–312

C

Continuous distribution, limit of, 69–71
Convective conditions, vertical turbulence component in, 125–130
Convective limit, value of σ_w in, 126–129
Convectively unstable boundary layer, thermodynamic model of, 111–123
Corrsin hypothesis, 149–152
Cromwell current, 360–361

D

Density interfaces, structure of, 433–443
Diffusion
 see also Turbulent diffusion
 absolute vs. relative, 324–326
 atmospheric, 1–21
 environmental conditions in, 322–323
 of floating pollutants, 371–381
 lateral, 316–321
 peak concentrations in, 321–322
 of turbidity, 331–337
 variability of, 326–328
 vertical, 321
Diffusion coefficient of thermal pollution, 383–390
 estimation of, 388–389
Diffusion equation limit, in random walks, 35
Diffusivity, law for, 20

458 SUBJECT INDEX

Dissipation equations, in turbulent transport, 172–175, 182
Distribution, continuous, 69–71

E

Eddy diffusion, 5
  model of, 170
Entrainment equation, 123
Entrainment process, in unstable boundary layer, 115–121
Environmental conditions, diffusion and, 322–323
Eulerian field
  generation of, 153–156
  results of, 144–145
Eulerian flow field simulation, 143
Eulerian-Lagrangian relationship, in turbulence model, 165–168
Eulerian-Lagrangian turbulent diffusion problem, 62, 160–161
Eulerian-Lagrangian vertical-velocity spectra, 422–423
Eulerian space-time correlation, for two-dimensional numerically generated flow, 150

F

Filtered momentum and continuity equations, 238–241
FLIP measurements, of oceanic heat and momentum, 358–364
Floating pollutants
  see also Pollution
  diffusion and deposition of on beaches, 371–381
  model for, 379–381
Fluctuating quantity, heat transfer measurements and, 99–102
Fluid dynamic turbulence, diffusive character of, 141–142

G

Geostrophic drag law, 199
Geostrophic wind shear, atmospheric boundary layer and, 204

Gradient transport, directional constraint in, 47–53
Gradient transport approximations, mathematical restrictions leading to, 27–29
Gradient transport models, limitations of, 25–54
  see also Transport model
Great Lakes, *Alosa pseudoharengus* deposition in, 377–379

H

Harmonic analysis, in turbulent atmospheric diffusion, 17
Heat, turbulent diffusion of in ocean, 353–369
Heat transfer
  equations and definitions in, 94
  experimental arrangements in, 95–97
  mean velocity and temperature distribution in, 97–98
  measurement of fluctuating quantities in, 99–102
Heat transport, across turbulent mixing layer, 93–108

I

I.M.S.T. (Institut de Mecanique Statistique de la Turbulence), 391–404
Instantaneous velocity and length scales, in turbulent shear flow, 213–223
Intermediary distribution, in turbulent regime, 103
Internal boundary layer
  experimental conditions in, 264–265
  propagation velocities and spectra of u and w in, 272–283
  skewness and flatness factors in, 265–271
  turbulence within, 263–284
Irish Sea, diffusion experiments in, 318

K

Kinetic theory model, simplest, 29–31
Kinetic theory/random walk model, 29–31
  time restriction in, 31

SUBJECT INDEX

Kramers-Moyal expansion, 34
K-theory model, 170, 183

L

Lagrangian autocorrelation, Corrsin hypothesis and, 149–152
Lagrangian equations, for diffusion in stationary isotropic field, 146–149
Lagrangian-Eulerian time-scale relationship for balloon flights, 419–430
  variations from wind speed in, 426–429
Lagrangian-Eulerian vertical-velocity spectra, 422–423
Lagrangian field, generation of, 146
Lagrangian information, in three-dimensional calculation, 156–160
Lagrangian quantities, particle trajectories and, 157
Lagrangian Reynolds stress, 149
Lagrangian turbulence, two-particle separation in, 152
Lagrangian turbulent quantities, numerical simulation of, 141–162
Langmuir circulation
  diffusive power of, 381
  floating pollutants and, 372–373
Large-eddy simulation
  filtered momentum and continuity equations in, 238–241
  numerical simulation in, 238
Large-scale turbulence, energy loss in, 241–246
Lateral diffusion, from source at sea, 316–321
Length scales, in turbulent shear flow, 216–218

M

Markov chain, 62
Markovian binary velocity fields, random walks in, 61–71
Mean free path, size limitation in, 29–31
Mixed layer, height of in planetary boundary layer, 73–91
Modeling, 77–88
  computational, 169–191
Molecular diffusion, effect of in turbulent density interface structure, 433–443
  see also Diffusion; Turbulent diffusion

Momentum, turbulent diffusion of in ocean, 343–369
Momentumless wake, in turbulent shear flow, 232–234

N

Navier-Stokes equations, 3–4, 15, 63
  in Lagrangian turbulence, 153
  in turbulent shear flow, 225–226
  Wiener-Hermite statistical turbulence and, 445
North Sea, turbidity diffusion in, 331–337
No-slip boundary, in turbulent shear flow, 230–231

O

Ocean
  electronic circuits for turbulence signals in, 358–359
  heat measurement in, 355–360
  Cromwell current in, 360–361
  interactions of with atmosphere, 391–404
  mixed layers in, 364–368
  surface temperature distribution in, 385–388
  turbulence sources for, 360
  turbulent diffusion of heat and momentum in, 353–369
  universal similarity theory for, 355–358
Ocean surface waves, turbulence generation by, 368–369

P

Particle statistics, for arbitrary P, Q, 66
Peclet number, density interfaces and, 441–443
Planetary boundary layer
  baroclinic cases in, 412
  barotropic, 410–412
  eddy diffusion coefficient in, 407–417
  length of mixed layer in, 73–91
  maximum ground concentration error in, 416–417
  nonstationarity of, 413–417

Pollution
see also Thermal pollution
  dispersion of through density differences, 433
  disposal of, 442
  floating, 371–381
  thermal, 383–390
Prandtl number, turbulent, 93
Probability density distribution analysis, in wall turbulence, 253–254

R

Random walks
  binary, 61, 63
  continuous limits of, 35–40
  Eulerian and Lagrangian statistics for, 39–40
  finite, 63–67
  gradient transport model limitations in, 25–54
  with large steps, 40–42
  in Markovian binary velocity fields, 61–71
  simplest model of, 29–31
  telegraph equation for, 35, 40
Random walk transport rate, power series for, 55–57
Rayleigh numbers, in unstable conditions, 131–132
Reynolds equations, 3–4, 15
Reynolds number, 2
  high approximation of in turbulent transport, 172–175
  in unstable conditions, 131
Reynolds/Peclet number, order-of-magnitude analysis of in turbulent transport, 176
Reynolds stress
  bursts/sweeps and, 306–311
  conditionally sampled and sorted measurements in, 291–294
  distribution of across boundary layer, 299–304
  experimental apparatus and methods in, 289
  experimental measurements in, 289–311
  large individual contributions to near wall, 297-299
  structure of, 287–304
Richardson number, 439–442

S

Salinity interfaces, molecular diffusion and, 437–440
Sewage, double-diffusion convection, and, 442
Shear effect, turbidity diffusion by, 331–337
Skewness/flatness factors, in internal boundary layer, 265–271
Southern Bight (North Sea), turbidity diffusion in, 331–337
Statistical turbulence, renormalization from Wiener-Hermite representation of, 445–455
Stratification effects, in turbulent processes, 400
Stratified waters
  layer types in, 346–351
  temperature structure and mixing in, 339–351
Surface layer
  new theory in, 136–138
  Rayleigh and Reynolds numbers in, 131–132
  similarity theories in, 133–136

T

Taylor's diffusion law, 146–149, 157–158
Telegraph equation, for random walks, 35, 40
Temperature field
  self-preserving region of, 96–97
  in turbulent regime, 103–105
Temperature fluctuation, variance in, 103–105
Temperature structure
  in stratified waters, 339–351
  tracer distribution and, 342–346
Tetroon (balloon) tracking radar, 419–422
Thermal models, for correcting unstable boundary layers, 111–123
Thermal pollution
  coefficient of, 383–390
  instrument and observation for measurement of, 383–385
  lateral spread of, 387
Tracer distribution, temperature structure and, 342–346

Transport mechanism, inhomogeneity and nonstationarity of, 32–33
Transport model
 for gradient modeling of turbulent transport, 43–47
 inhomogeneity and nonstationarity of transport mechanism in, 32–33
 transition probability and diffusion equation in, 33–34
Transport terms, in turbulent transport, 180–181
Turbidity, diffusion of by shear effect, 331–337
Turbulence
 Burgers' model of, 447–450
 diffusive character of, 141
 energy loss in large-scale type, 241–246
 experimental data on, 4–5
 fluid dynamic, 141
 generation of by ocean surface waves, 368–369
 gradient transport model limitations in, 25–54
 high Reynolds number in, 401–402
 within internal boundary layer, 263–284
 Lagrangian quantities in, 141–162
 mathematical analysis of, 4
 "noise" as, 367
 phenomenological studies of, 289
 scale of, 4
 second-order, 170
 stratification effects in, 400
 theory of, 2–3
 vertical component of, in convective conditions, 125–130
 wall, see Wall turbulence
Turbulence distributions, in atmospheric boundary layer, 199–203
Turbulence field, diffusion in, 327–328
Turbulence intensity, as pure Eulerian quantity, 166
Turbulence model, Eulerian-Lagrangian relationship in, 165–168
Turbulence signals, electronic circuits for, 358–364
Turbulent atmospheric diffusion, 1–21
 see also Diffusion; Turbulence; Turbulent diffusion
 cause of, 7–8
 generalized harmonic analysis in, 17

Turbulent boundary layer
 bursts in, 306–313
 bursts near wall in, 295–297, 304–305
 low-momentum fluid from wall in, 311
 "sweep" in, 287, 295–297, 304–313
Turbulent density interface, molecular diffusion effect in, 433–443
Turbulent diffusion, 315–329
 see also Diffusion
 beach pollution and, 371–381
 from continuous source at sea, 315–329
 defined, 2
 equation for, 61
 Euler-Lagrangian problem of, 62–63
 of passive scalar, 246
 simulation of from two-dimensional flow, 142–152
 "wavelike," 37–38
Turbulent diffusion coefficient, 12
Turbulent diffusivity tensor, 47–53
Turbulent flows
 see also Turbulent shear flow
 "Burgerlence" in, 244
 energy cascade in large-eddy simulations of, 327–347
 energy loss in large-scale turbulence, 241–246
 filtered momentum and continuity equations in, 238–241
 large-eddy simulations of, 237–247
 linear homogeneous, 16
 turbulent diffusion of passive scalar in, 246
Turbulent kinematic viscosity, 48 $n.$
Turbulent mixing layers
 self-preserving region and virtual origin in, 96–97
 temperature field characteristics in, 103
 temperature fluctuations in, 103–106
 transport of heat across, 93–108
Turbulent motion, continuous interaction in, 54
Turbulent Prandtl number, 93
Turbulent shear flow
 angular velocity in, 218–220
 equations of motion in, 226–228
 free-slip boundaries in, 228–230
 instantaneous velocity and length scales in, 213–223
 initial conditions in, 231–232
 momentumless wake in, 232–234

no-slip boundaries in, 230–231
numerical computation and methods in, 225–236
plume dispersion model of, 222
time scales in, 220–222
Turbulent stable boundary layer, model of, 77–88
Turbulent transport
  coefficients in, 182–184
  computational models of, 169–191
  dissipation equations in, 172–175, 182
  gradient models in, 25–54
  high Reynolds number approximation in, 172–175
  mean field homogeneity condition in, 45
  mean field time stationary condition in, 45
  modeling of third moments in, 176–179
  reversed kinetic energy transformation in, 42
  Reynolds/Peclet number order of significance in, 176
  weak anisotropy in, 177
  third moments in, 176–179
  Townsend's data in, 186–190
  transport model for, 43–47
  transport terms in, 180–181
  two-dimensional isothermal wake in, 184–191
Turbulent viscosity coefficient, 8
Two-dimensional flow
  field equation for, 143–144
  simulation of turbulent diffusion for, 142–152
Two-dimensional isothermal wake, in turbulent transport, 184–191
Two-dimensional vorticity field, numerically generated, 145

U

Unstable boundary layer
  entrainment process in, 115–121
  numerical integration procedure in, 121–123
  thermodynamic model for, 111–123
Unstable conditions
  similarity theories in, 133–136
  surface layer of atmosphere in, 131–138

V

Velocity isopleths, in channel flow, 155
Vertical mixing, in stably stratified waters, 339–351
Vorticity field, numerically generated, 145

W

Wall turbulence
  conditional averaging analysis in, 254–259
  experimental technique in, 251
  probability density distributions in, 253, 260
  statistical analysis of, 249–260
Wave equation limit, in random walks, 35
Waves, wind-generated, 395–397
Weak anisotropy, in turbulent transport, 177
Wiener-Hermite representation, for statistical turbulence, 445–455
Windrows, floating pollution and, 372–377

QC
806
A3
v.18A
1974

JAN 9 1974